A Course on Queueing Models

STATISTICS: Textbooks and Monographs

Recent Titles

Probability and Statistical Inference, *Nitis Mukhopadhyay*

Handbook of Stochastic Analysis and Applications, *edited by D. Kannan and V. Lakshmikantham*

Testing for Normality, *Henry C. Thode, Jr.*

Handbook of Applied Econometrics and Statistical Inference, *edited by Aman Ullah, Alan T. K. Wan, and Anoop Chaturvedi*

Visualizing Statistical Models and Concepts, *R. W. Farebrother and Michaël Schyns*

Financial and Actuarial Statistics: An Introduction, *Dale S. Borowiak*

Nonparametric Statistical Inference, Fourth Edition, Revised and Expanded, *Jean Dickinson Gibbons and Subhabrata Chakraborti*

Computer-Aided Econometrics, *edited by David E.A. Giles*

The EM Algorithm and Related Statistical Models, *edited by Michiko Watanabe and Kazunori Yamaguchi*

Multivariate Statistical Analysis, Second Edition, Revised and Expanded, *Narayan C. Giri*

Computational Methods in Statistics and Econometrics, *Hisashi Tanizaki*

Applied Sequential Methodologies: Real-World Examples with Data Analysis, *edited by Nitis Mukhopadhyay, Sujay Datta, and Saibal Chattopadhyay*

Handbook of Beta Distribution and Its Applications, *edited by Arjun K. Gupta and Saralees Nadarajah*

Item Response Theory: Parameter Estimation Techniques, Second Edition, *edited by Frank B. Baker and Seock-Ho Kim*

Statistical Methods in Computer Security, *edited by William W. S. Chen*

Elementary Statistical Quality Control, Second Edition, *John T. Burr*

Data Analysis of Asymmetric Structures, *Takayuki Saito and Hiroshi Yadohisa*

Mathematical Statistics with Applications, *Asha Seth Kapadia, Wenyaw Chan, and Lemuel Moyé*

Advances on Models, Characterizations and Applications, *N. Balakrishnan, I. G. Bairamov, and O. L. Gebizlioglu*

Survey Sampling: Theory and Methods, Second Edition, *Arijit Chaudhuri and Horst Stenger*

Statistical Design of Experiments with Engineering Applications, *Kamel Rekab and Muzaffar Shaikh*

Quality by Experimental Design, Third Edition, *Thomas B. Barker*

Handbook of Parallel Computing and Statistics, *Erricos John Kontoghiorghes*

Statistical Inference Based on Divergence Measures, *Leandro Pardo*

A Kalman Filter Primer, *Randy Eubank*

Introductory Statistical Inference, *Nitis Mukhopadhyay*

Handbook of Statistical Distributions with Applications, *K. Krishnamoorthy*

A Course on Queueing Models, *Joti Lal Jain, Sri Gopal Mohanty, and Walter Böhm*

A Course on Queueing Models

Joti Lal Jain
University of Delhi
India

Sri Gopal Mohanty
McMaster University
Ontario, Canada

Walter Böhm
University of Economics
Vienna, Austria

Chapman & Hall/CRC
Taylor & Francis Group
Boca Raton London New York

Chapman & Hall/CRC is an imprint of the
Taylor & Francis Group, an informa business

Chapman & Hall/CRC
Taylor & Francis Group
6000 Broken Sound Parkway NW, Suite 300
Boca Raton, FL 33487-2742

Printed in the United States of America on acid-free paper
10 9 8 7 6 5 4 3 2 1

International Standard Book Number-10: 1-58488-646-3 (Hardcover)
International Standard Book Number-13: 978-1-58488-646-4 (Hardcover)

Library of Congress Cataloging-in-Publication Data

Jain, Joti Lal.
A course on queueing models / Joti Lal Jain, Sri Gopal Mohanty, Walter Böhm.
p. cm. -- (Statistics, textbooks and monographs ; v. 189)
Includes bibliographical references and index.
ISBN 1-58488-646-3 (acid-free paper)
1. Queueing theory--Textbooks. I. Mohanty, S.G. II. Böhm, Walter, 1957- III. Title.
IV. Series: Statistics, textbooks and monographs ; v. 189.

QA274.8.J35 2006
519.8'2 --dc22 2006045490

Visit the Taylor & Francis Web site at
http://www.taylorandfrancis.com

and the CRC Press Web site at
http://www.crcpress.com

To our wives

Kusum, Shanti and Julia

For their love and patience

Preface

We have decided to join the queue in writing a textbook on queues—not to ask for any service as the usual case is, but possibly to provide service to a community interested in learning about its theory and its applicability aspects.

There has been a growing interest in the subject due to its applications in fields as divergent as management science and communication. At the same time, several approaches have advanced both in theory and computation. Thus our decision is guided by three factors: to include recent advancements in the fields, to include different methodological tools, some of which not available elsewhere, and more importantly to include computational techniques for the purpose of application. Therefore, a compromise has to be made and that too without sacrificing the content. We believe we have achieved a balance.

We belong to very distinct teams and yet, as providence dictated, our paths converged to play the game together not one against the other but as members of the same team. Our opponent is nobody and that is why our joy is boundless. Each of us has taught 'Queueing Theory' at our natural domains and the experience of each is varied. The convergence of our collective experience, we believe, has enriched us. We have coached each other and learnt a lot.

In our endeavour to reach the goal, we have been very fortunate to receive enormous encouragement from Professor N. Balakrishnan, McMaster University, Canada; experienced technical support from Mrs. Debbie Iscoe and significant computational assistance from Ayekpam Meitei. We owe our gratitude to all of them. We are particularly thankful to Kevin Sequeira for taking special interest in this project. Our wives have very patiently kept us going over these years. The book is dedicated to them.

Joti Lal Jain
Sri Gopal Mohanty
Walter Böhm

Contents

Part 1

CHAPTER 1

Queues: Basic Concepts

1.1. Introduction

Lining up for some form of service is a common phenomenon, be it visible or invisible, be it by human beings or by inanimate objects. It is more organized or sometimes is made to be so in the modern world and therefore a systematic study of a lineup or equivalently a queueing process is instinctively more attractive. The concept of a queue is well recognized in the field of management science and lately is drawing attention in communication systems.

From very mundane understanding of a queue in daily life, we have progressed through to establish a sophisticated body of knowledge called theory of queues. In this context, lineup is visualized as time wasted and hence waiting time is uncomfortable in a society that sees idleness as socially counterproductive. Yet there are occasions when it becomes a sharing experience and provides time for people to chat with friends and companions. Nevertheless a reduced waiting time is always welcome.

Clearly a queue is a manifestation of congestion in flow of objects through a system or a network consisting of many systems. The interest being to reduce congestion in whatever situation it arises, led to studies to be called "congestion theory." In spite of its supposedly broad nature of coverage to include any flow with objectives of all kinds, we restrict ourselves to the study of congestion of certain types of flows and term it as the theory of queues.

In this chapter, we try to understand a queue in an intuitive manner through simple graphs and then develop the theory by adopting various models in later chapters.

1.2. Queues: Features and Characteristics

Whenever the available resource is not sufficient to satisfy the demands, then the demanding units have to wait and are subjected to the queueing analysis. Some examples are: customers arrive at a bank and wait to have certain monetary transactions; in any busy airport, planes are waiting for their turn to land; patients have to wait for an operation because either the surgeon is busy or there are not enough hospital beds; goods to be transported from city to city wait due to lack of enough transport vehicles; signals are sent through switches and wait to be transmitted. All these are systems of flow or moving processes and the congestion arises when the available facilities to clear the flow are not adequate. A few other situations are traffic flow through network of roads, flow of objects along a production line for completion of various tasks and flow of computer programs through time-sharing computer systems.

These examples have the following three basic features:

(i) A unit arrives to receive the service and is called a *customer* or an *arrival.* Customers may arrive in batches and are designated as bulk arrivals.

(ii) A system provides the service which may consist of a single *server* or many servers. The service may be done in batches.

(iii) Arriving customers wait to be served, if the service facility is busy and form a line what is known to be a *queue.*

The process consisting of arrivals, formation of a queue or queues, performance of service and departures is a *queueing process*, or a *queueing system.* It is illustrated in Figure 1.1.

Arriving Customer → Queue → Service facility → Departure

FIGURE 1.1. Queueing process

Now we present two examples, one from a communication system and the other from a manufacturing system.

(a) In a communication system, a portion of the frame capacity is allocated to the voice traffic and the rest to data traffic. For voice traffic there is no waiting space. If the allocated channels are busy, an incoming voice call is lost. For the data traffic, there is an infinite waiting space. This problem leads to two queues, one for the voice call and the other for the data transmission. A voice call is an arrival which is either being served immediately if it gets connected or is lost so that virtually no queue is formed. In this case, the loss of a call is of interest. For the transmission situation, data are sent in packets and these packets are transmitted each taking a constant length of time. Here packets are customers and form a queue to wait until ready to be transmitted and thus our interest will be the waiting time of a packet in the queue. Ultimately, the problem is how to divide the system into two in order to optimize the loss on either side.

(b) In a multi-location inventory system, the number of an item at a location is the inventory and each item is given to a customer on demand. Thus the inventory level is nothing but the number of servers to provide service to the same number of customers. Any demand of the item represents an arrival and an unfulfilled demand becomes a backlog and waits in the queue. Our interest is to reduce the backlog.

We consider a few variations in a queueing system, such as a customer's behavior and *queue discipline.* A customer may balk (i.e., upon arrival does not join the queue), renege (i.e., leaves the queue before being served) or jockey (i.e., switches from one queue to another). The queue discipline means the order in which customers are served. The most common situation is the one in which customers are served on a first-come first-served (FCFS) basis; also known as first-in first-out (FIFO) discipline. Some other queue disciplines are last-come first-served (LCFS), random service and service given on a priority basis. Another feature of a queue is the capacity of the waiting space which may be finite or infinite. Infinite waiting space is often not stated explicitly. In communication systems a switch, which acts as a waiting space (called buffer), has finite capacity.

Although in principle we could study all kinds of characteristics of any such process, it would be expedient to focus on those which suit to our purpose. The very fact that a queue is a waiting line, immediately invokes the natural curiosity to learn about the *queue length* or *number in the system* and *waiting time* of a customer. The management which provides the service facility has to design and control a queueing system by keeping two objectives in view, viz., to minimize the cost of providing the service and to minimize the discomfort to the customer. From these points of view the suggested characteristics are very appropriate. In addition, the management would like to keep servers busy and thus is interested in the length of a busy period (i.e., when a server is continuously serving) and *length of an idle period* (i.e., when a server is not serving). These characteristics are usually random variables and what we mean by studying them is to determine their probability distributions (see Section A.2 in Appendix A) or probabilities of events defined on them (see Section A.1 in Appendix A) either at any fixed time point (i.e., *transient analysis*) or after a long time has elapsed (i.e., *steady state* or *stationary analysis*). In many situations their means and variances, called *measures of performances*, are adequate for the purpose of the design and control of a queueing system. Without any ambiguity, we may not often distinguish between characteristics and measures of performance. Occasionally, we will introduce others, such as loss of customers, which is of significant importance in communication systems since this loss implies complete loss of information that is supposed to be originally transmitted.

1.3. Graphical Methods

Any knowledge about the characteristics or measures of performance would need information on the customers' arrival instants and departure instants. Consider a one server queueing system with FCFS queueing discipline. Let $A(t)$ and $D(t)$, respectively, represent the cumulative arrivals and cumulative departures during time $(0, t)$. Note that $A(t)$ and $D(t)$ are determined by the arrival and departure instants. For the observed interval (a, b) these are represented in Figure 1.2.

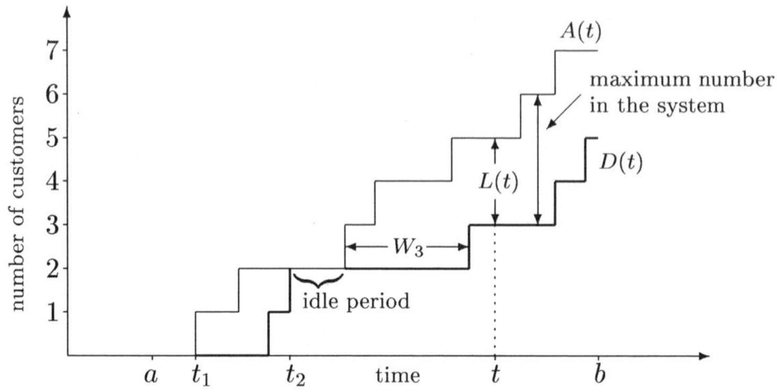

FIGURE 1.2. Graph of $A(t)$ and $D(t)$

Denoting by $L(t)$ and $L_q(t)$ the number in the system and the number in the queue at time t, we obtain from the graph that

$$L(t) = A(t) - D(t), \tag{1.1}$$

and

$$L_q(t) = \begin{cases} A(t) - D(t) - 1 & \text{if } A(t) - D(t) - 1 \geq 0, \\ 0 & \text{if } A(t) - D(t) - 1 < 0. \end{cases} \tag{1.2}$$

In the above graph, $L(t)$ is indicated. Observe that (i) there is an idle period after the departure of the second customer, (ii) the first busy period is between t_1 and t_2 and consists of serving two customers and (iii) the maximum number in the system during (a, b) is 3. Letting W_n to represent the waiting time in the system of the nth customer, we have indicated W_3 in the graph. We can compute performance measures, such as the mean and variance of such characteristics. The mean and variance of the number of customers in the system are given by

$$L = \frac{1}{b-a}\int_a^b L(t)\,dt \quad \text{and} \quad \frac{1}{b-a}\int_a^b (L(t) - L)^2\,dt \tag{1.3}$$

respectively. Similarly, other performance measures are the average waiting time in the system and the variance as given by

$$W = \frac{1}{N}\sum_{n=1}^{N} W_n \quad \text{and} \quad \frac{1}{N}\sum_{n=1}^{N}(W_N - W)^2 \tag{1.4}$$

where N is the number of customers in the sample. The information on characteristics and measures of performance to be computed from observed data have been asked in Exercises 1 and 2. In Exercise 1(ii), an alternative graphical representation is suggested.

In fact the above expressions are sample values of the same quantities in the population (i.e., the entire process if it could be hypothetically visualized) and in that sense are called the *estimates*. Common sense dictates the use of these values or estimates for all future purposes. If the interval (a, b) and N are large, then this procedure is justified under certain conditions and it is a well-known practice of drawing inferences in the statistical world. However, we would ask: "How large should be the sample to be large enough?" or "How do we check whether the conditions are valid?" Quite often no satisfactory answer is available and therefore this approach of studying the queueing system through graphical methods is less effective.

1.4. Modelling

Instead of depending on assumptions which are difficult to check we may make assumptions on the original processes, say, on arrival instants and service durations on the basis of past data and derive expressions for various characteristics under study. What is being suggested is to construct a model which prescribes a precise description of various features of the queueing process, in particular, the stochastic nature of the arrival process and of the service-time. Once the model is checked through statistical techniques, the distributions of characteristics and the measures of performance are obtained by applying probability theory.

Note that the approach in the previous section is strongly of statistical nature in the sense that the sample is a reflection of the population and thus the sample values are used for prediction in the population. In contrast, the modelling approach, although uses statistical procedures to check certain model assumptions, heavily relies on probability theory for the derivation of explicit mathematical expressions. The analysis of queues through models is common in the literature on theory of queues because (i) the stochastic nature of the arrival process and service-times is observable, (ii) the necessary statistical procedures are straightforward and (iii) tractable results are derivable by the application of probability theory and mathematical tools. In other words, the modelling technique examines the system at micro-level to construct a model and utilizes the emerging theory for deriving results at macro-level; in contrast, the graphical method produces macro-level results from the observed data without looking at the structure of the system. At this point, a word of caution is necessary. The real life queues are more complex than the constructed models. If we increase complexities, we do not know how to analyze them. A model should be made as simple as possible but should also be close to reality. Out of our excessive enthusiasm to utilize the attractive looking theories we may neglect to check the model assumptions or stretch a queueing process to fit a model without appropriate justification. In case we are in serious doubt, we should remind ourselves that the graphical method is still available as a last resort, how so ever imperfect it might be.

In this book, we assume a model for each queueing process and analyse it by studying the desired characteristics. It is realized from the graph that a queueing model is primarily defined by three basic properties: the nature of arrival process (also called input process) or the probability distribution of interarrival times denoted by the random variable T, the probability distribution of service-times denoted by the random variable S, and the number of servers. In addition, information on the relation between interarrival times and service-times (assumed to be independent in this book) queue discipline, capacity of waiting space and batch size are pertinent for a complete specification of most queueing models. There is a convention to denote a model by its three basic properties as $\cdot/\cdot/\cdot$, the first $(\cdot)$ describing the nature of the arrival process, the second $(\cdot)$ being on service-time distribution and the third $(\cdot)$ giving the number of servers. Whenever this notation is used, it is tacitly accepted that the arrival process and service-times are independent of each other. For example if we consider a model with one server in which the interarrival times and the service-times are fixed constants, then it is denoted by $D/D/1$ where D stands for the word "deterministic." Another model is $M/M/1$, where M refers to "Markovian," which means both interarrival times and service-times are independently and identically distributed (briefly, i.i.d.) random variables each having exponential distribution. (Strictly speaking, the word "Markovian" should imply that the underlying process is a Markov process. However, in the context of queueing theory it has been the convention to make it a Poisson process which is indeed a special Markov process. The Poisson process is described through an equivalent characterization, viz., through interarrival times and service-times (see Section A.7 in Appendix A). Other properties will be stated for each model without being a part of the notation. In succeeding chapters, a model will be completely described, whenever it is introduced.

In example (a), it is customary to assume arrival times of voice traffic form a Poisson process and call duration has exponential distribution. For data traffic, the assumption is that arrival times also form a Poisson process but the transmission time of a packet is a fixed constant.

The model $D/D/1$ is a trivial one to analyse since there is no randomness in the process. Denote by λ and μ the arrival rate and the service rate, respectively. In that case, the fixed interarrival time is $1/\lambda$ and the fixed service-time is $1/\mu$. If $\lambda > \mu$, then the number of customers in the system continues to grow indefinitely. In fact, assuming that initially the system is empty and letting $N(t)$ to be the number of customers in the system at time t, we have

$$\begin{aligned} N(0) &= 0, \qquad\qquad (1.5)\\ N(t) &= \{\text{number of arrivals during } (0,t)\} \\ &\quad - \{\text{number of services completed during } (0,t)\} \\ &= \left[\frac{t}{1/\lambda}\right] - \left[\frac{t-1/\lambda}{1/\mu}\right] \\ &= [\lambda t] - \left[\mu t - \frac{\mu}{\lambda}\right], \end{aligned}$$

where $[x]$ is the greatest integer $\leq x$. Observe that $N(t) > 0$ and it increases as t increases. Clearly for $\lambda = \mu$, either there is no queue if the system is empty at $t = 0$ or the length of the queue remains the same otherwise. A similar analysis can be done for $\lambda < \mu$.

For models featuring randomness, mathematical methods will be used to analyse them. But it turns out that these tools are not helpful for every model to yield explicit results on characteristics. In such cases, we may resort to the method of solution by simulation. It is a technique usually adapted for the usage of computers in which a programme consisting of data generation, organization and analysis is executed. The last two parts more or less follow steps similar to those in the graphical analysis except that the steps are implemented through computers but not through graphs. Regarding the generation of data, unlike recording the actual observations of a queueing system as would be done in the graphical method, it involves producing artificial observations from probability distributions that are specified by the model. Therefore, the assumption of a model is absolutely essential in simulation, whereas we do not assume any model in the graphical method. The real challenge in simulation is to create samples without actually observing them and it is done through what is commonly known as Monte Carlo method. The method is nothing but a procedure to generate observations from the specified distribution and therefore forms the basis of simulation. Recall that in the graphical method we are confronted with a serious limitation. The collection of data is constrained by the cost, procedure of collection, time lag, loss of information and other factors. We often ask: "Are the sample size and the number of samples large enough so that the estimated value is a representation of the future or in general of the population?" This is where the Monte Carlo simulation method comes to our rescue, since we can produce many samples of very large size with the aid of computers.

Let us briefly explain the Monte Carlo procedure and illustrate it in a simple but important case. To begin with, let us remember the following elementary but remarkable result in probability distribution (see Section A.4 in Appendix A):

If X is a continuous random variable with its distribution function $F_X(x)$ at x, then $Y = F_X(X)$ as a function of X has uniform distribution over $(0, 1)$.

Its implication for the Monte Carlo method is that a random number y_1 selected from the uniform distribution over $(0, 1)$ leads to an observation $x_1 = F_X^{-1}(y_1)$ from the distribution of X (see Figure 1.3), where F_X^{-1} is the inverse transform of the probability distribution function.

With the help of a computer we can easily generate many pseudorandom numbers which almost behave like random numbers from the uniform distribution over $(0, 1)$. Suppose we have selected $y_1, \ldots, y_n$ (n being large enough). Their inverse values form a random sample of observations from the original population of X. As an illustration, suppose we need a random sample from

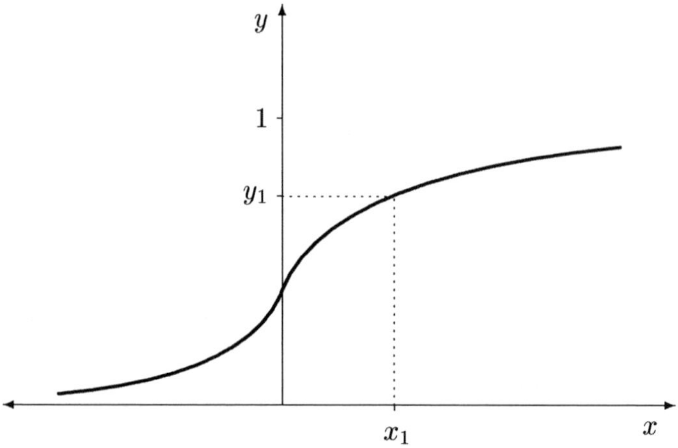

FIGURE 1.3. Graph of $y = F_X(x)$

the exponential distribution with density function

$$f_X(x) = \mu\, e^{-\mu x}, \quad x > 0 \text{ and } \mu > 0$$

and the distribution function

$$F_X(x) = 1 - e^{-\mu x}, \quad x > 0.$$

In this case the inverse function has an explicit form. Writing

$$y = 1 - e^{-\mu x}$$

we find the sample $x_1, \ldots, x_n$ to be

$$x_i = -\frac{1}{\mu} \ln(1 - y_i), \quad i = 1, \ldots, n. \tag{1.6}$$

If, however, an explicit inverse function is difficult to derive, then one uses the graph of F_X and numerically determine the inverse values (see Figure 1.3).

Remark: We are led to believe that all our observational data can be comfortably obtained only by working with a computer and thereby perhaps the actual collection of the real data could be avoided. Here, let us remember that the generation of data by a computer solely depends on the assumed model – a model which is postulated through some experience of handling the original data. Thus we cannot ignore the understanding of real observations. We only understand them better with the help of computers.

In the context of modelling of queues, one may like to ask: "Should time always be treated to be a continuous variable, as it has been obsessively accepted almost everywhere in the past?" Surely flow of time is continuous. However, when it is being measured, it is almost always an integer multiple of some unit and when it comes to data collection, the recorded time points at which observations are taken are frequently discrete. Integers and the associated mental process of counting them are conceptually simpler for us than anything related to continuous variables. Yet we construct continuous-time models since in that case we are able to use powerful tools from mathematical analysis for obtaining solutions. These models become surrogate for discrete-time models in the sense that the results for continuous-time models were substituted as an approximation for their counterparts in discrete-time models. As time progressed, the existence of discrete-time models is strongly experienced, especially in communication systems where the service-time is actually measured as an integer multiple of some fixed duration. This has given a legitimacy to the discrete-time models to be treated independently. Interestingly, the new approach of examining these models on their own has sometimes the advantage of the so-called "killing two birds in one shot." It implies that if we derive a probability distribution (say) of a characteristic in discrete-time model, then we may be able to derive the same distribution (which is not an approximation) in the analogous continuous-time model by a limiting process. But the reverse treatment is not possible, i.e., we cannot get a result for a discrete-time model from the corresponding result for the continuous-time model. Thus discrete-time models will appear several times in the book.

1.5. Scope and Organization

No doubt the model $D/D/1$ is elementary enough not to need any special attention beyond what has been said in the previous section. Among the models displaying randomness, the simplest one is $M/M/1$ for having the Markovian nature which only enunciates the memoryless property of the Poisson process or equivalently of the exponential distribution (see Section A.7 in Appendix A). Starting from the model, we proceed through different chapters to partially Markovian models or non-Markovian models of special types (Chapters 2, 3 and 6) and to general models (Chapter 7) for analytically deriving results on characteristics and performance measures. The model variations such as, many servers, batch arrivals, batch departures and network of servers are also discussed. In dealing with mathematical analysis of models, three methods, namely analytic, algebraic and combinatorial methods are used.

The derived exact results are not always in a convenient form for computation and in such cases numerical techniques may find a way out to provide solutions. Furthermore, for some models no explicit results are available but the solutions are numerically computable. The values are usually approximations and are computed with a certain degree of accuracy. Chapters 4 and 8 are exclusively devoted to discuss quite a few computational methods for finding solutions. Yet there is another approach to approximation in which the discrete state space (i.e., the number of customers in the system) is approximated by a continuous state space in order to use the existing rich theory of diffusion approximation. This is dealt with in Chapter 7. The difference between the two approaches is that in the former we approximate the results derived from exact models, whereas in the latter we derive exact results from approximated models.

Model assumptions need checking. This is usually done by statistical analysis which is discussed in Chapter 5. When mathematical analysis and numerical techniques do not take us far, we resort to simulation to guide us which is also included in the same chapter.

Discrete-time models are interspersed with their continuous counterparts in different chapters. The treatment of these models is done by enumeration of possible realizations of a process which belongs to the domain of combinatorics. That is how these are dealt with in Chapter 9. In fact, combinatorial methods provide an alternative to analytical methods in successfully dealing with transient analysis (see Section 1.2) including the analysis of busy periods. This becomes a subject of discussion in Chapter 6.

In Chapter 2, we realize very soon that the study of transient behaviour of a system is more involved than the steady-state behaviour (see Section 1.2) even in the simplest model $M/M/1$. This is one reason why traditionally the treatment of transient analysis is almost limited to the $M/M/1$ model. However, we devote a section in Chapter 4 for its study by numerical methods. Later it is discussed in Chapter 6 for continuous-time models and in Chapter 9 for discrete-time models.

Then there are topics which are related to different chapters or to the subject as a whole but themselves need not be connected, such as priority queues as a queue discipline, design and control of queues and statistical methods used in queues. These are lumped together in Chapter 10.

The book has ten chapters, equally divided between two parts and Appendices. Part I is intended for a beginner in the subject and consists of Markovian queues (Chapter 2), partially Markovian queues or non-Markovian queues of a special kind (Chapter 3) and simpler but basic computational methods (Chapter 4) and statistical inference and simulation (Chapter 5). Large parts of Chapter 2 and 3 are standard classical material invariably emphasizing the study of steady state behaviour. However, we have added a section on Markovian networks and a section on optimization in Chapter 2, so as to appreciate the purpose of studying the theory of queues in the context of communication systems and management problems. Similarly, a section on steady state solutions of discrete-time queues is included in Chapter 3. Anyone, after finishing Part I, should have a fairly good understanding of the subject by being exposed to all basic ideas, some theory and their applications to practical problems.

Part II provides more of the same but with a refinement and enrichment. It starts with Chapter 6 which is an extension of Chapter 3 and consists mostly of transient analysis. We discuss general queueing models in Chapter 7. Chapter 8 deals with recent computational approaches and Chapter 9 on discrete-time queues, whereas Chapter 10 consists of miscellaneous topics.

Whereas the main text comprising of these chapters is devoted strictly for the development of theory of queues, any other relevant definitions and results in probability theory and other branches of Mathematics are relegated to Appendices which also include a glossary of notations. In this way, the flow of the subject moves smoothly without interruption. Moreover, the reader can conveniently refer to the necessary materials for guidance and understanding.

In our presentation, we have adopted a few unusual features. To begin with, we prefer the continuity of an idea or of a technique of solution applied to different models, over the complete treatment of models individually and comprehensively. For example, in Chapter 2 the balance equation technique which emerges from finding the stationary distribution of number of customers in the $M/M/1$ system is repeatedly used for other models just to demonstrate its effectiveness without completing all aspects of the $M/M/1$ model. In this way, the emphasis is laid on the comprehension and reaffirmation of a notion rather than on the completeness of discussion of a given model.

Second, the new material in the theory of queues has increased steadily and should find a place in this book. In order to permit entry for new topics such as computational methods, discrete-time models and combinatorics, we have rearranged the classical material by discussing only the basic models and putting special but relevant ones as exercises. Take for instance, the case of the birth-death model which is a simple generalization of the $M/M/1$ model. We deal with the birth-death model in the body of the text of Chapter 2, leaving quite a few of its special cases to be included in exercises.

Third, we intend to introduce most of basic ideas on the subject in Part I. It is achieved by splitting up a topic (e.g., discrete-time queues, networks of queues and computational methods) into parts for inclusion in different chapters. We believe that this method of presentation allows the readers to gain a wider coverage of the subject at the end of Part I and at the same time helps them to build up a progressive comprehension of a topic.

Another feature is the addition of a section "Discussion" at the end of each chapter except this one. Besides summarizing the content of a chapter, it dwells on related issues whenever possible. It also concludes with relevant reference materials that have been used elsewhere and to put them together at one place in the chapter so as to permit the development of topics with least dependence on them. The overall purpose is to persuade the valued readers for reflection at the end of a day's journey.

Notations, which are common in the theory of queues and which appear at different places in the text, are presented in Appendix C for convenience. In Appendix D, some key formulas are provided for easy reference.

The references at the end of this chapter are mainly for graphical methods.

1.6. Exercises

(1) A bank starts functioning at 9 a.m. with one teller. The data given below represent the arrival time instants and service durations of 8

customers.

Customer	Arrival instant	Service duration (in minutes)
1	9:05	15
2	9:12	12
3	9:20	9
4	9:30	10
5	9:36	6
6	9:45	7
7	9:55	8
8	10:00	10

(i) For the above data, plot $A(t)$ and $D(t)$ – the graphics for the cumulative arrivals and cumulative departures. Using the graphical method, determine the number in the systems at time 9:50. Find the waiting time in the queue of each customer and determine the average waiting time. Compute the mean and variance of the number of customers in the system.

(ii) For the same data, plot a graphic by representing x-axis for time and by drawing a unit of vertical upward step at an arrival instant and a unit of vertical downward step at a departure instant. What does the y-coordinate represent in this graph?

(2) Suppose there are two tellers operating in place of one. For the data in Exercise 1, determine the average waiting time and the variances by using the graphical method.

(3) (i) Assuming that the interarrival times are i.i.d. (independently and identically distributed) exponentially distributed with parameters $\lambda = 8$, generate a sample of size 30 arrival instants. (Hint: start selecting 30 random numbers from uniform distribution over (0,1).) Compute the mean and variance of interarrival times and check their closeness to 1/8. What do you expect and why?

(ii) In addition to assumption (i), suppose service-times, independent of arrival times, are i.i.d. exponential random variables with mean 0.1. Consider 30 customers who arrive at generated random instants in (i) and generate their service-times under the present assumption. By graphical methods, compute performance measures (1.3) and (1.4).

The following exercises are on Poisson arrival process and exponential service-time (Chapter 2 starts with these assumptions).

(4) A bank opens at 9 a.m. Customers arrive at a Poisson rate of λ, that is, the p.f. of any interarrival time (in minutes) is $f_T(t) = \lambda e^{-\lambda t}$. The service-time of each customer is c minutes (i.e., it has

a deterministic distribution). Find the probability that the second arriving customer will not have to wait for service.

(5) The duration (in minutes) of a telephone conversation is exponentially distributed with parameter $\mu = 1/4$.
   (i) What is the average duration of a call?
   (ii) Suppose you arrive at a telephone booth two minutes after a call started. How long should you expect to wait?

(6) Arrivals of passengers and taxis at a particular taxi stand form two independent Poisson processes at rates λ and μ, respectively. Assume that the loading time is negligible. Let $X(t)$ denote the number of passengers or of taxis at time t. It takes on positive values if the number is of passengers otherwise takes on negative values. Find the distribution of $X(t)$.

(7) If customers arrive according to a Poisson process with rate λ and the service-times are exponentially distributed with mean μ^{-1}, show that the probability of an arrival occurring before a service completion is $\lambda/(\lambda+\mu)$. What is the probability that a service is completed before a new arrival?

References

Hall, R. W. (1991). *Queueing Methods for Service and Manufacturing*, Prentice-Hall, N.J.

Newell, G. F. (1982). *Applications of Queueing Theory*, Chapman & Hall, London.

Panico, J. A. (1969). *Queueing Theory: A Study of Waiting Lines for Business Economics and Science*, Prentice-Hall, N.J.

CHAPTER 2

Markovian Queues

2.1. Introduction

After introducing the subject of queue and a common sense based and pedestrian type analysis, we would aspire for a systematic development of the theory, especially keeping in view the abundant growth and maturity of the subject during a short period. As a starting point, we think of models associated with queueing processes. It is inspiring to know that there is a substantial class of queueing models, the so-called Markovian models whose analysis has been standardized to be dealt with routinely. In this chapter, it is our intention to expose the reader to the study of this class for its well-formulated structure yet quite suitable for many applications.

A Markovian queueing model is nothing but a Markov process (see Section A.6 in Appendix A) characterized by (A.61). It has an essential property that the time spent in any given state is an exponential random variable. As stated in Chapter 1, our attention will be focused on three characteristics, viz., the number of customers in the system, the waiting time of a customer and the busy period. In dealing with the first one, we will find that the Chapman-Kolmogorov equation (see (A.75)) and the stationary distribution (see (A.78)) would play a fundamental role in Markovian queues.

In order to ease our understanding, we proceed step by step by beginning with a simple Markovian model known as $M/M/1$, then moving to more comprehensive models called the birth-death models and finally ending with Erlangian models in which Markovian property which is originally nonexisting is induced suitably. In addition, it is decided to develop and derive the results from the first principle without directly depending on the Appendix. For completeness, the other two characteristics are discussed for the $M/M/1$ model. In case of other models, these are studied through general models, for example, $M/G/1$ and $G/M/1$ in Chapter 3.

We have taken the attitude that once some concepts and techniques are explained in a few models, the reader should try to solve similar problems in other situations. Therefore there is no need to present results on models which for example are special cases of birth-death models but to include them as exercises.

The world of queues is more complex than a simple independent system. A customer may require service at more than one service counter which are interconnected. Such a system is known as a queueing network. In this chapter, we introduce some simple networks for gaining a wider perspective on the subject. Similarly, for a better appreciation of the theory, we also introduce the idea of design and control of queues even at an early stage.

2.2. A Simple Model: Steady-State Behaviour

Consider a single-server queueing system with the following properties:

(i) The arrival or input process is Poisson with rate λ;

(ii) The service-times are independently and identically distributed (i.i.d.) exponential random variables with parameter μ;

(iii) Service-times are independent of arrival times;

(iv) Infinite waiting space is available;

(v) Customers are served in a first-come first-served (FCFS) fashion (in short, FCFS queue discipline).

This model is denoted by $M/M/1$ (the first 'M' refers to the Markovian input, the second one to the Markovian service completion time and *1* to the single server). It is the simplest nontrivial model because of the interconnection between the Poisson process, the exponential distribution and the structural simplicity of the Poisson process (see Section A.7 in Appendix A). We know that the interarrival times T are i.i.d. exponential (λ).

Some immediate consequences of (i) and (ii) are:

$$\begin{aligned}
P(\text{an arrival occurs in } (t, t+\Delta t)) &= \lambda\Delta t + o(\Delta t),\\
P(\text{no arrival occurs in } (t, t+\Delta t)) &= 1 - \lambda\Delta t + o(\Delta t),\\
P(\text{a service is completed in } (t, t+\Delta t)) &= \mu\Delta t + o(\Delta t),\\
P(\text{no service is completed in } (t, t+\Delta t)) &= 1 - \mu\Delta t + o(\Delta t),
\end{aligned}$$

and

$$\begin{aligned}
f_T(t) &= \lambda e^{-\lambda t}, \qquad t \geq 0,\\
f_S(s) &= \mu e^{-\mu s} \qquad s \geq 0,
\end{aligned}$$

where f_T and f_S are the probability density functions (p.d.f.) of interarrival time T and service-time S, respectively.

We will study most of the characteristics of interest and the corresponding performance measures, for the $M/M/1$ system. The number in the system being of primary interest, we look for an expression of the probability that there are n customers in the system at time t.

Notation:

$X(t)$: the number in the system at time t
$P_n(t)$: $P(X(t) = n)$

A standard technique is put forward what is known as the differential-difference equations for $P_n(t)$ which is described hereafter. By the well-known result on total probability (see (A.4)) we can write for $n \geq 1$,

$$\begin{aligned}P_n(t+\Delta t) =\ & P_{n+1}(t)P(\text{no arrival and one service completion in } (t, t+\Delta t))\\ &+ P_n(t)P(\text{no arrival and no service completion or one arrival}\\ &\quad \text{and one service completion in } (t, t+\Delta t))\\ &+ P_{n-1}(t)P(\text{one arrival and no service completion in } (t, t+\Delta t))\\ &+ \text{terms containing probability of more than one arrival or}\\ &\quad \text{service completion in } (t, t+\Delta t)\\ =\ & P_{n+1}(t)[(1-\lambda\Delta t + o(\Delta t))(\mu\Delta t + o(\Delta t))]\\ &+ P_n(t)[(1-\lambda\Delta t + o(\Delta t))(1-\mu\Delta t + o(\Delta t))\\ &+ (\lambda\Delta t + o(\Delta t))(\mu\Delta t + o(\Delta t))]\\ &+ P_{n-1}(t)[(\lambda\Delta t + o(\Delta t))(1-\mu\Delta t + o(\Delta t))] + o(\Delta t).\end{aligned}$$

by using (iii) and consequences of (i) and (ii). When we remember $(\Delta t)^2$ is $o(\Delta t)$, the equation simplifies to

$$P_n(t+\Delta t) = P_{n+1}(t)\mu\Delta t + P_{n-1}(t)\lambda\Delta t + P_n(t)(1-\lambda\Delta t - \mu\Delta t) + o(\Delta t),\ n \geq 1. \tag{2.1}$$

A similar argument shows that

$$P_0(t+\Delta t) = P_1(t)\mu\Delta t + P_0(t)(1-\lambda\Delta t) + o(\Delta t). \tag{2.2}$$

Notice that (2.1) is true for $n \geq 1$ due to the property (iv), and property (v) or in general the queue discipline has no impact on $P_n(t)$. However, the usefulness of queue discipline will be seen in the discussion of waiting time of a customer. We may rewrite (2.1) and (2.2) as

$$\frac{P_n(t+\Delta t) - P_n(t)}{\Delta t} = \mu P_{n+1}(t) + \lambda P_{n-1}(t) - (\lambda+\mu)P_n(t) + \frac{o(\Delta t)}{\Delta t}, \quad n \geq 1$$

and

$$\frac{P_0(t+\Delta t) - P_0(t)}{\Delta t} = \mu P_1(t) - \lambda P_0(t) + \frac{o(\Delta t)}{\Delta t}.$$

When $\Delta t \to 0$, these become

$$\frac{dP_n(t)}{dt} = \mu P_{n+1}(t) + \lambda P_{n-1}(t) - (\lambda+\mu)P_n(t), \quad n \geq 1$$

and

$$\frac{dP_0(t)}{dt} = \mu P_1(t) - \lambda P_0(t) \tag{2.3}$$

because

$$\lim_{t\to 0} \frac{o(\Delta t)}{\Delta t} = 0.$$

These happen to be a well-known form of Kolmogorov's forward equations.

Let the initial condition be $P_i(0) = 1$, which implies that there are i customers at the beginning. Equations (2.3) are differential-difference equations (differential equations in t and difference equations in n) that describe the transient or time-dependent behaviour of the system which when solved subject to the initial condition will provide the desired probability distribution of $X(t)$. It turns out that the solution is not elementary and will be dealt with later in Section 2.5. However, we may direct our attention to evaluate $\lim_{t\to\infty} P_n(t)$, although one would naturally ask whether it has any meaningful implication and whether it forms a probability distribution. These questions will be answered shortly.

Notation:

$$P_n \quad : \quad \lim_{t\to\infty} P_n(t)$$

Taking limits on both sides of (2.3) as $t \to \infty$, we obtain

$$\begin{aligned} &P_{n+1} = (\rho+1)P_n - \rho P_{n-1}, \quad n \geq 1 \qquad \text{and} \\ &P_1 = \rho P_0, \end{aligned} \tag{2.4}$$

where $\rho = \lambda/\mu$. Equations (2.4) express the steady-state behaviour of the number in the system and may be called *stationary equations.* Note that these types of equations are also known as difference equations or recurrence relations. In general, the ratio of the rate of input and the rate of service is known as the *traffic intensity.* For the $M/M/1$ model obviously the traffic intensity is ρ. The recursive approach leads to

$$P_n = \rho^n P_0, \quad n \geq 0. \tag{2.5}$$

The sequence $\{P_n\}$ forms a probability distribution (see (A.78)), if

$$\sum_{n=0}^{\infty} P_n = 1$$

which is called the *normalizing equation.* In general, for obtaining $\{P_n\}$ this equation forms an integral part of the system of equations. The normalizing equation along with (2.5) leads to

$$P_0 = \frac{1}{\sum_{n=0}^{\infty} \rho^n} = 1 - \rho, \tag{2.6}$$

if and only if $\rho < 1$. Therefore

$$P_n = \rho^n(1-\rho), \quad \rho < 1, n \geq 0 \tag{2.7}$$

which is the probability function (briefly p.f., see Sections A.2 and A.4 in Appendix A) of a geometric distribution.

Regarding the implication of $\{P_n\}$, it forms a probability distribution where P_n represents the probability that there are n in the system when the system is in operation for a long period of time and the transient behaviour subsides and the conditions for stationarity are satisfied. In other words, as $t \to \infty$, that the system is in a particular state does neither depend on t nor on the initial state. The sequence $\{P_n\}$ is called the *steady-state* or *stationary* distribution.

Since (2.4) is a set of homogeneous difference equations with constant coefficients, we may find the solution by the use of characteristic equation of the difference equations. For this purpose, set $P_n = \alpha^n$ in the first set of equations in (2.4). Then we have

$$\alpha^2 - (\rho + 1)\alpha + \rho = 0,$$

which is called the characteristic equation of

$$P_{n+1} = (\rho + 1)P_n - \rho P_{n-1}, \quad n = 1, 2, \ldots$$

Let the two roots of the characteristic equation be α_1 and α_2 which in our case are $\alpha_1 = 1$ and $\alpha_2 = \rho$. Then the general solution is

$$P_n = c_1\alpha_1^n + c_2\alpha_2^n,$$

where c_1 and c_2 are constants to be determined from the boundary conditions

$$P_1 = \rho P_0 \quad \text{(last equation in (2.4))}$$

and

$$\sum_{i=0}^{\infty} P_i = 1 \quad \text{(normalizing equation).}$$

In our present case, the first boundary condition gives rise to

$$c_1 + c_2\rho = \rho(c_1 + c_2),$$

which leads to $c_1 = 0$. From the second boundary condition it follows that $c_2 = 1 - \rho$ provided $\rho < 1$. Thus we get (2.7). Note that $c_1 = 0$ follows also from the convergence of $\sum_{i=0}^{\infty} P_i$.

We solve (2.4) by yet another method, viz., the method of generating function or the probability generating function (briefly p.g.f., see Section A.5 in Appendix A) which is often a standard technique for solving a set of difference equations. The p.g.f. of $\{P_n\}$ is given by $\sum_{n=0}^{\infty} P_n z^n$, where z is complex with $|z| \leq 1$.

Notation:

$$P(z) \;:\; \sum_{n=0}^{\infty} P_n z^n, \quad |z| \leq 1$$

Multiplying the first equation in (2.4) by z^n and summing from $n = 1$ to ∞, we get

$$z^{-1}\left[\sum_{n=-1}^{\infty} P_{n+1}z^{n+1} - P_1 z - P_0\right] = (\rho+1)\left[\sum_{n=0}^{\infty} P_n z^n - P_0\right] - \rho z \sum_{n=1}^{\infty} P_{n-1}z^{n-1}$$

which with the help of the second equation in (2.4) simplifies to

$$z^{-1}\left[P(z) - (\rho z + 1)P_0\right] = (\rho + 1)\left[P(z) - P_0\right] - \rho z P(z).$$

This yields

$$P(z) = \frac{P_0}{1 - z\rho}.$$

But $P(1) = 1$, therefore $P_0 = 1 - \rho$ provided $\rho < 1$ which was otherwise established in (2.6). $P(z)$ when expanded as a power series in z, becomes

$$P(z) = \sum_{n=0}^{\infty}(1-\rho)\rho^n z^n$$

which checks that

$$P_n = (1-\rho)\rho^n.$$

Also noting that $P_n = \frac{1}{n!}\frac{d^n P(z)}{dz^n}|_{z=0}$, one can obtain P_n by differentiation. We observe that the generating function technique suggests (i) to obtain an explicit expression for $P(z)$ from the difference equations and (ii) to invert $P(z)$ in order to get the desired probability distribution. In our case, the power series expansion has achieved the inversion of $P(z)$.

Remark: That $\rho < 1$ is necessary and sufficient for the steady state solution is intuitively justified, otherwise the size of queue will tend to increase without limit as time advances and therefore no time independent solution will exist. Also it makes sense that P_n decreases as n increases.

A simple but useful technique to write Equations (2.3) or (2.4) is by considering the system as a dynamic flow. For example, $\frac{dP_n(t)}{dt}$ in (2.3) represents the rate of change of flow into state n, which should be equal to the difference between the flow rate into state n (i.e., $\mu P_{n+1}(t)+\lambda P_{n-1}(t)$) and the flow rate out of state n (i.e., $(\lambda+\mu)P_n(t)$). Similarly the stationary Equations (2.4) are written by remembering that the flow must be conserved (i.e., $\frac{dP_n(t)}{dt} = 0$) that is, inflow rate equals outflow rate (see (A.87)).

For $M/M/1$, the following *flow diagram* in Figure 2.1 displays the input and output of any state:

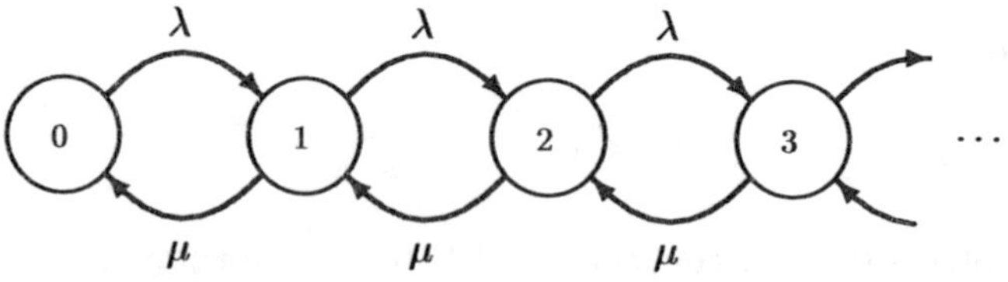

FIGURE 2.1. Flow diagram for the $M/M/1$ model

Here the state is represented by the number enclosed in a circle. Regarding the arrows, an example will be sufficient for explanation; say, the arrow from 1 to 2 with λ indicates the transition from state 1 to state 2 at the rate λ. The rate of transition into state $n (n \neq 0)$ is $\mu P_{n+1} + \lambda P_{n-1}$ and the rate of transition out of the same state is $(\lambda+\mu)P_n$. When equated we get the first equation in (2.4) and similarly the second one for $n = 0$. Indeed, in the future we suggest using this simple principle (i.e., rate in = rate out) for obtaining the stationary equations, also called *balance equations* due to the present approach.

Because of the simplicity of the above technique to obtain the limiting distribution of the number in the system in the $M/M/1$ model, we postpone the investigation of the other characteristics and continue the application of this technique to other models in Sections 2.3 and 2.4.

While we are studying the number of customers present in the system, it is worthwhile to find its mean and variance which are the main performance measures. Often these quantities may provide adequate information regarding the system, whereas the exact distribution being at times unsuitable for computational purposes may be of little practical value.

Notation:

L : expected number in the system
L_q : expected number in the queue

By remembering from (2.7) that P_n is the p.f. of a geometric distribution or by deriving directly, we have,

$$L = \sum_{n=0}^{\infty} nP_n = (1-\rho)\sum_{n=0}^{\infty} n\rho^n = \frac{\rho}{1-\rho} \tag{2.8}$$

and

$$L_q = \sum_{n=1}^{\infty}(n-1)P_n = L - \rho = \frac{\rho^2}{1-\rho}. \tag{2.9}$$

The variance of the number in the system is

$$\sum_{n=0}^{\infty}(n-L)^2 P_n,$$

which can be simplified and written as

$$\sigma^2 = \frac{\rho}{1-\rho} + \frac{\rho^2}{(1-\rho)^2}. \tag{2.10}$$

Although in the present case, the derivation of L_q and L is straightforward, we will see that the situation is not so in general (see Erlangian models, Section 2.4). Instead of evaluating the mean, variance and other moments directly from the definition, we may obtain them from the p.g.f. $P(z)$ by differentiation (see Section A.5 in Appendix A).

2.3. Birth-Death Models: Steady-State Behaviour

The method of obtaining steady-state probabilities in the last section is seen to be elementary and can be effective for models other than $M/M/1$. The idea of flow conservation which leads to the stationary equations can in fact be applied to a more general setting, namely, any continuous time Markov process (see Section A.8 in Appendix A). Although there exists a mathematical justification of these equations (see (A.87)), we simply illustrate the use of the technique of flow conservation for a few more models.

Of special interest are those models which are particular cases of the so-called birth-death process (see Sections A.6 and A.7). A birth-death process is one in which the transition takes place from a given state only to the nearest neighbours (i.e., either a 'birth' or a 'death' occurs). A birth or a death may respectively be considered as an arrival or a departure in a queueing system. In this process, the notion of a birth rate λ_n and a death rate μ_n is introduced which generalizes λ and μ of Section 2.2 and has the probability interpretation in the language of queues as follows:

$$P(\text{an arrival occurs in } (t, t+\Delta t) | n \text{ in the system}) = \lambda_n \Delta t + o(\Delta t)$$
$$P(\text{a service is completed in } (t, t+\Delta t) | n \text{ in the system}) = \mu_n \Delta t + o(\Delta t).$$

The flow diagram for the birth-death process is shown in Figure 2.2.

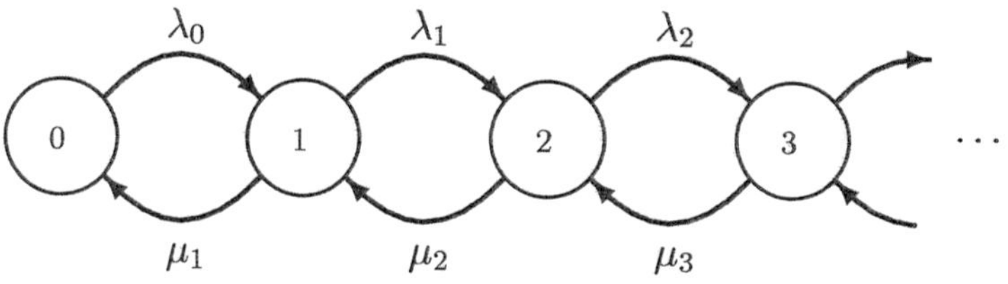

FIGURE 2.2. Flow diagram for birth-death process

Observe that in the $M/M/1$ model $\lambda_n = \lambda$ and $\mu_n = \mu$ for all n. From the diagram we get the balance equations as

$$\lambda_{n-1}P_{n-1} + \mu_{n+1}P_{n+1} - (\lambda_n + \mu_n)P_n = 0, \quad n \geq 1, \qquad (2.11)$$
$$\mu_1 P_1 - \lambda_0 P_0 = 0.$$

Also the normalizing equation is

$$\sum_{n=0}^{\infty} P_n = 1. \qquad (2.12)$$

By iteration, it is easy to check from (2.11) that

$$P_n = P_0 \prod_{i=0}^{n-1} \frac{\lambda_i}{\mu_{i+1}}, \quad n \geq 1. \qquad (2.13)$$

Evaluation of P_0 is done by substituting P_n from (2.13) in (2.12), which leads to

$$P_0 = \frac{1}{1+\sum_{n=1}^{\infty}\prod_{i=0}^{n-1}\frac{\lambda_i}{\mu_{i+1}}}. \tag{2.14}$$

Expressions (2.13) and (2.14) constitute the solutions of (2.11) and (2.12), provided the denominator in P_0 converges. Observe that we have the following from (2.13):

$$P_n = \frac{\lambda_{n-1}}{\mu_n} P_{n-1}, \quad n \geq 1.$$

A few models which are special cases of the birth-death process are listed below along with the solutions.

1. Discouraged arrivals

Arrivals get discouraged when more and more customers are present in the system. A possible model is given by

$$\lambda_n = \frac{\lambda}{n+1} \quad \text{for every } n \geq 0,$$
$$\mu_n = \mu \quad \text{for every } n \geq 1.$$

Then

$$P_n = \frac{\rho^n e^{-\rho}}{n!}, \quad \rho < 1 \tag{2.15}$$

which is the p.f. of the Poisson distribution with parameter ρ. Therefore

$$L = \rho, \tag{2.16}$$

which is also equal to the variance of the number in the system.

2. M/M/c

The system has c servers. The second 'M' which refers to the service-time distribution implies that the service-times for each server are i.i.d. exponential variables with parameter μ and these service-times are independent for different servers. Thus the model parameters λ_n and μ_n are:

$$\lambda_n = \lambda \quad \text{for every } n,$$

$$\mu_n = \begin{cases} n\mu & \text{for } 0 \leq n \leq c \\ c\mu & \text{for } n \geq c. \end{cases}$$

Then

$$P_n = \begin{cases} \dfrac{P_0\rho^n}{n!} & 0 \leq n \leq c, \\ \dfrac{P_0\rho^n}{c!c^{n-c}} & n \geq c. \end{cases} \tag{2.17}$$

It can be verified that

$$L_q = \frac{\rho^{c+1}}{(c-1)!(c-\rho)^2} P_0. \tag{2.18}$$

It may be noted that a direct derivation of L is possible and yet becomes simpler by a result developed in Section 2.6.

3. $M/M/\infty$

$$\lambda_n = \lambda \qquad \text{for every } n,$$
$$\mu_n = n\mu \qquad \text{for every } n.$$
$$P_n = \frac{\rho^n e^{-\rho}}{n!} \tag{2.19}$$

Observe that Models 1 and 3 are not the same, but their solutions are identical.

4. $M/M/c$ with finite customer population and limited waiting space

In this model, we have the following two additional constraints on Model 2:

(i) An arrival is coming from a source which is finite, say consists of J customers,

(ii) The waiting space is limited to hold at most K customers in the system ($c \leq K \leq J$).

Moreover, the first "M" for finite number of customers has the interpretation that each customer's arrival is governed by the Poisson process with rate λ and the processes are independent. Therefore, when there are n customers in the system, any one of the remaining $J-n$ customers is to arrive at the rate $(J-n)\lambda$, since the arrival process at this point is the superimposition of $J-n$ independent Poisson processes. Another assumption is that when $n = K$ (i.e., waiting space is full), the arriving customer returns to the source as if it has just completed the service. In this sense, the customer is considered to be "lost". Thus

$$\lambda_n = \begin{cases} \lambda(J-n) & 0 \leq n \leq K-1, \\ 0 & \text{otherwise}, \end{cases}$$

$$\mu_n = \begin{cases} n\mu & 1 \leq n \leq c, \\ c\mu & c \leq n \leq K, \end{cases}$$

and

$$P_n = \begin{cases} P_0 \rho^n \binom{J}{n} & 0 \leq n \leq c, \\ P_0 \rho^n \binom{J}{n} \dfrac{n!}{c! c^{n-c}} & c \leq n \leq K. \end{cases} \tag{2.20}$$

When $c = K$,

$$P_n = P_0 \binom{J}{n} \rho^n, \quad n = 0, 1, \ldots, c. \tag{2.21}$$

Application:

An example of this model is that of machine maintenance. The population is a pool of J machines which are repaired by c repairmen when out of order. We assume $J = K$; otherwise, we face the absurd situation when a machine which becomes inoperative and does not find waiting space has to go back for operation. The Poisson nature of the arrival process presupposes the operation time being exponentially distributed, which seems to be a reasonable assumption.

Remarks:

1. The model is general enough to cover many interesting variations and special cases.
2. It is possible to get the infinite population model from the finite one by assuming $\lim_{J\to\infty} \lambda J = \lambda^*$ (a constant) as $J \to \infty$ and $\lambda \to 0$. In other words, the arrival rate for an individual customer is small enough such that the aggregate rate is a constant.
3. When J is large, the binomial coefficient in expression (2.20) becomes computationally unmanageable because it involves $J!$. In order to avoid this difficulty, one can develop a recursive relation between P_n and P_{n-1} as

$$P_n = f_{n-1}P_{n-1} \quad 1 \le n \le J$$

with

$$P_0 \left[1 + \sum_{n=1}^{J} \prod_{i=0}^{n-1} f_i\right]^{-1}$$

where

$$f_n = \begin{cases} \dfrac{J-n}{n+1}\rho & 0 \le n \le c-1, \\ \dfrac{J-n}{c}\rho & c-1 \le n \le J-1. \end{cases}$$

2.4. Erlangian Models: Steady-State Behaviour

The birth-death process which is typified by the nearest-neighbour transition, enables us not only to easily write down stationary or balance equations but also to obtain a simple explicit expression for P_n. Now we consider models called 'Erlangian models' (to be explained below) in which non-neighbour transition takes place. For these models the stationary equations are easily derived by following the flow conservation technique. However, the explicit solution for P_n may no longer be simple.

2.4.1. The $M/E_k/1$ model. The service-time distribution 'E_k' stands for the Erlangian distribution, with k stages (or phases) as suggested by A. K. Erlang (see (A.48) in Appendix A). What it means is that a customer's service-time may be viewed as the sum of k i.i.d. exponential service periods each having the parameter $k\mu$. Any exponential period is considered to be a stage and the service is complete when the customer goes through all stages. This assumption seems to be reasonable when a server performs several duties of various nature which are required by a customer (for example, at a bus or a railway station where the server issues the ticket and collects the luggage).

From the assumptions, it is clear that

$$f_T(t) = \lambda\, e^{-\lambda t}, \quad t \geq 0, \tag{2.22}$$

$$f_S(t) = \frac{(k\mu)^k}{(k-1)!} t^{k-1} e^{-k\mu t}, \quad t \geq 0,$$

since the sum of k i.i.d. exponential $(k\mu)$ variables is a gamma variable with parameters $(k,\, k\mu)$. Thus an Erlangian distribution is a gamma distribution with integer-valued first parameter. Observe that the rate of service is μ (same as that in the $M/M/1$ model). The advantage of decomposing the service-time into k successive stages is realized when the number of stages to be completed is considered as a state of a Markov process. This particular procedure of splitting the entire duration which is nonexponential, into several stages each of which is exponential (so that the Markovian property is retained) may be called the *method of stages.* More specifically, if there are r customers in the system and the customer in service is in the ith stage, then we say that the process is in state $(r-1)k + (k-i+1) = rk - i + 1$, which is the number of stages to be completed by the customers in the system. Using this convention, the flow diagram is given in Figure 2.3.

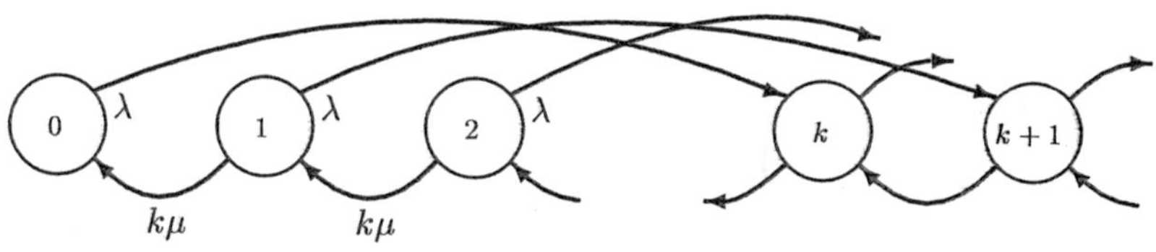

FIGURE 2.3. Flow diagram for the $M/E_k/1$ model

Suppose the process is in state $n-k$. It changes either to state $n-k-1$ if the service in the particular stage is complete or to state n if a customer arrives. This explains the above diagram.

Notation:

$p_n \quad : \quad P(\text{system in state } n)$

Using the technique of the flow diagram, we obtain

$$\begin{aligned} p_n &= 0, \quad n < 0 \\ \lambda\, p_0 &= k\mu p_1 \\ (\lambda + k\mu)p_n &= \lambda p_{n-k} + k\mu p_{n+1}, \quad n = 1, 2, \ldots . \end{aligned} \tag{2.23}$$

The simplest way to find p_n is to solve (2.23) recursively and to use $\sum_{n=0}^{\infty} p_n = 1$. However, it is instructive to look into the generating function approach and without any ambiguity let us use the same notation $P(z)$ for the generating function of $\{p_n\}$. As in the $M/M/1$ case, we multiply z^n to the last equation in (2.23) and add over all n to get

$$P(z) = \frac{p_0(\lambda + k\mu - \frac{k\mu}{z}) - k\mu p_1}{\lambda + k\mu - \lambda z^k - \frac{k\mu}{z}}$$

which by the second equation in (2.23) becomes

$$P(z) = \frac{p_0 k\mu(1-z)}{k\mu + \lambda z^{k+1} - (\lambda + k\mu)z}. \tag{2.24}$$

Since $P(1) = 1$, we use L'Hôpital's rule to obtain

$$\frac{p_0 k\mu}{k\mu - \lambda k} = 1$$

and finally derive

$$p_0 = 1 - \rho, \quad \rho < 1$$

which is the same as (2.6).

Substitution of p_0 in (2.24) leads to

$$P(z) = \frac{k\mu(1-\rho)(1-z)}{k\mu + \lambda z^{k+1} - (\lambda + k\mu)z}. \tag{2.25}$$

In its present form the power series expansion is not apparent. However, when written as

$$P(z) = \frac{1-\rho}{1 - \frac{\rho}{k} z \frac{(1-z^k)}{1-z}}$$

the power series expansion ultimately becomes

$$P(z) = (1-\rho) \sum_{n=0}^{\infty} \sum_{j=0}^{\infty} \sum_{i=0}^{n} (-1)^i \binom{n}{i} \binom{n+j-1}{j} z^{j+ik+n} \left(\frac{\rho}{k}\right)^n. \tag{2.26}$$

Obviously, p_n is obtained as the coefficient of z^n in (2.26). It seems the explicit expression for p_n obtained in this manner is not very appropriate for computational purposes.

When a power series expansion is not conveniently available as in the present case, one adopts a standard method of inversion by getting a partial fraction expansion of $P(z)$ and then inverting each term, usually by inspection.

For the partial fraction expansion, it is necessary to find the roots of the polynomial in the denominator, which may often be difficult unless k is 1 or 2.

In our case, one checks by inspection that $z = 1$ is a root of the denominator. Let the remaining roots be distinct and denoted by $z_1, \ldots, z_k$. Then the partial fraction expansion gives

$$P(z) = (1-\rho)\sum_{i=1}^{k} \frac{A_i}{(1 - z/z_i)}$$

where

$$A_i = \prod_{\substack{j=1 \\ j \neq i}}^{k} \frac{1}{(1 - z_i/z_j)}.$$

The inversion of $\frac{1}{1-z/z_i}$ being $\frac{1}{z_i^n}$, the usual inversion technique gives rise to

$$p_n = (1-\rho)\sum_{i=1}^{k} \frac{A_i}{z_i^n}, \quad n = 1, 2, \ldots . \tag{2.27}$$

Here, we have merely demonstrated the technique of partial fraction expansion which could be used in many situations. We observe that the distribution of the number of stages is a weighted sum of geometric distribution. When $k = 1$, the system is $M/M/1$ and in that case

$$\begin{aligned} P(z) &= \frac{1-\rho}{1-\rho z} \\ &= (1-\rho)\sum_{j=0}^{\infty} \rho^j z^j. \end{aligned}$$

Thus

$$p_n = P_n = (1-\rho)\rho^n, \quad n = 0, 1, \ldots$$

which was observed previously. The reader is encouraged to find p_n for $k = 2$.

Our interest being in P_n, the following relations are useful.

$$P_n = \sum_{j=(n-1)k+1}^{nk} p_j, \quad n = 1, 2, \ldots$$

and

$$P_0 = p_0 = 1 - \rho. \tag{2.28}$$

Remarks:

1. A direct evaluation of P_n will be possible by another approach discussed in the next chapter.

2. Interesting to note that $M/E_k/1$ model is the same as $M/M/1$ model with bulk arrival of size k, since k customers being served one after the other is the same as one customer being served k times. In other words, a Poisson (λ) arrival consisting of k customers and an arrival being served means k customers being served each with rate $k\mu$.

3. Many distributions may be approximated by E_k for suitable selection of k. The mean and variance of the distribution E_k are $\frac{1}{\mu}$ and $\frac{1}{k\mu^2}$, respectively. Thus when $k \to \infty$, E_k in the limit becomes the degenerate distribution at $\frac{1}{\mu}$ (see the end in Section A.3 for the definition of a degenerate distribution).

Instead of a fixed bulk size we may consider a random size bulk arrival. If we assume the probability of arrival of size i to be α_i, the rate of arrival of size i becomes $\lambda\alpha_i$. For simplicity we keep the service rate to be μ. Then the stationary equations are

$$\lambda P_0 = \mu P_1,$$

$$(\lambda + \mu)P_n = \mu P_{n+1} + \sum_{i=0}^{n-1} P_i \lambda \alpha_{n-i}, \quad n \geq 1. \tag{2.29}$$

Again the application of the generating function technique gives us

$$(\lambda + \mu)(P(z) - P_0) = \frac{\mu}{z}[P(z) - P_1 z - P_0] + \lambda \sum_{n=1}^{\infty}\sum_{i=0}^{n-1} P_i \alpha_{n-i} z^n. \tag{2.30}$$

Introducing the generating function for $\{\alpha_n\}$ as

$$G(z) = \sum_{n=1}^{\infty} \alpha_n z^n$$

we get

$$\begin{aligned}\sum_{n=1}^{\infty}\sum_{i=0}^{n-1} P_i \alpha_{n-i} z^n &= \sum_{i=0}^{\infty} P_i z^i \sum_{n=i+1}^{\infty} \alpha_{n-i} z^{n-i} \\ &= \sum_{i=0}^{\infty} P_i z^i \sum_{j=1}^{\infty} \alpha_j z^j = P(z)G(z).\end{aligned} \tag{2.31}$$

Therefore, from (2.30) and (2.31) and using $P(1) = 1$ through L'Hôpital's rule, one finally obtains

$$P(z) = \frac{\mu P_0(1-z)}{\mu(1-z) - \lambda z(1 - G(z))} \tag{2.32}$$

where P_0 can be checked to be $1 - \rho G'(1)$. In order to complete the discussion on the derivation of the distribution, it may be said that an expression for $G(z)$ is needed and then the problem of inversion is to be handled. As a special case, one notes that for fixed bulk size of k, $G(z) = z^k$ and (2.32) checks with (2.25) in which $k\mu$ is replaced by μ.

Now we return to the evaluation of L and L_q. From (2.23), it follows that

$$(\lambda + k\mu)\sum_{n=0}^{\infty} n^2 p_n = k\mu \sum_{n=0}^{\infty} n^2 p_{n+1} + \lambda \sum_{n=k}^{\infty} n^2 p_{n-k},$$

which may be written as

$$(\lambda + k\mu)\sum_{n=0}^{\infty} n^2 p_n = k\mu \sum_{n=1}^{\infty} (n-1)^2 p_n + \lambda \sum_{n=0}^{\infty} (n+k)^2 p_n.$$

On expanding and simplifying, we have

$$2(\mu - \lambda)\sum_{n=1}^{\infty} n p_n = \mu(1 - p_0) + \lambda k.$$

From (2.28) we know that $p_0 = 1 - \rho$. Therefore

$$\sum_{n=1}^{\infty} n p_n = \frac{\lambda(k+1)}{2(\mu - \lambda)},$$

which represents the average number of stages left in the system and is not the same as L. Next, let us find the average number of incomplete stages in service. Given that the system is not empty, it is equally likely that the service is in any one of k stages. Thus, the average number of incomplete stages in service is given by

$$\rho \frac{1}{k} \frac{k(k+1)}{2} = \frac{\lambda(k+1)}{2\mu}.$$

Thus the average number of stages in the queue is equal to

$$\frac{\lambda(k+1)}{2(\mu - \lambda)} - \frac{\lambda(k+1)}{2\mu} = \frac{\lambda^2(k+1)}{2\mu(\mu - \lambda)},$$

which when divided by k gives an expression for L_q. Therefore

$$L_q = \frac{\lambda^2(k+1)}{2k\mu(\mu - \lambda)} \tag{2.33}$$

and

$$L = L_q + \rho \qquad \text{(see (2.9))}.$$

Remark: The present situation is an example where L and L_q are derived in an indirect manner, not through the basic definition, since the expansion for the exact distribution is not simple (see (2.26)). We can also obtain the moments including the mean L from (2.25) by differentiation.

2.4.2. The $E_k/M/1$ model. Following an argument very similar to the $M/E_k/1$ model, we note that

$$f_T(t) = \frac{(k\lambda)^k}{(k-1)!} t^{k-1} e^{-k\lambda t} \quad t \geq 0 \tag{2.34}$$
$$f_S(s) = \mu\, e^{-\mu s} \quad s \geq 0.$$

Observe that the interarrival time and service-time of $M/E_k/1$ and $E_k/M/1$ are interchanged and in that sense one is the 'dual' of the other. In this case, the interarrival time is divided into k successive stages, the duration of each being i.i.d. exponential $(k\lambda)$. The system state is marked by the completed arrival stages. Thus if there are r customers in the system and the arriving customer is at the ith stage, then the system state is $rk + i - 1$. The flow diagram is given in Figure 2.4.

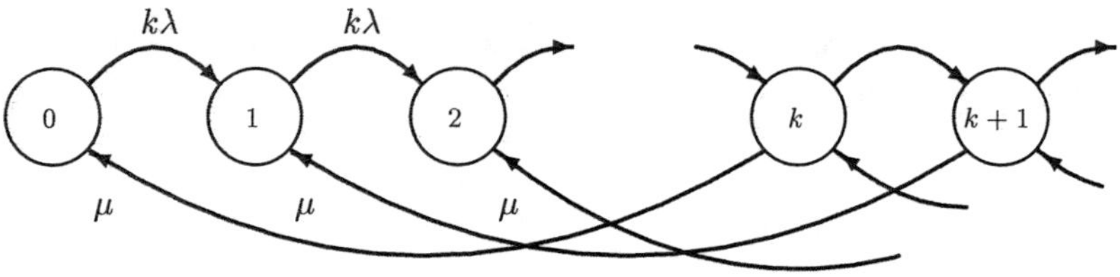

FIGURE 2.4. Flow diagram for the $E_k/M/1$ model

The following may be established routinely:

$$\begin{aligned} k\lambda p_0 &= \mu p_k \\ k\lambda p_n &= k\lambda p_{n-1} + \mu p_{n+k}, \qquad 1 \leq n \leq k-1 \\ (k\lambda + \mu) p_n &= k\lambda p_{n-1} + \mu p_{n+k}, \qquad k \leq n \end{aligned} \tag{2.35}$$

$$P(z) = \frac{(1 - z^k) \sum_{j=0}^{k-1} p_j z^j}{k\rho z^{k+1} - (1 + k\rho) z^k + 1}. \tag{2.36}$$

In the present situation, neither the direct power series expansion nor the partial fraction expansion is available. The main difficulty arises due to the factor $\sum_{j=0}^{k-1} p_j z^j$ in the numerator which contains unknown p_j 's. Therefore a slightly more subtle analysis is required.

Observe that $z = 1$ is a root in the denominator. It can be seen by Rouché's Theorem (see Section A.5) that $k - 1$ of the remaining roots are in $|z| < 1$ and one in $|z| > 1$. However $P(z)$ being a probability generating function must be bounded in $|z| \leq 1$. Hence in $|z| \leq 1$, the zeroes of the denominator must be the zeroes in the numerator. The zeroes of $(1 - z^k)$ in the numerator have absolute value equal to unity. Therefore, letting the root of the denominator in $|z| > 1$ be z_0, $P(z)$ in (2.36) can be expressed as

$$P(z) = \frac{C(1-z^k)}{(1-z)(z_0-z)} = \frac{C\sum_{j=0}^{k-1} z^j}{z_0 - z},$$

where C is a constant. Eventually with the help of $P(1) = 1$, we get

$$P(z) = \frac{(z_0-1)\sum_{j=0}^{k-1} z^j}{k(z_0-z)}.$$

Using the power series expansion for $(1-\frac{z}{z_0})^{-1}$ or the method of differentiation, we get

$$\begin{aligned} p_n &= \frac{1-{z_0}^{-n-1}}{k}, \qquad n \leq k-1, \\ p_n &= \frac{{z_0}^{-n-1}(z_0^k-1)}{k}, \quad k \leq n. \end{aligned} \tag{2.37}$$

From (2.37) we can compute P_n which has the following expression of a geometric distribution:

$$P_0 = 1 - \rho$$

and

$$P_n = \sum_{j=nk}^{nk+k-1} p_j = \rho {z_0}^{-(n-1)k}(1-{z_0}^{-k}), \quad n \geq 1. \tag{2.38}$$

Remarks:

1. Similar to $M/E_k/1$, we may consider here an arriving customer to pass through k exponential $(k\lambda)$ stages. As a model although it sounds somewhat artificial, we can use it in situations when customers are arriving at a counter in a Poisson process $(k\lambda)$ and only every kth arrival is admitted for service (thus a customer for service passes through k arriving stages).

2. Again the $E_k/M/1$ model is equivalent to the $M/M/1$ model with bulk service of fixed size k (i.e., customers are served collectively in batches of size k). This is so because the situation to have a customer going through k independent interarrival times before reaching the service counter is seen in another way to have k arrivals going together for service. Obviously, the server has to wait until there are k customers in the system.

A variation of the bulk service model is when the server upon becoming free accepts k or less number of customers. The stationary equations are

$$\begin{aligned} (\lambda+\mu)P_n &= \mu P_{n+k} + \lambda P_{n-1}, \quad n \geq 1 \\ \lambda P_0 &= \mu(P_1 + \ldots + P_k). \end{aligned}$$

By an argument similar to the one given for the $M/E_k/1$ model, we note that

$$P(z) = \frac{1 - 1/z_0}{1 - z/z_0}, \tag{2.39}$$

where z_0 is the root of

$$k\rho z^{k+1} = (1 + k\rho)z^k - 1 \tag{2.40}$$

such that $|z_0| > 1$. Finally, after the inversion of (2.39) we get

$$P_n = \left(1 - \frac{1}{z_0}\right)\left(\frac{1}{z_0}\right)^n, \quad n = 0, 1, \ldots \tag{2.41}$$

which is the p.f. of the geometric distribution. Observe that z_0 is the same as that in (2.37).

2.5. Transient Behaviour

The study of transient behaviour of Markovian queues is more complex than one would like to think, which will be demonstrated in the case of the $M/M/1$ model. Yet there exist several methods of solution for this simple model. Let us start with the $M/M/1$ model and assume there are initially i customers.

2.5.1. Method 1. The differential-difference equations (2.3) which reflect the transient (in contrast to steady state) behaviour of $M/M/1$ system have been delayed until now, because the procedure for solving the differential-difference equations is more complicated than solving the difference equations or balance equations. A standard technique to solve difference equations, in particular balance equations, is to use the p.g.f. which has been discussed in Section 2.2. As far as the differential part is concerned, the Laplace transform (L.T. in short, see Section A.5) which corresponds to the p.g.f. will be used. One may be reminded of the fact that the recursive method will not work because of the presence of the differential part.

Notation:

$$P(z,t) \quad : \quad \sum_{n=0}^{\infty} P_n(t)z^n$$

$$P^*(z,\theta) \quad : \quad \int_0^{\infty} e^{-\theta t} P(z,t)dt \quad \text{(L.T. of } P(z,t))$$

$$P_n^*(\theta) \quad : \quad \int_0^{\infty} e^{-\theta t} P_n(t)dt \quad \text{(L.T. of } P_n(t))$$

(Invariably the L.T. of any function $a(t)$ will be denoted by $a^*(\theta)$.)

In order to get a relation on the p.g.f. of $P(z,t)$ we start from (2.3) and end up with

$$z\frac{\partial P(z,t)}{\partial t} = (1 - z)[(\mu - \lambda z)P(z,t) - \mu P_0(t)] \tag{2.42}$$

by using the fact that

$$\sum_{n=1}^{\infty} \frac{dP_n(t)}{dt} z^n = \frac{\partial P(z,t)}{\partial t} - \frac{dP_0(t)}{dt}$$
$$= \frac{\partial P(z,t)}{\partial t} + \lambda P_0(t) - \mu P_1(t).$$

Equation (2.42) as it is, cannot be solved directly because of the existence of $P_0(t)$. Therefore, we apply the L.T. on both sides of (2.42) and use the relation

$$\int_0^{\infty} e^{-\theta t} \left(\frac{\partial P(z,t)}{\partial t} \right) dt = \theta P^*(z,\theta) - z^i$$

(which can be checked by using integration by parts and by remembering the initial condition that there are i customers at $t = 0$) to obtain

$$z(\theta P^*(z,\theta) - z^i) = (1-z)[(\mu - \lambda z)P^*(z,\theta) - \mu P_0^*(\theta)]$$

which leads to

$$P^*(z,\theta) = \frac{z^{i+1} - \mu(1-z)P_0^*(\theta)}{(\lambda+\mu+\theta)z - \mu - \lambda z^2}. \tag{2.43}$$

Now we want to find an expression for $P_0^*(\theta)$ by a well-known method. Notice that the denominator of $P^*(z,\theta)$ has two zeroes z_1 and z_2 given by

$$z_1 = \frac{\lambda+\mu+\theta - \sqrt{(\lambda+\mu+\theta)^2 - 4\lambda\mu}}{2\lambda}, \tag{2.44}$$
$$z_2 = \frac{\lambda+\mu+\theta + \sqrt{(\lambda+\mu+\theta)^2 - 4\lambda\mu}}{2\lambda}.$$

In addition to the fact that

$$|z_1| < |z_2|$$

the following properties are easily derived:

$$z_1 + z_2 = \frac{\lambda+\mu+\theta}{\lambda}, \quad z_1 z_2 = \frac{\mu}{\lambda}. \tag{2.45}$$

Using Rouché's Theorem (see Section A.5 in Appendix A) with the contour $C : |z| = 1$, $f(z) = (\lambda+\mu+\theta)z$ and $g(z) = -\mu - \lambda z^2$, we note that the denominator has only one zero in the region $|z| < 1$ and $\Re(\theta) > 0$ and this must be z_1 since $|z_1| < |z_2|$. But $P^*(z,\theta)$ is convergent in the same region and hence its numerator must vanish at z_1. This leads to

$$P_0^*(\theta) = \frac{z_1^{i+1}}{\mu(1-z_1)}. \tag{2.46}$$

Up to this point we focused our attention to derive an explicit expression for $P^*(z,\theta)$. What remains to be done is to go through the two-stage inversion, viz., first to obtain $P_n^*, (\theta)(n > 0)$ and then $P_n(t)$. Now we briefly discuss how to obtain $P_n^*(\theta)$.

The denominator of (2.43) with the help of (2.44) becomes

$$\lambda(z - z_1)(z_2 - z).$$

Thus (2.43) through (2.46) can be written as

$$\frac{z^{i+1} - \frac{(1-z)z_1^{i+1}}{1-z_1}}{\lambda(z - z_1)(z_2 - z)}$$

which can be reduced to

$$P^*(z, \theta) = \frac{1}{\lambda z_2(1 - \frac{z}{z_2})}\left[z^i(1 + \frac{z_1}{z} + \ldots + \left(\frac{z_1}{z}\right)^i) + \frac{z_1^{i+1}}{1 - z_1}\right].$$

With some further simplification by using (2.45), one can show that for $n \geq i$,

$$P_n^*(\theta) = \frac{1}{\lambda}\Big[z_2^{i-n-1} + \rho^{-1}z_2^{i-n-3} + \rho^{-2}z_2^{i-n-5} + \ldots + \rho^{-i}z_2^{-i-n-1} + \rho^{n+1}\sum_{j=n+i+2}^{\infty}(\rho z_2)^{-j}\Big]. \tag{2.47}$$

In order to get the expression for $P_n(t)$, the inverse L.T. of $P_n^*(\theta)$ is needed. For this purpose, we may use the fact that the inverse L.T. of z_2^{-n} equals

$$e^{-(\lambda+\mu)t}n\rho^{n/2}t^{-1}I_n(2\sqrt{\lambda\mu}\,t),$$

where $I_n(z)$ is the modified Bessel function of the first kind for which the expression is

$$I_n(z) = \sum_{j=0}^{\infty}\frac{(z/2)^{n+2j}}{j!(n+j)!}. \tag{2.48}$$

When some of the properties of $I_n(z)$, viz.,

$$\frac{2n}{z}I_n(z) = I_{n-1}(z) - I_{n+1}(z),$$

and

$$I_n(z) = I_{-n}(z), \tag{2.49}$$

are applied, one can finally obtain for $n \geq i$,

$$P_n(t) = e^{-(\lambda+\mu)t}\Big[\rho^{\frac{n-i}{2}}I_{n-i}(2\sqrt{\lambda\mu}\,t) + \rho^{\frac{n-i-1}{2}}I_{n+i+1}(2\sqrt{\lambda\mu}\,t) + (1-\rho)\rho^n\sum_{j=n+i+2}^{\infty}\rho^{-j/2}I_j(2\sqrt{\lambda\mu}\,t)\Big]. \tag{2.50}$$

It can be checked that (2.50) is true also for $n < i$. Using the asymptotic behaviour of the modified Bessel function, it can be proved that

$$\lim_{t\to\infty}P_n(t) = P_n = \rho^n(1-\rho), \qquad n \geq 0.$$

The proof is left as an exercise (Exercise 17) in order not to overshadow the basic subject matter with complex mathematical tricks.

2.5.2. Method 2. As observed earlier, the L.T. is required due to the appearance of $P_0(t)$ in (2.42), which arises from the second equation in (2.3). If the second equation can be converted into the first one with appropriate boundary condition, then the L.T. can be avoided. This idea leads us to an alternative method of solution in which we consider an extended system of equations

$$P'_n(t) = -(\lambda+\mu)P_n(t) + \lambda P_{n-1}(t) + \mu P_{n+1}(t) \qquad \text{for every } n, \tag{2.51}$$

subject to the condition

$$P_0(t) = \rho P_{-1}(t) \tag{2.52}$$

and

$$P_n(0) = \delta_{i,n} \tag{2.53}$$

where $\delta_{i,n}$ is the Kronecker delta as given by

$$\delta_{i,n} = \begin{cases} 1 & \text{if } n = i, \\ 0 & \text{otherwise} \end{cases}.$$

Note that under condition (2.52), Equations (2.51) become (2.3).

Now we look for the solution of the extended system. Using the extended p.g.f. (without any confusion we adopt the same notation),

$$P(z,t) = \sum_{n=-\infty}^{\infty} P_n(t) z^n,$$

we get

$$\frac{\partial P(z,t)}{\partial t} = (\lambda z - (\lambda+\mu) + \frac{\mu}{z})P(z,t),$$

which is an elementary differential equation. Its solution is

$$P(z,t) = \Phi(z)\exp\{-(\lambda+\mu)t + (\lambda z + \frac{\mu}{z})t\}, \tag{2.54}$$

where $\Phi(z)$ is an arbitrary function of z. When the right side is expressed in its power series expansion, (2.54) can be written as

$$P(z,t) = \Phi(z)e^{-(\lambda+\mu)t} \sum_{n=-\infty}^{\infty} I_n(2\sqrt{\lambda\mu}\,t) z^n \rho^{\frac{n}{2}}. \tag{2.55}$$

Let

$$\Phi(z) = \sum_{n=-\infty}^{\infty} \psi_{-n} z^n.$$

Then $P_n(t)$, being the coefficient of z^n in $P(z,t)$, has the expression from (2.55) as

$$P_n(t) = e^{-(\lambda+\mu)t} \sum_{j=-\infty}^{\infty} \psi_j \rho^{\frac{1}{2}(n+j)} I_{n+j}(2\sqrt{\lambda\mu}\,t) \tag{2.56}$$

where ψ's are determined through the initial conditions (2.52) and (2.53). When $t = 0$ in (2.56), it follows that

$$\psi_{-n} = \delta_{i,n}, \qquad n \geq 0 \tag{2.57}$$

by (2.53) and by $I_n(0) = \delta_{0,n}$, a property of the Bessel function. ψ_n for $n > 0$ is determined by (2.52). For convenience, introduce

$$\gamma_n = \psi_n \rho^{\frac{n}{2}} \tag{2.58}$$

and $I_n = I_n(2\sqrt{\lambda\mu}\, t)$. Then (2.52) with the help of (2.56) and (2.57) becomes

$$\rho^{-\frac{i}{2}} I_i + \sum_{j=1}^{\infty} \gamma_j I_j = \rho \left[\rho^{-\frac{i+1}{2}} I_{i+1} + \rho^{-\frac{1}{2}} \sum_{j=1}^{\infty} \gamma_j I_{j-1} \right]. \tag{2.59}$$

Comparing coefficients of I on both sides of (2.59), we can get

$$\gamma_j = \begin{cases} 0 & j = 1, 2, \ldots, i, \\ \rho^{-j/2} & j = i+1 \\ \rho^{-j/2}(1-\rho) & j = i+2, i+3, \ldots . \end{cases} \tag{2.60}$$

Finally, substituting values of ψ's from (2.57), (2.58) and (2.60) in (2.56) we get (2.50).

Because of the difficulties even in this simple model, we postpone further developments on transient solutions to Chapter 6 including the treatment of combinatorial methods (Section 6.3).

2.5.3. Some alternative formulas. It is instructive to mention a few alternative expressions for (2.50). By an independent probabilistic argument, we can show that

$$\begin{aligned} P_n(t) &= e^{-(\lambda+\mu)t} \rho^{(n-i)/2} (I_{n-i} - I_{n+i}) \\ &\quad + e^{-(\lambda+\mu)t} \rho^n \sum_{k=n+i+1}^{\infty} k \rho^{-\frac{k}{2}} (\mu t)^{-1} I_k \end{aligned} \tag{2.61}$$

(see Exercise 19) where $I_j = I_j(2\sqrt{\lambda\mu} t)$. The second term on the right side becomes

$$e^{-(\lambda+\mu)t} \rho^n \sum_{k=n+i+1}^{\infty} \rho^{-\frac{k-1}{2}} (I_{k-1} - I_{k+1})$$

when the first relation in (2.49) is applied, and this simplifies to

$$e^{-(\lambda+\mu)t} \{ \rho^{\frac{n-i}{2}} I_{n+i} + \rho^{\frac{n-i-1}{2}} I_{n+i+1} + (1-\rho)\rho^n \sum_{k=n+i+2}^{\infty} \rho^{-\frac{k}{2}} I_k \}.$$

Upon substitution in (2.61), we obtain the classical formula (2.50). Again, the second term can be rewritten as

$$e^{-(\lambda+\mu)t} \rho^n \rho^{-\frac{n+i}{2}} (I_{n+i} - I_{n+i+2}) + e^{-(\lambda+\mu)t} \rho^n \sum_{k=n+i+2}^{\infty} k \rho^{-\frac{k}{2}} (\mu t)^{-1} I_k,$$

which leads to

$$\begin{aligned}P_n(t) &= e^{-(\lambda+\mu)t}\rho^{\frac{n-i}{2}}I_{n-i} - e^{-(\lambda+\mu)t}\rho^{\frac{n-i}{2}}I_{n+i+2}\\ &+e^{-(\lambda+\mu)t}\rho^n \sum_{k=n+i+2}^{\infty} k\rho^{-\frac{k}{2}}(\mu t)^{-1}I_k. \end{aligned}\tag{2.62}$$

Now we split the solution into the stationary solution and the remainder. For this purpose, we use the identity

$$1-\rho = e^{-(\lambda+\mu)t}\sum_{k=-\infty}^{\infty} k\rho^{-\frac{k}{2}}(\mu t)^{-1}I_k$$

(to be proved in Exercise 19) in (2.61) and get

$$\begin{aligned}P_n(t) &= (1-\rho)\rho^n + e^{-(\lambda+\mu)t}\rho^{\frac{n-i}{2}}(I_{n-i}-I_{n+i})\\ &-e^{-(\lambda+\mu)t}\rho^n(\mu t)^{-1}\sum_{k=-\infty}^{n+i} k\rho^{-\frac{k}{2}}I_k \end{aligned}\tag{2.63}$$

in which the first term is the stationary solution. Now it can be seen that

$$\sum_{k=-\infty}^{n} k\rho^{-\frac{k}{2}}I_k = \sum_{m=0}^{\infty}\frac{(\lambda t)^m}{m!}\sum_{k=0}^{m+n}(k-m)\frac{(\mu t)^k}{k!}\ .$$

Thus (2.63) becomes

$$\begin{aligned}P_n(t) &= (1-\rho)\rho^n + e^{-(\lambda+\mu)t}\rho^n\sum_{m=0}^{\infty}\frac{(\lambda t)^m}{m!}\sum_{k=0}^{m+n+i}(m-k)\frac{(\mu t)^{k-1}}{k!}\\ &+e^{-(\lambda+\mu)t}\sum_{m=0}^{\infty}(\lambda t)^{m+n-i}\frac{(\mu t)^m}{(2m+n-i)!}\\ &\times\left[\binom{2m+n-i}{m}-\binom{2m+n-i}{m-i}\right]. \end{aligned}\tag{2.64}$$

It can be written in an alternative form as

$$\begin{aligned}P_n(t) &= (1-\rho)\rho^n + e^{-(\lambda+\mu)t}\rho^n\sum_{m=0}^{\infty}\frac{(\lambda t)^m}{m!}\sum_{k=0}^{m+n+i+1}(m-k)\frac{(\mu t)^{k-1}}{k!}\\ &+e^{-(\lambda+\mu)t}\rho^n\sum_{m=0}^{\infty}(\lambda t)^{m+1}\frac{(\mu t)^{m+V}}{m!}\\ &\times\left[\frac{(\lambda t)^{-U-1}}{(m+|i-n|)!}-\frac{(\mu t)^{U+1}}{(m+i+n+2)!}\right] \end{aligned}\tag{2.65}$$

where $U=\min(i,n)$ and $V=\max(i,n)$.

2.6. Waiting Time and Little's Formula

Besides the number of customers in the system, which is of importance from the management's point of view, it will be of interest from the customers' point of view to study another characteristic, namely, waiting time of a customer, for example, the time a customer upon arrival has to wait in the queue before being served, or the time spent in the system. Before the waiting time distribution is discussed, we heuristically develop an elegant-looking general formula, known as Little's formula (J.D.C. Little was responsible for bringing it out), which relates the average waiting time to the average number of customers.

Notation:

W : waiting time in the system of a customer
W_q : waiting time in the queue of a customer

Furthermore, let us denote, respectively, by $\lambda(t)$, $\beta(t)$, $L(t)$, $Y(t)$ and $W(t)$, the average arrival rate, the total time spent in the system by all customers, the average number of customers in the system, the number of arrivals and the average waiting time of a customer in the system, each during the time $(0, t)$. Then,

$$\lambda(t) = \frac{Y(t)}{t},$$
$$L(t) = \frac{\beta(t)}{t},$$

and

$$W(t) = \frac{\beta(t)}{Y(t)}.$$

Combining these three expressions, we have

$$L(t) = \lambda(t)W(t)$$

from which emerges *Little's formula*

$$L = \bar{\lambda}E(W) \tag{2.66}$$

when limits are taken on both sides, provided that all limits exist and where $\bar{\lambda}$ is the limiting average arrival rate.

Note that no assumption is made on the input distribution, the service-time distribution and the queue discipline. Also observe that the argument for (2.66) will lead to

$$L_q = \bar{\lambda}E(W_q)$$
$$L_s = \bar{\lambda}E(W_s) \tag{2.67}$$

where L_s and W_s, respectively, refer to the average number of customers in the service and the time spent in the service. In other words, the result (2.66) is true for any specific class of customers, regardless of what situation is under consideration and therefore each expression in (2.67) is also called Little's formula.

Usually the average waiting time of a customer is difficult to calculate directly. Relation (2.66) comes to our rescue, because the evaluation of the average number of customers in the system is relatively easy. As examples, $E(W)$ is derived from (2.66) for the following models:

1. $M/M/1$:

$$E(W) = \frac{1}{\lambda}\frac{\rho}{1-\rho}, \qquad \text{by using (2.8).}$$

$$E(W_q) = \frac{1}{\lambda}\frac{\rho^2}{1-\rho}, \qquad \text{by using (2.9).}$$

2. Discouraged arrivals:

$$\begin{aligned}\bar{\lambda} = \sum_{n=0}^{\infty} \lambda_n P_n &= \sum_{n=0}^{\infty} \frac{\lambda \rho^n e^{-\rho}}{(n+1)!} \quad \text{by using (2.15),}\\ &= \frac{\lambda}{\rho}\sum_{n=1}^{\infty}\frac{\rho^n e^{-\rho}}{n!}\\ &= \mu(1-e^{-\rho}).\end{aligned}$$

$$E(W) = \frac{\rho}{\mu(1-e^{-\rho})} \qquad \text{by using (2.16).}$$

3. $M/M/\infty$:

$$E(W) = \frac{1}{\mu}.$$

In another example, a simple expression for L in the $M/M/c$ model is obtained as

$$L = L_q + \rho$$

by utilizing the second expression of (2.67) in which $\bar{\lambda} = \lambda$ and $E(W_s) = 1/\mu$.

Next, we may consider the distribution of the waiting time.

Notation:

Q_n : steady-state probability of n units in the system at an arrival instant

Note that Q_n is not the same as P_n. However, for Markovian arrivals, we will show in the next chapter (see (3.74)) that $Q_n = P_n$ and, therefore, for the $M/M/1$ model, $Q_n = P_n$.

Assuming the queue discipline to be FCFS, the waiting period of an arrival equals the service completion period of customers ahead of the new arrival and, therefore, we have for $t > 0$

$$\bar{F}_{W_q}(t) = \sum_{n=1}^{\infty} P(\text{time for } n \text{ service completions} > t | \text{arrival found } n \text{ in the system}) Q_n$$

which holds good for any model. For the $M/M/1$ model, it becomes

$$\bar{F}_{W_q}(t) = \int_t^{\infty} (1-\rho) \left[\sum_{n=1}^{\infty} \rho^n \frac{\mu^n}{(n-1)!} x^{n-1} \right] e^{-\mu x} dx$$

when we recall that the sum of n i.i.d. exponential variables is the Erlangian variable $(n, 1/\mu)$. The expression simplifies to

$$\begin{aligned} (1-\rho)\rho \int_t^{\infty} \mu\, e^{-\mu x} \left[\sum_{n=1}^{\infty} \frac{(\rho \mu x)^{n-1}}{(n-1)!} \right] dx \\ = (1-\rho)\rho \int_t^{\infty} \mu e^{-\mu x(1-\rho)} dx \\ = \rho\, e^{-\mu t(1-\rho)}, \quad t > 0. \end{aligned} \tag{2.68}$$

Observe that $\bar{F}_{W_q}(0) = \rho \neq 1$ which is explained by the fact that

$$\begin{aligned} P(W_q = 0) &= P(\text{system empty at an arrival}) \\ &= P_0 = 1 - \rho. \end{aligned} \tag{2.69}$$

Remarks:

1. For other models with Markovian arrivals, we proceed similarly and obtain the distribution of W_q. For example, in case of the $M/M/c$ model, it is given by

$$\bar{F}_{W_q}(t) = P_0 \frac{\rho^c}{c!(1-\rho/c)} e^{-c\mu t(1-\rho/c)}. \tag{2.70}$$

   In general, a simple expression may not be produced if the expression for P_n is more complicated (e.g., which happens for the model $M/E_k/1$) or due to the inability to simplify the summation inside the integral.

2. In the next chapter, we develop a general functional relation between Q_n and P_n which will be useful to derive the distribution of W_q for models with non-Markovian arrivals.

2.7. Busy Periods and Idle Periods

A characteristic which is of concern to everyone who deals with a queueing system is a busy period. This can be defined as the time interval when the serving system (also called the server) continuously remains operative or simply busy. The word "operative" is rather not definite and may have a connotation depending on a particular situation. In the $M/M/1$ model, it begins when there are no customers in the system and a customer arrives, and it ends when for the first time the system becomes empty. A complementary characteristic is an idle period during which the system remains inoperative. The idle period in the $M/M/1$ model is the duration when there are no customers in the system.

Notation:

Δ : length of a busy period
I : length of an idle period

For the study of a busy period, both Δ and I are of common interest. By its very nature, the study is directly associated with the approach of the time-dependent solution. Let us consider the $M/M/1$ model. The distribution of Δ can be determined by following the technique for the transient behaviour analysis with an *absorbing barrier* at zero system size and with only one initial customer. Then the differential-difference equations (2.3) become the following:

$$\begin{aligned} P_0'(t) &= \mu P_1(t), \\ P_1'(t) &= -(\lambda+\mu)P_1(t) + \mu P_2(t), \\ P_n'(t) &= -(\lambda+\mu)P_n(t) + \lambda P_{n-1}(t) + \mu P_{n+1}(t), \qquad n \geq 2. \end{aligned} \tag{2.71}$$

Observe the changes in the first two equations which are due to the presence of the absorbing barrier. Similar to (2.43) the new L.T. of the p.g.f. is

$$P^*(z,\theta) = \frac{z^2 - (1-z)(\mu - \lambda z)(z_1/\theta)}{-\lambda(z-z_1)(z-z_2)}$$

where z_1 and z_2 are given by (2.44). Then

$$P_0^*(\theta) = P^*(0,\theta) = \frac{\mu}{\lambda\theta z_2}.$$

It is known that

$$\text{L.T.}\Big(P_0'(t)\Big) = \theta P_0^*(\theta) - P_0(0) = \theta P^*(0,\theta),$$

which when inverted and simplified leads to

$$P_0'(t) = \frac{\sqrt{\mu/\lambda}\, e^{-(\lambda+\mu)t} I_1(2\sqrt{\lambda\mu}\, t)}{t}. \tag{2.72}$$

Here we note that $P_0(t)$ being the probability of the system to be empty at time t, is also the probability that $\Delta \leq t$. Therefore, what is obtained in (2.72) is an expression for the p.f. of Δ. It can be shown that for $\lambda < \mu$,

$$E(\Delta) = \frac{1}{\mu - \lambda}. \tag{2.73}$$

When $\mu \leq \lambda$, Δ can be infinity with some positive probability.

As far as the length of an idle period I is concerned, it represents the period from the moment the system becomes empty to the arrival time. Since the interarrival times are exponential (λ) which has the memoryless property, the density of I is given by

$$f_I(t) = \lambda e^{-\lambda t}, \qquad t > 0. \tag{2.74}$$

Here we have illustrated the study of the busy period only for a simple Markovian model. However, in the next chapter the busy period will be more elaborately discussed for various general models.

2.8. Networks of Queues – I

A natural generalization of the birth-death model is a network of queues which is a collection of several service systems such that a customer after completing service at one system may join another including itself. In that case, an arrival may arise from two different sources, viz., external and internal. An external arrival is one coming from outside the system, whereas an internal arrival comes from any service system after its service completion. An internal arrival is governed by a set of rules, commonly known as a switching or routing process, that prescribes the transition probabilities of moving from one service system to another. Such a collection of service systems, along with respective input process and service-time distribution, waiting capacity and queue discipline is called a network of queues or queueing network. In a network, we identify a service system as a node. One can picture a job shop, a communication system, a maintenance facility, an air traffic control facility, and a medical care system as examples of queueing networks.

We consider a queueing network having J service systems with the following properties:

(i) The arrival process from the external source (without any confusion we call it node 0) to node i is Poisson with rate $\lambda_i, (i = 1, \ldots, J)$;

(ii) The service-times at node i, independent of arrival times, are i.i.d. exponential random variables with parameter $\mu_i, (i = 1, \ldots, J)$;

(iii) On completion of service at node i, a unit goes instantaneously to node j with probability $p_{ij}, (i, j = 1, \ldots, J)$ for its next service or leaves the network with probability p_{i0}, such that $\sum_{j=0}^{J} p_{ij} = 1$. These routing probabilities are independent of the history of the system.

(iv) Infinite waiting space is available at each node and the queue discipline is FCFS.

A network having these properties is called an *open Jackson network* (named after J.R. Jackson) – open because of the existence of the external source. To make these less superfluous, assume that $p_{i0} > 0$ for some i. In this section we derive only the joint steady-state probability distribution of the number of units at various nodes of the network.

Notation:

$X_i(t)$: the number in the system at node i $(i = 1, \ldots, J)$ at time t

$P_i(n_i)$: the steady-state probability of having n_i units at node i, $(i = 1, \ldots, J)$. (We say that the network is in state $\mathbf{n} = (n_1, \ldots, n_J)$ at time t, if $X_i(t) = n_i, i = 1, \ldots, J$.)

$P(\mathbf{n})$: the joint steady-state probability of the network in state $\mathbf{n}$

Letting α_i to represent the composite arrival rate (comprising of both external and internal) at node $i (i = 1, \ldots, J)$ we develop certain properties of α_i which will be used later for the derivation of a steady-state solution. It is easy to see that α_i's must satisfy

$$\alpha_i = \lambda_i + \sum_{j=1}^{J} \alpha_j p_{ji}, \qquad i = 1, \ldots, J. \tag{2.75}$$

These equations are called *traffic equations* which when expressed in matrix form become

$$\boldsymbol{\alpha} = \boldsymbol{\lambda} + \boldsymbol{\alpha}\mathbf{P},$$

where

$$\boldsymbol{\alpha} = (\alpha_1, \ldots, \alpha_J), \qquad \boldsymbol{\lambda} = (\lambda_1, \ldots, \lambda_J),$$

and

$$\mathbf{P} = (p_{ij}) \qquad 1 \leq i, j \leq J.$$

Note that $\mathbf{P}$ is a sub-stochastic matrix and therefore $(\mathbf{I} - \mathbf{P})^{-1}$ exists and has only non-negative elements (see Section A.10). Hence (2.75) has the unique solution

$$\boldsymbol{\alpha} = \boldsymbol{\lambda}(\mathbf{I} - \mathbf{P})^{-1}. \tag{2.76}$$

Next, if we sum both sides of (2.75) over i and use the fact that

$$1 - p_{j0} = \sum_{i=1}^{J} p_{ji},$$

we obtain

$$\sum_{i=1}^{J} \lambda_i = \sum_{i=1}^{J} \alpha_i p_{i0}. \tag{2.77}$$

Now let us formulate the appropriate balance equations for the network which will be referred to as *global balance equations* for reasons to be obvious later. For convenience, introduce the J-component vector $\mathbf{e}(i)$ consisting of all zeroes except that the ith component equals 1. Hence $\mathbf{n} + \mathbf{e}(i)$ and $\mathbf{n} - \mathbf{e}(i)$

are, respectively, equal to $\mathbf{n}$ with one unit more and one unit less in component i. Following the usual procedure, we write the global balance equations as

$$\begin{aligned}\left(\sum_i \lambda_i + \sum_i \mu_i I_{(n_i>0)}\right) P(\mathbf{n}) \\ = \sum_j \sum_i \mu_i p_{ij} I_{(n_j>0)} P(\mathbf{n}+\mathbf{e}(i)-\mathbf{e}(j)) \\ + \sum_i \lambda_i I_{(n_i>0)} P(\mathbf{n}-\mathbf{e}(i)) + \sum_i \mu_i p_{i0} P(\mathbf{n}+\mathbf{e}(i))\end{aligned} \tag{2.78}$$

where

$$I_{(n>0)} = \begin{cases} 1 & \text{if } n > 0, \\ 0 & \text{otherwise} \end{cases}$$

Direct solution of (2.78) is not easy. However, the solution is obtained through the so-called *local balance equations* which are balance equations for different nodes, each node being isolated from others. Indeed, the global balance equation is the sum of all local balance equation. Thus solution for the local equations is the solution for the global balance equation.
The local equations for $j = 1, 2, \ldots, J$ are

$$P(\mathbf{n}+\mathbf{e}(j))\mu_j = \sum_i P(\mathbf{n}+\mathbf{e}(i))\mu_i p_{ij} + P(\mathbf{n})\lambda_j, \tag{2.79}$$

and for $j = 0$ is

$$P(\mathbf{n})\sum_i \lambda_i = \sum_i P(\mathbf{n}+\mathbf{e}(i))\mu_i p_{i0}. \tag{2.80}$$

We can rewrite (2.79) as

$$1 = \sum_i \frac{P(\mathbf{n}+\mathbf{e}(i))\mu_i}{P(\mathbf{n}+\mathbf{e}(j))\mu_j} p_{ij} + \frac{P(\mathbf{n})}{P(\mathbf{n}+\mathbf{e}(j))\mu_j}\lambda_j. \tag{2.81}$$

If we set

$$\frac{P(\mathbf{n}+\mathbf{e}(i))}{P(\mathbf{n}+\mathbf{e}(j))} = \frac{\alpha_i \mu_j}{\mu_i \alpha_j} \tag{2.82}$$

and

$$\frac{P(\mathbf{n})}{P(\mathbf{n}+\mathbf{e}(j))} = \frac{\mu_j}{\alpha_j} \tag{2.83}$$

then the set (2.81) becomes the set of traffic equations (2.75) which is always true. Therefore the value of $P(\mathbf{n})$ that satisfies (2.82) and (2.83) will satisfy (2.81). From (2.82) and (2.83) we recursively obtain

$$P(\mathbf{n}) = c\prod_{i=1}^{J} \rho_i^{n_i}, \qquad \rho_i = \frac{\alpha_i}{\mu_i}$$

where c is the normalizing constant and is equal to $\prod_{i=1}^{J}(1-\rho_i)$.

Thus

$$P(\mathbf{n}) = \prod_{i=1}^{J} (1-\rho_i)\rho_i^{n_i}. \tag{2.84}$$

If (2.84) is the solution of all local equations then it must satisfy (2.79) and (2.80). On substitution of (2.84) in (2.80) we find it reduces to (2.77). Therefore the steady-state distribution of the open Jackson network is given by (2.84), which is called product form solution because of its form.

Remarks:

1. The implication of (2.84) is that one can decompose the steady-state solution of a network with J service nodes into solutions of J independent $M/M/1$ queueing systems. The queue length process at node i for every i imitates as if it were generated by a $M/M/1$ queueing system with parameters α_i and μ_i $(i = 1, \dots, J)$.

2. The form of (2.84) may mislead someone to believe that the input process at any individual node is Poisson. But it is not necessarily so.

3. Result (2.84) can be extended to the case where there are c_i servers at node i each having the same exponential service-time distribution with parameters μ_i $(i = 1, \dots, J)$. In that event,

$$P(\mathbf{n}) = \prod_{i=1}^{J} \frac{\rho_i^{n_i}}{a_i} p_{0i} \tag{2.85}$$

where

$$a_i = \begin{cases} n_i! & \text{if } n_i \le c_i \\ c_i^{n_i - c_i} c_i! & \text{if } n_i \ge c_i \end{cases}$$

and ρ_{0i} is such that

$$\sum_{n_i} p_{0i} \frac{\rho_i^{n_i}}{a_i} = 1.$$

An open Jackson network with

$$\lambda_i = \begin{cases} \lambda & \text{if } i = 1 \\ 0 & \text{otherwise} \end{cases}$$

and

$$p_{ij} = \begin{cases} 1 & \text{if } j = i+1, \quad 1 \le i \le J-1 \\ 1 & \text{if } i = J, \quad j = 0 \text{ (external node)} \\ 0 & \text{otherwise} \end{cases}$$

is called a series or tandem queue. From the steady-state solution the usual measures of performance can be computed.

Manufacturing – the job shop environment. Open Jackson networks have interesting applications in the theory of production and manufacturing. A typical setting is that of a job shop environment: the system consists of a number of manufacturing units each equipped with certain machines or stations of particular type and capability. Jobs are arriving randomly and each job has certain processing requirements. These requirements specify the routing of jobs through the network of stations. The performance of such a manufacturing system may be evaluated by means of the *mean work in process*, that is the mean number of jobs in the network, and the *mean lead time* of a job, which is defined as the mean time a job spends in the network, starting with its entry from outside and terminating when the job finally leaves the system.

For simplicity assume that each processing unit in the system can be modelled as a $M/M/1$ system. Then using (2.8) the mean number of jobs at unit i, L_i is given by

$$L_i = \frac{\rho_i}{1-\rho_i}, \qquad i = 1, 2, \ldots, J.$$

The mean work in process L is simply the sum:

$$L = \sum_{i=1}^{J} L_i.$$

To determine the mean lead time $E(W)$ of a job, we apply Little's formula (2.66) to the entire manufacturing system:

$$L = \bar{\lambda} E(W),$$

where $\bar{\lambda} = \sum_{i=1}^{J} \lambda_i$ is the total external arrival rate. Thus the mean lead time of a job is given by

$$E(W) = \frac{1}{\bar{\lambda}} \sum_{i=1}^{J} \frac{\rho_i}{1-\rho_i}.$$

Another important performance measure is the *total mean service* δ_i that a job requires from processing unit i during its stay in the system. This number is easily found by the following argument: every time a job visits unit i it requires processing time which equals $1/\mu_i$ in the mean. Hence

$$\delta_i = \frac{\omega_i}{\mu_i}, \qquad i = 1, 2, \ldots, J,$$

where ω_i is the mean number of visits at unit i. Now we know that the mean number of jobs entering the system from outside per unit of time equals $\bar{\lambda}$ and therefore the mean number of arrivals at unit i per unit of time equals $\bar{\lambda}\omega_i$. But that is just the composite arrival rate α_i. It follows that

$$\omega_i = \frac{\alpha_i}{\bar{\lambda}}, \qquad i = 1, 2, \ldots, J$$

Combining this with the previous result yields finally

$$\delta_i = \frac{\omega_i}{\mu_i} = \frac{\rho_i}{\bar{\lambda}}, \qquad i = 1, 2, \ldots, J. \tag{2.86}$$

Formula (2.86) when written as $\rho_i = \delta_i \bar{\lambda}$ has an interesting interpretation: the load ρ_i is just the mean amount of work destined for processing unit i that enters the manufacturing system from outside per unit of time.

2.9. Optimization

Thus far we have dwelt on the study of the characteristics of a queue, which may portray the descriptive nature of a queueing model. It merely helps us to understand the intrinsic properties of a queue but does not prescribe any action.

An elementary example for action is as follows: Suppose there are 5 different machines available for a given job. The average time needed by the jth machine is μ_j and the cost of the machine j is c_j per unit time, $j = 1, \ldots, 5$. The amount received per completed job is u and the cost of spending time in the system is c per job per unit time. The arrival rate of jobs is λ. Assume the model to be $M/M/1$ with $\lambda < \mu_j$. Let R_j be the profit per unit time from the jth machine. Then

$$R_j = \lambda u - cL - c_j = \lambda u - \frac{\lambda c}{\mu_j - \lambda} - c_j.$$

Thus machine j should be used which has minimum R_j, $j = 1, \ldots, 5$.

This section illustrates the applications for making an optimal decision by taking into account the economic measures of the system performance. In other words our focus will be on the prescriptive nature of a queue from the point of view of the management, more particularly on the system design and its control which are meant to be action oriented. It may be necessary to explain the difference between the two problems, viz., design and control. Given a cost structure, the design problem attempts to find optimal values of queue parameters, e.g., service rate and number of servers. Even in an optimal design (with optimal parameters) one may be able to utilize the service facility more effectively by imposing a policy called control policy, say, by removing the service facility at the beginning of an idle period either until the queue size reaches a certain level or for a certain duration. In a control problem, the optimal level or duration is ascertained. In a broader sense it is a design problem, except that in the latter the parameters (say, service-time) are fixed (static) and in the former these vary over time (dynamic).

2.9.1. Optimal designs. Returning to design policies for consideration by the management, we note that many queueing problems in practice require the decision on the service parameters since these are the ones which the decision maker can change. To be more specific, let us consider the $M/M/c$ model and determine c which minimizes the expected total cost in designing a system. Introduce the cost components as follows:

c_1 : cost per unit time of a customer while waiting in the queue,
c_2 : cost per unit time of a customer while being served,
c_3 : cost per unit time of a server during an idle period,
c_4 : cost per unit time of a server while serving.

Let TC, WC and SC denote the total cost, the waiting cost and the service cost per unit time, respectively. Then we have

$$\begin{aligned} E(TC) &= E(WC) + E(SC) \\ &= c_1 L_q + c_2(L - L_q) + c_3 \sum_{n=0}^{c} (c-n)P_n + c_4(c - \sum_{n=0}^{c}(c-n)P_n). \end{aligned} \tag{2.87}$$

Realizing that the average number in service is the same as the average number of service counters that are busy, (2.87) becomes

$$\begin{aligned} E(TC(c)) &= c_1 L_q + (c_2 + c_4)(L - L_q) + c_3(c - L + L_q) \\ &= c_1 L_q(c) + (c_2 + c_4)\frac{\lambda}{\mu} + c_3(c - \frac{\lambda}{\mu}), \qquad \text{by (2.67)}. \end{aligned} \tag{2.88}$$

We have written $TC(c)$ and $L_q(c)$ indicating that these are functions of c. c^* is an optimal value of c if

$$E(TC(c^*)) - E(TC(c^* - 1)) \leq 0$$

and

$$E(TC(c^*)) - E(TC(c^* + 1)) \leq 0.$$

hold good. Thus c^* must satisfy

$$L_q(c^*) - L_q(c^* + 1) \leq \frac{c_3}{c_1} \leq L_q(c^* - 1) - L_q(c^*). \tag{2.89}$$

As one would expect the optimality criterion depends on c_1, (a customer's waiting cost) and c_3, (a server's cost during an idle period).

For the next example, let us use the same model as before, but wish to optimize c and μ. The cost structure consists of two components, viz., WC, the waiting cost per unit time and SC, the service cost per unit time. In a general set-up, either we may have WC as a function of n (the number of customers in the system), say $g(n)$, or we may be given a customer's waiting cost as a function the waiting time W, say $h(W)$. This may be expressed as

$$E(WC) = \begin{cases} \displaystyle\sum_{n=0}^{\infty} g(n)P_n & \text{in the first case,} \\ \displaystyle\lambda \int_0^{\infty} h(w) f_W(w)\, dw & \text{in the second case.} \end{cases} \tag{2.90}$$

Obviously, if

$$g(n) = c_0 n \qquad \text{and} \qquad h(w) = c_0 w$$

where c_0 is the cost of waiting per unit time for each customer, then

$$E(WC) = c_0 L \tag{2.91}$$

in both cases, since $L = \lambda E(W)$ by Little's formula (2.66). Let us denote by $f(\mu)$ the cost of a server per unit time when μ is the service rate. The optimal values of c and μ are those which minimize

$$E(TC) = cf(\mu) + E(WC). \tag{2.92}$$

Assume that the number of feasible values of μ is finite which is usually the case. We adopt a two-stage optimization approach. First, for each μ find an optimal c, say $c^*(\mu)$, by using the procedure of the previous example. Next, select the overall optimal values (c^*, μ^*) for which $E(TC(c^*(\mu)))$ is minimized over all feasible values of μ.

2.9.2. Optimal control policies. After a brief discussion on design problems, let us turn to the control problem. In order to have a flavor of the problem, consider the $M/M/1$ model in which the server is removed when the system becomes empty and is brought back when the queue size is K. Such a policy is called a $(0, K)$-policy. Let $P_n(t;1)$ and $P_n(t;0)$ denote the following:

$$\begin{aligned}
P_n(t;1) &= P(X(t) = n \text{ when the system is busy}), \quad n = 1, 2, \ldots \\
P_n(t;0) &= P(X(t) = n \text{ when the system is idle}), \quad n = 0, 1, \ldots, K-1.
\end{aligned}$$

The associated differential-difference equations are:

$$\begin{aligned}
&\frac{d}{dt}P_0(t;0) = -\lambda\, P_0(t;0) + \mu\, P_1(t;1), \\
&\frac{d}{dt}P_n(t;0) = -\lambda\, P_n(t;0) + \lambda\, P_{n-1}(t;0), \qquad 1 \le n < K, \\
&\frac{d}{dt}P_1(t;1) = -(\lambda+\mu)P_1(t;1) + \mu\, P_2(t;1), \\
&\frac{d}{dt}P_n(t;1) = -(\lambda+\mu)P_n(t;1) + \lambda\, P_{n-1}(t;1) + \mu\, P_{n+1}(t;1), \quad n \ge 2, n \ne K, \\
&\frac{d}{dt}P_K(t;1) = -(\lambda+\mu)P_K(t;1) + \lambda(P_{K-1}(t;0) + P_{K-1}(t;1)) + \mu(P_{K+1}(t;1)).
\end{aligned} \tag{2.93}$$

Assuming that the stationary solution exists, and denoting by

$$P_n[i] = \lim_{t\to\infty} P_n(t;i), \qquad i = 0, 1,$$

the resulting balance equations are:

$$\begin{aligned}
&-\lambda P_0[0] + \mu P_1[1] = 0, \\
&-\lambda P_n[0] + \lambda P_{n-1}[0] = 0, \quad 1 \le n < K \\
&-(\lambda+\mu)P_1[1] + \mu P_2[1] = 0, \\
&-(\lambda+\mu)P_n[1] + \lambda P_{n-1}[1] + \mu P_{n+1}[1] = 0, \quad n \ge 2, n \ne K, \\
&-(\lambda+\mu)P_K[1] + \lambda(P_{K-1}[0] + P_{K-1}[1]) + \mu P_{K+1}[1] = 0.
\end{aligned} \tag{2.94}$$

Solving these equations recursively, we get

$$P_0[0] = P_1[0] = \ldots = P_{K-1}[0],$$

and

$$P_n[1] = \begin{cases} \dfrac{\rho(1-\rho^n)}{1-\rho} P_0[0], & 1 \le n \le K, \\ \\ \rho^{n+1-K} \dfrac{(1-\rho^K)}{1-\rho} P_0[0], & n \ge K+1. \end{cases} \tag{2.95}$$

Using the normalizing equation

$$\sum_{j=0}^{K-1} P_j[0] + \sum_{j=1}^{\infty} P_j[1] = 1,$$

we get

$$P_0[0] = \frac{1-\rho}{K}. \tag{2.96}$$

Thus we have

$$P_n[0] = \frac{1-\rho}{K}, \qquad 0 \le n \le K-1, \tag{2.97}$$

and

$$P_n[1] = \begin{cases} \dfrac{\rho(1-\rho^n)}{K}, & 1 \le n \le K \\ \rho^{n+1-K} \dfrac{(1-\rho^K)}{K}, & n \ge K+1. \end{cases} \tag{2.98}$$

Now L, the expected number of customers in the system, is given by

$$\begin{aligned} L &= \sum_{j=0}^{K-1} jP_j[0] + \sum_{j=1}^{\infty} jP_j[1] \\ &= \frac{\rho}{1-\rho} + \frac{K-1}{2}, \end{aligned} \tag{2.99}$$

the first term being the average queue length in the $M/M/1$ model without any restriction.

In order to find an optimal K, we introduce the cost of setting up and withdrawing the service facility from the system which is denoted by A. Letting T to represent the length of a cycle of an idle period followed by a busy period and using (2.73) and (2.74), we find

$$E(T) = \frac{K}{\lambda} + \frac{K}{\mu-\lambda} = \frac{\mu}{\lambda(\mu-\lambda)} K. \tag{2.100}$$

Now TC is a function of K and its expected value is

$$E(TC(K)) = \frac{A}{E(T)} + c_0 L. \tag{2.101}$$

It follows from (2.99) and (2.100) that

$$E(TC(K)) = \frac{A\lambda(\mu-\lambda)}{\mu K} + c_0\left(\frac{\rho}{1-\rho} + \frac{K-1}{2}\right). \tag{2.102}$$

When (2.102) is minimized with respect to K, the optimal value K^* is obtained as

$$K^* = \sqrt{\frac{2A\lambda}{c_0}\left(1-\frac{\lambda}{\mu}\right)}. \tag{2.103}$$

As far as the optimization is concerned, it is no surprise in the previous examples that two costs (WC and SC) are of significance. In general, costs which involve the variables to be optimized should be taken into account for the total cost. The expected total cost is then minimized with respect to the variables of interest. From these remarks it is quite obvious how we can deal with an optimization problem and thus its discussion is omitted in the present example.

Evidently, all our examples are directed to minimize the total cost which is primarily the objective of any management. However, this goal will not be shared by the customers and therefore a different objective function for optimization should be under consideration. Any policy resulting from such an optimization is called customer's optimization policy, as against the server's optimization policy that has been the topic of discussion in this section. Moreover, note that every model treated here is a Markovian one. In Chapter 10, the customer's optimization policy and optimization policy for non-Markovian systems will be studied.

At the end we discuss the impact of a situation in which the service facility is removed for a random duration. In such a situation we say that the server is on vacation. This time we may deal with the birth-death model (see Section 2.3). The server goes on vacation for a random duration whenever the system becomes empty. Upon returning from a vacation if the system is found to be empty, the server leaves for another vacation. Otherwise the service continues. The durations are i.i.d. exponential random variables with parameter δ. Using the notations $P_n[i]$, $(i = 0, 1)$, for this case also, we can derive the following balance equations:

$$\begin{aligned}
&-\lambda_0 P_0[0] + \mu_1 P_1[1] = 0, \\
&-(\lambda_n + \delta)P_n[0] + \lambda_{n-1}P_{n-1}[0] = 0, \quad n \geq 1, \\
&-(\lambda_1 + \mu_1)P_1[1] + \mu_2 P_2[1] + \delta\ P_1[0] = 0, \\
&-(\lambda_n + \mu_n)P_n[1] + \mu_{n+1}P_{n+1}[1] + \delta\ P_n[0] + \lambda_{n-1}P_{n-1}[1] = 0, \quad n \geq 2.
\end{aligned} \tag{2.104}$$

By iteration, it can be checked that

$$P_n[0] = P_0[0]\prod_{i=0}^{n-1}\frac{\lambda_i}{\lambda_{i+1}+\delta}, \qquad n \geq 1, \tag{2.105}$$

and

$$P_n[1] = P_0[0] \left(\prod_{i=0}^{n-1} \frac{\lambda_i}{\mu_{i+1}} \right) \left(1 + \sum_{k=1}^{n-1} \prod_{i=1}^{k} \frac{\mu_i}{\lambda_i + \delta} \right), \quad n \geq 1. \tag{2.106}$$

Compare the results with (2.13). As a special case consider the $M/M/1$ model. The corresponding results are:

$$P_n[0] = \left(\frac{\rho_1}{1 + \rho_1} \right)^n P_0[0], \qquad n \geq 0,$$

and

$$P_n[1] = \frac{1 + \rho_1}{1 + \rho_1 - \rho_2} \left(\rho^n - \left(\frac{\rho_1}{1 + \rho_1} \right)^n \right) P_0[0], \quad n \geq 1, \tag{2.107}$$

where $\rho_1 = \frac{\lambda}{\delta}$, $\rho_2 = \frac{\mu}{\delta}$.

From the normalizing equation we can show that

$$P_0[0] = \frac{1 - \rho}{1 + \rho_1}. \tag{2.108}$$

The average number in the system L can be found to be

$$L = \frac{\rho}{1 - \rho} + \rho_1. \tag{2.109}$$

Evidently, L has increased by ρ_1 due to random vacation. When $\delta \to \infty$, then there is no vacation time and we have the usual $M/M/1$ model. In that case,

$$\lim_{\delta \to \infty} \rho_1 = \lim_{\delta \to \infty} \rho_2 = 0, \tag{2.110}$$

which reduces (2.107), (2.108) and (2.109) to the known expressions (see (2.5), (2.6) and (2.8)).

2.10. Discussion

It is time to pause and have an overview of what has been covered so far. As a stepping stone to the theory of queues, a class of queueing models is treated in this chapter in order to exploit certain well defined basic features of Markov processes. The treatment is able to kill two birds in one shot in the sense that not only these features have remarkably simplified the structure of queues amenable to ready-made solutions, but also are exhibited in a fairly large variety of practical problems.

First of all, in the study of the distribution $\{P_n(t)\}$ of the number in the queueing system at time t, the Kolmogorov's equations of a Markov process lead us in $M/M/1$ to the set of differential-difference equations on $P_n(t)$ (Section 2.2) reflecting the transient behaviour in the queueing system. These are solved (Section 2.5) with the help of the p.g.f., L. T. and Rouché's Theorem (Method 1) or only of the p.g.f. through the extended system of variables (Method 2) so that the use of L. T. and Rouché's Theorem is avoided. In spite of the fact that these are standard techniques, the process of arriving at the solution has turned out to be lengthy without any simple looking result and involving Bessel functions. However, some numerical techniques for computation are discussed in Chapter 4.

The second but most striking feature of a Markov process is the existence of the stationary solution and the associated linear equations for deriving the solution (Section A.8 in Appendix A). It has made possible to derive the stationary distribution $\{P_n = \lim_{t\to\infty} P_n(t)\}$ which is the indicator of the steady-state behaviour of the number in the system. The overwhelming simplicity in which the system of linear equations (also popularly known as balance equations) can be obtained from the so-called principle of conservation of flow (i.e., rate in = rate out) has become the unifying force for several sections in this chapter (Sections 2.2–2.4).

Viewing the balance equations as difference equations in $\{P_n\}$, we have given three different approaches to solve them (Section 2.2). Of these the simplest one is the recursive or iterative method where P_n, $n = 1, 2, ...$ is successively expressed in terms of P_0 and P_0 in turn is derived from the normalizing equation. This elementary approach works out well for the birth-death models which encompass a considerable class of queueing systems (Section 2.3). The other two approaches, viz., solution through the characteristic function and through the generating function are well-known techniques for solving difference equations and are illustrated for $M/M/1$.

Because the balance equations are linear, we may solve them directly by the usual method (such as *Cramer's rule* or by finding the inverse of the coefficient matrix (see Section A.10 in Appendix A)) if the system of equations is finite. Let us consider the model $M/M/1$ with space only for K customers in the system (this is a special case of Model 4 in Section 2.3 with $J \to \infty$ and $c = 1$). It is easy to check that the balance equations are not linearly independent, but become so by dropping one of them. If we drop the last equation $\lambda P_{K-1} - \mu P_K = 0$, the system of equations can be written in matrix form as

$$\begin{bmatrix} \mu & & & & & \\ -(\lambda+\mu) & \mu & & & & \\ \lambda & -(\lambda+\mu) & \mu & & & \\ & & \ddots & & & \\ & & & \lambda & -(\lambda+\mu) & \mu \end{bmatrix} \begin{bmatrix} P_1 \\ \vdots \\ \\ \\ P_K \end{bmatrix} = \begin{bmatrix} \lambda P_0 \\ -\lambda P_0 \\ 0 \\ \vdots \\ 0 \end{bmatrix} \tag{2.111}$$

and if we drop the first one $\lambda P_0 - \mu P_1 = 0$, the matrix form is

$$\begin{bmatrix} -(\lambda+\mu) & \mu & & & & \\ \lambda & -(\lambda+\mu) & \mu & & & \\ & & \ddots & & & \\ & & & \lambda & -(\lambda+\mu) & \mu \\ & & & & \lambda & -\mu \end{bmatrix} \begin{bmatrix} P_1 \\ \vdots \\ \\ \\ P_K \end{bmatrix} = \begin{bmatrix} -\lambda P_0 \\ 0 \\ \vdots \\ \\ 0 \end{bmatrix} \tag{2.112}$$

The inverse of the extreme left matrix can be evaluated and thus every P_n, $n = 1, \ldots, K$ can be expressed in terms of P_0. Ultimately using the normalizing equation we obtain P_n's. This observation is also true for the birth-death process.

In the class of Markovian models including birth-death models and simple networks of queues, one may come across more queueing systems in real life situations than are discussed here. We contend ourselves by choosing only a few under a given unified treatment rather than trying to exhaust every feasible model in practice. A variety of interesting models will be included in the exercises. Nevertheless models with some novelty are discussed below.

Of particular interest is the effect of customer impatience which is usually of three forms, viz., balking, reneging and jockeying. The reluctance of a customer joining the system upon arrival is called balking, whereas the reluctance to remain in the queue until being served is reneging. If there are parallel service counters each having a separate queue, the customer's moving back and forth among different queues is known as jockeying.

For simplicity, consider the $M/M/1$ model. In case of balking, a customer may be discouraged to join the queue if the queue size is large. This fact is taken care of if $M/M/1$ is changed to a birth-death model with $\mu_n = \mu$ and $\lambda_n = b_n\lambda$ such that $0 \leq b_{n+1} \leq b_n \leq 1$. Another assumption may be of a constant balking rate irrespective of the queue size, i.e., $\lambda_n = p\lambda$. For reneging, we may define a reneging function $r(n)$ as

$$r(n) = \lim_{\Delta t \to 0} \{P(\text{unit reneges during } \Delta t \text{ when there are } n \text{ customers in the system})/\Delta t\}$$

The new model is still a birth-death one with $\lambda_n = \lambda$ and $\mu_n = \mu + r(n)$. An alternative postulate would be that the customer's waiting time before leaving the queue is a random variable. We can have both types of customer impatience in a given situation and still make it a birth-death model. The analysis of jockeying is more complicated at least when there are more than two channels.

A second case with a difference is when the service rate depends on the number in the system. Yet it is nothing but a birth-death model with $\lambda_n = \lambda$ and a general μ_n. A special case of this is the $M/M/\infty$ model where the number of servers is infinity. In any system which has a very large number of service channels (e.g., telephone exchange, a large parking lot, communication system via satellite) may be approximated by this model. Surprisingly, its transient solution is easy to study. With i initial customers, it can be shown that

$$P(z,t) = (q + pz)^i e^{-\rho q(1-z)} \tag{2.113}$$

where $p = e^{-\mu t}$ and $q = 1 - p$. From (2.113), one can conclude by the uniqueness theorem (see Section A.3 in Appendix A) that $X(t)$ is the sum of two independent variables, one having the binomial distribution with parameters

i and $e^{-\mu t}$ (see (A.32)) and the other having Poisson distribution (see (A.39)) with parameter $\rho(1-e^{-\mu t})$.

Among birth-death models, the models $M/M/c$ and $M/M/c$ with capacity c are of special interest in telephone congestion theory. The probability that an arriving customer finds all servers busy in the first model is given in Exercise 8(a) and is referred to as Erlang's C formula. The same probability in the second model is referred to as Erlang's B formula or Erlang's loss formula since an arriving customer is lost to the system. Its expression is given by P_c in Exercise 9(a) with $N = c$. These formulas also describe the fraction of time that all servers are busy.

It is observed in Section 2.3 that any Erlangian model or its equivalent bulk queue model (see Remark 2) is not a birth-death type but is a Markovian one by suitably choosing the state space. The main aim is to retain the Markovian nature of the queueing process by some suitable method. In that section it is achieved by the method of stages and the same theme will be repeated in Chapters 3, 6 and 7 through other methods. Another class of models is proposed here which are different from Markovian and Erlangian models. Let us closely examine these models before introducing the new one. The coefficient of variation (see Section A.3 in Appendix A) of an exponential distribution is one and of an Erlangian distribution with k stages and with the same mean for each stage is $1/\sqrt{k}$ which is less than one for $k > 1$. Therefore, we may like to consider the class of distributions whose coefficient of variation is greater than one. For a moment, let us view the situation from another angle and for simplicity we only discuss service-time distributions. In an Erlangian service time with k stages, the service mechanism is known to have k stages such that a service is considered complete when a customer passes through each stage one after the other in a given order. In other words, the stages are in series, which suggests us to think about service in stages which are in parallel. Let there be k stages of service, the service-time at the jth stage being exponential with mean $1/\mu_j$, $j = 1, \ldots, k$. A customer who is ready to be served enters the jth stage with probability α_j, $j = 1, \ldots, k$ and is out of the service mechanism when the service in that stage is completed. Then it can be shown that the overall service-time is hyper-exponential (see Section A.4 in Appendix A) with p.d.f.

$$\sum_{i=1}^{k} \alpha_i \mu_i e^{-\mu_i s}, \qquad s \geq 0. \tag{2.114}$$

Denote the distribution by HE_k. It turns out that its mean and variance, respectively, are as follows:

$$\mu_s = \sum_{i=1}^{k} \frac{\alpha_i}{\mu_i}, \tag{2.115}$$

and

$$\sigma_s^2 = 2\sum_{i=1}^{k} \frac{\alpha_i}{\mu_i^2} - \mu_s^2. \tag{2.116}$$

Using the *Cauchy-Schwarz inequality*, which is stated as

$$\left(\sum_i a_i b_i\right)^2 \leq \left(\sum_i a_i\right)^2 \left(\sum_i b_i\right)^2$$

for real a's and b's, we can show that the coefficient of variation $(\sigma_s/\mu_s) \geq 1$, the equality being true when $\mu_1 = \ldots = \mu_k$. Thus hyperexponential distributions are examples of those which have coefficients of variation that are greater than one.

In Section 2.2 we have not exactly eluded, but have not done proper justice to the question of the existence of $\lim_{t\to\infty} P_n(t) = P_n$ and that whether $\{P_n\}$ is a probability distribution. Without proving anything, the question has been dealt with in Section A.8 for general Markov chains and Markov processes. In spite of being repetitious, we prefer once again to bring the topic here for our special models. In this chapter we have been concerned with birth-death models and their special cases (Sections 2.2 and 2.3) or with some other types of Markov processes such as Erlangian models (Section 2.4) which do not belong to birth-death models. For the existence of the $\lim_{t\to\infty} P_n(t) = P_n$ and for $\{P_n\}$ being a probability distribution it is necessary that any Markov process and hence in particular any of these models must be irreducible and positive recurrent. The property of irreducibility that every state can be reached from any other one at some time or the other can be checked from the transition rates. For example, in the birth-death models there is a positive probability of having any number in the system at a future time given any fixed number in the system at present, and therefore the models are irreducible. However, in order to verify the second property we note that the condition under which a particular model is positive recurrent is the same as the condition for which the steady-state solution of the balance equations exists. Thus the birth-death model is positive recurrent provided

$$\sum_{n=1}^{\infty} \prod_{i=1}^{n-1} \frac{\lambda_i}{\mu_{i+1}} < \infty$$

(see (2.14) and the sentence following it). Similarly the condition for $M/M/1$ (see (2.7)) or Erlangian models (see the expression for p_0 after (2.24)) is $\rho < 1$.

It is true that the study of the number of customers in the system or in the queue has preoccupied us in most of the sections. Yet we have not quite neglected the investigation of the waiting time and the busy period. As far as the average waiting time is concerned, the enunciation of Little's formula (Section 2.6) on a general relation between the average number in the system and the average waiting time comes to our rescue. We are also delighted to learn that the formula is general enough to be applicable to almost any model and therefore will appear in subsequent chapters. On the other hand, the derivation of the exact distribution of the waiting time depends on the queue discipline and is illustrated for $M/M/1$ with FCFS discipline (other disciplines to be discussed in Section 10.2 in Chapter 10). The time-dependent nature of a busy period suggests that in its study too, the technique of transient behaviour analysis can be applicable. For the benefit of the readers, a simple concept of networks of queues is introduced in Section 2.8. More on networks is covered in Chapter 10, Section 10.5. Lastly, we touched upon simple problems in design and control of Markovian models by using some of their performance measures. Some more situations are dealt with in Chapter 10, Section 10.4.

The chapter is concluded with brief notes on departure process (i.e., distribution of inter-departure times) and estimation of parameters in the $M/M/1$ model.

(i) Departure process: It is known that in the $M/M/1$ model the distribution function of inter-departure times under steady state is given by

$$F_x(x) = 1 - e^{-\lambda x}, \qquad \geq 0 \tag{2.117}$$

which is interestingly the same as the distribution function of inter-arrival times (see Exercise 32).

(ii) Estimation of λ and μ : If all other assumptions in the model $M/M/1$ are correct except the knowledge of λ and μ then these two parameters are to be estimated. Suppose the queueing system is observed for a duration of t. Let n_a, n_s and t_b be the observed number of arrivals, number of departures and duration of all busy periods. Then simple estimates of λ and μ are given by

$$\hat{\lambda} = \frac{n_a}{t} \quad \text{and} \quad \hat{\mu} = \frac{n_s}{t_b}, \tag{2.118}$$

respectively. A detailed development of this problem will be part of Chapter 5.

Most of the material in this chapter being standard in the theory of queues, may be referred to any textbook on the subject (for example, Cox and Smith (1961), Gross and Harris (1985), Kleinrock (1975), Saaty (1961) Medhi (1991). For Erlangian models one may see Jackson (1956) and Jackson and Nickols (1956). Similarly, Method 1 of Section 2.5 is originally given in Bailey (1954). However, Method 2 is discussed in Cox and Smith (1961). For Section 2.5.3 see Conolly (1975), Conolly and Langaris (1993), Sharma (1997) and Jain, Mohanty and Meitei (2003). More particularly, formula (2.64) comes from Sharma (1997) and (2.65) from Conolly and Langaris (1993). The concept of open Jackson networks comes from Jackson (1963) and is included in most of the textbooks. Other references relate to exercises and applications.

2.11. Exercises

1. Customers arrive at a one-window drive-in of a bank according to the Poisson process with rate of arrival $\lambda = 10$ per hour. Service-times are independent of arrivals and are i.i.d. exponential random variables with service rate $\mu = 12$ per hour. Moreover, there is enough parking space to accommodate all arriving cars. Find the following:
   (a) The expected number of cars waiting in the queue.
   (b) The expected waiting time in the queue.
   (c) The minimum number of parking space to accommodate all arriving cars at least 80% of the time.

2. (a) Derive the steady-state equations for the $M/M/1$ system when the maximum number of units (customers) allowed in the system is N.
   (b) Show that
      (i) for $n = 0, 1, \ldots, N$,

$$P_n = \begin{cases} \left(\dfrac{1-\rho}{1-\rho^{N+1}}\right)\rho^n & \rho \neq 1, \\ \dfrac{1}{N+1} & \rho = 1. \end{cases}$$

      (ii)

$$L = \begin{cases} \rho\left(\dfrac{1-(N+1)\rho^N + N\rho^{N+1}}{(1-\rho)(1-\rho^{N+1})}\right) & \rho \neq 1 \\ \dfrac{N}{2} & \rho = 1. \end{cases}$$

      and
      (iii)

$$\lambda_{\mathit{eff}} = \lambda(1 - P_N) = \mu(L - L_q),$$

      where λ_{eff} is the rate of joining the system. (Remember that not every arrival can join.)

3. Consider a telephone system consisting of N lines. The arrival process is Poisson with rate λ, and call holding times are exponentially distributed with average $1/\mu$. An arriving call is lost if all lines are busy. Then show that the proportion of lost calls over a long period of time is

$$\frac{(\rho^N/N!)}{\sum_{n=0}^{N}(\rho^n/n!)}$$

4. (a) Consider the problem of repairing J machines by one operator. Any machine on failure is sent to the operator. After its repair it operates until it fails again. The failure time of a machine is

exponential with parameter λ and the repair time independent of failure times is also exponential with parameter μ. Draw the flow diagram and write the balance equations.
Let the probability that there are n machines which are out of order be P_n. Using balance equations show that:

$$P_n = \frac{J!}{(J-n)!}\left(\frac{\lambda}{\mu}\right)^n P_0, \quad n = 0, 1, 2, \ldots, J,$$

where

$$P_0 = \left[\sum_{n=0}^{J} \frac{J!}{(J-n)!}\left(\frac{\lambda}{\mu}\right)^n\right]^{-1}.$$

(b) Prove that

(i) The expected number of machines which are out of order is given by

$$L = J - \frac{\mu}{\lambda}(1 - P_0),$$

(ii) The expected number of machines waiting to be repaired is given by

$$L_q = J - \left(\frac{\lambda+\mu}{\mu}\right)(1 - P_0),$$

(iii) The expected time that a machine spends in the repair system is $L/\bar{\lambda}$, and

(iv) The expected time that a machine spends in the queue is $L_q/\bar{\lambda}$, where $\bar{\lambda} = \lambda(J - L)$.

5. In Exercise 4, let there be c operators. Write the balance equations and show that

$$P_n = \begin{cases} \binom{J}{n}\left(\frac{\lambda}{\mu}\right)^n P_0 & 0 \le n < c \\ \binom{J}{n}\frac{n!}{c^{n-c}c!}\left(\frac{\lambda}{\mu}\right)^n P_0 & c \le n \le J. \end{cases}$$

Obtain L and L_q.

6. Starting with an arbitrary value of P_0, determine P_n's in Exercise 4 for the case $J = 4$, $\lambda = 15$ and $\mu = 20$. (This exercise provides an alternative method of evaluating P_n's.)

7. For the $E_k/M/1$ queueing system,

(a) Show that

(i) $P_0 = (1 - \rho)$,

(ii) $P_n = \rho(1 - z_0^{-k})z_0^{-(n-1)k}$, $n \ge 1$,
where z_0 is the zero of $k\rho z^{k+1} - (1 + k\rho)z^k + 1$, lying outside the region $|z| < 1$.

(b) Let Q_n denote steady-state probability of n in the system at an arrival instant. Suppose the relationships

$$Q_0 = kp_{k-1},$$
$$Q_n = kp_{nk+k-1}, \qquad n \geq 1,$$

are true. (This will be discussed in Chapter 5.)
Show that
(i) $Q_n = (1 - z_0^{-k})z_0^{-nk}$, $n \geq 0$, and
(ii) The conditional waiting time in the queue is exponential with parameter $k\lambda(z_0 - 1)$.

8. Draw the flow diagram for a Markovian queueing system consisting of c servers, each with a rate μ (statistically independent of each other). Using the flow diagram, write the balance equations and show that

$$P_n = \begin{cases} \dfrac{(\lambda/\mu)^n}{n!} P_0 & 0 \leq n < c \\ \\ \dfrac{(\lambda/\mu)^n}{c!c^{n-c}} P_0 & n \geq c \end{cases}$$

where

$$P_0 = \left[\sum_{n=0}^{c-1} \frac{(\lambda/\mu)^n}{n!} + \frac{(\lambda/\mu)^c}{c!(1 - \frac{\lambda}{c\mu})} \right]^{-1},$$

and hence verify (2.18) and the following:
(a) The probability for all c servers to be occupied is given by

$$\frac{(\lambda/\mu)^c}{c!(1 - \frac{\lambda}{c\mu})} P_0,$$

(b) P(j units waiting in the queue | c servers are occupied) is equal to

$$\left(1 - \frac{\lambda}{c\mu}\right)\left(\frac{\lambda}{c\mu}\right)^j, j \geq 0,$$

(i.e., the conditional distribution of the number of units waiting in the queue is geometric with parameter $\frac{\lambda}{c\mu}$),
(c) The conditional waiting time distribution is exponentially distributed with parameter $(c\mu - \lambda)$,
(d) The expected number of busy servers is $\frac{\lambda}{\mu}$,
(e) (i) $\frac{P_n}{P_{n-1}} = \frac{\lambda}{c\mu}$ for $n \geq c$,
(ii) $P_{n+1} > P_n$ for $n + 1 < \frac{\lambda}{\mu}$,
(iii) $P_{n+1} < P_n$ for $n + 1 > \frac{\lambda}{\mu}$.

9. Consider the $M/M/c$ queueing model with N as the capacity of the system ($N \geq c$). Show that

(a)

$$P_n = \begin{cases} \dfrac{\rho^n}{n!} P_0 & 0 \le n < c \\[2ex] \dfrac{\rho^n}{c! c^{n-c}} P_0 & c \le n \le N. \end{cases}$$

(b)

$$P_0^{-1} = \begin{cases} \displaystyle\sum_{n=0}^{c-1} \frac{\rho^n}{n!} + \frac{\rho^c [1 - (\frac{\rho}{c})^{N-c+1}]}{c!(1 - \frac{\rho}{c})} & \frac{\rho}{c} \ne 1 \\[2ex] \displaystyle\sum_{n=0}^{c-1} \frac{\rho^n}{n!} + \frac{\rho^c}{c!}(N - c + 1), & \frac{\rho}{c} = 1. \end{cases}$$

(c)

$$L_q = \begin{cases} P_0 \dfrac{\rho^{c+1}}{(c-1)!(c-\rho)^2} \left[1 - \left(\dfrac{\rho}{c}\right)^{N-c} \right. \\ \qquad \left. -(N-c)\left(\dfrac{\rho}{c}\right)^{n-c} \left(1 - \dfrac{\rho}{c}\right) \right], & \frac{\rho}{c} \ne 1 \\[2ex] P_0 \dfrac{\rho^c (N-c)(N-c+1)}{2c!} & \frac{\rho}{c} = 1. \end{cases}$$

(d) $\lambda_{\mathit{eff}} = \lambda(1 - P_N) = \mu(c - \bar{c})$ (see Exercise 2 for definition),

(e) $L = L_q + (c - \bar{c}) = L_q + \frac{\lambda_{\mathit{eff}}}{\mu}$, where $\bar{c}$ is the expected number of idle servers.

10. Consider the $M/M/c$ queueing system with balking and reneging having the following description:
    (a) An arriving customer who finds all servers busy may join the system with probability p or balk (does not join) with probability $1 - p$.
    (b) After joining the queue, a customer may renege; he/she waits an exponential (α) length of time for service to begin, otherwise departs from the system.
    (c) Any customer who leaves the system and decides to return is a new arrival.
        (i) Find P_n, $n \ge 0$.
        (ii) Find P_n, $n \ge 0$, when no reneging is allowed.

11. Derive (2.19) from (2.17) as $c \to \infty$.

12. Verify (2.20) and find the mean and variance of the number of customers in the system.

13. In the $M/M/2$ model, verify the following:
    (a) $P_0 = \frac{2\mu - \lambda}{2\mu + \lambda}$,
    (b) $L_q = \frac{2(\frac{\lambda}{2\mu})^3}{1 - (\frac{\lambda}{2\mu})^2}$,

(c) $L = \frac{4\lambda\mu}{4\mu^2 - \lambda^2}$,
(d) $E(W) = \frac{4\mu}{4\mu^2 - \lambda^2}$.

14. Compare which one of the following two queueing systems with response time as the criterion is the best:
   (a) Two independent $M/M/1$ queue with $\frac{\lambda}{2}$ as the rate of arrival,
   (b) $M/M/2$ with λ as the rate of arrival.

15. Consider the $M/M/2$ queueing system where the service rates of the server are μ_1 and μ_2. Assume without loss of generality that $\mu_1 > \mu_2$. The state of the system is defined to be the tuple (n_1, n_2) where $n_1 \geq 0$ denotes the number of customers at the faster server, and $n_2 \geq 0$ denotes the number of customers at the slower server. When both servers are idle, the faster server is scheduled for service before the slower one. Draw the flow diagram and write down the balance equations. Find the values of P_0, L, and $E(W)$.

16. Show that the Laplace transform of p.g.f. for the system size at time t for the $M/M/1$ model with an absorbing barrier at zero and having initially i customers, is given by

$$P^*(z, s) = \frac{z^{i+1} - (1-z)(\mu - \lambda z)(z_1^i/s)}{-\lambda(z - z_1)(z - z_2)},$$

where z_1 and z_2 are as in (2.44).

17. For the $M/M/1$ model,
   (a) Check that (2.50) holds true for $n < i$;
   (b) Show that $\sum_{n=0}^{\infty} P_n(t) = 1$; and
   (c) Verify that

$$\lim_{t\to\infty} P_n(t) = (1-\rho)\rho^n, \qquad n = 0, 1, 2, \ldots,$$

   when $\rho < 1$.

18. In the $M/M/1$ model, let $\lambda(t)$ and $\mu(t)$ be the arrival and service rate at time t, respectively, (this is called time dependent Markovian queueing system). Denote by $L(t)$ the expected number of customers in the system at time t. Show that
   (a) $\frac{dL(t)}{dt} = \lambda(t) - \mu(t)[1 - P_0(t)]$;
   and
   (b) Obtain its solution as $L(t) = X(0) + \int_0^t \lambda(\tau)d\tau - \int_0^t \mu(\tau)d\tau + \int_0^t \mu(\tau)P_0(\tau)d\tau$.
   (See Rider (1976).)

19. For the $M/M/1$ model, using the relation

$P_n(t) = P(X(t) = n, X(s) > 0, 0 \leq s \leq t | X(0) = i) + P(X(u) = 0$ for the first time for some u, $0 \leq u \leq t$ and $X(t-u) = n | X(0) = i)$, prove (2.61). (See Jain et al. (2003).)

20. (a) Using the generating function formula for the modified Bessel function to be

$$\sum_{n=-\infty}^{\infty} y^n I_n(z) = \exp\left(\frac{z}{2}\left(y + \frac{1}{y}\right)\right),$$

and the properties (2.49), prove the following identity:

$$\sum_{k=-\infty}^{\infty} \frac{e^{-(\mu+\lambda)t}}{\mu t} k\rho^{-k/2} I_k(2\sqrt{\lambda\mu}\, t) = (1-\rho).$$

(b) Using the above identity, prove (2.63).

21. Prove (2.64) and (2.65). (See Conolly and Langaris (1993) and Sharma (1997).)

22. Let $P_{ij}(t)$ be the probability that there are exactly i arrivals and j services up to time t. For the $M/M/1$ model which is idle at time $t = 0$, obtain the differential-difference equations that are satisfied by $P_{ij}(t)$; i.e.,

$$\begin{aligned}
\frac{dP_{00}}{dt}(t) &= -\lambda P_{00}(t), \\
\frac{dP_{i0}}{dt}(t) &= -(\lambda+\mu)P_{i0}(t) + \lambda P_{i-1,0}(t) & i \geq 1, \\
\frac{dP_{ii}}{dt}(t) &= -\lambda P_{ii}(t) + \mu P_{i,i-1}(t) & i \geq 1 \\
\frac{dP_{ij}}{dt}(t) &= -(\lambda+\mu)P_{ij}(t) + \mu P_{i,j-1}(t) \\
&\quad + \lambda P_{i-1,j}(t) & i \geq 2 \\
&& \text{and } 1 \leq j < i.
\end{aligned}$$

Show that the solution of the differential-difference equations is given by

$$\begin{aligned}
P_{ij}(t) &= \left(\frac{\lambda}{\mu}\right)^i \frac{(\mu t)^j e^{-\lambda t}}{i!} \sum_{k=0}^{j} \frac{(i-k)}{k!} \sum_{m=0}^{j-k} (-1)^m \\
&\quad \times \frac{(m+i+k-1)!\left(1 - e^{-\mu t}\sum_{r=0}^{m+i+k-1} \frac{(\mu t)^r}{r!}\right)}{m!(j-k-m)!(\mu t)^{m+k}}
\end{aligned}$$

for $i \geq 1$ and $0 \leq j \leq i$;
(See Pegden and Rosenshine (1982)).

23. For the $M/M/1$ queueing system with i initial customers, use the transformation

$$q_n(t) = e^{(\lambda+\mu)t}\left[\mu P_n(t) - \lambda P_{n-1}(t)\right],$$

to prove that

$$q_n(t) = \mu\rho^{\frac{n-i}{2}}(1-\delta_{0i})\left[I_{n-i}(2\sqrt{\lambda\mu}\,t) - I_{n+i}(2\sqrt{\lambda\mu}\,t)\right] + \lambda\rho^{\frac{n-i-1}{2}}\left[I_{n+i+1}(2\sqrt{\lambda\mu}\,t) - I_{n-i-1}(2\sqrt{\lambda\mu}\,t)\right],$$

for $n = 1, 2, \ldots$.
Express $P_n(t)$ in terms of $q_n(t)$ as follows:

$$P_n(t) = \frac{e^{-(\lambda+\mu)t}}{\mu}\sum_{k=1}^{n}\rho^{n-k}q_k(t) + \rho^n P_0(t), \qquad n \geq 1,$$

and

$$P_0(t) = \int_0^t e^{-(\lambda+\mu)y}q_1(y)dy + \delta_{0i}.$$

(See Parthasarathy (1987).)

24. Consider the birth-death model with

$$\lambda_n = \begin{cases} (N-n)\lambda & n = 0, 1, \ldots, N-1, \\ 0 & n \geq N, \end{cases}$$

and

$$\mu_n = n\mu \qquad n = 1, 2, \ldots, N.$$

(This can be viewed as the machine maintenance model with N machines and N operators. See Exercise 5.)
Find the transient solution for $P_n(t)$ of the model. (See Giorno et al. (1985).)

25. For the $M/M/\infty$ model with i initial customers, establish the following:
   (a)
   $$P(z,t) = (q+pz)^i e^{-\frac{\lambda}{\mu}q(1-z)},$$
   where $p = e^{-\mu t}$ and $q = 1-p$.
   (b) Find the mean and variance of the system size at time t.

26. (a) Prove (2.73).
   (b) Show that
   $$P_0 = \frac{E(I)}{E(I)+E(\Delta)}$$
   and interpret it.

27. Consider the model in Exercise 20 when no reneging is allowed. Suppose the period during which all servers are busy is called a busy period. Let N and A, respectively, denote the average number lost and the average attended the service during a unit service-time within a busy period.

(a) Find $N/(N+A)$, the proportion of loss during a unit service-time within a busy period and hence find the average loss during a busy period.

(b) Show that the proportion of time the system remains busy is $\lambda P_0/(c\mu - \lambda)$. (Hint: use Exercise 25(b).)

(c) Use (a) and (b) and find the average loss of customers during any duration of time t.

28. Consider an open Jackson network with three nodes designated as 1,2 and 3. Nodes 1,2 and 3 have service-times with parameters μ_1, μ_2 and μ_3, respectively, and Poisson arrivals with rate λ only at node 1. Customers follow one of two routes through the network: node 1 to node 2 with probability p and node 1 to node 3 with probability $q = 1 - p$. Obtain the effective arrival rates α_i at node $i (i = 1, 2, 3)$ and the mean time spent by a customer in the network.

29. Consider a tandem queue with 2 nodes, where λ is the arrival rate at node 1, and μ_i is the service rate at the ith node ($i = 1, 2$). Write the local balance equations for the queueing system and check that

$$P(n_1, n_2) = \prod_{i=1}^{2} (1 - \rho_i)\rho_i^{n_i}$$

and

$$L = \frac{\rho_1}{(1 - \rho_1)} + \frac{\rho_2}{(1 - \rho_2)},$$

where $\rho_i = \frac{\lambda}{\mu_i}$, $i = 1, 2$.

30. Check (2.80).

31. For the $M/M/c$ queueing system with mean interarrival time equal to 0.25 hours and mean service-time equal to 0.20 hours, determine the number of servers that should be assigned to the system to minimize the expected total cost per hour, given that the cost of each server is \$20/hour and the estimated cost that is incurred on each customer waiting in the queue is \$180/hour.

32. For the $M/M/1$ system, denote by D the inter-departure time and by $N(t)$ the number of customers in the system at time t, measured from the previous departure. Let

$$F_n(t) = P(N(t) = n, D > t).$$

Show that

$$\frac{dF_n(t)}{dt} = -(\lambda + \mu)F_n(t) + \lambda F(t)_{n-1}, \quad n \geq 1,$$

$$\frac{dF_0(t)}{dt} = -\lambda F_0(t),$$

with the initial condition

$$F_n(0) = (1-\rho)\rho^n, \qquad n \geq 0.$$

Verify that

$$F_n(t) = (1-\rho)\rho^n e^{-\lambda t}, \qquad n \geq 0,$$

and hence under steady state, check that D follows the exponential distribution with parameter λ.

33. Suppose there is a one service center consisting of a single server. Let the external arrival process be Poisson with rate λ, the service times be exponential with mean service-time $1/\mu$, and p be the probability that a customer rejoins the queue after completing the service (in the notation of open network $p_{11} = p$, $p_{10} = 1-p$). Such a queueing system is called a $M/M/1$ with Bernoulli feedback. Check that
   (a) The effective arrival rate to the queue is $\frac{\lambda}{(1-p)}$;
   (b) The steady-state probability that the system is in state n is

   $$P_n = (1-\alpha)\alpha^n, \qquad n \geq 0,$$

   where $\alpha = \frac{\lambda}{\mu(1-p)}$; and
   (c) The interarrival times (external and internal) are not exponentially distributed (i.e., the effective arrival process is not Poisson).

34. Consider a *tandem production line*, that is a series of manufacturing and assembly stages, consisting of J processing units, each may be modelled as a $M/M/1$ system with mean processing time $1/\mu_i$, $i = 1, 2, \ldots, J$. The last unit in the line performs quality inspection. If an item does not satisfy quality requirements, it is immediately sent back to the first unit of the production line. It is known that a fraction π, $0 < \pi < 1$ of jobs requires such postprocessing. Find the mean lead time of a job in this system under the assumption that each processing unit has a buffer to store waiting jobs of unlimited size.

References

Bailey, N. T. J. (1954), A continuous time treatment of a simple queue using generating functions, *J. Roy. Statist. Soc.*, Ser. B, **16**, 288–291.

Conolly, B. W. (1975), *Lecture Notes on Queueing Systems*, Ellis Horwood, Chichester Ltd.

Conolly, B.W. and Langaris, C. (1993), On a new formula for the transient state probabilities for $M/M/1$ queues and computational implications, *J. Appl. Prob.*, **30**, 237–246.

Cox, D. R. and Smith, W. L. (1961), *Queues*, Methuen & Co., London.

Giorno, V., Negri, C. and Nobile, A.G. (1985), A solvable model for a finite capacity queueing systems, *J. Appl. Prob.*, **22**, 903–911.

Gross, D. and Harris, C. M. (1985), *Fundamentals of Queueing Theory*, Second Edition, John Wiley, New York.

Jackson, J.R. (1963), Job-shop-like queueing system, *Management Sci.*, **10**, 131–142.

Jackson, R. R. P. (1956), Queueing processes with phase-type service, *J. Roy. Statist. Soc.*, Ser. B, **18**, 129–132.

Jackson, R. R. P. and Nickols, D. J. (1956), Some equilibrium results for the queueing process $E_k/M/1$, *J. Roy. Statist. Soc.* Ser. B, **18**, 275–279.

Jain, J. L., Mohanty, S. G. and Meitei, A. J. (2003), Transient solution of $M/M/1$ queues: A probabilistic approach, *Operational Research and its Applications: Recent Trends: Proceedings of the APORS*, 74–81.

Kleinrock, L. (1975), *Queueing Systems*, Vol. 1, John Wiley, New York.

Medhi, J. (1991), *Stochastic Models in Queueing Theory*, Academic Press, New York.

Mitrani, I. (1987), *Modelling of Computer and Communication Systems*, Cambridge University Press.

Parthasarathy, P.R. (1987), A transient solution to an $M/M/1$ queue: A simple approach, *Adv. Appl. Prob.*, **19**, 997–998.

Pedgen, C.D. and Rosenshine, M. (1982), Some new results for the $M/M/1$ queue, *Management Sci.*, **28**, 821–828.

Rider, K.L. (1976), A simple approximation to the average queue size in the time-dependent $M/M/1$ Queue, *J. ACM*, **23**, pp. 361–367.

Saaty, T. L. (1961), *Elements of Queueing Theory with Applications*, McGraw-Hill, New York.

Sharma, O.P. (1997), *Markovian Queues*, Allied, New Delhi.

Suri, R., Sanders, J. L., Kamath, M. (1993), Performance Evaluation of Production Networks, in *Handbooks in Operations Reseach and Management Science*, S. C. Graces et al. Eds., Amsterdam.

CHAPTER 3

Regenerative Non-Markovian Queues – I

3.1. Introduction

In the last chapter, the study of various characteristics was done for models which are either Markovian or can be reformulated to be Markovian (such as the $M/E_k/1$ model). In such models, the number of customers $X(t_0)$ in the system at any instant t_0 provides sufficient information for the study of the number of customers $X(t_1)$ in the system at some future time t_1. This Markovian property has made the theory attractively elementary. Guided by the principle of maintaining the Markovian property, we find that some special non-Markovian models exhibit the Markovian property at certain points of time called regeneration points if either the interarrival times or the service times are i.i.d. exponential random variables. These models may be called regenerative non-Markovian models, which will be studied by the so-called *imbedded Markov chain technique.* The technique will be explained when we directly deal with specific models such as, $M/G/1$ and $G/M/1$ where G represents a general or arbitrary distribution.

The imbedded Markov chain technique certainly restricts the study of the number in the system to the regeneration points. This limitation, however, is overcome for the steady-state distributions by the development of functional relations linking the distribution at a regeneration point to the distribution at any point in time. As in the previous chapter, the waiting time and the busy period are also dealt with for the models introduced in this chapter.

Traditionally, the main focus of the theory of queues has been on continuous-time models. In this chapter, a section has been included, mainly to introduce discrete-time models and to present results parallel to those in continuous-time models, remembering that these models occur in the telecommunications field.

3.2. Markovian Input Models

We consider a model which may be called the $M/G/1$ model. As the notation suggests, it is a single-server system with the arrival and service-times having the following properties:

(i) The arrival process is Poisson with rate λ;

(ii) The service-times are i.i.d. with an arbitrary distribution function F_S with mean $1/\mu$.

In addition, properties (iii), (iv) and (v) of $M/M/1$ are also valid. Property (i) implies that the interarrival times are i.i.d. exponential (λ) random variables.

Similar to the last chapter, at first we look into the number in the system at any time. Because of the memoryless property of the exponential distribution, in any Markovian model the probability of an arrival or a departure in a small interval $(t,\ t + \Delta t)$ is independent of what happened up to time t. Thus as we noticed earlier, the number $X(t + \Delta t)$ in the case of a birth-death process (which is a Markov process) only depends on $X(t)$. However, in the $M/G/1$ model, since any arbitrary service-time distribution except the exponential distribution does not possess the memoryless property, $X(t+\Delta t)$ not only depends on $X(t)$ but also on how much service-time is elapsed for the customer in service at time t. In this sense, the model is non-Markovian. Instead of considering the process at all time points, suppose it is examined at a set of selected points of time, say just after the departure instants. At these instants, the elapsed service-time for the entering customer being zero which is so even when the system is empty, makes the number of customers at a departure instant dependent on the number at the previous departure instant and the number of arrivals during a service period which in any case does not depend on any departure instant.

Notation:

C_n	:	nth customer in the system
X_n	:	number of customers in the system at the departure of C_n
X_0	:	initial number of customers waiting for service
γ_n	:	number of arrivals during the service-time of C_n
Π_n	:	steady-state probability of n in the system at a departure instant

From our discussion it is clear that the sequence $\{X_n\}$, called the departure point process, forms a Markov chain which is imbedded over the original non-Markovian process at regeneration points such as service completion instants. That we study X_n in place of $X(t)$ is the basis of the imbedded Markov chain technique.

We first develop an important but simple relation between X_{n+1} and X_n. Notice that if C_n does not leave the system empty, then C_{n+1} will leave behind one customer less than those in the queue left by C_n (because C_{n+1} is departing) plus all the arrivals during the service of C_{n+1}. Thus, we have

$$X_{n+1} = X_n - 1 + \gamma_{n+1}, \qquad X_n > 0. \tag{3.1}$$

On the other hand, if C_n leaves the system empty then C_{n+1} will leave behind those who arrived during the service of C_{n+1}. Hence,

$$X_{n+1} = \gamma_{n+1}, \qquad X_n = 0. \tag{3.2}$$

Combining (3.1) and (3.2), one can write

$$X_{n+1} = X_n - \delta(X_n) + \gamma_{n+1} \tag{3.3}$$

where

$$\delta(X_n) = \begin{cases} 0 & \text{if } X_n = 0 \\ 1 & \text{if } X_n > 0 \end{cases}$$

Let us find the transition probabilities $P_{ij} = P(X_{n+1} = j | X_n = i)$ of the Markov chain. Denote by $\alpha_k = P(\gamma_n = k)$ which is clearly independent of n since γ_n's are i.i.d. random variables (r.v.). The arrivals having formed the Poisson process (λ), we have the conditional distribution of γ_{n+1} given the service-time $S = s$ as

$$P(\gamma_n = k | S = s) = \frac{(\lambda s)^k e^{-\lambda s}}{k!}$$

and therefore

$$\alpha_k = \int_0^\infty \frac{(\lambda s)^k e^{-\lambda s}}{k!} dF_S(s). \tag{3.4}$$

Now it is obvious from (3.1) and (3.2) that

$$P_{ij} = \begin{cases} \alpha_{j-i+1} & \text{if } i > 0, j \geq i-1 \\ \alpha_j & \text{if } i = 0, j \geq 0 \\ 0 & \text{if } i > 0, j < i-1 \end{cases} \tag{3.5}$$

Denoting by $\mathbf{P} = (P_{ij})$, we have

$$\mathbf{P} = \begin{bmatrix} \alpha_0 & \alpha_1 & \alpha_2 & \alpha_3 & \cdot & \cdot & \cdot \\ \alpha_0 & \alpha_1 & \alpha_2 & \alpha_3 & \cdot & \cdot & \cdot \\ 0 & \alpha_0 & \alpha_1 & \alpha_2 & \cdot & \cdot & \cdot \\ 0 & 0 & \alpha_0 & \alpha_1 & \cdot & \cdot & \cdot \\ \vdots & \vdots & \vdots & \vdots & \vdots & \vdots & \vdots \end{bmatrix},$$

which is of *upper Hessenberg form*, a well-known special type of matrix.

Let us assume that the service-time S has finite mean and variance as $\frac{1}{\mu}$ (so that the rate of service completion is μ which is the same as assumed in the $M/M/1$ model) and σ^2, respectively. Then, it is easy to verify that the mean and variance of γ_n are, respectively,

$$E(\gamma_n) = \rho \qquad \text{and} \qquad \text{Var}(\gamma_n) = \lambda^2\sigma^2 + \rho. \tag{3.6}$$

The investigation of the transient behaviour as observed in the last chapter being difficult (see Section 6.2), we only focus our attention on the evaluation Π_n, the existence of which for $\rho < 1$ will be justified in Section 3.8. However, what will be derived here is an explicit expression not for Π_n but for the generating function of $\{\Pi_n\}$. Following the principle of conservation of flow (see flow diagrams in Chapter 2 and replace average rates by appropriate probabilities) or referring to the properties of a Markov chain (see Section A.8), we can write the balance equations or stationary equations as

$$\Pi_j = \alpha_j \Pi_0 + \sum_{i=1}^{j+1} \alpha_{j-i+1} \Pi_i, \qquad j = 0, 1, 2, \ldots \tag{3.7}$$

such that

$$\sum_{j=0}^{\infty} \Pi_j = 1. \tag{3.8}$$

Notation:

$\Pi(z)$: $\sum_{j=0}^{\infty} \Pi_j z^j$

For solving (3.7) and (3.8), the usual recursive technique (see the $M/M/1$ model) does not lead to a straightforward explicit solution. We may use the method of generating function and after some simplification obtain

$$\Pi(z) = \frac{(z-1)\alpha(z)}{z-\alpha(z)}\Pi_0 \tag{3.9}$$

where $\alpha(z)$ is the generating function of $\{\alpha_j\}$ and is given by

$$\alpha(z) = \sum_{j=0}^{\infty} \alpha_j z^j.$$

Applying L'Hôpital's rule to (3.9) and realizing that $\Pi(1) = 1$, $\alpha(1) = 1$ and $\alpha'(1) = \rho$, we get

$$\Pi_0 = 1 - \rho. \tag{3.10}$$

Therefore,

$$\Pi(z) = \frac{(z-1)\alpha(z)}{z-\alpha(z)}(1-\rho) \tag{3.11}$$

which is known as the *Pollaczek-Khinchin (P-K) formula.*

The procedure we have just now used is a standard one. An alternative derivation of (3.11) by utilizing (3.3) but not (3.7) is presented below.
Let

$$G_n(z) = E(z^{X_n}) = \sum_{j=0}^{\infty} P(X_n = j)z^j.$$

Clearly

$$\Pi(z) = \lim_{n\to\infty} G_n(z).$$

From (3.3) it is evident that

$$\begin{aligned} G_{n+1}(z) &= E(z^{X_n-\delta(X_n)+\gamma_{n+1}}) \\ &= E(z^{X_n-\delta(X_n)})E(z^{\gamma_{n+1}}) \\ &= E(z^{X_n-\delta(X_n)})\alpha(z). \end{aligned} \tag{3.12}$$

Now

$$
\begin{aligned}
E(z^{X_n-\delta(X_n)}) &= P(X_n=0)+\sum_{j=1}^{\infty} P(X_n=j)z^{j-1} \\
&= P(X_n=0)+\frac{1}{z}(G_n(z)-P(X_n=0)).
\end{aligned}
$$

When this expression is substituted in (3.12) and the limit as $n \to \infty$ is taken, the derived expression is

$$\Pi(z)=\alpha(z)(\Pi_0+\frac{\Pi(z)-\Pi_0}{z})$$

leading to the desired P-K formula (3.11).

Let us establish an identity between $\alpha(z)$ and the Laplace-Stieltjes transform (L.S.T. in short, see Section A.5 in Appendix A) of $F_S(s)$ which will be useful in the context of the waiting time distribution.

Notation:

$$\Phi_X(\theta) \quad : \quad \int_0^{\infty} e^{-\theta t} dF_X(t) \quad \text{(L.S.T. of } F_X(t)\text{)}.$$

It is easy to check from (3.4) that

$$\alpha(z)=\Phi_S(\lambda-\lambda z) \tag{3.13}$$

The generating function when inverted would give Π_n, but the inversion is not always simple, even though one can succeed in finding it for the $M/M/1$ model. However, it readily provides the moments of the distribution of the number of customers in the system. (In fact, the knowledge of the mean and the variance of a characteristic in a queue is often sufficient in many situations.) For example, the expected number of customers in the system at the time of an arbitrary customer's departure is given by

$$\Pi'(1)=\rho+\frac{\rho^2}{2(1-\rho)}(1+\mu^2\sigma^2). \tag{3.14}$$

Expression (3.14) is called the *Pollaczek-Khinchin (P-K) mean value formula* (not the same as P-K formula).

The expected number of customers in the system at a departure point which is given in (3.14) can be evaluated by an indirect but interesting technique. Squaring both sides of (3.3) and then taking expectation, we get

$$
\begin{aligned}
E(X_{n+1}^2) =&E(X_n^2)+E(\delta(X_n))+E(\gamma_{n+1}^2) \\
&+2E(X_n)E(\gamma_{n+1})-2E(\delta(X_n))E(\gamma_{n+1})-2E(X_n) \qquad (3.15)
\end{aligned}
$$

since $\delta^2(X_n)=\delta(X_n)$ and $X_n\delta(X_n)=X_n$. When we take the limit as $n\to\infty$, we obtain (3.14) by using (3.6) and realizing that

$$\lim E(X_{n+1}^2)=\lim E(X_n^2)$$

and

$$\begin{aligned}\lim E(\delta(X_n)) &= \lim P(X_n > 0) \\ &= 1 - \Pi_0 = 1 - P_0 = \rho.\end{aligned}$$

Let the steady-state number of customers in the queue at an arrival instant be N. Using the fact that $Q_n = \Pi_n$ to be shown in (3.74), we get

$$E(N) = \sum_{n=1}^{\infty}(n-1)Q_n = \sum_{n=1}^{\infty}(n-1)\Pi_n = \sum_{n=1}^{\infty} n\Pi_n - (1 - \Pi_0). \tag{3.16}$$

On the right side, the first term represents the expected number of customers in the system at a departure instant, whereas the second equals ρ by (3.10). Hence from (3.14), we get

$$E(N) = \frac{\rho^2}{2(1-\rho)}(1 + \mu^2\sigma^2). \tag{3.17}$$

It may be noted but is not proved that (3.17) is true at any instant. Recalling Little's formula (2.67) which is true in general, we are able to evaluate the expected value of the time W_q of a customer to be waiting in the queue as

$$E(W_q) = \frac{1}{\lambda}E(N) = \frac{\rho}{2\mu(1-\rho)}(1 + \mu^2\sigma^2). \tag{3.18}$$

Because of the one-to-one relations among (3.14), (3.17) and (3.18), we may also call (3.17) and (3.18) P-K mean value formula, without ambiguity.

Remark: A particular model of practical significance is the one in which the service time is constant. It is written as $M/D/1$, D standing for 'deterministic' as opposed to random and referring to the service-times. In other words, the service-time is constant and in that sense it has degenerate distribution at its mean. Assume its duration is $1/\mu$ to be the same as the mean service time in $M/M/1$. A relative comparison of models $M/M/1$ and $M/D/1$ based on (3.17) (or (3.18)) shows that the mean queue length (or the mean waiting time in the queue) for the former (see (2.9)) is twice as large as that for the latter, by noting that the variance of exponential service-time is $\frac{1}{\mu^2}$ and that of constant service-time is zero.

Manufacturing – an automatic assembly station. Consider an automatic assembly station which is equipped with a single machine. It is assumed that the machine is working *jam-free*. The occurrence of a jam means that the work of the machine is interrupted and it requires a repair process, like replacing tools or other maintenance. It is typical for such stations that the machining times are known quite accurately and therefore may be assumed to be deterministic and equal to τ, say. The jobs being processed arrive from outside the station according to a Poisson process with rate λ. Based on these assumptions the assembly station may be modelled as a $M/D/1$ system with

$\tau = 1/\mu$. The mean waiting time of a job can be calculated using the P-K mean value formula (3.18)

$$E(W_q) = \frac{\rho\tau}{2(1-\rho)}.$$

Observe that if we had modelled the station as a $M/M/1$ system with $1/\mu = \tau$, we would have committed an error in the mean waiting time of 100%. Clearly this affects also the *mean lead time* of a job, which for this assembly station is given by

$$E(W) = \tau + \frac{\rho\tau}{2(1-\rho)} = \tau\frac{1+\rho}{1-\rho}.$$

In order to find other moments of W_q, it is helpful to have the knowledge about the distribution of W_q. As in the case of Π_n, we derive the L.S.T. of the distribution functions of W_q (similar to the p.g.f. of a distribution), rather than the distribution itself. The waiting time in the system being the sum of customer's waiting time in the queue and the service time, the L.S.T. can be expressed as

$$\Phi_{W_q+S}(\theta) = \Phi_{W_q}(\theta)\Phi_S(\theta),$$

since W_q and S are independent. Referring back to (3.13), we notice that $\Phi_S(\lambda - \lambda z)$ equals the generating function of $\{\alpha_j\}$, where α_j as given by (3.4) is the probability that there are j arrivals during the service of a customer. Analogously, $\Phi_{W_q+S}(\lambda - \lambda z)$, should be equal to the generating function of the probabilities of number of arrivals during a customer's waiting time in the system. Assuming FCFS queue discipline this turns out to be the generating function $\Pi(z)$ of the number in the system at the time of departure. Therefore,

$$\Pi(z) = \Phi_{W_q+S}(\lambda - \lambda z) = \Phi_{W_q}(\lambda - \lambda z)\Phi_S(\lambda - \lambda z).$$

Using (3.11) and (3.13), we find that

$$\Phi_{W_q}(\lambda - \lambda z) = \frac{(z-1)(1-\rho)}{z - \Phi_S(\lambda - \lambda z)}$$

which can be written as

$$\Phi_{W_q}(\theta) = \frac{\theta(1-\rho)}{\theta - \lambda(1 - \Phi_S(\theta))} \tag{3.19}$$

by changing $\lambda - \lambda z$ to θ. For obvious reasons, (3.19) is also called the P-K formula. Expression (3.19) has the following power series form:

$$\Phi_{W_q}(\theta) = \sum_{j=0}^{\infty}(1-\rho)\rho^j\left(\frac{\mu}{\theta}\right)^j(1-\Phi_S(\theta))^j. \tag{3.20}$$

If $\frac{\mu}{\theta}(1 - \Phi_S(\theta))$ happens to be the L.S.T. of the distribution of a particular random variable, say Y, then $(\frac{\mu}{\theta})^j(1-\Phi_S(\theta))^j$ is the L.S.T. of the jth convolution of the same distribution function (equivalently the distribution of the sum of j independent Y's). Actually, from renewal theory one can show that Y is the elapsed time from the arrival of a customer until the end of the service-time of the customer in service. Although (3.20) has a simple form, unfortunately no intuitive explanation even by using the above interpretation of Y exists so far. However, for $M/M/1$ the result can be easily justified.

3.3. Markovian Service-Time Models

The most natural model next to be looked into is the $G/M/1$ model, in which the service-times have the Markovian property. From the notation it is apparent that the model is a single-server system with the properties of the $M/G/1$ model except that properties (i) and (ii) are changed to the following:

(i) Interarrival times T_n are i.i.d. with an arbitrary distribution function F_T with mean $1/\lambda$;

(ii) Service-times are i.i.d. exponential(μ) random variables.

For the study of the number in the system it is evident from the $M/G/1$ model that in the present case we may look for regeneration points and use the imbedded Markov chain technique. Since the interarrival times have a general distribution, the natural selection for such points will be the set of instants just prior to the arrival instants.

Notation:

Y_n : number of customers in the system just before the arrival of C_n

γ_n^* : number of customers (real or virtual) served during the interarrival time between the arrivals of C_n and C_{n+1}

What is meant by real or virtual number of customers served is that the service process is hypothetically assumed to continue even if there are no customers in the queue.

Similar to (3.3), we have

$$Y_{n+1} = \begin{cases} Y_n + 1 - \gamma_n^* & \text{if } Y_n + 1 - \gamma_n^* \geq 0 \\ 0 & \text{otherwise,} \end{cases} \tag{3.21}$$

which can be rewritten as

$$Y_{n+1} = \max(Y_n + 1 - \gamma_n^*, 0). \tag{3.22}$$

Now we derive the transition probabilities $P_{ij} = P(Y_{n+1} = j | Y_n = i)$ to be needed in evaluating the steady-state probability Q_n of n in the system at an arrival instant. Since service-times are i.i.d. exponential (μ), the number of departures during the interarrival time T_n between the arrivals of C_n and C_{n+1} form the Poisson process with rate μ. Thus denoting by $\beta_k = P(\gamma_n^* = k)$, we have

$$P(\gamma_n^* = k | T_n = t) = e^{-\mu t} \frac{(\mu t)^k}{k!},$$

and

$$\beta_k = \int_0^\infty e^{-\mu t} \frac{(\mu t)^k}{k!} dF_T(t). \tag{3.23}$$

By the way, letting $\beta(z)$ to be the p.g.f. of $\{\beta_i\}$, we find that

$$\beta(z) = \Phi_T(\mu - \mu z) \tag{3.24}$$

by comparing with (3.13). The transition probabilities P_{ij} are given by

$$P_{ij} = \begin{cases} \beta_{i+1-j} & \text{if } i+1 \geq j \geq 1 \\ 1 - \sum_{k=0}^{i} \beta_k & \text{if } j = 0, i \geq 0 \\ 0 & \text{if } i < 0, \text{ or } j < 0 \text{ or } i+1 < j \end{cases} \tag{3.25}$$

which is similar to (3.5). These transition probabilities can be written in the following matrix form:

$$\begin{bmatrix} 1-\beta_0 & \beta_0 & 0 & 0 & 0 & \cdot & \cdot & \cdot \\ 1-\sum_{k=0}^{1}\beta_k & \beta_1 & \beta_0 & 0 & 0 & \cdot & \cdot & \cdot \\ 1-\sum_{k=0}^{2}\beta_k & \beta_2 & \beta_1 & \beta_0 & 0 & \cdot & \cdot & \cdot \\ \vdots & \vdots & \vdots & \vdots & \vdots & \vdots & \vdots & \vdots \end{bmatrix}.$$

It is instructive to explain the expression $P_{i0} = 1 - \sum_{k=0}^{i} \beta_k$ which is naturally obvious if we remember that $\sum_{j=0}^{\infty} P(Y_{n+1} = j | Y_n = i) = 1$. However, one may like to think that P_{i0} should be equal to $P(\gamma_n^* = i+1) = \beta_{i+1}$, since the event $\{Y_{n+1} = 0 | Y_n = i\}$ occurs when the event $\{\gamma_n^* = i+1\}$ occurs. But $\{Y_{n+1} = 0 | Y_n = i\}$ also occurs if $\{\gamma_n^* > i+1\}$ which is evident from (3.21) and therefore $P(Y_{n+1} = 0 | Y_n = i) \neq \beta_{i+1}$. There is a reason for relating Y_{n+1} to Y_n and γ_n^* in the form (3.21) or (3.22) although one may argue that by the very definitions of Y_n and γ_n^* the relation $\gamma_n^* \leq Y_n + 1$ is true and therefore is redundant. It is to allow γ_n^* to take any non-negative integer value permissible under the Poisson process, the consideration of which has led us to (3.23). Relations similar to (3.22) will have a crucial role to play in other models (for example, see Section 3.4).

We present a method to evaluate Q_n, the probability that an arrival finds n customers in the system. The question of the existence of Q_n will be dealt with in Section 3.8. The stationary equations are

$$Q_0 = \sum_{i=0}^{\infty} Q_i \left(1 - \sum_{k=0}^{i} \beta_k\right) \tag{3.26}$$

$$Q_j = \sum_{i=0}^{\infty} Q_{j+i-1} \beta_i, \qquad j = 1, 2, \ldots \tag{3.27}$$

with the normalizing equation

$$\sum_{j=0}^{\infty} Q_j = 1. \tag{3.28}$$

Instead of using the p.g.f. technique to solve the above equations, we may treat (3.27) as a set of difference equations as was done in the $M/M/1$ model. The characteristic equation of (3.27) is

$$z - \beta(z) = 0 \tag{3.29}$$

where the second term is the p.g.f. of $\{\beta_i\}$. Obviously $z = 1$ is a root of (3.29). Now we should look for the remaining roots only in the interval $(0, 1)$ because $\beta(z)$ is convergent for $|z| \leq 1$ and any root in $[-1, 0]$ will make some Q_j's negative. Observe that since $\beta_i > 0$ for every i, $\beta'(z)$ and $\beta''(z)$, the first and second derivatives of $\beta(z)$ are positive in $(0, 1)$ and therefore $\beta(z)$, which is continuous, is increasing and strictly convex. Moreover, $0 < \beta(0) = \beta_0 < 1$. Taking these facts into account, if we graph $y = z$ and $y = \beta(z)$ in a (z, y)-plane and watch for points of intersection in $(0, 1)$, then the roots are the z-coordinates of these points of intersection.

Two possible situations can arise which are demonstrated in Figures 3.1 and 3.2.

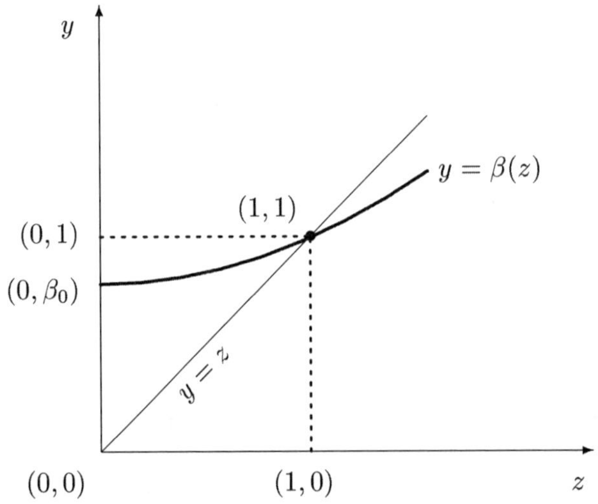

FIGURE 3.1. Roots of (3.29) for $\rho \geq 1$

In case (i) of Figure 3.1 there are no roots in $(0, 1)$ which happens when

$$\begin{aligned} \beta'(1) &= E[\text{number served during an interarrival time}] \\ &= \frac{\mu}{\lambda} \leq 1 \end{aligned}$$

i.e., when $\rho \geq 1$. Clearly when there are no roots other than 1, $\sum Q_j$ diverges. On the other hand, in case (ii) of Figure 3.2 there is exactly one root in $(0, 1)$. Let the root be r_0. Then

$$Q_j = c_1 + c_2 r_0^j$$

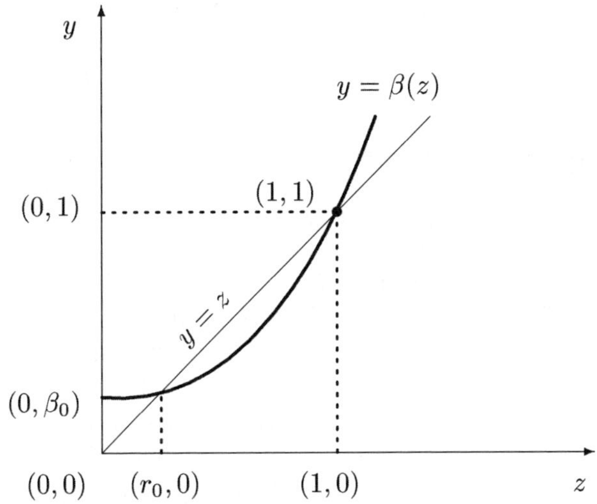

FIGURE 3.2. Roots of (3.29) for $\rho < 1$

where constants c_1 and c_2 are to be determined from the boundary conditions and in particular from (3.28). In order to make $\sum Q_j$ convergent, we find $c_1 = 0$ as was noted in the $M/M/1$ model. Again from (3.28) it follows that $c_2 = 1 - r_0$. Furthermore, we realize that case (ii) of the diagram is equivalent to $\rho < 1$. Hence $\rho < 1$ is a necessary and sufficient condition for the existence of the stationary solution which is given by

$$Q_j = (1 - r_0)r_0^j, \qquad j = 0, 1, \ldots . \tag{3.30}$$

Sometimes it may be worthwhile to check whether the solution satisfies the original equations and in our case we can easily verify that (3.30) does satisfy (3.26) and (3.27).

Remarks:

1. An important difference between (3.27) and its counterpart (3.7) for $M/G/1$ is that the summation in (3.27) extends up to infinity in contrast to (3.7). This difference has become an advantage in favour of $G/M/1$ in that the technique of difference equations could be applied with success, as the resulting characteristic equation (3.29) turns out to be a solvable functional equation.

2. In the system of equations (3.26) and (3.27) one equation may be considered redundant in view of (3.28). Thus in our case (3.26) has not been used.

3. The fact that $c_1 = 0$ which corresponds to the root at 1 is true holds good in general as long as we are dealing with a probability distribution such as $\{Q_n\}$.

4. That there is exactly one root of (3.29) in $(0, 1)$ whenever $\rho < 1$ can be derived by the application of Rouché's Theorem (see Section A.5 in Appendix A). This has been dealt with in Section 3.4.

5. Observe that (3.30) is exactly the same as (2.7) in the $M/M/1$ model except that ρ is replaced by r_0. Thus the expected number in the system, the expected number in the queue and other characteristics just before an arrival can be obtained from the corresponding results for $M/M/1$ by merely replacing ρ by r_0. Surprisingly, we also get the distribution of the waiting time W_q from the $M/M/1$ model ((2.68) and (2.69) in Section 2.6 in Chapter 2) by the same replacement. In fact, for deriving the distribution in Section 2.6, Q_n but not P_n has been used and therefore the replacement of ρ by r_0 is justified.

Although for simplicity we started with the single-server model, the study of the $G/M/c$ model having c servers is no more difficult than $G/M/1$ except that the analysis is more involved. We give below the expressions of P_{ij} for various regions and leave these for checking as an exercise.

$$P_{ij} = \begin{cases} 0 & i < 0 \text{ or } j < 0 \text{ or } i+1 < j, \\ \displaystyle\int_0^\infty \binom{i+1}{j} e^{-j\mu t}(1-e^{-\mu t})^{i+1-j}\, dF_T(t) & j \le i+1 \le c, \\ \displaystyle\int_0^\infty \frac{(c\mu t)^{i+1-j}}{(i+1-j)!} e^{-c\mu t}\, dF_T(t) & c \le j \le i+1, \\ \displaystyle\int_0^\infty \int_0^\infty \binom{c}{j} \frac{(c\mu y)^{i-c}}{(i-c)!} e^{-j\mu t} & \\ \qquad (e^{-\mu y} e^{-\mu t})^{c-j} c\mu \, dy \, dF_T(t) & j < c < i+1. \end{cases} \tag{3.31}$$

Denoting the 3rd expression in (3.31) by $\beta_{i+1-j}(c)$, we see from (3.23) that $\beta_i(1) = \beta_i$.

The stationary equations are

$$Q_j = \sum_{i=0}^{\infty} P_{ij} Q_i, \qquad j \ge 0. \tag{3.32}$$

However, these become

$$Q_j = \sum \beta_i(c) Q_{j+i-1} \quad \text{for } j \ge c \tag{3.33}$$

which are similar to (3.27) and therefore we should have

$$Q_j = C[r_0(c)]^j, \qquad j \ge c \tag{3.34}$$

where $r_0(c)$ is the only root of the equation

$$z = \sum_{n=0}^{\infty} \beta_n(c) z^n,$$

and C is a constant. We have to determine C and Q_j, $j < c$ from the first $c-1$ stationary equations in (3.32) and the normalizing equation $\sum_j Q_j = 1$. The solution is done recursively. From the normalizing equation, it is easy to see that

$$C = \frac{1 - \sum_{j=0}^{c-1} Q_j}{[r_0(c)]^c/(1 - r_0(c))}. \tag{3.35}$$

From (3.32), (3.34) and the first expression of (3.31), we have

$$Q_j = \sum_{i=j-1}^{c-1} P_{ij} Q_i + C \sum_{i=c}^{\infty} P_{ij} [r_0(c)]^i, \qquad 1 \le j \le c$$

which leads to the recursive relation

$$Q_{j-1} = \frac{Q_j - \sum_{i=j}^{c-1} P_{ij} Q_i - C \sum_{i=c}^{\infty} P_{ij} [r_0(c)]^i}{P_{j-1,j}}, \qquad 1 \le j \le c. \tag{3.36}$$

Letting $CQ_j^* = Q_j$, it becomes

$$Q_{j-1}^* = \frac{Q_j^* - \sum_{i=j}^{c-1} P_{ij} Q_i^* - \sum_{i=c}^{\infty} P_{ij} [r_0(c)]^i}{P_{j-1,j}}, \qquad 1 \le j \le c. \tag{3.37}$$

When $j = c$, we obtain

$$Q_{c-1}^* = \frac{[r_0(c)]^c - \sum_{i=c}^{\infty} \beta_{i+1-c}(c) [r_0(c)]^i}{\beta_0(c)} \tag{3.38}$$

By repeated use of (3.37) for $j = c-1$, $c-2$, ..., 1, one gets $Q_{c-2}^*, \ldots, Q_0^*$. Finally, rewriting (3.35) as

$$C = \frac{1}{\sum_{j=0}^{c-1} Q_j^* + \frac{[r_0(c)]^c}{1 - r_0(c)}}. \tag{3.39}$$

In order to obtain the waiting time distribution, we follow arguments similar to the $M/M/1$ model (see Section 2.6 in Chapter 2) and derive

$$F_{W_q}(t) = \frac{C[r_0(c)]^c}{1 - r_0(c)} \left[1 - e^{-\mu c(1 - r_0(c))t}\right], \qquad t \ge 0 \tag{3.40}$$

and

$$P(W_q = 0) = C \left[\frac{1}{C} - \frac{[r_0(c)]^c}{1 - r_0(c)}\right]. \tag{3.41}$$

The verification of (3.40) and (3.41) is left as an exercise (Exercise 20).

3.4. Bulk Queues

So far we have looked into a queueing system under the assumption that arrivals occur singly and the customers are served one at a time. There are, however, many queueing processes in which arrivals can occur and services can be performed in groups (i.e., in bulk or in batches). For example, several people may go to a restaurant together and may be served in batches. Similarly, elevators handle passengers in batches which is a case of bulk services. These systems in short may be called bulk queues. Recall that $M/E_k/1$ and $E_k/M/1$ models can be thought of as *bulk queues* (see Remark 2 in each model in Section 2.4 in Chapter 2).

3.4.1. The $M/G^{a,b}/1$ model. We consider the model $M/G^{a,b}/1$ which is a bulk service generalization of $M/G/1$ (Section 3.2). Here $G^{a,b}$ refers to the service mechanism and is defined as follows:

(i) As earlier the duration of service-time has an arbitrary distribution with mean service-time $1/\mu$;

(ii) The server takes up all customers for service if the number of customers waiting is at least a and at most b;

(iii) If there are fewer than a customers waiting, the service does not begin (i.e., the server waits until there are a customers). On the other hand, if there are more than b customers waiting, only b customers are taken up for service.

At first we look into the steady-state behaviour of the number in the system. As in Section 3.2, we easily see that the departure point process $\{X_n : n \geq 1\}$ forms a Markov chain which is imbedded over the non-Markovian process at regeneration points such as service completion instants of a batch. For convenience, we repeat the notations of Section 3.2. It can be seen by a similar argument that the transition probabilities $P_{ij} = P(X_{n+1} = j | X_n = i)$ of the Markov chain $\{X_n : n \geq 1\}$ are given by

$$P_{ij} = \begin{cases} \alpha_{j-i+b} & \text{if } i > b, j \geq i - b, \\ 0 & \text{if } i \geq b, j < i - b, \\ \alpha_j & \text{if } 0 \leq i \leq b, j \geq 0. \end{cases} \tag{3.42}$$

Clearly P_{ij}'s are independent of a and are reduced to (3.5) when $b = 1$. They are also independent of i for $0 \leq i \leq b$. Note that the first $b + 1$ rows of the transition matrix are identical. A simple examination of this matrix indicates that the Markov chain is irreducible and aperiodic. Moreover, it will be justified in Section 3.8 that when the traffic intensity $\frac{\lambda}{b\mu}$ is less than one, the chain is ergodic and the stationary departure point probabilities $\{\Pi_n : n \geq 0\}$ exist.

The balance equations can be written in a usual manner as

$$\Pi_j = \alpha_j \sum_{i=0}^{b-1} \Pi_i + \sum_{i=b}^{b+j} \alpha_{j-i+b}\Pi_i, \qquad j = 0, 1, \ldots \tag{3.43}$$

such that

$$\sum_{j=0}^{\infty} \Pi_j = 1.$$

Using the method of generating function and remembering that $\alpha(z)$ is the generating function of $\{\alpha_i\}$, we obtain

$$\begin{aligned}\Pi(z) &= \sum_{i=0}^{b-1} \Pi_i \sum_{j=0}^{\infty} \alpha_j z^j + \sum_{j=0}^{\infty} z^j \sum_{i=b}^{b+j} \alpha_{j-i+b}\Pi_i \\ &= \alpha(z) \sum_{i=0}^{b-1} \Pi_i + z^{-b} \sum_{i=b}^{\infty} \Pi_i z^i \sum_{j=i-b}^{\infty} \alpha_{j-i+b} z^{j-i+b} \\ &= z^{-b}\alpha(z) \left[\sum_{i=0}^{b-1} \Pi_i z^b + \Pi(z) - \sum_{i=0}^{b-1} \Pi_i z^i \right].\end{aligned}$$

The solution for $\Pi(z)$ is

$$\Pi(z) = \frac{\sum_{i=0}^{b-1} \Pi_i (z^b - z^i)}{(z^b/\alpha(z)) - 1}. \tag{3.44}$$

Unfortunately (3.44) involves b unknown Π_i's, $i = 0, 1, \ldots, b-1$ whereas we have only one relation among Π_i's, viz., the normalizing equation $\Pi(1) = 1$. Recall that in (2.43) we have a similar situation with only one unknown to be determined. Thus a parallel treatment is suggested to deal with the present impasse.

Since $\Pi(z)$ is a p.g.f., it is absolutely convergent for $|z| \leq 1$. Therefore, in this region the zeroes of the numerator must coincide with those of the denominator which are the same as those of

$$z^b - \alpha(z) = 0. \tag{3.45}$$

Consider the contour $C : |z| = 1 + \epsilon$, $\epsilon > 0$. By Taylor's series expansion, it is evident that

$$\alpha(1+\epsilon) = 1 + \epsilon\rho + o(\epsilon)$$

since $\alpha'(1) = \rho = \lambda/\mu$. Using the information that $\{\Pi_n,\ n \geq 0\}$ exist under the condition $\rho < b$, we may take ϵ small enough such that

$$\alpha(1+\epsilon) < 1 + \epsilon b.$$

But

$$1 + \epsilon b < (1+\epsilon)^b.$$

Thus on C, for small enough ϵ one has

$$|\alpha(z)| < |z^b|.$$

Applying Rouché's theorem (see Section A.5 in Appendix A) we deduce that the number of roots of (3.45) inside C is b, ϵ being extremely small. From these discussions, it is clear that all b zeroes of the numerator in (3.44) must be the roots of (3.45) that are inside C. Obviously one of them is 1. Let the remaining zeroes be $r_1, \ldots, r_{b-1}$. Then we can write the numerator of (3.44) as

$$\sum_{i=0}^{b-1} \Pi_i(z^b - z^i) = A(z-1) \prod_{i=1}^{b-1} (z - r_i), \tag{3.46}$$

where A is a constant to be evaluated from $\Pi(1) = 1$ and we find that

$$A = \frac{b - \rho}{\prod_{i=1}^{b-1}(1 - r_i)}, \tag{3.47}$$

when we use L'Hôpital's rule, $\alpha(1) = 1$ and $\alpha'(1) = \rho$. Finally,

$$\Pi(z) = \frac{(b-\rho)(z-1)\prod_{i=1}^{b-1}(z - r_i)/(1 - r_i)}{(z^b/\alpha(z)) - 1}. \tag{3.48}$$

When $b = 1$, (3.48) reduces to (3.11). This provides a check for (3.48). The next obvious question one would like to ask is how to derive r_i's. This question is dealt with in Chapter 4.

3.4.2. The $M/E_k^{a,b}/1$ model. As a special case, let the service-times have the k-stage Erlangian distribution with p.f. (see Section A.4 (A.48) in Appendix A).

$$f_S(t) = \frac{(k\mu)^k}{(k-1)!} t^{k-1} e^{-k\mu t}, \qquad t \geq 0. \tag{3.49}$$

Using (3.13) for the present case, we have

$$\alpha(z) = \left(1 + \frac{\rho}{k}(1 - z)\right)^{-k}, \tag{3.50}$$

which upon substitution in (3.48) gives the appropriate expression. We may get an alternative answer with some advantage. To do this, the denominator of (3.48) is expressed with the help of (3.50) as

$$z^b \left(1 + \frac{\rho}{k}(1 - z)\right)^k - 1 \tag{3.51}$$

which being a polynomial of degree $b + k$ has $b + k$ zeroes. Those inside C are $1, r_1, \ldots, r_{b-1}$ and let the remaining zeroes which are on or outside C be $r_1^*, \ldots, r_k^*$. It is easy to see that (3.44) can now be written as

$$\Pi(z) = \frac{B}{\prod_{i=1}^{k}(z - r_i^*)}. \tag{3.52}$$

The evaluation of B is done as usual through the normalizing equation. Finally we get

$$\Pi(z) = \prod_{i=1}^{k} \frac{1 - r_i^*}{z - r_i^*}. \tag{3.53}$$

Remarks:

1. Suppose $b > k$. Then we need a lesser number of zeroes to determine in (3.52) than in (3.48).
2. It can be shown that β_i's are distinct. Then one can apply the technique of partial fraction expansion as demonstrated in (2.27) and obtain Π_j's.
3. The β_i's become inside zeroes of the polynomial obtained from (3.51) by replacing z by $1/z^*$. The computational technique for deriving inside zeroes is discussed in Chapter 4.

3.4.3. The $M/D^{a,b}/1$ model. For a second special case of practical importance, let the service-times be a fixed constant for each batch. This model without any batch has appeared in Section 3.2 in a remark. Again, using (3.13) we find

$$\alpha(z) = e^{-\rho(1-z)}$$

which is substituted in (3.48).

Remarks:

1. The moment generating function (see Section A.3 in Appendix A) briefly m.g.f. of E_k is $(1 - \frac{t}{k\mu})^{-k}$, by considering the p.f. in (3.49). Since its limit as $k \to \infty$ is $e^{-t/\mu}$ which is the m.g.f. of the deterministic distribution D with fixed service-time $1/\mu$, we note that D is obtained as the limiting distribution for E_k, by applying the uniqueness theorem. Thus, we observe that formula above could be obtained from (3.50) when $k \to \infty$.
2. Because of the previous remark, we have not attempted to derive a solution similar to (3.52) or (3.53) as the number of r_i^* has to be infinite for $k \to \infty$.

Next, in the study of the waiting time W_q, we assume $a = 1$ and the queue discipline to be FCFS as before. Let $c_r(x)$ be the probability that a customer who is the rth unit in a batch that is ready to be served (i.e., at a departure instant) and henceforth is referred to as a test customer, has waiting time $\leq x$. Furthermore, let p_r be the probability of a customer to be a test customer. It should be proportional to the probability that there are at least r customers in the queue at the time of the departure of a batch.

Moreover, the conservation of flow implies the equality of the average inflow and the average outflow which leads to

$$\lambda E(S) = \sum_{j=1}^{b-1} j\Pi_j + b\sum_{j=b}^{\infty} \Pi_j.$$

Using this and the fact that $\sum_{r=1}^{b} p_r = 1$, we get

$$p_r = \frac{1}{\lambda E(S)} \sum_{i=r}^{\infty} \Pi_i, \qquad r = 1, \ldots, b. \tag{3.54}$$

Obviously, the distribution function (d.f.) of W_q, the waiting time in the queue is given in terms of our above notation as

$$F_{W_q}(x) = \sum_{r=1}^{b} p_r c_r(x). \tag{3.55}$$

Now we establish an identity which is to be used in the sequel. Suppose a test customer has waiting time x. Let there be j arrivals during this time interval x, the probability for which is $e^{-\lambda x}(\lambda x)^j/j!$. Then

$$\int_0^{\infty} e^{-\lambda x} \frac{(\lambda x)^j}{j!}\, dc_r(x)$$

represents the probability that there are j arrivals during the waiting time of a test customer. On the other hand, the same event may be viewed differently. Realize that the end of the waiting period of a test customer is a departure instant. Therefore, j arrivals during the waiting period of a test customer is equivalent to the event that there are $j + r$ customers at the above mentioned departure instant given that there are at least r customers at that instant (because the test customer's position is the rth in the queue). The probability of the latter event is

$$\frac{\Pi_{j+r}}{\sum_{i=r}^{\infty} \Pi_i}.$$

Hence we have the identity

$$\int_0^{\infty} e^{-\lambda x} \frac{(\lambda x)^j}{j!}\, dc_r(x) = \frac{\Pi_{j+r}}{\sum_{i=r}^{\infty} \Pi_r} \qquad j = 0, 1, \ldots, \quad r = 1, \ldots b. \tag{3.56}$$

Similar to the derivation of (3.19), we proceed to develop a relation between $\Pi(z)$ and $\Phi_{W_q}(\theta)$ in order to eventually express $\Phi_{W_q}(\theta)$ in terms of $\Pi(z)$. Thus

$$\begin{aligned}
\Pi(z) &= \sum_{i=0}^{r-1} \Pi_i z^i + \sum_{i=0}^{\infty} \Pi_{i+r} z^{i+r} \\
&= \sum_{i=0}^{r-1} \Pi_i z^i + \left(\sum_{i=r}^{\infty} \Pi_i\right) z^r \int_0^{\infty} e^{-\lambda x} \sum_{j=0}^{\infty} \frac{(\lambda x z)^j}{j!}\, dc_r(x)
\end{aligned}$$

due to (3.56). Denoting by

$$M_r(\theta) = \int_0^\infty e^{-\theta x}\, dc_r(x)$$

and using (3.54), we get

$$\Pi(z) = \sum_{i=0}^{r-1} \Pi_i z^i + \rho z^r p_r M_r(\lambda(1-z)). \tag{3.57}$$

On multiplying by z^{b-r} and summing over $1 \le r \le b$, (3.57) becomes

$$\begin{aligned} \Pi(z)\frac{1-z^b}{1-z} &= \sum_{r=1}^{b} z^{b-r} \sum_{i=0}^{r-1} \Pi_i z^i + \rho z^b \sum_{r=1}^{b} p_r M_r(\lambda(1-z)) \\ &= \sum_{i=0}^{b-1} \Pi_i \frac{z^i - z^b}{1-z} + \rho z^b \Phi_{W_q}(\lambda(1-z)) \end{aligned} \tag{3.58}$$

because of (3.55). From (3.44) we get

$$\sum_{i=0}^{b-1} \Pi_i (z^i - z^b) = -\Pi(z)\left(\frac{z^b}{\alpha(z)} - 1\right).$$

Therefore, (3.58) simplifies to

$$\Pi(z)(1-z^b) = \rho z^b (1-z) \Phi_{W_q}(\lambda(1-z)) - \Pi(z)\left(\frac{z^b}{\alpha(z)} - 1\right)$$

which finally yields the desired expression

$$\Phi_{W_q}(\theta) = \left(\frac{1}{\Phi_S(\theta)} - 1\right) \frac{\Pi(1-\frac{\theta}{\lambda})}{\rho}. \tag{3.59}$$

3.4.4. The $G^k/M/1$ model. As the notation suggests, the model is a bulk arrival generalization of $G/M/1$ (Section 3.3) in which each arrival consists of a batch of a fixed size k and the interarrival time T has an arbitrary distribution with mean duration $1/\lambda$.

For continuity we use the same notations as in Section 3.3. Our starting point is the generalization of (3.22) to bulk arrivals as

$$Y_{n+1} = \max(Y_n + k - \gamma_n^*, 0). \tag{3.60}$$

Notation:

$$Q(z) \;:\; \sum_{n=0}^{\infty} Q_n z^n$$

Unlike earlier models, the present model is dealt with using an innovative approach directly based on (3.60) but totally different from earlier methods.

Note from (3.60) that Y_n's cannot be negative, but the way γ_n^* is defined $Y_n + k - \gamma_n^*$ can be. Therefore, associated with $Y_n + k - \gamma_n^*$ we may consider

$$c_j = \lim_{n\to\infty} P(Y_n + k - \gamma_n^* = j), \qquad j = 0, \pm 1, \pm 2, \ldots \tag{3.61}$$

as a new but fundamental notion in our subsequent development. It is obvious from (3.60) that

$$Q_j = c_j, \qquad j = 1, 2, \ldots \tag{3.62}$$

and

$$Q_0 = \sum_{j=-\infty}^{0} c_j. \tag{3.63}$$

Let η be the r.v. corresponding to the probability distribution $\{c_j\}$. Suppose $Y = \lim Y_n$ and $\gamma^* = \lim \gamma_n^*$ (actually $\gamma^* = \gamma_n^*$ since γ_n^* are i.i.d. r.v.'s). Then (3.61) is stated in terms of r.v.'s as

$$\eta = Y + k - \gamma^*.$$

Recalling the notation $\beta(z)$ as the p.g.f. of $\{\beta_i\}$ we can express the p.g.f. of η as

$$\sum_{j=-\infty}^{\infty} c_j z^j = E[z^\eta] = E[z^{Y+k-\gamma^*}] = Q(z) z^k \beta(z^{-1}). \tag{3.64}$$

But from (3.63), it follows that

$$\begin{aligned}\sum_{j=-\infty}^{\infty} c_j z^j &= \sum_{j=-\infty}^{0} c_j z^j + Q(z) - \sum_{j=-\infty}^{0} c_j \\ &= Q(z) - \sum_{j=0}^{\infty} c_{-j}(1 - z^{-j}).\end{aligned}$$

Therefore,

$$Q(z) = \frac{\sum_{j=0}^{\infty} c_{-j}(1 - z^{-j})}{1 - z^k \beta(z^{-1})}, \tag{3.65}$$

$c_0, c_{-1}, \ldots$ being constants to be determined.

At this point we move to look into the roots of the equation

$$z^k = \beta(z). \tag{3.66}$$

Noting that $\beta'(1) = 1/\rho > k$ and following an argument similar to the one applied in determining the roots of (3.45) but using the contour $|z| = 1 - \epsilon$, $\epsilon > 0$, we can ascertain that there are exactly k roots of (3.66) inside $|z| = 1$. Denote the roots by r_j, $j = 1, \ldots, k$. Clearly, there are exactly k zeroes of the denominator in (3.65) outside the unit circle and these are $1/r_j$, $j = 1, \ldots, k$. The root finding technique is discussed in Chapter 4.

Consider the function

$$A(z) = Q(z)\prod_{j=1}^{k}(1 - r_j z) \tag{3.67}$$

which is regular for $|z| \leq 1$. Moreover, since in the region $|z| > 1$ the zeroes of the denominator in (3.67) are zeroes of its numerator, $A(z)$ is regular in $|z| > 1$. Therefore, by analytic continuation $A(z)$ is regular everywhere. Furthermore $A(z) = o(|z|)$ as $|z| \to \infty$. Hence by Liouville's theorem (see Section A.5 in Appendix A) $A(z)$ must be a constant, say, A. From (3.67) and the condition $Q(1) = 1$, we get

$$Q(z) = \prod_{j=1}^{k}\left(\frac{1 - r_j}{1 - r_j z}\right). \tag{3.68}$$

3.4.5. $M^k/M/1$**.** As an illustration of the general result, we consider T to be exponentially distributed. From (3.24) we have

$$\beta(z) = \frac{\rho}{\rho + 1 - z} \tag{3.69}$$

in our special case. It is known that the denominator $1 - z^k\beta(z^{-1})$ in (3.65) has k zeroes $1/r_j$, $j = 1, \ldots, k$ outside $|z| = 1$. In addition $z = 1$ is also a zero. In our model, the equation $1 - z^k\beta(z^{-1}) = 0$ simplifies to

$$\rho z^{k+1} - z(1 + \rho) + 1 = 0,$$

which has exactly $k+1$ roots. As a consequence, it is evident that these roots are 1, $\frac{1}{r_j}$, $j = 1, \ldots, k$. Thus we may write

$$\rho z^{k+1} - z(1 + \rho) + 1 = c(1 - z)\prod_{j=1}^{k}(1 - r_j z) \tag{3.70}$$

where c is a constant. From (3.68) we get

$$Q(z) = \frac{c(1 - z)\prod_{j=1}^{k}(1 - r_j)}{\rho z^{k+1} - z(1 + \rho) + 1}$$

with the help of (3.70). The fact that $Q(1) = 1$ leads to

$$c\prod_{j=1}^{k}(1 - r_j) = 1 - k\rho.$$

Hence

$$Q(z) = \frac{(1 - k\rho)(1 - z)}{\rho z^{k+1} - z(1 + \rho) + 1}. \tag{3.71}$$

Remarkably, (3.71) does not involve any of the roots r_j.

Let us now turn to W_q and let it refer to the waiting time of the first unit in an arbitrary arriving batch. Using the usual conditional argument that the arriving batch finds j customers in the system, we find

$$F_{W_q}(x) = \int_0^x \sum_{j=0}^{\infty} Q_j e^{-\mu t} \frac{(\mu t)^{j-1}}{(j-1)!} \mu \, dt.$$

Therefore

$$\Phi_{W_q}(\theta) = \sum_{j=0}^{\infty} Q_j \left(\frac{\mu}{\mu+\theta}\right)^j = Q\left(\frac{\mu}{\mu+\theta}\right). \tag{3.72}$$

It is left as an exercise to prove that the waiting time of a random unit in a batch has L.S.T. equal to

$$\prod_{j=1}^{k} \frac{(1-r_j)}{\left(1-r_j \frac{\mu}{\mu+\theta}\right)} \frac{1}{k} \sum_{j=0}^{k-1} \left(\frac{\mu}{\mu+\theta}\right)^j. \tag{3.73}$$

3.5. Functional Relations: A Heuristic Approach

By inducing an imbedded Markov chain over a non-Markovian process, what has been achieved so far is the possibility of studying the number in the system only at regeneration points. However, we desire to learn as in Chapter 2 about the same number at any arbitrary instant — a mathematically intractable problem. Yet it is solved by a functional relation which is developed here by a strikingly simple but heuristic argument.

Let us begin with the $M/G/1$ model. Denote by M_j the number of transitions from state j to state $j+1$ (briefly, $j \to j+1$) and by N_j the number of transitions from state $j+1$ to state j (briefly, $j+1 \to j$), respectively, over a long period of time. Observe that a single movement $j \to j+1$ of the chain must be compensated by a movement $j+1 \to j$ so that another movement $j \to j+1$ would be possible. In other words, $|M_j - N_j| \leq 1$. We may also understand that Q_j and Π_j, respectively, represent the proportion of transitions $j \to j+1$ among all transitions $i \to i+1$, $i = 0, 1, \ldots$ and $j+1 \to j$ among all $i+1 \to i$, $i = 0, 1, \ldots$ over a long period of time. Thus it is evident that

$$Q_j = \Pi_j.$$

Moreover, due to the memoryless property of the input process, the interval between an arbitrary instant and the arrival instant is exponential and therefore we may regard the arbitrary instant as an arrival instant. This implies that

$$P_j = Q_j$$

which in combination with the previous one yields an important functional relation for $M/G/1$ as

$$P(z) = Q(z) = \Pi(z), \qquad \rho < 1. \tag{3.74}$$

The same elementary justification can be extended to the $M/G^k/1$ leading to the first part of the relation. For a possible relation between $\Pi(z)$ and $Q(z)$ some further structural investigation is needed. We see that the process can have $j \to j+1$ if and only if it has reached state $i_0, i_0 \le j$ from a state $i_1, i_1 > j$. Without any ambiguity, let us denote by M_j and N_j the number of transitions $j \to j+1$ and $j+k \to j$ over a long period of time. Then M_j is almost equal to

$$\begin{cases} N_0 + \ldots + N_j, & \text{if } j < k \\ N_{j-k+1} + \ldots + N_j & \text{if } j \ge k. \end{cases} \tag{3.75}$$

Thus

$$\sum_{j=0}^{\infty} M_j = k \sum_{j=0}^{\infty} N_j,$$

which leads to

$$Q_j = \begin{cases} \frac{1}{k}(\Pi_0 + \ldots + \Pi_j) & j < k \\ \frac{1}{k}(\Pi_{j-k+1} + \ldots + \Pi_j) & j \ge k. \end{cases} \tag{3.76}$$

Finally, combining the p.g.f. of (3.76) and the earlier relation between $P(z)$ and $Q(z)$, we have for $M/G^k/1$,

$$P(z) = Q(z) = \frac{1}{k}\frac{1-z^k}{1-z}\Pi(z), \qquad \rho < k. \tag{3.77}$$

Remark: Since we have made no assumptions on the distributions in justifying the last part of (3.77), the relation

$$Q(z) = \frac{1}{k}\frac{1-z^k}{1-z}\Pi(z), \qquad \rho < k, \tag{3.78}$$

is true for the $G/G^k/1$ model.

By using a parallel argument for $G^k/M/1$, it can be established that

$$\Pi(z) = \frac{1}{k}\frac{1-z^k}{1-z}Q(z), \qquad k\rho < 1. \tag{3.79}$$

This is true even for $G^k/G/1$ by the above remark.

Relation $P(z) = Q(z)$ for $M/G^k/1$ depends on the Markovian property of the input process and therefore may not be true in general for $G^k/M/1$. An appropriate relation between $P(z)$ and $\Pi(z)$ which corresponds to $Q(z)$ in $M/G^k/1$ is presented below.

A starting point is to fully utilize the Markovian property in the model. Since the service-time distribution is exponential, an arbitrary instant can be regarded as a departure instant, provided the departure instants form a Poisson process. This is possible when the service completion process is allowed to continue without interruption, irrespective of the fact whether there is a customer present to be served. In that case, a service end point is real (i.e., a real departure point) if an actual customer's service is completed, otherwise it is considered to be virtual (i.e., a virtual departure point). The notion of real or virtual departures has been introduced in Section 3.3.

Now we have to determine the probability of a real departure point. Recall that in $M/M/1$, $\rho = \frac{\lambda}{\mu}\,(0 \leq \rho < 1)$ happens to be equal to $1 - P_0$ (see (2.6)), the probability of the system being busy at any arbitrary instant. Indeed, it turns out that it is true for any one-server general model with λ and μ, respectively being the average number of arrivals (i.e., arrival rate) and average number of completed services (i.e., rate of service completion) during a unit of time. Here we sketch an intuitive justification of this result. Let t be an arbitrarily long time interval. We expect λt number of arrivals during this time interval. During the same interval, since the expected length of busy period is $(1 - P_0)t$, the average number of customers served is $\mu(1 - P_0)t$. For large t, we expect the number of arrivals and the number of customers served to be equal and therefore

$$1 - P_0 = \frac{\lambda}{\mu}. \tag{3.80}$$

In our case the average number of arrivals is $k\lambda$ and average number of customers served μ. Therefore the departure point to be real (which is equivalent to the system being busy at that instant), has the probability $k\lambda/\mu = k\rho$. Moreover, if there are j customers in the system at an arbitrary instant which coincides with a real departure instant (i.e., just after the departure), then we have $j - 1$ customers just after the departure instant. All these lead to

$$P_j = k\rho\Pi_{j-1}, \qquad j \geq 1.$$

It follows that

$$P_0 = 1 - k\rho.$$

Therefore, for $G^k/M/1$ we have

$$P(z) = (1 - k\rho) + k\rho z\Pi(z), \qquad k\rho < 1. \tag{3.81}$$

Remarks:

1. This section plays a role analogous to Section 2.6 on Little's formula. Little's formula helped us to directly derive the average waiting time painlessly from the average number in the system. The formula was also arrived at by a heuristic argument. The situation in this section is parallel in the sense that through functional relations we are able to determine P_j from either Q_j or Π_j and these relations are established heuristically. In Section 6.4 in Chapter 6, we will deal with them somewhat more rigorously.

2. Two models where input and service mechanism characteristics are interchanged may be known as dual models, for example $M/G/1$ and $G/M/1$ (to be discussed in Section 7.7 in Chapter 7). Results such as

(3.78) for $G/G^k/1$ and (3.79) for $G^k/G/1$ are called dual relations. Observe that in dual relations $\Pi(z)$ and $Q(z)$ are interchanged. Another way of obtaining them is to replace $\Pi(z)$ and $Q(z)$ in one by their reciprocals in order to get the other. A further observation is that the relations between $P(z)$ and $\Pi(z)$ (or $Q(z)$) are not dual. An intuitive explanation for the discrepancy is that $Q(z)$ and $\Pi(z)$ in a sense represent dual characteristics of dual models ($Q(z)$ refers to the number just before an arrival and $\Pi(z)$ to the number just after a departure), whereas $P(z)$ and $\Pi(z)$ (or $Q(z)$) do not.

3.6. Busy Periods

In Section 2.7 in Chapter 2, the concept of a busy period has been introduced and the distribution of Δ, (i.e., the length of a busy period), has been studied for the $M/M/1$ model. Here we derive the distributions of Δ and the number served during a busy period for $M/G/1$ and $G/M/1$ models by using various approaches including the combinatorial one.

At first we take up the $M/G/1$ model and attempt to derive the distribution of Δ as was done for $M/M/1$ (Section 2.7 in Chapter 2). Realize that in the absence of the nicety of Markovian nature inclusive of Chapman-Kolmogorov equation which was readily available for $M/M/1$, it is necessary to look for an alternative approach. For that purpose we bring in the concept of first passage time. The duration of a process starting from state i to reach state j for the first time is called the first passage time from i to j and is denoted by $\Delta_{i,j}$, i.e., if the process is denoted by $\{\xi(t), t \geq 0\}$ then

$$\Delta_{i,j} = \inf\{t : \xi(t) = j|\xi(0) = i\}.$$

Similarly, let

$$N_{i,j} = \inf\{n : X_n = j|X_0 = i\}$$

where

$$X_0 = \xi(0).$$

Notation:

Δ_i	:	length of a busy period with i initial customers
N_i	:	number of customers served during a busy period with i initial customers

Clearly $\Delta_{i,0} = \Delta_i$ and $\Delta_{1,0} = \Delta_1 = \Delta$. Also $N_{i,0} = N_i$ and $N_{1,0} = N_1 = N$.

3.6.1. Analytic methods. Supposing that there are k, $k = 0, 1, \ldots$ arrivals during the first service period and noting that after completion of the first service period a new busy period starts with k initial customers, we obtain for $t \geq 0$,

$$F_\Delta(t) = \int_0^\infty \left[\sum_{k=0}^\infty e^{-\lambda y} \frac{(\lambda y)^k}{k!} F_{\Delta_k}(t-y)\right] dF_S(y). \tag{3.82}$$

From the notations, it follows that

$$\Delta_k = \sum_{j=1}^{k} \Delta_{j,j-1}.$$

Evidently, $\Delta_{1,0}, \Delta_{2,1}, \ldots, \Delta_{k,k-1}$ are i.i.d. r.v.'s.
Therefore

$$\Phi_{\Delta_k}(\theta) = [\Phi_\Delta(\theta)]^k, \qquad k = 0, 1, \ldots \tag{3.83}$$

Taking the L.S.T. of (3.82) and using (3.83), we find

$$\begin{aligned}\Phi_\Delta(\theta) &= \int_0^\infty e^{-y(\theta+\lambda)} \sum_{k=0}^{\infty} \frac{(\lambda y)^k}{k!} [\Phi_\Delta(\theta)]^k \, dF_S(y) \\ &= \Phi_S(\theta + \lambda - \lambda\Phi_\Delta(\theta)).\end{aligned}$$

Thus $\Phi_\Delta(\theta)$ satisfies the functional equation

$$\Phi_\Delta(\theta) = \Phi_S(\theta + \lambda - \lambda\Phi_\Delta(\theta)), \qquad \Re(\theta) \geq 0, \tag{3.84}$$

which is known as the *Kendall-Takács functional equation.*

We may attempt to derive something similar for the distribution of N. Let the p.g.f. of N be denoted by $G(z)$. Then our earlier analysis leads to

$$\begin{aligned}G(z) = E(z^N) &= \int_0^\infty \left[\sum_{k=0}^{\infty} e^{-\lambda y} \frac{(\lambda y)^k}{k!} E(z^{1+N_k})\right] dF_S(y) \\ &= z \int_0^\infty \left[\sum_{k=0}^{\infty} e^{-\lambda y} \frac{(\lambda y)^k}{k!} (G(z))^k\right] dF_S(y)\end{aligned}$$

from which follows the functional equation

$$G(z) = z\Phi_S(\lambda - \lambda G(z)), \qquad |z| \leq 1. \tag{3.85}$$

This is called the *Good-Takács functional equation.*

With simple differentiation of (3.84) and (3.85) we are able to derive

$$\mu_\Delta = \frac{1}{\mu - \lambda} \quad \text{and} \quad \mu_N = \frac{1}{1-\rho}.$$

Let us consider the solution of (3.84) for $M/M/1$. We find in this case, (3.84) becomes

$$\Phi_\Delta(\theta) = \frac{\mu}{\mu + \theta + \lambda - \lambda\Phi_\Delta(\theta)}$$

which is the quadratic equation

$$\lambda[\Phi_\Delta(\theta)]^2 - (\lambda + \mu + \theta)\Phi_\Delta(\theta) + \mu = 0.$$

Among the two roots of the equation, only

$$\Phi_\Delta(\theta) = \frac{\lambda + \mu + \theta - \sqrt{(\lambda + \mu + \theta)^2 - 4\lambda\mu}}{2\lambda}$$

is admissible, because it satisfies $\Phi_\Delta(0) = 1$ when $\lambda < \mu$, whereas the other root leads to $\Phi_\Delta(0) > 1$. Clearly,

$$\Phi_\Delta(\theta) = \frac{\mu}{\lambda z_2}$$

z_2 being given by (2.44) and its inversion is the expression (2.72) in Section 2.7 in Chapter 2, note that L.T. $(P_0'(t)) = \mu/\lambda z_2$).

Closed form solutions of (3.84) and (3.85) can be derived by using Lagrange's series expansion. Let us consider (3.85) which may be rewritten as

$$G(z) = z\alpha(G(z)), \qquad |z| \leq 1, \tag{3.86}$$

by using (3.13). By Lagrange's series expansion (see Section A.5 in Appendix A), we obtain

$$G(z) = \sum_{n=1}^{\infty} \frac{z^n}{n!} \left(\frac{d^{n-1}}{d\omega^{n-1}} \alpha^n(\omega) \right)_{\omega=0}. \tag{3.87}$$

For specific service-time distribution F_S, we may be able to get an expression for $\alpha(z)$ through α_k in (3.4) and thus an expression for $P(N = n)$.

It can be shown that the p.g.f. of N_k is

$$\sum_{n=k}^{\infty} P(N_k = n) z^n = G^k(z) \tag{3.88}$$

and its Lagrange's series expansion simplifies to

$$G^k(z) = \sum_{n=k}^{\infty} \frac{kz^n}{n(n-k)!} \left(\frac{d^{n-k}}{d\omega^{n-k}} \alpha^n(\omega) \right)_{\omega=0}. \tag{3.89}$$

This gives

$$P(N_k = n) = \frac{k}{n(n-k)!} \left(\frac{d^{n-k}}{d\omega^{n-k}} \alpha^n(\omega) \right)_{\omega=0}. \tag{3.90}$$

From (3.84) an analogous series expansion will give the distribution of Δ_k.

We realize that the distributions are derived by analytic methods. Following the same approach, the joint distribution of N_k and Δ_k can be obtained. Let

$$G(n, t : k) = P(N_k = n, \Delta_k \leq t). \tag{3.91}$$

Suppose

$$H(z, \theta) = E\left(z^N e^{-\theta\Delta}\right)$$

which exists for $|z| \leq 1$ and $\Re(\theta) \geq 0$. If $|z| \leq 1$ and $\Re(\theta) \geq 0$, we can prove that $\omega = H(z, \theta)$ is the root with smallest absolute value in ω of the equation

$$\omega = z\Phi_S(\theta + \lambda - \lambda\omega), \tag{3.92}$$

and hence

$$G(n, t : k) = \frac{k}{n} \int_0^t e^{-\lambda y} \frac{(\lambda y)^{n-k}}{(n-k)!} \, dF_S^{(n)}(y), \tag{3.93}$$

by Lagrange's expansion.

Next we present an alternative procedure for deriving the joint distribution of Δ_k and N_k. Here instead of having some initial customers, we consider a busy period with an initial work load (i.e., preoccupation time before the service begins, say for example, the time to set up the counter or to complete the unfinished work).

Notation:

$V(t)$	:	unfinished work (i.e., cumulative service load or work load), in the system at time t
$\Delta(x)$	:	length of a busy period when the initial work load is x (i.e.,$V(0) = x$)
$N(x)$	:	number of customers served during a busy period when the initial work load is x

If the queue discipline is FCFS, then $V(t)$ represents the waiting time in the queue of an arrival at time t and therefore is called the virtual waiting time. Remember that the waiting time variable W_q (also W) is a steady-state variable with $\lim_{t\to\infty} V(T) = W_q$ under FCFS discipline.

The busy period ends when for the first time in addition to the initial work load, all those arrived are served. This means,

$$\Delta(x) = \inf\{t : t \geq x + Y(t)\} \tag{3.94}$$

where $Y(t)$ represents the service-time of customers who arrive during the interval $(0, t]$. Similar to $G(n, t : k)$, let

$$G(n, t; x) = P(N(x) = n, \Delta(x) \leq t). \tag{3.95}$$

We derive an expression for the joint p.f. $dG(n, t; x)$ of $N(x)$ and $\Delta(x)$ through a recurrence relation. Clearly,

$$dG(0, t; x) = \begin{cases} e^{-\lambda t} & t = x, \\ \\ 0 & \text{otherwise}, \end{cases} \tag{3.96}$$

since $n = 0$ and $t = x$ imply no arrivals during $[0, t]$ and either $t < x$ is impossible or $n = 0$ is impossible when $t > x$. When $n \geq 1$, the p.f. of a new arrival at time $\eta (0 < \eta \leq x)$ is $\lambda e^{-\lambda\eta}$. Suppose y is the service-time of the new arrival. Consider the busy period starting at time η. Then the initial work load should be $x - \eta + y$ and during the remaining interval (η, t), $n - 1$ customers should arrive and be served. Thus, if $t \geq x$

$$dG(n, t; x) = \int_0^x \int_0^{t-x} [\lambda e^{-\lambda\eta}\, dG(n - 1, t - \eta; x - \eta + y)]\, dF_S(y)\, d\eta \tag{3.97}$$

and

$$dG(n, t; x) = 0, \qquad \text{if } t < x,$$

which is a recurrence relation in n. Using (3.96) and (3.97) we find for $n = 1$,

$$dG(1,t;x) = \int_0^x \lambda e^{-\lambda\eta} e^{-\lambda(t-\eta)}\, dF_S(t-x)d\eta$$
$$= \lambda x e^{-\lambda t}\, dF_S(t-x). \tag{3.98}$$

For $n > 1$, we need the identity

$$\int_0^t y\, dF_X^{(n)}(y)\, dF_X^{(m)}(t-y) = \frac{nt}{m+n} dF_X^{(m+n)}(t) \tag{3.99}$$

(see (A.19) for notation). This can be proved by taking L.T. on both sides. An inductive argument easily leads to the general formula

$$dG(n,t;x) = e^{-\lambda t}\frac{(\lambda t)^{n-1}}{n!}\lambda x\, dF_S^{(n)}(t-x). \tag{3.100}$$

It may seem to be surprising that we will be able to find an expression for $G(n,t:k)$ from $G(n,t;x)$. The illustration for $k = 1$ is perhaps sufficient for our purpose, as the general case is only a routine extension.

If there is initially one customer, the initial work load equals the service time of one customer which is a r.v. with d.f. $F_S(s)$. Therefore,

$$\begin{aligned}
G(n,t) = G(n,t:1) &= \int_0^t G(n-1,t;x)\, dF_S(x) \\
&= \int_0^t \left[\int_x^t e^{-\lambda y}\frac{(\lambda y)^{n-2}}{(n-1)!}\lambda x\, dF_S^{(n-1)}(y-x)\right] dF_S(x) \\
&= \int_0^t e^{-\lambda y}\frac{(\lambda y)^{n-2}}{(n-1)!}\lambda \int_0^y x\, dF_S^{(n-1)}(y-x) dF_S(x) \\
&= \frac{1}{n!}\int_0^t e^{-\lambda y}(\lambda y)^{n-1}\, dF_S^{(n)}(y)
\end{aligned} \tag{3.101}$$

by using the identity (3.99).

Remarks:

1. The technique just now described is an isolated one and does not seem to apply to other situations because it depends on induction and a specific identity, namely, (3.99). However, $G(n,t;x)$ can be obtained by Lagrange's expansion, which is given as an exercise.

2. If $\{\gamma_1, \gamma_2, \ldots\}$ is any set of i.i.d. r.v.'s not necessarily arising from the Poisson input, it can be shown that (3.86) is true (to be proved in an exercise) and hence $P(N_k = n)$ is given by (3.90) through Lagrange's expansion. However, without the knowledge of $\alpha(z)$ we cannot proceed further.

3.6.2. Combinatorial methods. Let us examine (3.90). Noting that

$$P(\gamma_1+\ldots+\gamma_n=n-k)=\frac{1}{(n-k)!}\left(\frac{d^{n-k}}{d\omega^{n-k}}\alpha^n(\omega)\right)_{\omega=0}$$

we observe from (3.90) that

$$P(N_k=n|\gamma_1+\ldots+\gamma_n=n-k)=\frac{k}{n}. \tag{3.102}$$

But, as a first passage problem the event $\{N_k=n\}$ is equivalent to

$$\left\{\sum_{i=1}^{r}\gamma_i>r-k, k\leq r<n \text{ and } \sum_{i=1}^{n}\gamma_i=n-k\right\}.$$

It therefore follows from (3.102) that

$$P\left(\sum_{i=1}^{r}\gamma_i>r-k, k\leq r<n|\sum_{i=1}^{n}\gamma_i=n-k\right)=\frac{k}{n}. \tag{3.103}$$

Since $\gamma_1,\ldots,\gamma_n$ are i.i.d. r.v.'s, we can replace them by $\gamma_n,\gamma_{n-1},\ldots,\gamma_1$ and obtain

$$P\left(\sum_{i=1}^{r}\gamma_i<r, 1\leq r<n|\sum_{i=1}^{n}\gamma_i=n-k\right)=\frac{k}{n} \tag{3.104}$$

which is remarkable since it does not depend on the distribution of γ_i.

Next we state the solution of an *urn problem*. Consider an urn with n cards marked with non-negative integers $k_1,\ldots,k_n$ such that $\sum_{i=1}^{n}k_i=k\leq n$. Cards are selected one by one at random from the urn without replacement until the last card is drawn. Let ν_r be the number on the rth card drawn, $r=1,\ldots,n$. Then

$$P(\nu_1+\ldots+\nu_r<r, r=1,\ldots,n)=1-\frac{k}{n}. \tag{3.105}$$

What is of immediate interest to us is to note the closeness between (3.104) and (3.105) and that the urn problem is a combinatorial one. The proofs and detailed developments on combinatorial methods are postponed to Section 6.3 in Chapter 6. Now assuming that (3.103) and (3.104) are true for any set of i.i.d. r.v.'s we derive the joint distribution $G(n,t)$ for the $G/M/1$ model for a change instead of the $M/G/1$ model.

Suppose that the busy period starts when a customer arrives. Then the number of arrivals during the busy period is $n-1, n>1$. Recall that γ_i^* represents the number of services completed during T_i. Therefore during a busy period $\sum_{i=1}^{r}\gamma_i^*<r$, $r=1,\ldots,n-1$ must be satisfied. Denote by T^* the duration from the last arrival instant to the termination of the busy period and by γ^* the number of services completed during T^*. Then $G(n,t)$ can be

expressed as

$$G(n,t) = \sum_{j=2}^{n} P\Bigg(\sum_{i=1}^{r} \gamma_i^* < r, r = 1,\ldots,n-1, \sum_{i=1}^{n-1} \gamma_i^* = n-j,$$

$$\gamma^* = j, T_1 + \ldots + T_{n-1} + T^* \leq t\Bigg)$$

$$= \int_0^t \sum_{j=2}^{n} P\left(\sum_{i=1}^{r} \gamma_i^* < r, r = 1,\ldots,n-1 | \sum_{i=1}^{n-1} \gamma_i^* = n-j, \sum_{i=1}^{n-1} T_i = u\right)$$

$$\times P\left(\sum_{i=1}^{n-1} \gamma_i^* = n-j | \sum_{i=1}^{n-1} T_i = u\right) P\left(T^* \leq t-u, \gamma^* = j\right) dF_T^{(n-1)}(u), \tag{3.106}$$

since T^* and γ^* are independent of $T_1,\ldots,T_{n-1}$ and of $\gamma_1^*,\ldots,\gamma_{n-1}^*$ and the density function (p.d.f.) of $\sum_{i=1}^{n-1} T_i$ is the $(n-1)$-fold convolution of the p.d.f. of T. If we fix $T^* = v$, then the event $\{T^* = v, \gamma^* = j\}$ means that there are j services completed during the interval $(0, v)$ including the one which takes place at time v. Thus

$$P\left(T^* \leq t-u, \gamma^* = j\right) = \int_0^{t-u} e^{-\mu v} \frac{(\mu v)^{j-1}}{(j-1)!} (1 - F_T(v))\mu \, dv. \tag{3.107}$$

Also

$$P\left(\sum_{i=1}^{n-1} \gamma_i^* = n-j | \sum_{i=1}^{n-1} T_i = u\right) = e^{-\mu u} \frac{(\mu u)^{n-j}}{(n-j)!}. \tag{3.108}$$

Furthermore, with the help of (3.104) we have

$$P\left(\sum_{i=1}^{r} \gamma_i^* < r, r = 1,\ldots,n-1 | \sum_{i=1}^{n-1} \gamma_i^* = n-j, \sum_{i=1}^{n-1} T_i = u\right) = \frac{j-1}{n-1} \tag{3.109}$$

since the conditions on γ_i^*'s are met. When (3.106), (3.107), (3.108) and (3.109) are combined, $G(n,t)$ takes the form

$$G(n,t) = \frac{\mu^n}{(n-1)!} \int_0^t \int_0^{t-u} \left(\sum_{j=2}^{n} \binom{n-2}{j-2} u^{n-j} v^{j-1}\right) \times$$

$$\times e^{-\mu(u+v)} (1 - F_T(v)) \, dv \, dF_T^{(n-1)}(u)$$

which simplifies to

$$G(n,t) = \frac{\mu^n}{(n-1)!} \iint_R v(u+v)^{n-2} e^{-\mu(u+v)} (1 - F_T(v)) \, dF_T^{(n-1)}(u) \, dv \tag{3.110}$$

where $R = \{(u,v) : u \geq 0, v \geq 0, u + v \leq t\}$.

Let us conclude with a brief discussion of the $M/G/1$ model. It is in some way natural to consider two sets of random variables, the service-times and the number of arrivals during each service-time which are counterparts or dual variables (for duality see Section 7.7) of the interarrival times and the number of services completed during each interarrival time in the $G/M/1$ model. For a busy period, we derive $G(n,t)$ for $M/G/1$ by proceeding in a similar fashion as in $G/M/1$ and using (3.103). It can be seen that the distribution so derived is in agreement with (3.101) (see Exercise 32).

3.7. Discrete-Time Queues

So far we have focused our attention on continuous-time queues in which time t is a non-negative real number and interarrival times and service-times are non-negative real valued random variables. However, in practice, time is actually treated as a discrete variable and it is almost always expressed as an integer multiple of some convenient unit, like seconds, minutes, hours, etc. For instance, execution times of jobs on a computer may be measured in milliseconds, the time tankers need for unloading at a terminal may be measured in hours. In fact there are only a few instruments used in practice to control and process jobs continuously. A particular situation arises in the telecommunication area where discrete-time queues are natural to occur. Recently, BISDN (Broadband Integrated Service Digital Network) has been of significant interest because it can provide a common interface for future communication needs including video, data and speech. For this purpose, the standard transport vehicle is chosen to be ATM (Asynchronous Transfer Mode) in which messages are segmented into small fixed-size transmission units of 53 bytes, called cells. The transmission time of a cell may be taken as a discrete-time interval. Here the time axis is divided into a sequence of intervals of unit duration (transmission time of a cell) called slots. Transmission of a message (service) starts only at the beginning of a slot and therefore the service duration is recorded in terms of number of slots. This feature distinguishes the present case from a continuous-time model. It is therefore natural to consider discrete-time models in which interarrival times and service-times are measured in units of time (or slots) and thus are discrete random variables.

After choosing some time unit, we divide the time axis into a sequence of contiguous slots $(0,1], (1,2], \ldots, (n-1,n], \ldots$. Different assumptions can be made on arrival times and departure times in relation to slot boundaries. One assumption is that arrivals join the system immediately at the beginning of a slot and departures are recorded immediately before the end of a slot. Thus a customer completing service at the end of slot j leaves behind those arrived in the jth slot and those waiting at the beginning of the slot. A customer starts being served always at the beginning of a slot. This assumption is called the *early arrival system*. It is important to know exactly when the system is noticed, whether exactly at the point of division j or a little before at $j-$ or a little after at $j+$. Let $X_n = X(n)$, $Y_n = X(n-)$ and $Z_n = X(n+)$. For the same realization each situation gives rise to a different sample path. Consider the example with arrival time instants 1, 2, 4, 6, 7 and with service-times 3, 1, 1, 2, 1. In Figure 3.3 X_n, Y_n and Z_n are recorded for the above example:

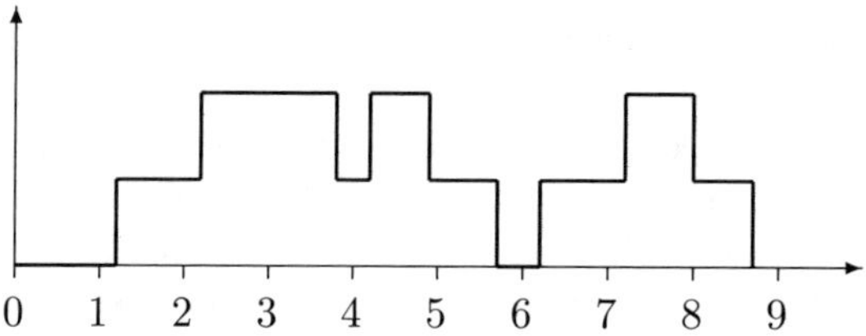

X_n:	0	0	1	2	1	1	0	1	1	0
Y_n:	0	0	1	2	2	2	1	1	2	1
Z_n:	0	1	2	2	2	1	1	2	1	0

FIGURE 3.3. Early arrival system

Alternatively, we may have another assumption, called the *late arrival system*. In this case, arrivals appear just prior to the end of a slot and departures at the beginning of a slot. The diagram for the above example under the late arrival system is given in Figure 3.4.

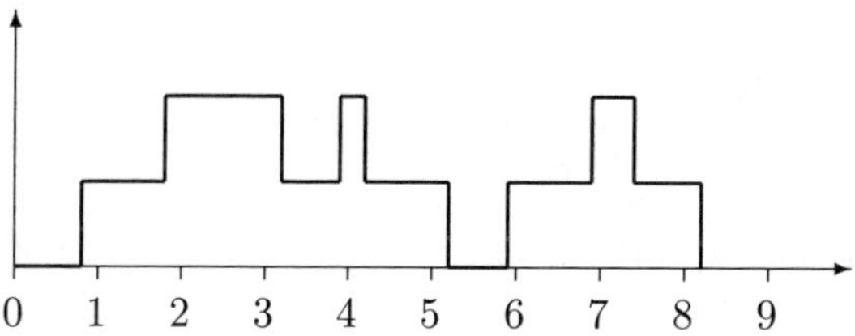

X_n:	0	1	2	2	2	1	1	2	1	0
Y_n:	0	0	1	2	1	1	0	2	1	0
Z_n:	0	1	2	1	1	0	1	1	0	0

FIGURE 3.4. Late arrival system

Observe that $Z_n = Y_{n+1}$.

Now we begin with the simplest discrete-time queueing model, known as Geometric/Geometric/1 (briefly, Geo/Geo/1) model, which is analogous to the $M/M/1$ model in continuous time. It is characterized by the following assumptions:

1. The arrivals are governed by the Bernoulli process with parameter λ, (see Section A.6 in Appendix A), i.e., the interarrival times are i.i.d. geometric random variables with parameter λ;

2. The service-times, independent of arrivals, are i.i.d. geometric random variables with parameter μ;

3. Arrivals and departures are according to the early arrival system.

In addition, the remaining assumptions of the $M/M/1$ model on system capacity and queue discipline hold good. Note that the geometric distribution is the discrete-time counterpart of the continuous exponential distribution and has the same memoryless property. Therefore the number in the system $X(t)$ forms a Markov chain.

Throughout this section, we adopt earlier notations from Chapter 2 and Chapter 3 except that $t = 0, 1, 2, \ldots$. To maintain uniformity, symbol t is retained for representing time, even though it is discrete. In the future it may be changed as the situation demands. Consider the case of viewing the system exactly at the point of division. Observe that

$$X(t+1) = X(t) + N(1), \tag{3.111}$$

where $N(1)$ represents the net addition to the system during the interval $(t, t+1]$. Evidently, $N(1)$ is independent of t and has the following distribution: for $X(t) \geq 1$,

$$N(1) = \begin{cases} 1 & \text{with probability } \lambda(1-\mu), \\ 0 & \text{with probability } (1-\mu)(1-\lambda) + \lambda\mu, \\ -1 & \text{with probability } (1-\lambda)\mu; \end{cases} \tag{3.112}$$

and for $X(t) = 0$,

$$N(1) = \begin{cases} 1 & \text{with probability } \lambda(1-\mu), \\ 0 & \text{with probability } 1 - \lambda(1-\mu) \end{cases} \tag{3.113}$$

In (3.112), $N(1)$ is possible when either no arrival and no departure or one arrival and one departure occur. It is important to note that the second situation contributes $\lambda\mu$ which did not exist in the continuous-time case. Observe the expressions for probabilities in (3.113) which are so due to the early arrival system.

Remembering that $X_n = X(n)$, we find $\{X_n, n \geq 0\}$ is a homogeneous Markov chain with the transition matrix

$$\begin{bmatrix} 1-\lambda(1-\mu) & \lambda(1-\mu) & 0 & 0 & \cdots \\ (1-\lambda)\mu & 1-(1-\lambda)\mu-\lambda(1-\mu) & \lambda(1-\mu) & 0 & \cdots \\ 0 & (1-\lambda)\mu & 1-(1-\lambda)\mu-\lambda(1-\mu) & \lambda(1-\mu) & \cdots \\ \vdots & \vdots & \vdots & \vdots & \vdots \end{bmatrix}.$$

Similar to (2.1) and (2.2), we can establish the time dependent equations as

$$\begin{aligned} P_n(t+1) =& \lambda(1-\mu)P_{n-1}(t) + ((1-\mu)(1-\lambda) + \lambda\mu)P_n(t) \\ & + (1-\lambda)\mu \, P_{n+1}(t) \qquad \text{for } n \geq 1, \end{aligned}$$

and

$$P_0(t+1) = (1-\lambda(1-\mu))P_0(t) + (1-\lambda)\mu \, P_1(t). \tag{3.114}$$

The transient behaviour of the system which is the solution of (3.114), will be discussed in Chapter 9. It can be shown that the steady-state solution

$P_n = \lim_{t\to\infty} P_n(t)$ exists when $\lambda < \mu$, since in that case the Markov chain is irreducible and ergodic. The stationary equations are:

$$\mu(1-\lambda)P_{n+1} - \lambda(1-\mu)P_n = \mu(1-\lambda)P_n - \lambda(1-\mu)P_{n-1} \qquad \text{for } n \geq 1,$$

and

$$\mu(1-\lambda)P_1 - \lambda(1-\mu)P_0 = 0. \tag{3.115}$$

These can be interpreted as balance equations in which the probability of entering into a state equals the probability of getting out of it. To solve (3.115), use the last one in the right side of the first one for $n = 1$ and recursively obtain

$$\mu(1-\lambda)P_{n+1} = \lambda(1-\mu)P_n, \qquad n \geq 0.$$

This leads to

$$P_n = \alpha^n P_0, \qquad n \geq 1$$

where

$$\alpha = \frac{\lambda(1-\mu)}{\mu(1-\lambda)}. \tag{3.116}$$

With the help of the normalizing equation (i.e.,$\sum_{n=0}^{\infty} P_n = 1$) we obtain

$$P_0 = 1 - \alpha$$

provided $\alpha < 1$ which is equivalent to $\rho < 1$. Therefore,

$$P_n = (1-\alpha)\alpha^n, \qquad n = 0, 1, 2, \ldots, \tag{3.117}$$

for $\alpha < 1$ (i.e., $\rho < 1$). This implies that the stationary distribution of the number in the system is geometric which is exactly the same as in the continuous-time case except that the parameter has changed from ρ to α.

Instead of viewing the system exactly at the points of division, suppose it is done just prior to it. In other words, we are interested in studying $\{Y(t), t \geq 0\}$. This changes (3.113) to

$$N(1) = \begin{cases} 1 & \text{with probability } \lambda, \\ 0 & \text{with probability } 1-\lambda. \end{cases} \tag{3.118}$$

Denoting by $\tilde{P}_n = \lim_{t\to\infty} P(Y(t) = n)$, we get the balance equations as

$$\mu(1-\lambda)\tilde{P}_{n+1} - \lambda(1-\mu)\tilde{P}_n = \mu(1-\lambda)\tilde{P}_n - \lambda(1-\mu)\tilde{P}_{n-1} \qquad \text{for } n \geq 2,$$

$$\mu(1-\lambda)\tilde{P}_2 - \lambda(1-\mu)\tilde{P}_1 = \mu(1-\lambda)\tilde{P}_1 - \lambda\tilde{P}_0, \tag{3.119}$$

and

$$\mu(1-\lambda)\tilde{P}_1 - \lambda\tilde{P}_0 = 0$$

It can be shown that

$$\tilde{P}_n = \begin{cases} \rho(1-\alpha)\alpha^{n-1}, & n \geq 1, \\ 1-\rho, & n = 0 \end{cases} \tag{3.120}$$

provided $\rho < 1$. Thus the distribution of the number in the system in this case is not geometric, but conditionally so given that the system is not empty.

Remark: The stationary distribution for continuous-time is derivable if λt and μt in the limit become constants as $t \to \infty$, in which case both (3.117) and (3.120) approach $(1-\rho)\rho^n$, the solution for continuous-time $M/M/1$ model.

What about the case of viewing the system just after the points of division? It does not need any new treatment and the stationary distribution of the number in the system is the same as given by (3.120) since $Z_n = Y_{n+1}$. The case of the late arrival system is left to the reader for completion.

We may extend the $Geo/Geo/1$ model to the discrete-time birth-death model. The only changes to be introduced are λ and μ to become state dependent λ_n and μ_n. In other words, if there are n customers in the system, the next interarrival time and the next service-time are geometric random variables with parameters λ_n and μ_n, respectively. For the early arrival system, we would find the stationary solution to be given by

$$P_n = P_0 \prod_{i=0}^{n-1} \frac{\lambda_i(1-\mu_i)}{\mu_{i+1}(1-\lambda_{i+1})} \qquad \text{for } n = 1, 2, \ldots \tag{3.121}$$

with

$$P_0^{-1} = 1 + \sum_{n=1}^{\infty} \prod_{i=0}^{n-1} \frac{\lambda_i(1-\mu_i)}{\mu_{i+1}(1-\lambda_{i+1})}. \tag{3.122}$$

The summation in (3.122) must converge for the solution to exist. These have the same form as in (2.13) and (2.14) where $\frac{\lambda_i}{\mu_{i+1}}$ is replaced by $\frac{\lambda_i(1-\mu_i)}{\mu_{i+1}(1-\lambda_{i+1})}$.

Next we consider the $Geo/G/1$ model under the assumption of late arrival. This being the discrete-time version of the $M/G/1$ model has the same description except that the arrivals form a Bernoulli process with parameter λ (i.e., the interarrival times have geometric distributions with parameter λ) and the service-times have an arbitrary discrete distribution with finite mean $1/\mu$. Let us assume the p.f. of the service-time S to be

$$P(S = k) = g_k, \qquad k = 0, 1, 2, \ldots . \tag{3.123}$$

Assume the arrivals and departures to follow the late arrival system. Then the probability distribution of γ, the number of arrivals during the service-time of a customer, is identical for every customer and is given by

$$\alpha_k = P(\gamma = k) = \sum_{i=k}^{\infty} g_i \binom{i}{k} \lambda^k (1-\lambda)^{i-k}, \qquad k = 0, 1, 2, \ldots . \tag{3.124}$$

Following the same steps as in the continuous-time case, we get the p.g.f. of the distribution for the number in the system just after a departure to be

$$\Pi(z) = \frac{(1-z)\alpha(z)}{\alpha(z) - z}(1-\rho) \tag{3.125}$$

(which is exactly the same as (3.11)), where $\alpha(z)$ and the p.g.f. $G_S(z)$ of service-time distribution $\{g_k\}$ are related through

$$\alpha(z) = G_S(1 - \lambda + \lambda z), \tag{3.126}$$

due to (3.124). Expression (3.126) is the discrete-time analogue of (3.13). For the stationary solution to exist the conditions are $\rho < 1$ and $g_1 < 1$.

Unfortunately this pattern of similarity between discrete-time and continuous-time cases is no longer true if we have the early arrival system. Here, the distribution of γ depends on whether the system was empty or not when the customer entered into the service facility and this changes $\Pi(z)$ to $\Pi_1(z)$ as given by

$$\Pi_1(z) = \frac{(1-z)G_S(1-\lambda+\lambda z)}{(1-\lambda+\lambda z)(G_S(1-\lambda+\lambda z)-z)}(1-\rho). \tag{3.127}$$

Now let us consider the model dual to $Geo/G/1$, which is nothing but the $G/Geo/1$ model with arbitrarily distributed interarrival times with finite mean $1/\lambda$ and geometrically distributed service-times with parameter μ. If it follows the early arrival system then the stationary distribution for the number in the system prior to an arrival can be seen to be given by

$$Q_j = (1 - r_0)r_0^j, \qquad j = 0, 1, \ldots \tag{3.128}$$

where r_0 is the unique root of the equation $z = G_T(1-\mu+\mu z)$ that lies in the region $(0,1)$ and $G_T(z)$ is the p.g.f. of the interarrival times. On the other hand, for the late arrival system we can obtain

$$Q_j = \begin{cases} 1 - \dfrac{r_0}{1-\mu+\mu r_0} & \text{if } j = 0, \\ \dfrac{r_0}{1-\mu+\mu r_0}(1-r_0)r_0^{j-1} & \text{if } j \geq 1. \end{cases} \tag{3.129}$$

We conclude this section with the evaluation of waiting time distribution and start with the $Geo/G/1$ model. Assume FCFS queue discipline and the late arrival system. Note that the number of customers a departing customer leaves behind is the same as the number of arrivals during the customers' waiting time in the system (i.e., including the service-time). Therefore,

$$\Pi_k = \sum_{i=k}^{\infty} w_i \binom{i}{k} \lambda^k (1-\lambda)^{i-k}, \qquad k = 0, 1, \ldots \tag{3.130}$$

where $w_j = P(W = j)$, W being the steady-state waiting time of a customer in the system (compare (3.130) with (3.124)). Denoting by $G_W(z)$ the p.g.f. of $\{w_j\}$ we obtain from (3.130)

$$\Pi(z) = G_W(1 - \lambda + \lambda z) \tag{3.131}$$

which is similar to (3.126). From (3.125), (3.126) and (3.131) it follows that

$$G_W(1-\lambda+\lambda z) = \frac{(1-\rho)(1-z)G_S(1-\lambda+\lambda z)}{G_S(1-\lambda+\lambda z)-z}.$$

When $1-\lambda+\lambda z$ is replaced by z we get the p.g.f. of W as

$$G_W(z) = \frac{(1-\rho)(1-z)G_S(z)}{\lambda G_S(z) - z + 1 - \lambda}, \tag{3.132}$$

(compare it with the continuous-time result (3.19)).

Next we turn to the $G/Geo/1$ model and assume FCFS queue discipline with early arrival system. Obviously,

$$P(W_q = 0) = Q_0 = 1 - r_0 \tag{3.133}$$

from (3.128). Realize that the waiting time of an arriving customer in the queue is the same as the service-time of those customers ahead of the new arrival. Therefore

$$\begin{aligned} P(W_q = j) = \sum_{n=1}^{j} & P(\text{time for service completion of } n \text{ customers} \\ & = j | \text{arriving customer found } n \text{ customers in the system}) Q_n. \end{aligned}$$

That geometric variable has memoryless property and that the sum of geometric variables is a negative binomial variable lead to

$$\begin{aligned} P(W_q = j) &= (1-r_0)(1-\mu)^j \sum_{n=1}^{j} \binom{j-1}{n-1} \left(\frac{\mu\, r_0}{1-\mu}\right)^n \\ &= (1-r_0)(1-\mu)^j \sum_{n=0}^{j-1} \binom{j-1}{n} \left(\frac{\mu\, r_0}{1-\mu}\right)^{n+1} \\ &= (1-r_0)\mu\, r_0 (1-\mu)^{j-1} \left(1 + \frac{\mu\, r_0}{1-\mu}\right)^{j-1} \\ &= \mu\, r_0 (1-r_0)(1-\mu+\mu r_0)^{j-1}, \qquad j = 1, 2, \ldots. \end{aligned} \tag{3.134}$$

Compare (3.132) with Remark 5 in Section 3.3. From the distribution of W_q and by the use of p.g.f. we can show that

$$P(W = j) = \mu(1-r_0)(1-\mu+\mu r_0)^{j-1}, \qquad j = 1, 2, \ldots. \tag{3.135}$$

Remark: Although the moments of W or W_q can be obtained directly from the distribution or from the p.g.f., we may not forget Little's formula when the derivation of only the mean is concerned.

3.8. Discussion

If we look back at our progress made in this chapter, we realize that the models under consideration are no longer fully in the secure lap of Markov processes as in Chapter 2 but are provided the support of the imbedded Markov chain technique which helps their development yet in a Markovian environment. The models have either Markovian input or Markovian service-time and in general possess some non-Markovian nature. Yet at some selected points of time called regeneration points, the number in the queueing system which is our focal point of study except in Section 3.6 forms a Markov chain. The regeneration points for Markovian input models are customers' departure instants and for Markovian service-time models are customers' arrival instants. Thus what the imbedded Markov chain technique does, is to restrict our study of the system only to these regeneration points.

Leaving the study of transient behaviour (which is more theoretical and involved) to be done in Chapter 6, we deal with stationary distributions of the number of customers in the system at the regeneration points (i.e., $\{\Pi_n\}$ for the Markovian input models and $\{Q_n\}$ for the Markovian service-time models) in Sections 3.2, 3.3 and 3.4.

In each case, the starting point is to derive the transition probabilities of the imbedded Markov chain from recurrence relations on the original variables (e.g., in X_n in $M/G/1$, Y_n in $G/M/1$). This facilitates us to write down the balance equations which are solved either through the generating function (e.g., $M/G/1$ and $M/G^{a,b}/1$) or through the characteristic function (e.g., in $G/M/1$). However, the $G^k/M/1$ model is treated in a different way. Most often Rouché's theorem comes out to be handy.

Interesting to note is that the departure instant is in general not a regeneration point in the $M/G/c$ model. Therefore the imbedded Markov chain technique cannot be applied. Nevertheless it works for the special model $M/D/c$, which is given as an exercise.

As pointed out in the remarks, we observe that Q_j in $G^k/M/1$ is of simpler form than Π_j in $M/G^k/1$, which is a special case of $M/G^{a,b}/1$. The p.g.f. of $\{\Pi_n\}$ is expressed as the well-known Pollaczek-Khinchin (P-K) formula.

In all these models batch sizes are fixed, say, equal to k. Suppose a random batch size is involved such that the size varies with a probability distribution. In that case the superscript k changes to X implying that it is a random variable. For example $G^k/M/1$ becomes $G^X/M/1$, if arrivals come in random batch sizes. In addition, the p.f. of X has to be specified.

Our prime interest being the stationary distribution $\{P_n\}$ at any arbitrary point, it is extremely satisfying to locate functional relations among $\{P_n\}$, $\{\Pi_n\}$ and $\{Q_n\}$ which remarkably happened to be based on an elementary level crossing approach (Section 3.5). These functional relations are similar in spirit as that of Little's formula. The relations are established heuristically and will be dealt with formally in Chapter 6.

As usual, the waiting time distribution is simultaneously treated along with the distribution of the number of customers in the system. It is often very helpful to remember that the waiting time of a customer joining at time t is nothing but $V(t)$, under the FCFS queue discipline. In contrast to the waiting time, a complete section (Section 3.6) is devoted on the busy period, carefully delineating both analytic (viz., functional equations, Lagrange's expansion) and combinatorial methods except that the proofs in the latter method are shifted to Chapter 6. In fact, Chapter 6 serves as a continuation of this chapter without hampering its readability. The concept of the first passage time is crucial in the development of busy period distribution.

Needless to emphasize that the discrete-time models have their rightful place in the theory of queues. Although the treatment seems to be analogous to that of continuous-time models, there does arise a difference when we deal with an early arrival system versus a late arrival system. The computational and theoretical aspects of transient analysis will be handled in Chapters 4 and 8, respectively.

To conclude this section we verify that Π_n and Q_n exist if and only if $\rho < 1$, which was left out in Sections 3.2 and 3.3. First we do it for Π_n. Suppose $\rho < 1$. Letting $y_j = j(1-\rho)^{-1}$ we find that $\{y_j\}$ satisfies the criterion in Result 1, Section A.8 in Appendix A, and therefore the Markov chain $\{X_n, n = 0, 1, \ldots\}$ is ergodic and hence Π_n exists. On the other hand, if the Markov chain is ergodic, then the mean first passage time from state i to state j denoted by μ_{ij} can be seen to satisfy

$$\mu_{i,i-1} = \mu_{10}, \qquad i \neq 0. \tag{3.136}$$

Moreover, it is true that

$$\begin{aligned} \mu_{i0} &= \mu_{i,i-1} + \mu_{i-1,0} \\ &= i\mu_{10} \quad \text{by (3.136).} \end{aligned} \tag{3.137}$$

Thus by Result 2, Section A.8 in Appendix A, we have

$$\mu_{10} \sum_j j\alpha_j = \mu_{10} - 1, \tag{3.138}$$

by using (3.5) and (3.137). When we use (3.6) that

$$E(\gamma_n) = \sum_j j\alpha_j = \rho,$$

the condition (3.138) becomes

$$\rho = 1 - \frac{1}{\mu_{10}} < 1 \qquad (\text{since } \mu_{10} > 0)$$

which was to be proved (see Foster (1953)). It is easy to check that the Markov chain is irreducible and aperiodic. This completes the proof of the existence of Π_n when $\rho < 1$. Also see Karlin and Taylor (1975) Chapter 3, Section 5.

The question of the existence of Q_n in Section 3.3 is simpler to answer. In (3.30) we have seen that the stationary solution Q_n exists (not the same as the existence of Q_n as a limiting distribution) if and only if $\rho < 1$. Then due to a result in Section A.8 in Appendix A, the Markov chain $\{Y_n, n = 0, 1, \ldots\}$ is clearly ergodic for $\rho < 1$. Again the properties of irreducibility and aperiodicity are easy to establish. Hence Q_n exists for $\rho < 1$.

The concept of imbedded Markov chain was introduced by Kendall in (1951) and (1953). The materials in Sections 3.2 and 3.3 being classical may be found in most standard textbooks, such as Gross and Harris (1985), Cooper (1981) and Prabhu (1965). The references for Section 3.4 on bulk queues are Saaty (1961), Bailey (1954), Downton (1955) and (1956), Foster (1961) and Medhi (1984). Any reader interested in the subject may refer to the book by Chaudhry and Templeton (1983). Most of the materials on functional equations are derived from papers by Foster et al. (1961, 1964 and 1965). The section on the busy period is based on Takács' work, more specifically on (1961, 1962, 1967 and 1975), except that the alternative analytic technique is referred to Prabhu (1960). For discrete-time queues one may refer to Hunter (1983).

3.9. Exercises

1. Check (3.6).

2. Show that if the distribution of S, the service-time is exponential with rate μ, then $\Pi(z)$ as given by (3.11) is simplified to

$$\Pi(z) = \frac{1-\rho}{1-\rho z}$$

(Note that $\Pi(z) = P(z)$.)

3. Consider a single server queueing system having Poisson arrivals with rate λ. Let the service-times be i.i.d. random variables having p.d.f.

$$f_S(t) = \mu(\mu t)e^{-\mu t}.$$

Under the condition $\frac{2\lambda}{\mu} < 1$, show that

$$\Pi(z) = \frac{\Pi_0 \mu^2}{(\mu^2 - 2\lambda\mu z - \lambda^2 z(1-z))}.$$

Calculate Π_0.

4. For the queueing system $M/E_k/1$, with λ as the arrival rate and μ as the service rate, show that

$$\Pi(z) = \frac{(1-\rho)(1-z)}{1 - z(1 + \frac{\rho}{k}(1-z))^k}.$$

5. For the queueing model $M/D/1$ with λ as the arrival rate and μ as the service rate, show that

$$\Pi(z) = \frac{(1-\rho)(1-z)}{1 - ze^{\rho(1-z)}},$$

and verify that

$$\Pi_0 = 1 - \rho \quad \text{and} \quad \Pi_1 = (1-\rho)(e^{\rho} - 1).$$

6. Obtain the Pollaczek-Khinchin mean value formula given by (3.14) directly from (3.11).

7. Referring to the Pollaczek-Khinchin formula (3.19), show that

$$(1-\rho)n\, E\left[W_q^{n-1}\right] = \lambda \sum_{k=2}^{n} \binom{n}{k} E\left[W_q^{n-k}\right] E\left[S^k\right],$$

$n = 2, 3, \ldots$, and in particular

$$2(1-\rho)E[W_q] = \lambda\, E(S^2),$$
$$3(1-\rho)E\left[W_q^2\right] = 3\lambda E[W_q]\, E\left[S^2\right] + \lambda E\left[S^3\right].$$

(This gives a relationship among moments of W_q.)

8. For the queueing model $M/G/1$ discussed in Section 3.2, use the transformation

$$g_n = \left(1 - \sum_{j=0}^{n} \alpha_j\right)\rho^{-1}, \qquad n = 0, 1, 2, \ldots$$

and prove that (3.11) can be expressed as

$$\Pi(z) = \alpha(z)G(z),$$

where

$$G(z) = \frac{1-\rho}{1 - \rho\sum_{n=0}^{\infty} g_n z^n}$$

(Note that $G(z)$ is the p.g.f. of a compound geometric distribution.) Hence verify that

$$\Pi_n = \sum_{j=0}^{n} q_j \alpha_{n-j}, \qquad n = 0, 1, \ldots,$$

where q_j is the coefficient of z^j in $G(z)$. (This is an explicit expression for Π_n.) (See Gordon (1988))

9. Show that on the average $M/D/1$ system has $\frac{\rho^2}{(1-\rho)}$ fewer customers than $M/M/1$ model.

10. Consider the model $M/HE_2/1$ with

$$f_S(t) = \alpha\mu_1 e^{-\mu_1 t} + (1-\alpha)\mu_2 e^{-\mu_2 t}, \qquad t \geq 0.$$

Find $\Pi_j, j = 0, 1, 2, \ldots$, for $\alpha = \frac{1}{4}$, $\mu_1 = \lambda$, $\mu_2 = 2\lambda$, and also obtain the value of L.

11. Find the queueing model $M/D/c$, where each customer is served for exactly one unit of time, $X(t)$ is a non-Markovian process. By considering $X(t)$ over the consecutive intervals $(0, t]$, $(t, t+1]$, obtain the following relation:

$$P_n(t+1) = \sum_{i=0}^{c} P_i(t)e^{-\lambda}\frac{\lambda^n}{n!} + \sum_{i=c+1}^{c+n} P_i(t)e^{-\lambda}\frac{\lambda^{n-i+c}}{(n-i+c)},$$

$n = 1, 2, \ldots,$

$$P_0(t+1) = \sum_{i=0}^{c} P_i(t)e^{-\lambda}.$$

Under the condition $\lambda < c$, show that

$$\Pi(z) = \frac{\sum_{i=0}^{c-1} \Pi(z^c - z^i)}{1 - z^c e^{\lambda(1-z)}},$$

and hence the following expression:

$$\Pi(z) = \frac{(c-\lambda)(1-z)\prod_{i=1}^{c-1}(z-z_i)}{\left(\prod_{i=1}^{c-1}(1-z_i)\right)\left(1-z^c e^{\lambda(1-z)}\right)},$$

where $z_1, z_2, \ldots, z_{c-1}$ are the zeroes of $1 - z^c e^{\lambda(1-z)}$ inside the unit circle.

12. Prove (3.20) and simplify it for the $M/M/1$ model.

13. For the $M/G/1$ model, let the service-time of a customer follow an Erlangian distribution with parameters k and μ. Show that as k increases, the mean waiting time in the queue decreases monotonically to the case where the service-time is constant.

14. In the $G/M/1$ model, let the interarrival times follow an Erlangian distribution with parameters k and λ. Show that

$$E(W_q) = \frac{r_0}{\mu(1-r_0)},$$

where r_0 is the least positive real root of

$$z = \left(\frac{k\lambda}{k\lambda + \mu - \mu z}\right)^k.$$

15. Considering $M/M/1$ as a particular case of the $G/M/1$ model, show that $Q_j = (1-\rho)\rho^j$.

16. In the $G/M/1$ model, let the Laplace transform of the p.d.f. of any interarrival time be given by

$$f_T^*(\theta) = \frac{2\mu^2}{(\theta+\mu)(\theta+2\mu)}.$$

Show that $r_0 = 2 - \sqrt{2}$.

17. In the $G/M/1$ model, let the p.d.f. of the interarrival time be

$$\alpha\lambda_1 e^{\lambda_1 t} + (1-\alpha)\lambda_2 e^{-\lambda_2 t}.$$

For the following values of the parameters $\alpha = \frac{5}{8}$, $\lambda_1 = 2$, $\lambda_2 = 1$, and $\mu = 2$, find r_0.

18. Obtain Q_j for the $E_k/M/1$ model by using Section 3.3.

19. Consider the $M^X/M/1$ model (see Section 3.8 for definition), with the p.g.f. of batch sizes to be $C(z)$. Show that $\Pi(z)$ is given by

(3.11) with $\rho = \frac{\lambda E(X)}{\mu}$ and

$$\alpha(z) = \int_0^\infty e^{-\lambda(1-C(z))t} \mu e^{-\mu t}\, dt.$$

20. Verify (3.40) and (3.41).

21. Show that for the queueing model $M/M^k/1$,

$$\Pi_n = (1 - r_0) r_0^n$$

where r_0 is the unique positive root less than 1 of the equation

$$z^{k+1} - (1-\rho)z + \rho = 0.$$

22. Derive the differential-difference equations for the $M^X/M/1$ system with i initial customers, where customers join in batches of random side with distribution $a_j = P(X = j)$, $j = 1, 2, \ldots$
Show that

$$P^*(z,\theta) = \frac{z^{i+1} - \mu(1-z)P_0^*(\theta)}{(\lambda + \mu + \theta)z - \mu - \lambda z \sum_{j=1}^\infty a_j z^j},$$

$$L(t) = i + (\lambda E(X) - \mu)t + \mu \int_0^t P_0(\tau)d\tau.$$

23. Same model as in Exercise 22. Using Lagrange's expansion, prove that

$$\begin{aligned} P_0(t) &= \frac{e^{-(\lambda+\mu)t}}{\mu t} \\ &\times \sum_{m=i+1}^\infty m \left[\frac{(\mu t)^m}{m!} + \sum_{n=1}^\infty \frac{(\lambda t)^n}{n!} \sum_{j=0}^\infty b_{nj} \frac{(\mu t)^{j+n+m}}{(j+n+m)!} \right] \end{aligned}$$

where

$$\left(\sum_{j=1}^\infty a_j z^{j-1} \right)^n = \sum_{j=0}^\infty b_{nj} z,$$

and for the special case when

$$a_j = \begin{cases} 1 & j = k \\ 0 & \text{otherwise} \end{cases}$$

verify that

$$P_0(t) = \frac{e^{-(\lambda+\mu)t}}{\mu t} \sum_{m=i+1}^\infty m \left[\frac{(\mu t)^m}{m!} + \sum_{n=1}^\infty \frac{(\lambda t)^n}{n!} \frac{(\mu t)^{m+nk}}{(m+nk)!} \right].$$

24. Show that under the condition $\frac{\lambda E(S)}{b} < 1$, the inside roots of (3.51) are distinct.

25. Show that for the model $M/E_k^{a,b}/1$, the mean and variance of the number of customers in the system, respectively, are

$$\sum_{i=1}^{k}(r_i^* - 1)^{-1},$$

and

$$\sum_{i=1}^{k} r_i^*(r_i^* - 1)^{-2},$$

where $r_i^* (i = 1, 2, \ldots, k)$ are the zeroes of (3.51) outside C, where $C : |z| < 1$.

26. Derive all results for the model $M/D^{a,b}/1$ as a special case of the model $M/G^{a,b}/1$.

27. Prove (3.73).

28. Verify (3.77) from (3.76).

29. Prove (3.93) and hence show that $P(\Delta < \infty) = 1$, if and only if $\rho < 1$.

30. Consider the model $M/M/1$.
    (i) Check (2.72) from (3.93) as a particular case.
    (ii) Using (3.85), show that

$$G(z) = \frac{(1+\rho)}{2\rho}\left[1 - (1 - \frac{4\rho z}{(1+\rho)^2})^{1/2}\right]$$

   and hence

$$P(N = n) = \frac{1}{n}\binom{2n-2}{n-1}\rho^{n-1}(1+\rho)^{1-2n}, \qquad n = 1, 2, \ldots.$$

   (iii) If $\rho < 1$, show that

$$\sum_{n=1}^{\infty} P(N = n) = 1,$$

   and

$$E(N) = \frac{1}{1-\rho}.$$

31. For the $M/D/1$ model with constant service-time equal to $\frac{1}{\mu}$, show that (3.85) is reduced to

$$G(z) = ze^{-\rho}e^{\rho G(z)}.$$

Putting $y = \rho e^{-\rho} z$, and $x = \rho G(z)$, verify that it is simplified to

$$y = xe^{-x}.$$

Using Lagrange's expansion, prove that

$$G(z) = \sum_{n=1}^{\infty} \frac{(n\rho)^{n-1}}{n!} e^{-n\rho} z^n.$$

32. Using (3.103), rederive (3.101).

33. Using (3.103), prove (3.93).

34. For the model $M/M/1$, derive the joint distribution of Δ and N from (3.110) and show that

$$dG(n,t) = e^{-(\lambda+\mu)t} \frac{\mu^n \lambda^{n-1}}{n!(n-1)} t^{2n-2}\, dt$$

35. For the $M/G/1$ model, show that

$$E(\Delta) = \frac{E(S)}{1-\rho},$$
$$E(\Delta^2) = \frac{E(S^2)}{(1-\rho)^3},$$
$$E(\Delta^3) = \frac{E(S^3)}{(1-\rho)^4} + \frac{3\lambda[E(S^2)]}{(1-\rho)^5},$$
$$E(\Delta^4) = \frac{E(S^4)}{(1-\rho)^5} + \frac{10\lambda E(S^2)E(S^3)}{(1-\rho)^6} + \frac{15\lambda^2[E(S^2)]^3}{(1-\rho)^7}.$$

36. For the $G/M/1$ system, show that

$$P_0 = \frac{E(I)}{E(I)+E(\Delta)},$$
$$E(N) = \frac{1}{1-r_0},$$
$$E(I)+E(\Delta) = \frac{1}{\lambda(1-r_0)},$$
$$E(\Delta) = \frac{1}{\mu(1-r_0)},$$
$$E(I) = \frac{\mu-\lambda}{\lambda\mu(1-r_0)}.$$

37. Show that the average length of a busy period for the $M/G/1$ model with one initial customer is equal to the average time a customer spends in the $M/M/1$ system.

38. For the $M/G/1$ model, show that

$$E(N_j) = \frac{j}{1-\rho}.$$

39. Consider the queueing model $M/D/1$ with constant service-time equal to τ. Letting $f_n^i = P(N_i = n)$, establish the following:

$$f_n^j = \begin{cases} e^{-\lambda j \tau} & \text{if } n = j, \\ \displaystyle\sum_{k=1}^{n-j} e^{-\lambda j \tau} \frac{(\lambda j \tau)^k}{k!} f_{n-j}^k & \text{if } n > j. \end{cases}$$

$$f_n^j = \frac{j}{n} e^{-\lambda n \tau} \frac{(\lambda n \tau)^{n-j}}{(n-j)!}, \qquad n = j, j+1, \ldots$$

40. (i) In Section 3.7, show that

$$Y_{n+1} = \begin{cases} X_n & \text{with probability } 1 - \lambda \\ X_n + 1 & \text{with probability } \lambda \end{cases}$$

and hence obtain (3.120) from (3.117).

(ii) Derive (3.117) from (3.120) by establishing a relation between X_n and Y_n.

41. Derive results for the $Geo/Geo/1$ model with the late arrival systems similar to the early arrival system.

42. Prove (3.125).

References

Bailey, N. T. J. (1954) On queueing processes with bulk service, *J. Roy. Statist. Soc. Ser. B*, **16**, 80–87.

Chaudhry, M. L. and Templeton, J. C. C. (1983) *A First Course in Bulk Queues*, John Wiley, New York.

Cooper, R. B. (1981) *Introduction to Queueing Theory*, Second Edition, North-Holland, New York.

Downton, F. (1955) Waiting times in bulk service queues, *J. Roy. Statist. Soc. Ser. B*, **17**, 256–261.

Downton, F. (1956) On Limiting distributions arising in bulk service queues, *J. Roy. Statist. Soc. Ser. B*, **18**, 265–274.

Foster, F. G. (1953) On Stochastic matrices associated with certain queueing processes, *Ann. Math. Statist.*, **24**, 355–360.

Foster, F. G. (1961) Queues with batch arrivals I, *Acta Math. Acad. Sci. Hungar.*, **12**, 1–10.

Foster, F. G. and Nyunt, K. M. (1961) Queues with batch departures I, *Ann. Math. Statist.*, **32**, 1324–1332.

Foster, F. G. and Perera, A. G. A. D. (1964) Queues with batch departures II, *Ann. Math. Statist.*, **35**, 1147–1156.

Foster, F. G. and Perera, A. G. A. D. (1965) Queues with batch arrivals II, *Acta Math. Acad. Sci. Hungar.*, **16**, 275–287.

Gordon, W. (1988) A Note On the Equilibrium $M/G/1$ Queue Length, *J. Appl. Prob.*, **25**, 228–239.

Gross, D. and Harris, C. M. (1985) *Fundamentals of Queueing Theory*, Second Edition. John Wiley, New York.

Hunter, J. J. (1983) *Mathematical techniques of Applied Probability*, **2**, Academic Press, New York.

Karlin, S. and Taylor, H. M. (1975) *A First Course in Stochastic Processes*, 2nd ed., Academic Press, New York.

Kendall, D. G. (1951) Some problems in the theory of queues, *J. Roy. Statist. Soc. Ser. B*, **13**, 151–185.

Kendall, D. G. (1953) Stochastic processes occurring in the theory of queues and their analysis by the method of imbedded Markov chains, *Ann. Math. Statist.*, **24**, 338–354.

Medhi, J. (1984) *Bulk Queueing Models*, Wiley Eastern, New Delhi.

Prabhu, N. U. (1960) Some results for the queue with Poisson arrivals, *J. Roy. Statist. Soc. Ser. B*, **22**, 104–107.

Prabhu, N. U. (1965) *Queues and Inventories*, John Wiley, New York.

Saaty, T. L. (1961) *Elements of Queueing Theory with Applications*, McGraw-Hill, New York.

Takács, L. (1961) The probability law of the busy period for two types of queueing processes, *Oper. Res.*, **9**, 402–407.

Takács, L. (1962) A combinatorial method in the theory of queues, *J. Soc. Indust. Appl. Math.*, **10**, 691–694.

Takács, L. (1962) A single-server queue with recurrent input and exponentially distributed service-times, *Oper. Res.*, **10**, 395–399.

Takács, L. (1967) *Combinatorial methods in the theory of stochastic processes*, John Wiley, New York.

Takács, L. (1975) Combinatorial and analytic methods in the theory of queues, *Adv. Applied Prob.*, **7**, 607–635.

CHAPTER 4

Computational Methods – I

4.1. Introduction

We have presented the theoretical aspects of Markovian models and models which exhibit Markovian nature in some form of the other. Quite often, the solutions are left in a state that needs further computational tools for completion so that these are usable by practitioners. Sometimes numerical techniques are available even when no other forms of solution are known. This chapter is completely devoted to discuss a few techniques which enable us to compute solutions numerically.

Among a good number of results on stationary behaviour in queues, the final expressions are left in terms of zeroes of certain polynomials. In Section 4.2, we describe some root finding techniques that could help us in this direction. It is known that the stationary distribution of a finite Markov chain can be obtained by solving a finite system of linear equations. If the number of equations is not small, it may be difficult to solve them by hand. An algorithm to solve them is provided in Section 4.3. This one does not need roots of any equation.

When we come to transient solutions, the situation to get a nicely computable expression is worse. We resort to some numerical methods and in Section 4.4, present four techniques for birth-death models. There is also a method included in this section just for the $M/M/1$ model, since it seems to have some computational advantage over the classical expression.

4.2. Root Finding Methods

In Chapter 3, we have encountered several situations in which the steady-state probabilities depend on roots of some equation. For example, Q_n in the $G/M/1$ model is expressed in terms of r_0 (see (3.30)), the only root in (0,1) of Equation (3.29). Also recall that in bulk queues the expressions for $\Pi(z)$ in (3.44) and $Q(z)$ in (3.65) needed the roots of Equations (3.45) and (3.66), respectively, which are of the form

$$z^c = k(z) \tag{4.1}$$

where $k(z)$ is a p.g.f. We may call (4.1) the characteristic equation of the corresponding queueing model (different from the characteristic equation of a set of difference equations). These roots help in determining the unknown constants involved in $\Pi(z)$ or $Q(z)$. It may be visualized that there are other models for which the characteristic equation is of the form (4.1). In this section, we describe some numerical techniques for finding the roots of (4.1).

Among the classical techniques that one finds in most textbooks, we will discuss *Muller's method* (see Section A.11 in Appendix A), which has been used on computers with remarkable success, more particularly when applied to queueing models. Suppose we are interested in finding the roots of the equation $f(z) = 0$. Let us start with three distinct arbitrary values x_1, x_2, x_3 of z. Then fit the unique parabola passing through $(x_1,\ f(x_1))$, $(x_2,\ f(x_2))$ and $(x_3,\ f(x_3))$. Its equation happens to be

$$g(x) = f(x_1) + f(x_1, x_2)(x - x_1) + f(x_1, x_2, x_3)(x - x_1)(x - x_2) \tag{4.2}$$

where

$$f(x_i, x_j) = \frac{f(x_i) - f(x_j)}{x_i - x_j}$$

and

$$f(x_1, x_2, x_3) = \frac{f(x_1, x_2) - f(x_2, x_3)}{x_1 - x_3}.$$

Suppose we write $g(x)$ as

$$g(x) = a_0 + a_1 x + a_2 x^2.$$

Its zeroes can be expressed as

$$x = \frac{2a_0}{-a_1 \pm (a_1^2 - 4a_0 a_2)^{1/2}}. \tag{4.3}$$

The first approximation of a root of $f(x) = 0$ is given by (4.3), after selecting the sign so that the denominator is largest in magnitude. Denote this approximated value by x_4. The process of approximation is repeated with x_2, x_3, x_4. The sequence of approximation to the first root stops when it converges. Let the final approximated value of the first root be z_1.

Remark: Although three initial values are needed in Muller's method, it has been customary only to select one, say x_0, and take the other two as $x_0 + 0.5$ and $x_0 - 0.5$.

Muller's method is iterative. Once z_1 is determined, the next approximated root is found similarly by working with the function $f(z)/(z - z_1)$. If r roots $z_1, \ldots, z_r$ have already been found, the next root is approximated by considering $f(z)/\prod_{i=1}^{r}(z - z_i)$. In this manner the function $f(z)$ is getting deflated before finding the other roots.

Although this method behaves better than other well-known methods, experience suggests that when applied as such it is confronted with some problems. Returning to (4.1), in a queueing situation we are often interested in roots inside and on the unit circle which are usually c in number, whereas the number of roots of (4.1) could be infinite. Thus for large c the process is tedious to search for the right roots and therefore becomes impractical. It is also prone to an accumulation of error. In order to improve over Muller's method, Powell suggested a modification to replace (4.1) by an equivalent system of c equations.

Clearly, (4.1) can be written as

$$z^c = k(z)e^{2n\pi i} \tag{4.4}$$

where $i = \sqrt{-1}$. Then (4.4) is equivalent to a set of c equations which are as follows:

$$z = |k(z)|^{1/c} \exp\left(\frac{(\phi + 2n\pi)i}{c}\right), \qquad n = 1, \ldots, c \tag{4.5}$$

where $\phi = \arg(k(z))$. Applying Rouché's theorem to (4.5), one can check that each equation has exactly one root inside or on the unit circle. Therefore, instead of solving Equation (4.1) for its c roots inside or on the unit circle, we can find them by solving each equation in (4.5) for a single root inside or on the unit circle. These single roots may be approximated by using Muller's method and then tested for accuracy by substituting in (4.1). In this way, the deflation technique is bypassed and the required roots can be found even for large value of c. We call this the *modified Muller's method.*

The following observations are useful hints while one actually computes the roots:

(i) The characteristic equation has all real coefficients and hence the complex roots must appear in conjugate pairs. It is known that one root is 1 which may be denoted by z_c.

(ii) Sometimes the required roots of an equation may not be inside the unit circle. However, a suitable transformation of the equation will possibly bring the roots inside the unit circle. Therefore, one may have a computational algorithm only for finding roots inside the unit circle.

(iii) Inside roots generally lie on a smooth curve and in some cases this curve approximates a circle with center at the origin.

(iv) Besides z_c, one may look for other real roots mostly in the interval $(-1, 0)$.

(v) For any given precision for roots of (4.1), one should compute the roots of (4.5) with double the original precision.

(vi) Powell's modification is computationally more stable than Muller's.

The last four observations have not been theoretically established but are merely based on enough computational experience.

For the purpose of illustration, let us examine two particular cases of the bulk queue model $M/G^{a,b}/1$ (see Section 3.4 in Chapter 3).

1. $M/D^{a,b}/1$. In this model, the service-time is a constant, say, $1/\mu$. This implies that

$$k(z) = \alpha(z) = e^{-\rho(1-z)}$$

and the characteristic equation is

$$z^b = e^{-\rho(1-z)}. \tag{4.6}$$

We have to find b roots of (4.6) inside and on the unit circle. Powell's modification leads to the following b equations:

$$z = \exp\left(\frac{\rho(z-1) + 2n\pi i}{b}\right), \qquad n = 1, \ldots, b. \tag{4.7}$$

Besides $z_b = 1$, other roots are approximated by solving each of (4.7) through Muller's method. Because of the remark we are to select one initial approximation for each equation. Denote by $x_0^{(n)}$ the initial approximation for the nth equation, $n = 1, 2, \ldots, b-1$. Furthermore, let z_n be the nth approximated root, $n = 1, 2, \ldots, b-1$. It has been found empirically that the initial approximations as given by

$$x_0^{(1)} = 1 - \frac{\rho}{b}\exp\left(\frac{2\pi i}{b}\right) \tag{4.8}$$

and

$$x_0^{(n)} = |z_{n-1}|\exp\left(\frac{2n\pi i}{b}\right), n = 2, \ldots, b-1,$$

provide good result in the sense of faster convergency and better accuracy.

2. $M/E_r^{a,b}/1$. From (3.51), we have the characteristic equation of the model as

$$z^b\left(1 + \frac{\rho}{r}(1-z)\right)^r - 1 = 0 \tag{4.9}$$

which has $b+r$ roots. Under the condition $\rho < b$, it has b roots inside and on the unit circle and therefore has r roots outside the unit circle. If r is very small in comparison with b (i.e., symbolically $r \ll b$), then it is often faster to solve for r outside roots. In order to use the algorithm to find inside roots, we use the transformation

$$\tilde{z} = 1 + \frac{\rho}{r}(1-z)$$

transforming (4.9) into

$$\tilde{z}^r = \left(1 + \frac{r}{\rho}(1-\tilde{z})\right)^{-b} \tag{4.10}$$

which has r roots inside the unit circle. Once the roots are found, the p.g.f. of the stationary distribution under consideration is completely determined (see (3.48) and (3.68)).

Based on the modified Muller's method, a software package known as QROOT (see Chaudhry (1993)) has been developed for finding roots of characteristic equations.

4.2.1. The factorization method. For deriving the stationary distribution $\{Q_N\}$ (see (3.68)) in the $G^k/M/1$ model, the inside roots of (3.66) or the outside roots of the denominator of (3.65) are needed and thereafter $\{Q_N\}$ is obtainable from (3.68).

We consider another model where roots have to be determined to find the distribution. Suppose we are interested in the stationary distribution $\{\pi_j\}$ of a special type of a Markov chain on state space $\{0, 1, \ldots\}$. Denoting by $\{\pi_{ij}\}$ the transient probabilities of the Markov chain, the special chain should have the condition $\pi_{ij} = q_{j-i}$ when $j \geq c$ for some fixed c where the sequence $\{q_i\}$ forms a distribution, say of a random variable U.

Evidently, the condition leads to

$$\pi_j = \sum_{i=0}^{\infty} \pi_i \pi_{ij} \qquad \text{for } j < c, \tag{4.11}$$

$$\pi_j = \sum_{i=0}^{\infty} \pi_i q_{j-i} \qquad \text{for } j \geq c. \tag{4.12}$$

Note that the $G^k/M/1$ model satisfies these conditions with $c = 1$ and $q_j = \alpha_{k-j}, j \leq k$ and $q_{k+1} = q_{k+2} = \ldots = 0$. In this example clearly U has the largest value k. In fact, for many problems the assumption of

$$-g \leq U \leq h \tag{4.13}$$

with probability one where g and h are positive integers is a reasonable one, at least for computational purposes. It essentially means that although theoretically U can vary from $-\infty$ to $+\infty$, computationally $P(U < -g)$ and $P(U > h)$ are so small that such values can be neglected.

The following facts may be checked and are left as exercises:

(i) Under (4.12),

$$\pi_{ij} = 0 \qquad \text{when} \quad i \geq c+g \quad \text{and} \quad j < c \tag{4.14}$$

(ii) Let $\pi(z)$ be the p.g.f of $\{\pi_j\}$. Then

$$\pi(z) = \frac{V(z)}{1 - U(z)}, \tag{4.15}$$

where $U(z)$ is the p.g.f. of U and $V(z)$, known as the compensation function, is a polynomial of degree $c + g - 1$. Observe (4.15) is in the form of (3.65).

(iii) Since $U \leq h$, under the condition that $E(U) < 0$, we have

$$1 - U(z) = A(z)B(z), \tag{4.16}$$

where $A(z)$ is a polynomial of degree g containing all zeroes of $1 - U(z)$ inside and on the unit circle and $B(z)$ of degree h containing zeroes outside the unit circle.

The possibility of the factorization (4.16) is fundamental to the development of the numerical method in this section. Up to this point, we have practically advanced nothing new. Now, let us write $B(z)$ as

$$B(z) = -1 + \sum_{j=1}^{h} B_j z^j, \tag{4.17}$$

where without loss of generality B_0 is assumed to be -1. Next, we intend to express each π_j as a function of B_i's. Since $\pi(z)$ is convergent inside the unit circle, $V(z)$ must be divisible by $A(z)$ and therefore we may write

$$\frac{V(z)}{A(z)} = C(z) \qquad \text{(say)}. \tag{4.18}$$

From (ii) and (iii) we may deduce that $C(z)$ is a polynomial of degree $c-1$ and thus may be expressed as

$$C(z) = \sum_{j=0}^{c-1} C_j z^j. \tag{4.19}$$

Combining (4.15), (4.16) and (4.18), we have

$$\pi(z)B(z) = C(z). \tag{4.20}$$

Note that $A(z)$ in (3.67) has similar structure as (4.20). If we use expansions (4.17) and (4.19) and collect appropriate coefficients, we get

$$\sum_{i=0}^{j} B_i \pi_{j-i} = \begin{cases} C_j & \text{when } j < c \\ 0 & \text{when } j \geq c \end{cases} \tag{4.21}$$

where $B_i = 0$ if $i > h$ (see (4.17)). The second one leads to the basic result

$$\pi_j = \sum_{i=1}^{j} B_i \pi_{j-i}, \qquad j \geq c. \tag{4.22}$$

The remarkable feature of (4.22) is that it is a finite sum.

In order to ascertain them, not that (4.11) is reduced to

$$\pi_j = \sum_{i=0}^{c+g-1} \pi_i \pi_{ij}, \quad j = 0, 1, \ldots, c-1, \tag{4.23}$$

due to (4.14). In (4.23), it gives the impression that $c + g$ π_i's are unknown, to be determined from c equations. However, once B_i's are determined, one can express $\pi_c, \ldots, \pi_{c+g-1}$ in terms of $\pi_0, \ldots, \pi_{c-1}$ from (4.22). Now from (4.23) the initial probabilities $\pi_{c-1}, \ldots, \pi_0$ can be determined. Thus what is of immediate interest is to evaluate the B_j's. One way of doing so is to obtain them through the zeroes of $1 - U(z)$ outside the unit circle.

4.3. The State Reduction Method

We describe below an algorithmic procedure for obtaining stationary solutions for any finite Markov chain. This procedure is called the state reduction method. It avoids the use of transforms and is easily usable.

Let the Markov chain have a finite state space $\{0, 1, \ldots, N\}$. Then the stationary equations are:

$$\pi_j = \sum_{i=0}^{N} \pi_i \pi_{ij}, \qquad j = 0, \ldots, N. \tag{4.24}$$

We apply the age old Gaussian elimination procedure by eliminating the Nth variable first, $(N-1)$th variable next and so on. The expression for π_N from (4.24) is

$$\pi_N = \frac{\sum_{i=0}^{N-1} \pi_i \pi_{iN}}{1 - \pi_{NN}}. \tag{4.25}$$

By substituting (4.25) in the remaining equations, we have

$$\pi_j = \sum_{i=0}^{N-1} \pi_i \left(\pi_{ij} + \frac{\pi_{iN}\pi_{Nj}}{1 - \pi_{NN}} \right), \qquad j = 0, \ldots, N-1. \tag{4.26}$$

Equations (4.26) may be viewed as the stationary equations of a reduced Markov chain with states $0, \ldots, N-1$ such that its (i,j)th transition probability becomes

$$\pi_{ij} + \frac{\pi_{iN}\pi_{Nj}}{1 - \pi_{NN}}. \tag{4.27}$$

In fact, it is easier to prove that this is the transition probability from i to j in a new Markov chain formed by collapsing state N in the original Markov chain. Proceeding in this manner, we see that the elimination of a variable corresponds to the reduction of a state in the associated chain. This *Gaussian elimination* has the above elegant probabilistic interpretation and may be called the *state reduction method.*

In a natural way, (4.27) is denoted by $\pi_{ij}^{(N-1)}$, and so π_{ij} should be denoted by $\pi_{ij}^{(N)}$. By repeating the procedure, we may represent the transition probabilities of the successive reduced Markov chains by $\{\pi_{ij}^{(N)}\}$, $\{\pi_{ij}^{(N-1)}\}$, $\ldots$, $\{\pi_{ij}^{(1)}\}$, where

$$\pi_{ij}^{(N)} = \pi_{ij}$$

and

$$\pi_{ij}^{(n-1)} = \pi_{ij}^{(n)} + \frac{\pi_{in}^{(n)}\pi_{nj}^{(n)}}{S_n} \qquad n = N, N-1, \ldots, 2 \tag{4.28}$$

with

$$S_n = 1 - \pi_{nn}^{(n)} = \sum_{j=0}^{n-1} \pi_{nj}^{(n)}.$$

In order to avoid negative signs for computation, we use the last expression. In this notation, (4.25) is written as

$$\pi_N = \frac{\sum_{i=0}^{N-1} \pi_i \pi_{iN}^{(N)}}{S_N}.$$

Similarly, we can write

$$\pi_{N-1} = \frac{\sum_{i=0}^{N-2} \pi_i \pi_{i,N-1}^{(N-1)}}{S_{N-1}}$$

and in general

$$\pi_j = \frac{\sum_{i=0}^{j-1} \pi_i \pi_{ij}^{(j)}}{S_j}, \qquad j = 1, \ldots, N. \tag{4.29}$$

Express π_j, $j = 1, \ldots, N$, in terms of π_0 recursively starting from π_1. Let

$$\pi_j = r_j \pi_0, \qquad j = 0, 1, \ldots, N \tag{4.30}$$

obviously $r_0 = 1$ and other r_j's are determined. From the normalizing equation and from (4.30), it is clear, that

$$\pi_0 = \frac{1}{\sum_{j=0}^{N} r_j}.$$

Thus the evaluation of π_j's is complete by computing

$$\pi_j = \frac{r_j}{\sum_{i=0}^{N} r_i}, \qquad j = 0, 1, \ldots, N. \tag{4.31}$$

This completes the development of the state reduction method to determine π's in a finite Markov chain.

Alternatively, we set $\pi_0 = 1$ and call it π'_0. Then compute π_1 from (4.29) with $\pi_0 = \pi'_0$ and call it π'_1. Continue to compute $\pi'_2, \ldots, \pi'_N$ recursively in this manner. Then

$$\pi_j = \frac{\pi'_j}{\sum_{i=0}^{N} \pi'_i} \tag{4.32}$$

where $\sum_{i=0}^{N} \pi'_i$ is called the normalizing constant.

Remarks:

1. Here we have solved

$$\boldsymbol{\pi}(\mathbf{I} - \mathbf{P}) = \mathbf{0} \qquad \text{(see (A.79))}$$

subject to the normalizing condition

$$\sum_{j=0}^{\infty} \pi_j = 1$$

for a finite $\mathbf{P}$. The same method can be used to solve the steady-state equations

$$\boldsymbol{\pi}\mathbf{Q} = \mathbf{0} \qquad \text{(see (A.88))}$$

subject to the above normalizing condition, when $\mathbf{Q}$ is finite.

2. The state reduction method is an exact but algorithmic approach. Similar exact methods (i) by using Cramer's rule and (ii) by finding the inverse of a matrix arising from $\mathbf{Q}$ and the normalizing condition, can give the solution. There are well-known iterative methods of solution such as *Jacobi iteration* and *Gauss-Seidel iteration* (see Section A.11) that can be used. However, the state reduction method is simple and behaves better even for large matrix, especially when $\mathbf{P}$ is banded on both sides, as in the case of the transition matrix of a birth-death chain. It retains the banded property throughout.
3. Suppose $\mathbf{P}$ (or $\mathbf{Q}$) is infinite. A natural thing to do is to truncate the matrix to a size, say N. This is reasonable since π_j's are small enough for large j to be negligible. When the matrix is truncated, it may no longer be a stochastic matrix. Augment the last column to make it stochastic and then solve. Let $\mathbf{P}$ (or $\mathbf{Q}$) satisfy the following properties:
   (i) $\mathbf{P}$ has repeating rows. For some fixed numbers k and c,

   $$p_{ij} = p_{i+k,j+k}, \quad i, j \geq c$$

   and c is the minimum among such values.
   (ii) $\mathbf{P}$ is banded. There are positive numbers g and h such that

   $$p_{ij} = 0 \text{ whenever } j - i < -g \text{ or } j - i > h,$$

   and g and h are the minimum among such values.

   If $\mathbf{P}$ has these properties, the augmentation procedure is a convenient one to use and may produce better approximations for a given truncation level.

4.4. Transient Behaviour: Numerical Approaches

Being in an applied field, we may often wonder how to utilize the exact formulas for computation if these exist and how to arrive at numerical solutions if the explicit expressions are not known. Sometimes it turns out that a numerical technique as against the exact formula may result in obtaining the solution with desired accuracy with relative ease.

Although the treatment of stationary solution has produced satisfactory results, we remember from Section 2.5 in Chapter 2 that the transient solution does not come out in an obvious manner and even in the simplest $M/M/1$ model the exact solution (2.50) is not attractive for computational purposes.

It is here in Markovian models that we may like to explore the applicability of numerical methods, possibly with the help of a computer, in order to derive a solution. We discuss below a few such methods.

4.4.1. The eigenvalue method. To begin with let us write the differential-difference equations for $P_n(t)$ in the birth-death model which is initially in state i. These are as follows:

$$\frac{dP_0(t)}{dt} = -\lambda_0 P_0(t) + \mu_1 P_1(t),$$

and

$$\frac{dP_n(t)}{dt} = \lambda_{n-1}P_{n-1}(t) - (\lambda_n + \mu_n)P_n(t) + \mu_{n+1}P_{n+1}(t), \qquad n \geq 1 \quad (4.33)$$

which can be written in a similar way as in (2.3) or from the flow diagram. By taking the L.T. of (4.33), we obtain

$$(\lambda_0 + \theta)P_0^*(\theta) - \mu_1 P_1^*(\theta) = \delta_{i0}$$

and

$$-\lambda_{n-1}P_{n-1}^*(\theta) + (\lambda_n + \mu_n + \theta)P_n^*(\theta) - \mu_{n+1}P_{n+1}^*(\theta) = \delta_{in}, \qquad n \geq 1, \quad (4.34)$$

(δ_{in} is the Kronecker delta) which is nothing but an infinite system of linear equations in L.T.'s of the state probabilities $P_n(t)$. Let us change the infinite system to a finite system (i.e., with finite capacity) for which solution techniques are well known. Consider a birth-death model with the state space $\{0, 1, \ldots, N\}$. A queue with limited waiting space (see Model 4 in Section 2.3 in Chapter 2) is indeed of this type. Then Equations (4.34) are modified to

$$\begin{aligned}(\lambda_0 + \theta)P_0^*(\theta) - \mu_1 P_1^*(\theta) &= \delta_{i0},\\ -\lambda_{n-1}P_{n-1}^*(\theta) + (\lambda_n + \mu_n + \theta)P_n^*(\theta) - \mu_{n+1}P_{n+1}^*(\theta) &= \delta_{in},\\ 1 \leq n \leq N-1,\end{aligned}$$

and

$$-\lambda_{N-1}P_{N-1}^*(\theta) + (\mu_N + \theta)P_N^*(\theta) = \delta_{iN}. \quad (4.35)$$

Once we solve the system (4.35), then we have to take the inverse L.T. of $P_n^*(\theta)$ in order to get $P_n(t)$.

Denote by $A_N(\theta)$ the determinant of the coefficient matrix of (4.35). Let $T_r(\theta)$ and $B_r(\theta)$ be the determinants of $r \times r$ matrix formed at the top left corner and at the bottom right corner of $A_N(\theta)$, respectively. Set $T_0(\theta) =$

$B_0(\theta) = 1$. Using Cramer's rule, the solution of (4.35) is given below.

$$P_n^*(\theta) = \begin{cases} \dfrac{T_n(\theta)B_{N-i}(\theta)}{A_N(\theta)} \displaystyle\prod_{j=n+1}^{i} \mu_j & \text{for } 0 \le n \le i, \\ \dfrac{T_i(\theta)B_{N-n}(\theta)}{A_N(\theta)} \displaystyle\prod_{j=i}^{n-i} \lambda_j & \text{for } i+1 \le n \le N, \end{cases} \tag{4.36}$$

where $\prod_{j=k}^{n} \mu_j = 1$ if $n < k$. Note that we want to express (4.36) as a partial fraction expansion for the purpose of inversion. First we want to identify the zeroes of the denominator $A_N(\theta)$. It can be seen either by direct verification or by a similarity transformation that $A_N(\theta)$ is equal to $C_N(\theta)$ which is given by

$$C_N(\theta) = \\ = \begin{vmatrix} \lambda_0+\theta & \sqrt{\lambda_0\mu_1} & & & & \\ \sqrt{\lambda_0\mu_1} & \lambda_1+\mu_1+\theta & \sqrt{\lambda_1\mu_2} & & & \\ & \sqrt{\lambda_1\mu_2} & \lambda_2+\mu_2+\theta & \sqrt{\lambda_2\mu_3} & & \\ & & & & \ddots & \sqrt{\lambda_{N-1}\mu_N} \\ & & & & \sqrt{\lambda_{N-1}\mu_N} & \mu_N+\theta \end{vmatrix}. \tag{4.37}$$

Notice that θ_0 is a zero of $C_N(\theta)$ if and only if $-\theta_0$ is an *eigenvalue* (see Section A.10) of the matrix E_N where

$$E_N = \begin{vmatrix} \lambda_0 & \sqrt{\lambda_0\mu_1} & & & & \\ \sqrt{\lambda_0\mu_1} & \lambda_1+\mu_1 & \sqrt{\lambda_1\mu_2} & & & \\ & \sqrt{\lambda_1\mu_2} & \lambda_2+\mu_2 & \sqrt{\lambda_2\mu_3} & & \\ & & & & & \sqrt{\lambda_{N-1}\mu_N} \\ & & & & \sqrt{\lambda_{N-1}\mu_N} & \mu_N \end{vmatrix}. \tag{4.38}$$

It can be verified that $\theta = 0$ is an eigenvalue of E_N. Moreover, since E_N is a tridiagonal symmetric matrix with nonzero sub-diagonal elements, it follows by *Sturm sequence* property (see Section A.10) that other eigenvalues are positive and distinct. Hence $A_N(\theta)$ has exactly $N+1$ distinct zeroes one of which is zero and the rest are negative eigenvalues of E_N. It is this property which prompts up to call this method the eigenvalue method. For specific parametric values, the zeroes of $A_N(\theta)$ are determined when we realize that computer packages are available to evaluate the eigenvalues.

Let the zeroes of $A_N(\theta)$ be z_k, $k = 0, 1, \ldots, N$ with $z_0 = 0$. Writing $P_n^*(\theta)$ as

$$P_n^*(\theta) = \frac{G_n(\theta)}{A_N(\theta)}, \qquad n = 0, 1, \ldots, N, \tag{4.39}$$

and observing that the degree of $G_n(\theta)$ is less than that of $A_N(\theta)$ for every n, we have the following *partial fraction expansion* form:

$$P_n^*(\theta) = \sum_{k=0}^{N} \frac{\beta_{n,k}}{\theta - z_k}, \qquad n = 0, 1, \ldots, N \tag{4.40}$$

where

$$\beta_{n,k} = \frac{G_n(z_k)}{H_N(z_k)}, \qquad k = 0, 1, \ldots, N \tag{4.41}$$

and

$$H_N(z_k) = \prod_{\substack{j=0 \\ j \neq k}}^{N} (z_k - z_j). \tag{4.42}$$

The transient distribution of the process can now be obtained by inverting (4.40) which leads to

$$P_n(t) = \beta_{n,0} + \sum_{k=1}^{N} \beta_{n,k} e^{z_k t}, \qquad n = 0, 1, \ldots, N. \tag{4.43}$$

Remarks:

1. If $t \to \infty$, then $\lim_{n\to\infty} P_n(t) = \beta_{n,0}$ which is the stationary probability of state n. In fact it can easily be verified that

$$\beta_{n,0} = \beta_{0,0} \prod_{i=0}^{n-1} \frac{\lambda_i}{\mu_{i+1}} \tag{4.44}$$

   which checks with (2.13).

2. The transient solution of the finite capacity process may be considered as an approximation to the solution of the infinite capacity process and as N becomes larger the approximation should be better.

3. Let $\mathbf{x}_k = (x_{0k}, \ldots, x_{Nk})'$ and $\mathbf{y}_k = (y_{0k}, \ldots, y_{Nk})'$ be the right-hand and left-hand eigenvectors corresponding to the eigenvalue z_k, $k = 0, 1, \ldots, N$. Then it can be seen that

$$\beta_{n,k} = x_{ik} y_{kn} \text{ and } P_n(t) = \sum_{k=0}^{N} e^{z_k t} x_{ik} y_{kn} \quad n = 0, 1, \ldots, N. \tag{4.45}$$

4.4.2. The continued fraction method. Another approximation to the transient solution of the infinite capacity process is provided by the continued fraction approach which is discussed below.

Let us go back to (4.34) and find a solution for $P_n^*(\theta)$.

For simplicity assume $i = 0$. Denote by

$$f_n = (-1)^n M_n P_n^*(\theta) \tag{4.46}$$

where

$$f_0 = P_0^*(\theta) \qquad \text{and} \qquad M_n = \prod_{j=1}^{n} \mu_j.$$

Then (4.34) is reduced to the following recurrence relations:

$$f_1 = 1 - (\lambda_0 + \theta) f_0$$

and

$$f_{n+1} = -\lambda_{n-1}\mu_n f_{n-1} - (\lambda_n + \mu_n + \theta) f_n, \qquad n \geq 1. \tag{4.47}$$

The second set in (4.47) can be expressed as

$$\frac{f_{n+1}}{f_n} = -(\lambda_n + \mu_n + \theta) - \frac{\lambda_{n-1}\mu_n}{f_n/\, f_{n-1}}, \qquad n \geq 1$$

from which it follows that

$$\frac{f_n}{f_{n-1}} = \frac{-\lambda_{n-1}\mu_n}{\lambda_n + \mu_n + \theta + \frac{f_{n+1}}{f_n}}, \qquad n \geq 1.$$

By iteration, we have

$$\frac{f_n}{f_{n-1}} = \cfrac{-\lambda_{n-1}\mu_n}{\lambda_n + \mu_n + \theta - \cfrac{\lambda_n \mu_{n+1}}{\lambda_{n+1} + \mu_{n+1} + \theta - \dots}}, \qquad n \geq 1.$$

which is also written as

$$\frac{f_n}{f_{n-1}} = \frac{-\lambda_{n-1}\mu_n}{\lambda_n + \mu_n + \theta} - \frac{\lambda_n \mu_{n+1}}{\lambda_{n+1} + \mu_{n+1} + \theta} - \dots, \qquad n \geq 1. \tag{4.48}$$

Similarly considering the first relation in (4.47), we obtain

$$f_0 = \frac{1}{\lambda_0 + \theta} - \frac{\lambda_0 \mu_1}{\lambda_1 + \mu_1 + \theta} - \frac{\lambda_1 \mu_2}{\lambda_2 + \mu_2 + \theta} - \dots. \tag{4.49}$$

The right sides in (4.48) and (4.49) are nothing but *continued fractions.*

For simplicity, the recurrence relations (4.47) may be written in the following general form:

$$f_1 = a_1 - b_1 f_0$$

and

$$f_{n+1} = a_{n+1} f_{n-1} - b_{n+1} f_n, \qquad n \geq 1. \tag{4.50}$$

Then

$$\frac{f_n}{f_{n-1}} = \frac{a_{n+1}}{b_{n+1}} + \frac{a_{n+2}}{b_{n+2}} \dots, \qquad n \geq 1 \tag{4.51}$$

and

$$f_0 = \frac{a_1}{b_1} + \frac{a_2}{b_2} + \dots. \tag{4.52}$$

In the continued fraction

$$\frac{a_1}{b_1 +}\ \frac{a_2}{b_2 +}\ \cdots,$$

$$\frac{A_n}{B_n} = \frac{a_1}{b_1 +}\ \frac{a_2}{b_2 +}\ \cdots\ \frac{a_n}{+\ b_n}$$

is called the nth convergent of the continued fraction. Observe that A_n and B_n satisfy the recurrence relation

$$U_n = a_n U_{n-2} + b_n U_{n-1} \tag{4.53}$$

where $A_0 = 0$, $A_1 = a_1$, $B_0 = 1$ and $B_1 = b_1$. From (4.53) we can express A_n, $n = 2, 3, \ldots$ and B_n, $n = 1, 2, \ldots$ as the following determinants:

$$A_n = a_1 \begin{vmatrix} b_2 & 1 & & & & & \\ -a_3 & b_3 & 1 & & & & \\ & -a_4 & b_4 & 1 & & & \\ & & & \ddots & & & \\ & & & & -a_{n-1} & b_{n-1} & 1 \\ & & & & & -a_n & b_n \end{vmatrix} \tag{4.54}$$

and

$$B_n = \begin{vmatrix} b_1 & 1 & & & & & \\ -a_2 & b_2 & 1 & & & & \\ & -a_3 & b_3 & 1 & & & \\ & & & \ddots & & & \\ & & & & -a_{n-1} & b_{n-1} & 1 \\ & & & & & -a_n & b_n \end{vmatrix} \tag{4.55}$$

Now we are in a position to give an expression for f_n (or equivalently $P_n^*(\theta)$) in terms of a continued fraction. First of all, in (4.50) when (4.51) is used, we obtain

$$\begin{aligned} f_1 &= a_1 - b_1 f_0 = A_1 - B_1 f_0 = -(B_1 f_0 - A_1), \\ f_2 &= a_2 f_0 - b_2 f_1 = a_2 f_0 - b_2(A_1 - B_1 f_0) \\ &= B_2 f_0 - A_2. \end{aligned}$$

By iteration, it can be established that

$$(-1)^n f_n = B_n f_0 - A_n, \qquad \text{or} \qquad (-1)^n \frac{f_n}{B_n} = f_0 - \frac{A_n}{B_n}. \tag{4.56}$$

On the other hand, (4.52) with the help of (4.51) becomes

$$f_0 = \frac{a_1}{b_1 +}\ \cdots\ + \frac{a_{n+1}}{b_{n+1} + \dfrac{f_{n+1}}{f_n}} = \frac{A'_{n+1}}{B'_{n+1}} \quad \text{(say)}$$

where $A'_{n+1} = A_{n+1}$ and $B'_{n+1} = B_{n+1}$ with b_{n+1} replaced by $b_{n+1} + f_{n+1}/f_n$. When the recurrence relation (4.53) is used, we can express f_0 as

$$f_0 = \frac{A'_{n+1}}{B'_{n+1}} = \frac{A_{n+1} + \dfrac{f_{n+1}}{f_n} A_n}{B_{n+1} + \dfrac{f_{n+1}}{f_n} B_n}. \tag{4.57}$$

Letting $\alpha_r = \prod_{i=1}^{r} a_i$, it can be checked by induction that

$$A_{r+1}B_r - A_r B_{r+1} = (-1)^r \alpha_{r+1}. \tag{4.58}$$

Now

$$\begin{aligned} f_0 - \frac{A_n}{B_n} &= \frac{A_{n+1} + \dfrac{f_{n+1}}{f_n} A_n}{B_{n+1} + \dfrac{f_{n+1}}{f_n} B_n} - \frac{A_n}{B_n} \\ &= \frac{A_{n+1}B_n - A_n B_{n+1}}{B_n \left(B_{n+1} + \dfrac{f_{n+1}}{f_n} B_n \right)} \\ &= \frac{(-1)^n \alpha_{n+1}}{B_n \left(B_{n+1} + \dfrac{f_{n+1}}{f_n} B_n \right)} \qquad \text{by using (4.58)} \\ &= \frac{(-1)^n \alpha_{n+1}}{B_n B_{n+1}} + \frac{B_n^2 a_{n+2}}{b_{n+2}} + \frac{a_{n+3}}{b_{n+3}} + \cdots \end{aligned} \tag{4.59}$$

by using (4.51). Thus combining (4.56) and (4.59), we get the following continued fraction for f_r:

$$f_r = \frac{\alpha_{r+1}}{B_{r+1}} + \frac{B_r a_{r+2}}{b_{r+2}} + \frac{a_{r+3}}{b_{r+3}} + \cdots \tag{4.60}$$

Because of (4.46), this in turn yields

$$P_r^*(\theta) = \frac{(-1)^r}{M_r} f_r. \tag{4.61}$$

Let the nth convergent of (4.60) be denoted by $A_n^{(r)}/B_n^{(r)}$ and that of $P_r^*(\theta)$ be $P_{r,n}^*$. Observe that $B_n^{(r)} = B_{n+r}$. Then

$$P_{r,n}^* = \frac{(-1)^r}{M_r} \frac{A_n^{(r)}}{B_{n+r}}. \tag{4.62}$$

In our problem,

$$a_1 = 1, \quad b_1 = \lambda_0 + \theta,$$
$$a_{n+1} = -\lambda_{n-1}\mu_n \quad \text{and} \quad b_{n+1} = \lambda_n + \mu_n + \theta, \qquad n \geq 1.$$

Thus from (4.54) and (4.55), it is evident that B_{n+r} happens to be a polynomial in θ of degree $n+r$, whereas $A_n^{(r)}$ is a polynomial in θ of lower degree.

In fact, (4.62) is of the form (4.39) and therefore the inverse transform of $P^*_{r,n}$ is obtained as was done in (4.40), (4.41), (4.42) and (4.43).

In order to find the zeroes of B_{n+r}, we approach the same way as was done under the eigenvalue method. First realize that B_n in our special case is equal to the following determinant:

$$\begin{vmatrix} \lambda_0+\theta & \sqrt{\lambda_0\mu_1} & & & & \\ \sqrt{\lambda_0\mu_1} & \lambda_1+\mu_1+\theta & \sqrt{\lambda_1\mu_2} & & & \\ & \sqrt{\lambda_1\mu_2} & \lambda_2+\mu_2+\theta & \sqrt{\lambda_2\mu_3} & & \\ & & & & & \sqrt{\lambda_{n-2}\mu_{n-1}} \\ & & & & \sqrt{\lambda_{n-2}\mu_{n-1}} & \lambda_{n-1}+\mu_{n-1}+\theta \end{vmatrix}. \tag{4.63}$$

which is checked either directly or by a similarity transformation. Notice that (4.63) and (4.37) are almost the same but for the last diagonal element and that $\theta = 0$ is not a zero of (4.63). Moreover, its zeroes are the negative eigenvalues of a matrix which is real symmetric positive definite tridiagonal with nonzero subdiagonal elements. Thus the eigenvalues are real, positive and distinct. Suppose the partial fraction expansion of (4.62) is

$$P^*_{r,n} = \sum_{j=1}^{n+r} \frac{H_{r,j}}{\theta - \theta_j} \qquad r = 0, 1, \ldots$$

where $\theta_1, \ldots, \theta_{n+r}$ are zeroes of B_{n+r} (which are negative). Then its inversion is written as

$$\sum_{j=1}^{n+r} H_{r,j} e^{\theta_j t}, \qquad r = 0, 1, \ldots \tag{4.64}$$

which is an approximation of $P_r(t)$. As commented earlier, θ_i's are evaluated by using computer packages.

Remark: In the eigenvalue approach the infinite system is truncated to a finite system and its exact solution is used as an approximation. On the other hand, in the continued fraction approach the solution of the infinite system is obtained as a continued fraction which is truncated to provide an approximation.

For general i, we only state the result without proof.

$$P^*_r(\theta) = \begin{cases} \dfrac{(-1)^i}{L_{i-1}M_r} B_r f_i, & r \le i, \\ \dfrac{(-1)^r}{L_{i-1}M_r} B_i f_r, & r \ge i, \end{cases} \tag{4.65}$$

and

$$P^*_{r,n} = \begin{cases} \dfrac{(-1)^i}{L_{i-1}M_r} B_r \dfrac{A_n^{(i)}}{B_{n+r}}, & r \leq i, \\ \\ \dfrac{(-1)^r}{L_{i-1}M_r} B_i \dfrac{A_n^{(r)}}{B_{n+r}}, & r \geq i. \end{cases} \tag{4.66}$$

where $L_k = \prod_{j=0}^{k} \lambda_j$, $k = 0, 1, \ldots$ and $L_{-1} = 1$.

4.4.3. The power series method. Next we turn to a classical method of approximation. Observe that in matrix form (4.33) becomes

$$\frac{d}{dt}\mathbf{P}(t) = \mathbf{P}(t)\mathbf{Q} \tag{4.67}$$

where

$$\mathbf{P}(t) = (P_0(t), P_1(t), \ldots),$$

and

$$\mathbf{Q} = \begin{bmatrix} -\lambda_0 & \lambda_0 & & \\ \mu_1 & -(\lambda_1 + \mu_1) & \lambda_1 & \\ & \mu_2 & -(\lambda_2 + \mu_2) & \mu_2 \\ & & & \ddots \end{bmatrix}. \tag{4.68}$$

Its formal solution can be written as (see Section A.10)

$$\mathbf{P}(t) = \mathbf{P}(0)\exp(\mathbf{Q}t). \tag{4.69}$$

The Taylor series expansion gives

$$P_n(t+h) = P_n(t) + \frac{h}{1!}\frac{d}{dt}P_n(t) + \frac{h^2}{2!}\frac{d^2}{dt^2}P_n(t) + \ldots \qquad n \geq 0,$$

which again in a matrix form is

$$\mathbf{P}(t+h) = \mathbf{P}(t) + \frac{h}{1!}\frac{d}{dt}\mathbf{P}(t) + \frac{h^2}{2!}\frac{d^2}{dt^2}\mathbf{P}(t) + \ldots \tag{4.70}$$

When (4.67) is repeatedly used, (4.70) leads to

$$\mathbf{P}(t+h) = \mathbf{P}(t) + \mathbf{P}(t)\mathbf{Q}\frac{h}{1!} + \mathbf{P}(t)\mathbf{Q}^2\frac{h^2}{2!} + \tag{4.71}$$

which may be abbreviated as

$$\mathbf{P}(t+h) = \mathbf{P}(t)e^{\mathbf{Q}h}. \tag{4.72}$$

The importance of (4.71) is in its application to approximate $P_r(t)$. This is done by applying two standard techniques for numerical approximations. First, the time is discretized by letting $t = nh$. Given the initial number of customers, i.e., $\mathbf{P}(0)$, compute $\mathbf{P}(h)$ by using

$$\mathbf{P}(h) = \mathbf{P}(0) + \frac{h}{1!}\mathbf{P}(0)\mathbf{Q} + \frac{h^2}{2!}\mathbf{P}(0)\mathbf{Q}^2 + \ldots$$

and continue recursively to compute $\mathbf{P}(2h), \ldots, \mathbf{P}(nh) = \mathbf{P}(t)$. Second, for small enough h, neglect higher order terms of h in the right side of (4.71) and thus obtain an approximation. In this manner, we have

$$\mathbf{P}(nh) = \mathbf{P}((n-1)h) + \frac{h}{1!}\mathbf{P}((n-1)h)\mathbf{Q} + \ldots \\ + \frac{h^k}{k!}\mathbf{P}((n-1)h)\mathbf{Q}^k + O(h^{k+1}). \tag{4.73}$$

Realize that

$$\frac{h^j}{j!}\mathbf{Q}^j = \left(\frac{h^{j-1}\mathbf{Q}^{j-1}}{(j-1)!}\right)\mathbf{Q}\frac{h}{j} \tag{4.74}$$

which means the successive terms on the right side can be calculated recursively.

The method may be called the power series method of approximation. It is essentially a generalization of *Runge-Kutta method* (see Section A.11).

Remarks:

1. A common method of computing $e^{\mathbf{Q}h}$ is based on the spectral representation of $\mathbf{Q}$ which involves the computation of eigenvalues and eigenvectors. This is being avoided by the present procedure.
2. Just like the earlier procedures, the approximation is achieved by a truncation, in this case the truncation being of a series solution. However, it is taking place in each series for $\mathbf{P}(h), \ldots, \mathbf{P}(nh)$.
3. The technique of discretization is quite common in numerical methods. The estimation of the discretization error is difficult.
4. The method is applicable to a general infinitesimal generator matrix $\mathbf{Q}$. For computation, $\mathbf{Q}$ has to be a finite matrix and is thus truncated.

4.4.4. The randomization method. We would like to describe another numerical approach called *randomization* or *uniformization* technique. Assume that $X(t)$ is uniformizable, i.e., the diagonal elements of $\mathbf{Q}$ are uniformly bounded. Let

$$q \geq \sup_i |q_{ii}|.$$

and

$$\mathbf{P} = \frac{1}{q}\mathbf{Q} + \mathbf{I}. \tag{4.75}$$

It can be shown that $\mathbf{P}$ is a transition matrix of a Markov chain. In the randomization construction, replace $\mathbf{Q}$ in (4.69) by $\mathbf{P}$ through (4.75) and obtain as an expansion (see Section A.10)

$$\mathbf{P}(t) = \mathbf{P}(0)e^{-qt}e^{\mathbf{P}qt} = \mathbf{P}(0)e^{-qt}\sum_{n=0}^{\infty}\frac{(qt)^n}{n!}\mathbf{P}^n. \tag{4.76}$$

In particular

$$P_n(t) = e^{-qt} \sum_{j=0}^{\infty} \frac{(qt)^j}{j!} P_n^{[j]}, \tag{4.77}$$

where $P_n^{[j]}$ is the probability that the Markov chain in (4.75) is in state n at the jth step.

Remarks:

1. For the algorithmic implementation of the formula given by (4.77), one has to truncate the infinite series at some point, say m, so that for computational purpose $P_j(t)$ is approximated by

   $$\sum_{n=0}^{m} P_j(n) e^{-qt} \frac{(qt)^n}{n!}. \tag{4.78}$$

   It can be checked that

   $$P_j(t) - \sum_{n=0}^{\infty} P_j^{[n]} e^{-qt} \frac{(qt)^n}{n!} \leq 1 - e^{-qt} \sum_{n=0}^{m} \frac{(qt)^n}{n!}.$$

   Thus one can set m to bound the error due to truncation by satisfying

   $$\left[1 - e^{-qt} \sum_{n=0}^{m} \frac{(qt)^n}{n!}\right] \leq \epsilon,$$

   where ϵ is the desired amount of error.
2. The expansion of $\exp(\mathbf{P}qt)$ usually converges faster than the expansion of $\exp(\mathbf{Q}t)$. Thus for numerical computation (4.76) is preferable to (4.69).
3. Observe that $\mathbf{Q}$ has both positive and negative elements where $\mathbf{P}$ has no negative elements. Therefore, computation based on $\mathbf{P}$ is expected to have less round-off errors.
4. In certain special circumstances there are some problems in applying the algorithm. The most serious one is that it cannot be used to find the transient solutions for Markov processes with time-varying transition matrices. Also it is not an efficient one in finding the solution of so-called stiff systems, i.e., systems in which the diagonal elements vary in size considerably.

4.4.5. The Q-function method. Until now, we have discussed four numerical techniques to evaluate $P_n(t)$ in the birth-death queue. Here we turn our attention to two formulas computationally suitable for evaluating the same but just for the $M/M/1$ model. For easy reference, the classical solution (2.50) is restated below:

$$\begin{aligned} P_n(t) &= e^{-(\lambda+\mu)t}\{\rho^{\frac{n-i}{2}}I_{n-i}(2\sqrt{\lambda\mu}t)+\rho^{\frac{n-i-1}{2}}I_{n+i+1}(2\sqrt{\lambda\mu}t) \\ &\quad +(1-\rho)\rho^n\sum_{k=n+i+2}^{\infty}\rho^{-\frac{k}{2}}I_k(2\sqrt{\lambda\mu}t)\}. \end{aligned}$$

From a computational point of view, the infinite sum of modified Bessel functions may cause a large increase in central processing unit (CPU) time. With an objective to avoid these functions, alternative expressions are developed in Section 2.5.3 in Chapter 2. However, if the infinite sum is a problem, one may resort to something having a finite sum of these functions. Keeping this in mind, two formulas are presented mainly for numerical computation.

Introducing *Marcum's Q-function* defined as

$$Q(a,b)=\int_b^{\infty}\exp\left(-\frac{a^2+x^2}{2}\right)I_0(ax)x\,dx,\quad a,b\geq 0 \tag{4.79}$$

which can be shown to be

$$Q(a,b)=e^{-\frac{a^2+b^2}{2}}\sum_{k=0}^{\infty}\left(\frac{a}{b}\right)^k I_k(ab) \tag{4.80}$$

we are able to establish that

$$\begin{aligned} P_n(t) &= (1-\rho)\rho^n Q(\sqrt{2\mu t},\sqrt{2\lambda t}) \\ &\quad +e^{-(\lambda+\mu)t}\left\{\rho^{\frac{n-i}{2}}I_{n-1}(2\sqrt{\lambda\mu}t)+\rho^{\frac{n-i-1}{2}}I_{n+i+1}(2\sqrt{\lambda\mu}t)\right. \\ &\quad \left.-(1-\rho)\rho^n\sum_{k=0}^{n+i+1}\rho^{-\frac{k}{2}}I_k(2\sqrt{\lambda\mu}t)\right\}. \end{aligned} \tag{4.81}$$

Note that the infinite sum is absorbed in Q-function, leaving a finite sum in the last term. The remarkable feature of the Q-function is its easy computability at least through MATLAB.

The second formula arises from the *generalized Q-function* which is defined as

$$Q_m(a,b)=\frac{1}{a^{m-1}}\int_b^{\infty}\exp\left(-\frac{a^2+x^2}{2}\right)I_{m-1}(ax)x^m\,dx \tag{4.82}$$

for $a,b\geq 0$ and $m=0,1,\ldots$. As in the previous situation, we can show that

$$1-Q_m(a,b)=e^{-\frac{a^2+b^2}{2}}\sum_{k=m}^{\infty}\left(\frac{b}{a}\right)^k I_k(ab), \tag{4.83}$$

from which the expression for $P_n(t)$ emerges as

$$\begin{aligned} P_n(t) &= (1-\rho)\rho^n(1-Q_{n+i+2}(\sqrt{2\lambda t},\sqrt{2\mu t})) \\ &\quad +e^{-(\lambda+\mu)t}\left\{\rho^{\frac{n-i}{2}}I_{n-i}(2\sqrt{\lambda\mu}t)+\rho^{\frac{n-i-1}{2}}I_{n+i+1}(2\sqrt{\lambda\mu}t)\right\}. \end{aligned} \tag{4.84}$$

On comparing (4.81) and (4.84), observe that the first term having a Q-function and last term having a finite sum in (4.81) are replaced by the first term merely having a Q-function in (4.84). It is expected the second will be superior from a computational point of view. A word of explanation on (4.84) may be helpful. Given n and i, the lower limit of k in the last sum of classical formula is fixed to be $n+i+2$. This prompts $m=n+i+2$ to be fixed in (4.83) and m cannot be equal to 1. Therefore, the two formulas are different.

We may like to see how these new formulas (4.81) and (4.84) would fare in computability against (2.50). We may also consider expressions (2.63), (2.64) and (2.65) in Section 2.5.3. An examination of (2.63) and (2.64) shows that these two do not vary much to give a significant computational advantage of one over the other. Thus, expression (2.64) may represent them for our comparison. With the help of MATLAB 6.0 package on a Celeron 1.0 GHz, the CPU times (in seconds) for computing $P_n(t)$ for model $M/M/1$ by using (2.50), (2.63), (2.64), (4.81) and (4.84) are obtained. The computer program for each formula and the tables of CPU times to compute the distribution $\{P_n(t)\}$ at t for parameter sets $(\lambda=3,\mu=4)$, $(\lambda=9,\mu=10)$, $(\lambda=9,\mu=12)$, $(\lambda=10,\mu=10)$ and $(\lambda=6,\mu=4)$ given $i=5$ and $t=2$ and sets $(\lambda=0.75,\mu=1)$, $(\lambda=0.9,\mu=1)$, $(\lambda=1,\mu=1)$ and $(\lambda=1.5,\mu=1)$ given $i=5$ and $t=4$ are presented below. From Table 4.1, it is revealed that the computational formula (4.84) based on generalized Q-function has minimum execution time, except once.

Computer program:

```
% This program is to compute the transient probability distribution
% for the number of units in the system for M/M/1 Queueing system
% using five different algorithms:

% Algo1 : Using (2.50)
% Algo2 : Using (2.64)
% Algo3 : Using (2.65)
% Algo4 : Using (4.81)
% Algo5 : Using (4.84)

lamda = input('Enter the value of lamda :'); %rate of arrivals
mu = input('Enter the value of mu :'); %rate of service
ii = input('Enter the value of i :'); %initial no. of units
%in the system
t = input('Enter the value of t :'); %the system time

% Algo1
tic; % start monitoring CPU time.
sum = 0.0; % initialization of the sum of probabilities.
rho = lamda/mu; % the traffic intensity.
x1 = exp(-(1+rho) * mu * t);
x2 = 2*(sqrt (lamda * mu)*t);
for n=0:10000 % number of units in the system
```

```
  a1 = (rho^ ((n-ii)/2))* besseli(n-ii,x2);
  a2 = (rho^((n-ii-1)/2))* besseli(n+ii+1,x2);
  % Note : "besseli" is the Modified Bessel function.
  sum1 = 0;
  for k = (n+ii+2):10000
    b1 = ((mu/lamda)^(k/2))* besseli(k,x2);
    if b1<0.000001
      break;
    end
    sum1 = sum1 + b1;
  end
  a3 = (1 - rho) * (rho^n) * sum1;
  prob = x1 * (a1 + a2 + a3);
  sum = sum + prob;
  if prob < 0.000001
    break;
  end
  if sum >= 0.999999
    break;
  end
  fprintf('prob=%1.6f, n=%d\n', prob,n); %system probability at t
end
toc;
% stop monitoring CPU time
% and print total CPU time taken from start.
fprintf ('Algo1 : sum=%f\n', sum); % sum of the probabilities
fprintf('\n%s\n','====================================');
% Algo1 end

% Algo2
tic; % start the CPU time.
rho = lamda/mu; % the traffic intensity
a1 = 1 - rho;
a2 = (lamda + mu) * t;
a3 = exp(-a2);
a4 = (lamda * t);
a5 = (mu * t);
sum = 0.0; % initialization of the sum of probabilities.
for n=0:200
  b1 = (rho^n);
  sum1 = 0.0;
  for m = 0 : 500
    c1=(a4^m)/factorial(m);
    sum2 = 0.0;
    for k = 0 : m+n+ii
      d1 = ((m-k) * (a5^(k-1)))/factorial(k);
      if k > 170
        break;
      end
```

```
    sum2 = sum2 + d1;
  end
  c2 = c1 * sum2;
  if c1 < 0.0000001
     break;
  end
  sum1 = sum1 + c2;
end
sum3 = 0;
for r = 0:500
  e1 = (a4)^(r + n - ii);
  e2 =(2 * r + n - ii);
  e3 = (r - ii);
  if e2 > 0
    e4 = factorial(e2);
    % factorial, function program for finding factorial of n
  else
    e4 = 1;
  end
  e5 = (e1 * (( a5) ^ r )) / e4 ;
  if e2 < 0 | e2 < r
    y1 = 0;
  else
    y1 = nchoosekm (e2,r);
    % nchoosekm is a function program to compute nCx
    % this is obtained by removing the warning message
    % from nchoosek (matlab library function to compute nCx)
  end
  if e2 < 0 | e2 < e3 | e3 < 0
    y2 = 0;
  else
    y2 = nchoosekm (e2,e3);
  end
  e6 = e5 * (y1-y2);
  sum3 = sum3 + e6;
  if e5 <0.1e-10
    break;
  end
end
x1 = a1 * b1;
x2 = a3 * b1 * sum1 ;
x3 = a3 * sum3;
prob = x1 + x2 + x3 ;
sum = sum + prob;
if prob < 0.000001
  break;
end
if sum >= 0.999999
  break;
```

```
  end
  fprintf('prob=%1.6f, n=%d\n',prob,n); % system probability at t
end
toc;
% stop monitoring CPU time
% and print total CPU time taken from start.
fprintf ('Algo2 :sum=%f\n', sum); %system probability at time t
fprintf('\n%s\n','========================================');
% Algo2 end

% Algo3
tic; % start the CPU time.
sum = 0.0;
rho = lamda/mu;
x1 = 1 - rho;
v = (lamda + mu);
x2 = exp(- v * t);
for n = 0:1000 %'n' is the number of units in the system
  x3 = rho ^ n;
  sum1 = 0;
  for k = 0:500
    a1 = ((lamda*t)^k)/factorial(k);
    % factorial, function program for finding factorial of n
    sum2 = 0;
    for m = 0:k+n+ii+1
      b1 = (((k-m)*((mu*t)^(m-1)))/factorial(m));
      sum2 = sum2+b1;
    end
    U = min(n,ii);
    % min, function program for finding the minimum of n and ii
    V = max(n,ii);
    % max, function program for finding the maximum of n and ii
    b21 = ((lamda * t) ^ (k + 1));
    b22 = ((mu * t) ^ (k + V))/factorial(k);
    b3 = ((lamda * t) ^ (- (U + 1)))/factorial(k+abs(ii-n));
    b4 = ((mu * t)^(U + 1))/factorial(k+ii+n+2);
    b5 = (b3 - b4);
    b6 = a1 * sum2;
    b7 = b5 * b21 * b22;
    b8 = b6 + b7;
    sum1 = sum1 + b8;
    if abs(b7) < 1.0e-36 | abs(b5) < 1.0e-105 | abs(b8) < 1.0e-8
      break;
    end
  end
  prob = x1 * x3 + x2 * x3 * sum1;
  if prob < 0.000001
    break;
  end
```

```
  if sum >= 0.999999
    break;
  end
  sum = sum + prob;
  fprintf(' prob=%1.6f, n=%d\n',prob,n); % system probability at t
end
toc;
% stop monitoring CPU time
% and print total CPU time taken from start
fprintf('Algo3 :sum=%f\n',sum);% sum of the probabilities
fprintf('\n%s\n','=====================================');
% Algo3 end

% Algo4
tic; % start the CPU time.
sum = 0.0; % initialization of the sum of probabilities.
rho = (lamda / mu); % the traffic intensity
x1 = exp(-(lamda + mu) * t);
x2 = 2 * (sqrt(lamda * mu)) * t;
x3 = sqrt(2 * mu * t);
x4 = sqrt(2 * lamda *t);
x5 = (1 - rho) * marcumq(x3,x4);
for n = 0:10000 % e number of units in the system at time t
  a1 = x5 * (rho^n);
  b1 = (rho^((n - ii)/2)) * besseli(n - ii,x2);
  b2 = (rho^((n - ii - 1)/2)) * besseli(n + ii + 1,x2);
  % Note : "marcumq" is the Marcum Q-function
  % Note : "besseli" is the Modified Bessel function.
  sum1 = 0.0;
  for k = 0 : n+ii+1
    c1 = (rho^(- k/2))* besseli(k,x2);
    sum1 = sum1 + c1;
  end
  b3 = (1 - rho) * (rho^n) * sum1;
  a2 = x1 * (b1 + b2 - b3);
  prob = a1 + a2;
  sum = sum + prob;
  if prob < 0.000001
    break;
  end
  if sum >= 0.999999
    break;
  end
fprintf('prob=%1.6f , n=%d\n',prob,n); %system probability at t
end
toc;
% stop monitoring CPU time
% and print total CPU time taken from start.
fprintf ('Algo4 : sum=%f\n', sum); % sum of the probabilities
```

```
fprintf('\n%s\n','=======================================');
% Algo4 end

% Algo5
tic; % start the CPU time.
sum = 0.0; % initialization of the sum of probabilities.
rho = (lamda / mu); % the traffic intensity
x1 = exp(-(1 + rho) * mu * t);
x2 = sqrt(2 * lamda * t);
x3 = sqrt(2 * mu * t);
x4 = 2*(sqrt(rho)) * mu * t;
for n = 0:10000 % number of units in the system
  a1 = (rho^((n-ii)/2))*besseli(n-ii,x4);
  a2 = (rho^((n-ii-1)/2))*besseli(n+ii+1,x4);
  % Note : "besseli" is the Modified Bessel function
  b1 = x1 * (a1 + a2);
  b2 = (1 - rho) * (rho^n) * (1 - (marcumq(x2,x3,n + ii + 2)));
  % Note : "marcumq" is the Marcum Q-function
  prob = b1 + b2;
  sum = sum+prob;
  if prob < 0.000001
    break;
  end
  if sum >= 0.999999
    break;
  end
  fprintf (' prob=%1.6f, n=%4d \n',prob,n); %system probability at t
end
toc;
% stop monitoring CPU time and
% print total CPU time taken from start.
fprintf ('Algo5 : sum=%f\n', sum); % sum of the probabilities
% Algo5 end
```

4.5. Discussion

With a desire to tackle some real life problems leading to line-ups and congestion which bear the generic name 'queues,' we pass through an era of steady progress on the theoretical aspects of the subject. In Chapters 2 and 3, most of the core developments are recorded. Structurally, a model is assumed over the dynamics of the queueing system and the characteristics and the performance measures are studied as consequences of the model assumptions. At a first glance of the results, one may have the satisfaction that there is the availability of solutions to the posed problems, provided the assumed models are close to the real situation. Yes, it has been so, for example, when it is revealed in Chapter 2 that the steady-state behaviour of the number of units in the system in the $M/M/1$ model is geometric – this is explicit. Yet there is an apparent deception. Many of them are not quite so; these are not in the

TABLE 4.1. Comparison of CPU times (in seconds) for computing transient probability distribution of $M/M/1$ queue, using different formulas

Case $i = 5, t = 2$

(λ, μ)	$(3, 4)$	$(9, 10)$	$(9, 12)$	$(10, 10)$	$(6, 4)$
Algo1	0.2800	0.4900	0.5500	0.5410	0.2400
Algo2	1.3520	7.1400	6.3690	9.3130	4.4260
Algo3	1.8830	6.3190	6.0890	8.2920	5.1570
Algo4	0.2100	0.2810	0.3110	0.3910	0.2810
Algo5	0.1100	0.2800	0.2700	0.3200	0.2200

Case $i = 5, t = 2$

(λ, μ)	$(0.75, 1)$	$(0.9, 1)$	$(1, 1)$	$(1.5, 1)$
Algo1	0.0900	0.1300	0.1310	0.1150
Algo2	0.6310	0.8010	0.9010	1.6830
Algo3	0.8610	1.0110	1.1720	1.7920
Algo4	0.1000	0.1310	0.1800	0.1650
Algo5	0.0800	0.1500	0.0900	0.0950

Algo1: Using the expression (2.50)
Algo2: Using the expression (2.64)
Algo3: Using the expression (2.65)
Algo4: Using the expression (4.81)
Algo5: Using the expression (4.84)

Note: The above computations are done using the MATLAB 6.1 package on a Celeron 1.0 GHz.

form, convenient for application by the practitioners. In this chapter, we have discussed several techniques in an attempt to close the gap between theory and practice.

Among "not-quite-so-explicit" solutions, there is one class in which the solution, for example in (3.30), is a function of a not explicitly known root (sometimes roots) of a polynomial equation. It arises due to the commonly used transform techniques to arrive at steady-state distributions. So, in the absence of an explicitly expressed root, we obtain it numerically for which some numerical root finding techniques are described in Section 4.2. In fact, M. L. Chaudhry (1993) has refined the techniques to develop the QROOT Software Package for computer implementation. It is also possible to find roots by using other packages, such as MATLAB. However, QROOT Software Package is more efficient with higher accuracy, even when the number of roots is very large. We have also discussed the steady-state solution of a special class of Markov chain models. Interestingly, by the use of the factorization method, the solution happens to involve roots which may be obtainable by these techniques. Yet, the steady-state solution of finite Markov chain models are numerically derivable without involving roots and this is achieved by the state reduction method. The method is simply the Gaussian method of elimination of variables one by one, in a system of linear equations. Although, an explicit set of solutions for steady-state probabilities can be readily available by applying Cramer's rule but their computation may not be as efficient as the method of state reduction since the former involves evaluating determinants. The state reduction methods is also applicable to infinite Markov chains under some restriction. It has been tested for large N up to 4900 states (see Grassmann (1990)). Other computational approaches for obtaining steady-state solutions are presented in Chapter 8.

Another type of hurdle the practitioner can confront with is in the computability of an explicit solution, for example, the transient solution (2.50) of the $M/M/1$ model. This formula has an infinite sum of modified Bessel functions. Thinking that these functions are problem makers, expressions avoiding them are given in Section 2.5.3. Yet, as an alternative way out, we have presented in Section 4.4, four numerical techniques to compute the transient solutions of the birth-death model of which $M/M/1$ is a special case. Remembering the remarks at the end of Section 4.4.4 that the randomization method is usually better than the power series method, it is instructive to compare the remaining three methods. Also in the randomization method, that the solution is in terms of $P_j^{[n]}$, the nth step transition probability (see 4.77) of a Markov chain which becomes a random walk in a queueing process implies that some combinatorial methods may be useful to determine $P_j^{[n]}$. This is discussed in Section 6.3.

If we have a close look at the four methods in Section 4.4, we may see the starting point is the system differential equation (4.67) whose solution is (4.69). The method of treatment is algebraic in contrast to the usual transform method in the last two chapters. The task is to compute it, at least numerically. In the power series method, the time is discretized and a recursive procedure has been applied to compute $P_n(t)$ starting from $t = 0$, through the usual power series expansion of $\exp(\mathbf{Q}t)$. Noting that the procedure has several stages of approximations, one replaces $\mathbf{Q}$ by $\mathbf{P}$ in the randomization method, by using the relation (4.75) only if $\mathbf{Q}$ satisfies the uniformization condition and uses the same type of series expansion for $\exp(\mathbf{P}qt)$. It has been seen that this approach improves the approximation. As our development suggests, these two approaches apply not only to the birth-death process but also to a general class of Markov processes. Coming back to the first two methods, the tridiagonal nature of $\mathbf{Q}$ and therefore the assumption of birth-death process is essential and the assumption of finiteness of the matrix is necessary in order to obtain a solution. In both cases, we have truncated the matrix and in addition, in the first case have modified the bottom-right element. Furthermore, note that in the continued fraction method, we could have started from (4.63) by avoiding the preceding discussion except that we see the motivation for arriving at (4.63). Another point to note is that we have used L.T. whose solutions involve eigenvalues which are distinct due to

$\mathbf{Q}$ being tridiagonal. For an interested reader, the study of this type is done under spectral methods for which we suggest Ledermann and Reuter (1954). These approaches may be termed as the *algebraic method.* A general reference on the topic of numerical transient analysis of Markov models is Reibman and Trivedi (1988).

Let us now digress a little bit. Consider the eigenvalue method in Section 4.4.1 and apply it to the $M/M/1$ model with system capacity N. Let i be the number of customers initially waiting. In this particular case, we will be able to express the eigenvalues explicitly as

$$\begin{aligned} z_j &= 2\sqrt{\lambda\mu}\cos\frac{j\pi}{N+1} - (\lambda+\mu), \quad j = 1, \ldots, N, \text{ and} \\ z_0 &= 0. \end{aligned} \tag{4.85}$$

By using Remark 3 at the end of Section 4.4.1, it can be shown that (see Takács (1962), Chapter 1 for details) for $\rho \neq 1$,

$$\begin{aligned} &P_n(t) = \\ &= \frac{1-\rho}{1-\rho^{N+1}}\rho^n + \frac{2}{N+1}\sum_{j=1}^{N}\left[\frac{\rho^{\frac{n-i}{2}}\exp(-(\lambda+\mu)t + 2t\sqrt{\lambda\mu}\cos\frac{j\pi}{N+1})}{1+\rho-2\sqrt{\rho}\cos\frac{j\pi}{N+1}}\right] \\ &\quad \times\left[\sin\frac{ij\pi}{N+1} - \rho^{\frac{1}{2}}\sin\frac{(i+1)j\pi}{N+1}\right]\left[\sin\frac{nj\pi}{N+1} - \rho^{\frac{1}{2}}\sin\frac{(n+1)j\pi}{N+1}\right], \end{aligned} \tag{4.86}$$

and for $\rho = 1$,

$$\begin{aligned} P_n(t) &= \frac{1}{N+1} + \frac{1}{N+1}\sum_{j=1}^{N}\left[\frac{\exp(-2\lambda t + 2\lambda t\cos\frac{j\pi}{N+1})}{1-\cos\frac{j\pi}{N+1}}\right] \\ &\quad \times\left[\sin\frac{ij\pi}{N+1} - \sin\frac{(i+1)j\pi}{N+1}\right]\left[\sin\frac{nj\pi}{N+1} - \sin\frac{(n+1)j\pi}{N+1}\right]. \end{aligned} \tag{4.87}$$

Let $N \to \infty$ which means there is no limit on the system capacity. Then the transient solution to the $M/M/1$ model by this approach gives rise to

$$P_n(t) = p_n(t) + \begin{cases} (1-\rho)\rho^n & \text{if } \rho < 1, \\ 0 & \text{if } \rho \geq 1, \end{cases} \tag{4.88}$$

where

$$\begin{aligned} p_n(t) &= \frac{2}{\pi}e^{-(\lambda+\mu)t}\rho^{\frac{n-1}{2}}\int_0^{\pi}\left[\frac{\exp(2\sqrt{\lambda\mu}t\cos y)}{1+\rho-2\rho^{\frac{1}{2}}\cos y}\right] \\ &\quad \times[\sin iy - \rho^{\frac{1}{2}}\sin(i+1)y][\sin ny - \rho^{\frac{1}{2}}\sin(n+1)y]dy. \end{aligned} \tag{4.89}$$

This formula could have a computation advantage over others for obtaining numerical results because of its trigonometric integral representation rather than power series expression. Furthermore, the expected system size has also a tractable integral representation. Table 4.2 below shows CPU times (in seconds) for computation of transient probability distribution for $M/M/1$ model by using (4.86)/(4.87) which has trigonometric functions and (2.50) which has Bessel functions. It seems by using (2.50), CPU time is more if ρ is small and decreases to be better as ρ approaches 1. The Takács expression fails to produce any result when the value of $\rho > 1$. When it gives a better result, CPU times are recorded in bold letters in these tables. Sharma and Tarabia (2000) provide an alternative expression involving recurrence relations and binomial coefficients. Apparently it does not have computational advantages against the formula using Bessel functions.

While we dwell on the computability of transient solution of the model $M/M/1$, we refer to another approach in Section 4.4.5. (Not again, the same old model!) The major consideration here is to avoid the infinite sum of modified Bessel functions, a cause for concern from a computation point of view. This is achieved by introducing the so-called Q-function and it seems there is an improvement. Again it is claimed that this approach is more efficient. It is worthwhile to compare all these formulas and we have done that for a few in Section 4.4.5.

A final remark on the transient behavior of the $M/M/1$ is to refer to Abate and Whitt (1989) who have noted the importance of integral representations, more particularly the trigonometric integral representations which they found to be superior for obtaining numerical results. Morse (1955) suggested this representation in queues for the first time.

We conclude by mentioning that the chapter complements well to the previous ones. A reader should feel somewhat comfortable that there are at least some tools that can be fiddled around in the presence of a computer and that too with a great deal of satisfaction.

On references, for root finding methods in Section 4.2, see Conte and Deboor (1972), Baiere and Chaudhry (1987), Chaudhry (1993), Powell (1981, 1985), Grassmann and Chaudhry (1982) and Grassmann (1985). For Section 4.3, see Grassmann, Taksar and Heyman (1985), Grassmann and Heyman (1993) and Grassmann and Stanford (2000). One may refer to Sharma (1990), and Mohanty, Montazer-Haghighi and Trueblood (1993) for Section 4.4.1, to Murphy and O'Donohue (1975) for Section 4.4.2, to Grassmann (1977a, 1977b) and Gross and Miller (1984) for Section 4.4.4 and to Marcum (1960), Jones, Cavin and Johnston (1980), Cantrell (1986) and Cantrell and Ojha (1987) for Section 4.4.5. The material in Section 4.4.3 is classical.

TABLE 4.2. CPU times (in seconds) for the computing transient probability distribution by using trigonometric and Bessel functions

$i = 5$ and $t = 2$

	$\lambda = 3$ $\mu = 4$	$\lambda = 3$ $\mu = 5$	$\lambda = 3$ $\mu = 6$	$\lambda = 3$ $\mu = 7$	$\lambda = 3$ $\mu = 8$
Takács	2.333	1.733	1.522	0.872	0.711
Bessel	0.150	0.200	0.201	0.270	0.341

$i = 5$ and $t = 2$

	$\lambda = 3$ $\mu = 9$	$\lambda = 3$ $\mu = 10$	$\lambda = 3$ $\mu = 11$	$\lambda = 3$ $\mu = 12$
Takács	0.550	0.521	**0.371**	**0.210**
Bessel	0.321	0.510	**0.601**	**0.371**

$i = 5$ and $t = 2$

	$\lambda = 9$ $\mu = 9$	$\lambda = 9$ $\mu = 10$	$\lambda = 9$ $\mu = 11$	$\lambda = 9$ $\mu = 12$
Takács	5.819	5.097	**4.036**	**3.795**
Bessel	3.064	4.086	**4.897**	**4.417**

$i = 5$ and $t = 2$

	$\lambda = 10$ $\mu = 10$	$\lambda = 10$ $\mu = 11$	$\lambda = 10$ $\mu = 12$	$\lambda = 10$ $\mu = 13$	$\lambda = 10$ $\mu = 14$
Takács	6.369	**5.408**	**4.356**	**4.126**	**3.615**
Bessel	4.596	**5.458**	**6.660**	**7.130**	**7.311**

$\lambda = 0.75, i = 5$ and $t = 4$

	$\mu = 1$	$\mu = 2$	$\mu = 3$	$\mu = 4$	$\mu = 5$	$\mu = 6$	$\mu = 7$	$\mu = 8$
Takács	1.412	0.430	0.350	0.201	0.180	0.130	**0.100**	**0.09**
Bessel	0.020	0.101	0.100	0.120	0.100	0.100	**0.120**	**0.161**

$\lambda = 0.9, i = 5$ and $t = 4$

	$\mu = 1$	$\mu = 2$	$\mu = 3$	$\mu = 4$	$\mu = 5$	$\mu = 6$	$\mu = 7$	$\mu = 8$
Takács	2.263	0.781	0.391	0.230	0.170	0.120	**0.121**	**0.100**
Bessel	0.050	0.070	0.100	0.101	0.130	0.110	**0.150**	**0.181**

$\lambda = 1, i = 5$ and $t = 4$

	$\mu = 1$	$\mu = 2$	$\mu = 3$	$\mu = 4$	$\mu = 5$	$\mu = 6$	$\mu = 7$	$\mu = 8$
Takács	2.283	1.432	0.671	0.280	0.230	0.231	**0.190**	**0.170**
Bessel	0.040	0.130	0.161	0.131	0.180	0.230	**0.311**	**0.270**

Note: Takács: Using formula (4.86)/(4.87); Bessel: Using (2.50).

4.6. Exercises

1. In the queueing model $M/E_{10}^{a,50}/1$, determine the average queue length and average waiting time by numerically computing r_i^* (see (3.51)), $i = 1, \ldots, 10$. (Hint: use a computer package.)

2. Consider the $G^h/M/1$ queueing system. Specifically, let the inter-arrival time distribution be deterministic and let $h = 2$, $E(T) = 1$ and $E(S) = 0.4$. Hence the number of customers served between two successive arrivals has a Poisson distribution with parameter 2.5. Determine the steady-state distribution using the factorization method.

3. Check (4.14), (4.15) and (4.16).

4. Show that
$$\sum_{j=1}^{h} B_j < 1,$$
where B_j's are given by (4.17).

5. For a finite number of states in a Markov chain with transition matrix $\mathbf{P}$, let us assume that $\mathbf{P}$ is banded, that is there are positive numbers g and h such that $p_{ij} = 0$ whenever $j - i < g$ or $j - i > h$ and g and h are the minimum among such values. Show that under the state reduction procedure, $p_{ij}^{(n)}$ inherits this property (see Grassmann, Taksar and Heyman (1985)).

6. Consider the same model as in Exercise 10 in Chapter 2 with system capacity N, when no reneging is allowed. Note that in this situation, a customer may balk if on arrival finds all servers are busy but waiting space is not full. Using the eigenvalue method, find that distribution of the number of customers in the system at time t.

7. Consider the birth-death model with the state space $\{0, 1, \ldots, N\}$. Using the eigenvalue method, show that the distribution of the length of a busy period is hyperexponential. (See Mohanty, Montazer-Haghighi and Trueblood (1993).)

8. Consider the taxi stand problem as given in Exercise 6, Chapter 1. This is called a double-ended queueing system, since there can be a negative state, say $-k$, when there are k taxis waiting for customers. Let the state space $S = (-M, -M+1, \ldots, 0, 1, \ldots, N)$ be bounded on both sides.
   (i) Write the set of differential-difference equations for the system.
   (ii) Obtain the transient solution.
   (iii) Show that the steady-state distribution is given by

$$\begin{aligned} P_n &= \tfrac{(1-\rho)\rho^{M+n}}{1-\rho^{M+N+1}} \quad \text{for } -M \leq n \leq N,\ \rho \neq 1, \\ &= \tfrac{1}{M+N+1} \quad \text{for } -M \leq n \leq N,\ \rho = 1. \end{aligned}$$

(See Sharma (1990).)

9. Prove (4.63).

10. Consider the same model as in Exercise 10 in Chapter 2, when no reneging is allowed. For $i = 1$, $\lambda = 2$, $\mu = 2$, $c = 3$ and $p = 0.5$, find the transient distribution $\{P_n(10)\}$ numerically by using the eigenvalue method, continued fraction method and randomization method. Compare performance of these three methods.

11. Check (4.76).

References

Abate, J. and Whitt, W. (1989). Calculating time-dependent measures for the $M/M/1$ queue, *IEEE Trans. Commun.*, **37**, 1102–1104.

Baiere, G. and Chaudhry, M. L. (1987). Computational analysis of single-server bulk arrival queue $G_I^K/M/1$, *Queueing Systems*, **2**, 173–185.

Cantrell, P. E. (1986). Computation of the transient $M/M/1$ queue cdf, pdf, and mean with generalized Q-functions, *IEEE Trans. Commun.*, **34**, 814–817.

Cantrell, P. E. and Ojha, A. K. (1987). Comparison of generalized Q-function algorithms, *IEEE Trans. Inform. Theory*. **3**, 591–596.

Chaudhry, M. L. (1993). *QROOT-Software Package*, A & A Publications, Kingston, Ontario.

Conte, S. D. and Deboor, C. (1972). *ElementaryNumerical Analysis*, McGraw-Hill, New York.

Grassmann, W. K. (1977a). Transient solutions in Markovian queueing systems, *Comput. Oper. Res.*, **4**, 47–53.

Grassmann, W. K. (1977b). Transient solutions in Markovian queues, *Europ. J. Oper. Res.*, **1**, 396–402.

Grassmann, W. K. and Chaudhry, M. L. (1982). A new method to solve steady-state queueing equations, *Naval Res. Log. Quart.*, **29**, 461–473.

Grassmann, W. K. (1985). The factorization of queueing equations and their interpretation, *J. Operational Res. Soc.*, **36**, 1041–1050.

Grassmann, W. K. (1990). Computational methods in probability theory, *Handbooks in OR and MS, Vol. 2*, Heyman, D.P. and Sobel, M.J., Eds., North-Holland, Amsterdam, Chapter 5, 199–254.

Grassmann, W. K. and Heyman, D. P. (1993). Computation of steady-state probabilities for infinite-state Markov chains with repeating rows, *ORSA J. Computing*, **5**, 292–303.

Grassmann, W. K., Taksar, M. I. and Heyman, D. R. (1985). Regenerative analysis and steady state distributions for Markov chains, *Oper. Res.*, **33**, 1107–1116.

Grassmann, W. K. and Stanford, D. A. (2000). Matrix analytic methods, *Computational Probability*, Grassmann W.K., Ed., Kluwer Academic Publisher, Boston, Chapter 6, 153–202.

Gross, D. and Miller, D. R. (1984). The randomization technique as a modeling tool and solution procedure for transient Markov processes, *Oper. Res.*, **32**, 343–361.

Jones, S. K., Cavin, R. K. and Johnston, D. A. (1980). An efficient computational procedure for the evaluation of the $M/M/1$ transient state occupancy probabilities, *IEEE Trans. Commun.*, **28**, 2019–2020.

Lederman, W. and Reuter, G. E. H. (1954). Spectral theory for the differential equations of simple birth and death processes, *Philos. Trans. Roy. Soc. London Ser. A*, **246**, 321–369.

Marcum, J. I. (1960). A statistical theory of target detection by pulsed radar, *IRE Trans. Inform. Theory*, **6**, 59–267.

Mohanty, S. G., Montazer-Haghighi, A. and Trueblood, R. (1993). On the transient behavior of a finite birth-death process with an application, *Computers Oper. Res.*, **20**, 239–248.

Morse, P. M. (1955). Stochastic properties of waiting lines, *Oper. Res.*, **3**, 255–261.

Murphy, J. A. and O'Donohue, M. R. (1975). Some properties of continued fractions with applications in Markov processes, *J. Inst. Maths. Applics.*, **16**, 57–71.

Powell, W. B. (1981). Stochastic delays in transportation terminals: New results in the theory and applications of bulk queues. Ph.D. Dissertation, Dept. of Civil Engineering, MIT, Cambridge.

Powell, W. B. (1985). Analysis of vehicle holding and cancellation strategies in bulk arrival, bulk service queues, *Trans. Sci.*, **19**, 352–377.

Reibman, A. and Trivedi, K. (1988). Numerical transient analysis of Markov models, *Compute. Oper. Res.*, **15**, 19–36.

Sharma, O. P. (1990). *Markovian Queues*, Ellis Horwood, London.

Sharma, O. P. and Tarabia, A. M. K. (2000). A simple transient analysis of an M/M/1/N queue, *Sankhya*, Series A, **62**, 273–281.

Takács, L. (1962). *Introduction to the Theory of Queues*, Oxford University Press, New York.

CHAPTER 5

Statistical Inference and Simulation

5.1. Introduction

As the title suggests, this chapter consists of two topics, namely, statistical inference and simulation. Seemingly they are disjoint, but soon we may see that it is not so.

Invariably in all theoretical developments, model assumptions are made, such as the interarrival times are i.i.d. exponential(λ) variables. It is necessary to check the validity of exponentiality and to estimate λ, if unknown. This brings us to the field of statistical inference which is the theme of Section 5.2.

In the absence of theoretically possible solutions, simulation of models with the aid of computer may provide an understanding of the system characteristics. Once we have an accepted model, we generate a large number of samples from the model by using simulation methods. These samples provide estimates of desired queueing characteristics. Thus statistical tools are in fact needed to produce those estimates. In Section 5.3, this topic is discussed.

In Section 5.4, a practical application is included just to illustrate the use of some of the results developed so far.

For this chapter, one is advised to refer to Section A.9 in Appendix A.

5.2. Statistical Inference

So far, we have presented theoretical and computational aspects of queues. The first one is to deductively derive results on queueing characteristics from assumed models and in the second, the computability questions of derived results and numerical techniques are examined. In this section, a third aspect, namely, verification of model assumptions and related inferences, is addressed. It comprises of estimation of parameters in a given model, such as λ and μ, and testing of hypotheses, such as whether the interarrival times have an exponential distribution.

To begin with, let us assume the $M/M/1$ model for which we have to estimate λ and μ. This means that both interarrival times and service-times are exponential. Suppose a random sample of size n is taken from the exponential(θ) distribution with p.f. $\theta \exp(-\theta t)$, t being the time. This is a collection of n i.i.d. r.v.'s from exponential(θ). We want to estimate θ by the *maximum likelihood method* (see Section A.9). Let $t_1, \ldots, t_n$ be the random sample of observations. Here, t_i refers to the ith observed duration, for example, the ith interarrival time. Given $(t_1, \ldots, t_n)$, the likelihood function $L(\theta)$ at θ is given by

$$L(\theta) = \prod_{i=1}^{n} \theta e^{-\theta t_i} = \theta^n e^{-(t_1+\cdots+t_n)\theta}, \quad \theta > 0. \tag{5.1}$$

The maximum likelihood estimate (MLE in brief) $\hat{\theta}$ of θ is that value of θ which maximizes $L(\theta)$ or equivalently $\ln L(\theta)$. It is often convenient to work with $\ln L(\theta)$, which in our case is

$$\ln L(\theta) = n \ln \theta - \theta \sum_{j=1}^{n} t_j.$$

For maximization, we consider

$$\frac{\partial \ln L}{\partial \theta} = \frac{n}{\theta} - \sum_{j=1}^{n} t_j = 0.$$

Its only solution can be checked to maximize $\ln L(\theta)$ by using the second derivative criterion and thus the MLE $\hat{\theta}$ is given by

$$\hat{\theta} = \frac{n}{\sum_{j=1}^{n} t_j} = \frac{1}{\bar{t}}, \tag{5.2}$$

where $\bar{t}$ is the mean of $t_1, \ldots, t_n$. The mean of the distribution being $1/\theta$, its MLE is

$$\frac{1}{\hat{\theta}} = \bar{t}.$$

In the above example, the basic assumption is that $t_1, \ldots, t_n$ are independently observable. But can this be done in any queueing situation? Thus a primary concern is the procedure of data collection under a specific queueing structure. The question of what data can be collected and what should be collected so that statistical methods could be applicable, needs a preliminary investigation before the sampling plan for collection of data is formulated.

One convenient way is to collect data in a given time interval. Consider estimating λ where λ is the parameter of interarrival times and fix the duration to be $(0, T]$. Let there be n arrivals at time epochs $0 < t_1 < \cdots < t_n < T$. The contribution of $t_i - t_{i-1}$ (set $t_0 = 0$) to the likelihood function is $\lambda \exp(-\lambda(t_i - t_{i-1}))$, $i = 1, \ldots, n$ and of $T - t_n$ is $\exp(-\lambda(T - t_n))$, which is the probability that no event has occurred during that period. Thus the likelihood function has the expression,

$$L(\lambda) = e^{-\lambda(T-t_n)} \prod_{i=1}^{n} (\lambda e^{-\lambda(t_i - t_{i-1})}) = \lambda^n e^{-\lambda T}. \tag{5.3}$$

It can be checked that the MLE of λ is

$$\hat{\lambda} = \frac{n}{T}, \tag{5.4}$$

where n is the observed value of the random variable, the number of arrivals during $(0, T]$. This random variable, of course, has the Poisson distribution with mean and variance λT. Therefore, the mean and variance of λ are given by

$$E(\hat{\lambda}) = \lambda \quad \text{and} \quad V(\hat{\lambda}) = \frac{\lambda}{T} \tag{5.5}$$

implying that $\hat{\lambda}$ is an unbiased estimate of λ. Remember that an MLE is not necessarily unbiased.

Remark: Instead of observing for a fixed duration, suppose the process is observed until a fixed number of n arrivals occur. It can be seen that the MLE does not change. However, the unbiased estimate of λ is

$$\tilde{\lambda} = \frac{(n-1)\hat{\lambda}}{n} \quad \text{and its variance is} \quad V(\tilde{\lambda}) = \frac{\lambda^2}{n-2}, \; n > 2. \tag{5.6}$$

Again assume that the model is $M/M/1$ and we want to estimate both λ and μ. The system is observed for a fixed length of time t, which is sufficiently long so that the time spent in busy state is a preassigned value t_b. Let n_0 represent the observed initial queue length and n_a and n_s, respectively, the observed number of arrivals and number of service completions, during $(0, t]$. The likelihood function conditioned by these observations will have contributions from three parts. The contribution from arrivals to the likelihood function is $\lambda^{n_a} e^{-\lambda t}$. Here and elsewhere, we ignore factors independent of parameters since these do not affect the estimate. Similarly, we get the contribution from service completion times as $\mu^{n_s} e^{-\mu t_b}$. The contribution from the initial queue length has to be ascertained. Under the assumption that the system is in steady state, having been observed for a sufficiently long time, the probability that there are initially n_0 in the system is $(1-\rho)\rho^{n_0}$, $\rho < 1$, by (2.7). Finally, combining various contributions, we get the likelihood function as

$$L(\lambda, \mu) = \rho^{n_0}(1-\rho)\lambda^{n_a}\mu^{n_s}e^{-(\lambda t + \mu t_b)} \tag{5.7}$$

which is also expressed as

$$L(\lambda, \mu) = \rho^{n_0}(1-\rho)\lambda^{n_a}e^{-\lambda t_e}\mu^{n_s}e^{-(\lambda+\mu)t_b} \tag{5.8}$$

where t_e is the time spent in empty state. Differentiate $\ln L$ with respect to λ and μ and put them equal to zero. This yields

$$\lambda = (\mu - \lambda)(n_a + n_s - \lambda t) \text{ and } \mu = (\lambda - \mu)(n_s - n_a - \mu t_b). \tag{5.9}$$

Substituting the value of μ from the second in the first gives a quadratic equation in λ. Any negative solution is discarded and for each of the remaining values of λ, corresponding value of μ is obtained. Any pair (λ, μ) which does not satisfy $\mu > 0$ and $\rho < 1$ is rejected. If both solutions are valid, then the one which maximizes the likelihood function is the MLE $(\hat{\lambda}, \hat{\mu})$.

For large $n_s - n_0$, a simple approximation for $\hat{\lambda}$ and $\hat{\mu}$ can be

$$\hat{\lambda} = \frac{n_a + n_0}{t} \quad \text{and} \quad \hat{\mu} = \frac{n_s - n_0}{t_b}, \tag{5.10}$$

respectively. If, however, we ignore the initial queue size, which may be done when the observations are from a very long realization, then the estimates are

$$\hat{\lambda} = \frac{n_a}{t} \quad \text{and} \quad \hat{\mu} = \frac{n_s}{t_b}. \tag{5.11}$$

The above estimation procedure can be extended to the birth-death queue. Let n_{a_i}, n_{s_i} and t_i denote, respectively, the observed values of number of arrivals, number of service completions and the time spent in state i, $i = 0, 1, \ldots$ during the observational period $(0, t]$. Conditional upon these observations, the likelihood function is found to be

$$L = e^{-\Sigma(\lambda_i + \mu_i)t_i} \prod \lambda_i^{n_{a_i}} \mu_i^{n_{s_i}}, \tag{5.12}$$

in which the initial queue length is ignored. Observe the expression is similar to (5.7). For a finite state birth-death queue with $i = 0, 1, \ldots, M$, the MLEs are seen to be

$$\hat{\lambda}_i = \frac{n_{a_i}}{t_i}, \; i = 0, 1, \ldots, M-1 \text{ and } \hat{\mu}_i = \frac{n_{s_i}}{t_i}, \; i = 1, \ldots, M. \tag{5.13}$$

Now let us deal with the $M/G/1$ model and use a similar argument as before. Suppose we are taking observations during $(0, t]$. Recall that F_S represents the distribution function of the service-time S. Assume its p.f. f_S exists. As earlier, we have two components of L, if the initial queue size is ignored. The contribution from the arrivals is exactly the same as in (5.3) and from the service-times it is

$$\left(\prod_{i=1}^{n_s} f_S(s_i)\right)(1 - F(s^*)). \tag{5.14}$$

where s^* is the elapsed time from the epoch of last service completion till t. Therefore,

$$L = \lambda^{n_a} e^{-\lambda t} \left(\prod_{i=1}^{n_s} f_S(s_i)\right)(1 - F_S(s^*)) \tag{5.15}$$

leading to the estimate of λ as given in (5.11). When F_S is specified we may be able to estimate its parameters in a usual manner.

What about collecting data at departure points as induced by the imbedded Markov chain on $M/G/1$? In other words, we gather the number of arrivals during successive service-times. Suppose G is Erlangian with k phases. Then the number of arrivals during service intervals forms a sequence of i.i.d. random variables. If X_n denotes the number of arrivals during the service of C_n, then X_n has a negative binomial distribution with parameters k and ρ and is given by

$$P(X_n = x) = \binom{x+k-1}{x} \left(\frac{\rho}{\rho+k}\right)^x \left(\frac{k}{\rho+k}\right)^k, \quad x = 0, 1, \ldots. \tag{5.16}$$

Suppose our sample consists of $x_1, \ldots, x_n$, the number of arrivals during first n service-times. Then it can be shown that the MLE $\hat{\rho} = \Sigma x_i/n$.

Let us turn to the problem of testing of hypotheses. Because of our interest to know whether a particular distribution rightly describes the actual probabilistic nature of the interarrival times or service-times, we focus only on those types of hypotheses testing situations – testing probabilistic models. Otherwise, this subject extensively deals with testing problems related to parameters of distributions.

Before we apply some testing procedures to the data, it is advisable to start with graphical methods, just to adhere to the adage that a picture speaks more than a thousand words can do. A simple graphical method is known as a quantile-quantile plotting (briefly $Q-Q$ plotting). To describe it, let there be an assumed continuous distribution to be checked for its correctness, which is called the null distribution in statistical language. Suppose F_0 is the assumed distribution function. Then x is called the *population quantile* of order p if $F_0(x) = p$. For the observed sample $x_1, \ldots, x_n$ of size n, denote by y_k the kth ordered observation, $k = 1, \ldots, n$. We define the *empirical distribution function* F_n of the sample as

$$F_n(x) = \begin{cases} 0 & \text{for} \quad x < y_1, \\ \frac{k}{n} & \text{for} \quad y_k \leq x < y_{k+1}, \; k = 1, \ldots, n-1, \\ 1 & \text{for} \quad y_n \leq x. \end{cases} \tag{5.17}$$

By analogy, y_k represents the *sample quantile* of order k/n. Thus assuming F_0 to be the true distribution function, we expect y_k should be close to the population quantile $F_0^{-1}(k/n)$. In other words, if y_k's are plotted against $F_0^{-1}(k/n)$, $k = 1, \ldots, n$, then these points are expected to be on the line $x = y$. We use $(k - 0.5)/n$ instead of k/n in order to correct it for discontinuity of the empirical distribution function. A *Q-Q plot* is a diagram of plotting $(y_k, F_0^{-1}((k - 0.5)/n))$. Computation of F_0^{-1} depends on parameters of F_0, which are usually unknown. However, there is a way out at least in some cases. For example, in case of normal distribution, consider the standardized normal variable (see Section A.4) which is linear in the original normal variable. Therefore, if we plot y_k's against $\tilde{F}_0^{-1}((k-0.5)/n)$, $k = 1, \ldots, n$, where $\tilde{F}_0$ is the distribution function of the standardized normal variable, we expect them to lie on a straight line. Note that $\tilde{F}_0^{-1}$ does not depend on any unknown parameters and is tabulated. Similarly, if X is the exponential(μ) r.v., then μX is exponential(1) and therefore its distribution function is independent of the parameter μ. The inverse distribution function at $(k - 0.5)/n$ is $-\ln(1-(k-0.5)/n)$. Since μX is linear in X, points $(y_k, -\ln(1-(k-0.5)/n))$ $k = 1, \ldots, n$ are expected to be on a straight line. Thus, if an examination of the $Q - Q$ plot shows a straight line fit to be reasonable, then the assumed distribution is tenable. In case of doubt, we may resort to some statistical tests described below for further examination.

We illustrate the $Q - Q$ plot of the following data on interarrival times of 20 customers, the data being listed in ascending order.

TABLE 5.1. Interarrival times (in minutes)

2.80	3.20	3.95	4.36
2.88	3.38	4.04	4.39
2.94	3.47	4.11	4.45
2.98	3.62	4.16	4.57
3.06	3.84	4.25	4.63

Our interest is to test whether the underlying distribution of the above data is exponential. The table gives values of $y_1, \ldots, y_{20}$. We compute $-\ln\left(1 - \frac{k-0.5}{20}\right)$, $k = 1, \ldots, 20$. For example, when $k = 3$, the value is 0.134 which is plotted against $y_3 = 2.94$. In this way, the $Q - Q$ plot of the above data is drawn in Figure 5.1.

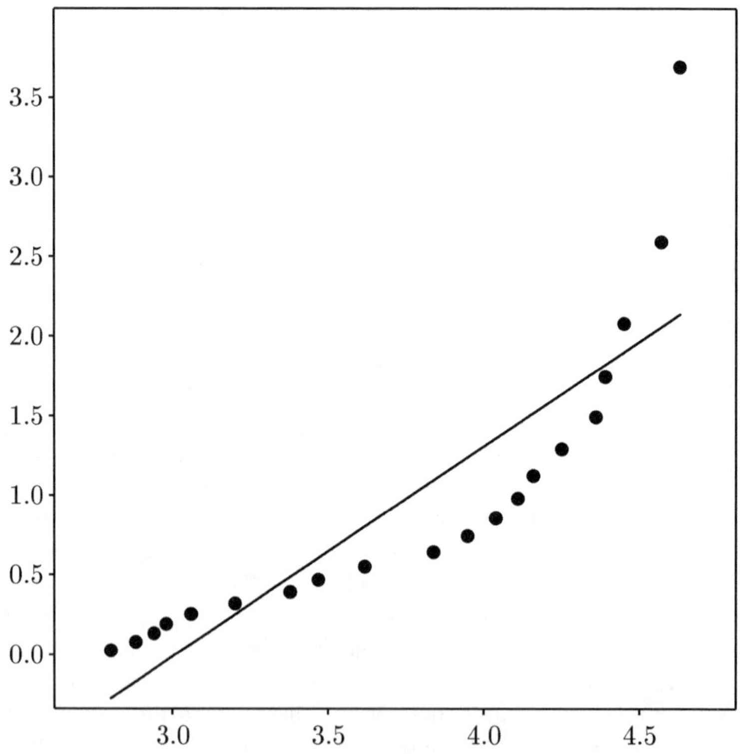

FIGURE 5.1. Q–Q plot

The points are not close to a straight line which implies the underlying distribution cannot be assumed to be exponential.

In $Q - Q$ plotting, we avoided parameters and their estimation. However, the parameters may be estimated by following the least square estimating procedure that is commonly used in fitting a regression line. If the fitted line is $y = a + bx$ over pairs $(x_1, y_1), \ldots, (x_n, y_n)$, then a and b are estimated as

$$\hat{a} = \bar{y} - \hat{b}\bar{x} \quad \text{and} \quad \hat{b} = \frac{\sum_{i=1}^{n} y_i(x_i - \bar{x})}{\sum_{i=1}^{n}(x_i - \bar{x})^2}$$

where $\bar{y} = \sum_{i=1}^{n} y_i/n$ and $\bar{x} = \sum_{i=1}^{n} x_i/n$. Consider F_0 to be the distribution function of a normal variable X with mean μ and variance σ^2. Then the standardized normal variable Z is known to be related to X by $X = \mu + \sigma Z$. But we know $x_k = F_0^{-1}(\frac{k}{n})$ and y_k, $k = 1, \ldots, n$ are expected to lie on $x = y$. Thus the fitted line is $y = \mu + \sigma z$ over $(z_1, y_1), \ldots, (z_n, y_n)$ where $z_k = \tilde{F}_0^{-1}((k - 0.5)/n)$, $k = 1, \ldots, n$. Using the above formula, the estimates of μ and σ are given by

$$\hat{\mu} = \bar{y} - \hat{\sigma}\bar{z} \text{ and } \hat{\sigma} = \frac{\sum_{i=1}^{n} y_i(z_i - \bar{z})}{\sum_{i=1}^{n}(z_i - \bar{z})^2} \tag{5.18}$$

where $\bar{z} = \sum_{i=1}^{n} z_i/n$. It is of interest to estimate μ if F_0 is exponential(μ) by this approach.

The most commonly used test is the *chi-square goodness-of-fit test.* In the language of testing, the null hypothesis H_0 is that the assumed distribution is true and the alternative hypothesis H_A is that it is not true. Let the unknown distribution function be F. Suppose the assumed distribution function is F_0, for instance, $F_0(t) = \lambda \exp(-\lambda t)$ for interarrival times. Then the two hypotheses are written as

$$H_0 : F = F_0 \quad \text{and } H_A : F \neq F_0. \tag{5.19}$$

The first step is to construct a *statistic*, a function of the random sample, which becomes an indicator to discriminate between two hypotheses. In case of our particular test, this statistic is formed on categorical data, the observations being the frequencies of certain categories or classes. Since we usually deal with continuous distributions, its observed values are classified into several intervals to form classes and the frequency in each class is noted. Let the observed frequency of the ith class be o_i. From F_0, we obtain the expected frequency of the ith class, denoted by e_i, by simply multiplying the probability that an observation lies in that class by total number of observations (this is justified by the fact that the expected value of the binomial distribution with parameters n and p is np). Often the distribution has unknown parameters which have to be estimated for computing the probabilities. The chi-square goodness-of-fit statistic, denoted by χ^2, is given by

$$\chi^2 = \Sigma \frac{(o_i - e_i)^2}{e_i} \tag{5.20}$$

where the summation is over all classes. It measures the deviation of observed frequencies from their expected values under H_0. Clearly, its smallness indicates the support for H_0, otherwise its rejection in favour of H_A. The question of how small it should be is decided by fixing the probability of *Type I error*, typically denoted by α, to be low, which is generally taken to be 0.05 or 0.01.

Type I error is committed by rejecting H_0 when it is true. Suppose we reject if the computed $\chi^2 > c$, for some constant c. Then $P(\chi^2 > c) = \alpha$ given that H_0 is true, where c is known as the *critical value* of the test for given α. In order to find that value, we realize that the statistic is a random variable and has, for a large sample size, approximately the chi-square distribution with k degrees of freedom when H_0 is true. The number of degrees of freedom k is equal to the total number of classes less one, minus the number of estimated parameters. The critical values of chi-square distribution for different degrees of freedom and different α's are tabulated and can be utilized for the test. For example, suppose we have $k = 5$, $\alpha = 0.05$. H_0 is rejected when the computed $\chi^2 > c$ so that $P(\chi^2 > c) = 0.05$. From Table B-2 we find $c = 11.07$. Thus if the observed value of $\chi^2 > 11.07$, then reject H_0 and infer that the assumed distribution is not appropriate. Otherwise, we accept the distribution under the present situation that there is no significant evidence to reject it. Applications of chi-square test can be seen in Section 5.4.

The sample size should be large enough for applying the chi-square test. For that reason, no expected frequency of any class should, as a rule of thumb, be less than 5. If this condition is not met, then the adjacent classes should be pooled so as to increase this value to become at least 5 or increase the sample size. Besides the limitation on sample size, this test leaves the choice of the number of classes and selection of class intervals to be arbitrary. Another drawback is that its probability of *Type II error* that occurs if H_0 is falsely accepted, possibly becomes high.

Even while applying the test, it is advisable to draw a diagram of relative frequencies and plot over it the hypothetical p.f. of F_0. It will visually show the closeness of fitting. This is done in Section 5.4.

The chi-square test is for categorical data. We introduce another one, called the *Kolmogorov-Smirnov test*, in which one does not need to lump the data into classes, but rather uses ordered observations $y_1, \ldots, y_n$ and the empirical distribution function F_n. The test statistic, denoted by D, compares the discrepancies between the empirical distribution function and the assumed distribution function. It is given by

$$D = \sup_x |F_n(x) - F_0(x)|. \tag{5.21}$$

Note that this test is applicable only for continuous F_0. It can be easily verified that the above can be written for computational purposes as

$$D = \max_j\{\max(F_0(y_j) - (j-1)/n,\ j/n - F_0(y_j))\}. \tag{5.22}$$

It is intuitively reasonable to reject H_0 for a large D. Given α, the critical value c such that $P(D > c) = \alpha$, can be determined if F_0 is completely specified. H_0 is rejected when the computed D from the sample is at least as large as c. Note that $F_0(X)$ has the uniform distribution over $(0, 1)$ and hence $F_0(y_k)$ is the kth order statistic of a sample of size n from the uniform distribution over (0,1) as long as F_0 is continuous. Therefore, the null distribution of D (i.e., under F_0) is independent of F_0, which makes it possible to tabulate critical values of this test. See Table B-4 in Appendix B.

Remark: We may compare both tests. The Kolmogorov-Smirnov test treats individual observations directly whereas the chi-square test discretizes the data by grouping and thereby may lose information. The former test is applicable even in case of small samples but the latter is essentially for large samples. The former assumes continuity of the distribution function which need not be so for the latter. Another important consideration is the χ^2 statistic still has a chi-square distribution even when the parameters are replaced by their estimates. This fact is not true for the other test.

In the context of queueing theory, it is of particular interest to check whether the assumed distribution is exponential. We describe two tests based on spacings. Let

$$S_i = (n - i + 1)(y_i - y_{i-1}), \quad i = 1, \ldots, n, \ y_0 = 0. \tag{5.23}$$

(For arrivals, $y_i - y_{i-1}$ represents the ith interarrival time.) Then for $r = [n/2]$ (the closest integer value of $n/2$),

$$U = \frac{\sum_{i=1}^{r} S_i/r}{\sum_{i=r+1}^{n} S_i/(n-r)} \tag{5.24}$$

turns out to have the F distribution with $(2r, 2(n-r))$ degrees of freedom (see Section A.9), under the assumption of exponentiality. A two-tailed F test is performed, rejecting H_0 when the calculated value is either small or large. The critical values c_1 and c_2 are determined from Table B-3 in Appendix B such that $P(U < c_1) = \alpha/2$ and $P(U > c_2) = \alpha/2$. For another test based on spacings, let $D_i = (n-i)(y_i - y_{i-1})$. The statistic

$$Z = 2\Sigma(n - 1 - i)D_i/(n-2)\Sigma D_i \tag{5.25}$$

can be used for testing. The null distribution of $2Z$ is the same as that of the mean of $(n-2)$ i.i.d. uniform random variables. It seems this test has often smaller probability of type II error among many of the competing tests.

5.3. Solving Queueing Problems by Simulation

The theoretical rather mathematical analysis is mostly concentrated on the study of probability distributions of queueing characteristics and their moments. We start with certain model assumptions and derive the distributions through our knowledge in probability theory. But if we understand through laws of large numbers that the probability of an event is the limit of sample proportions of the same event, then it is a matter of getting such samples in a large number and examining the sample proportions. In reality, this is not feasible and therefore mathematical tools have been very successfully used to obtain a vast number of elegant and useful results. Yet they are limited in their scope to cover an unending moor of applications. Thus arises the idea of simulation experiment as opposed to a real one, to artificially create, if possible, a large number of samples which otherwise would have hypothetically occurred in reality, provided the assumed models are true. The idea has become a real possibility through the advancement of computer technology. Strictly speaking, we will use the *Monte Carlo* method for generating random observations from a specified distribution and then apply our knowledge in probability and statistics to utilize the information existing in the artificially generated samples. We have briefly alluded to the Monte Carlo method in Section 1.4. This is another situation to solve problems by the use of computer, in addition to those utilized in deriving numerical solutions.

5.3.1. Discrete event simulation. When analyzing a queueing system by simulation, a special methodology is used: discrete event simulation. Roughly, this means the following: a prescribed stochastic system is observed by sampling, samples being collected at time points which form an increasing random sequence and the system may change its state only at those time points which are called event times.

There are some requirements which must be looked into for the simulation experiment to yield useful information about the system.

1. Designing the input of the simulation. Sufficient information is to be provided and maintained so that the state transitions and the times at which these transitions occur determine the system characteristics of interest. To start with, model specification and verification are necessary, such as arrival process and service-time distribution. The appropriateness of the model and estimation of parameters are done by statistical techniques as discussed in Section 5.2. A good input information will lead to good output data; otherwise we are reminded of the saying "garbage in, garbage out." Regarding maintenance of information, for instance, in simulating a $G/G/c$ queue, it is not sufficient to know the number of customers in the system at any time. We require also information about the residual service-times, the time to the next arrival and the status of the servers, whether they are idle or busy.

2. Generating the sample paths of the system. Basically (except for certain aspects of computer implementation) this means that we require algorithms which provide us with pseudo-random variates of a prespecified distribution. The generation of pseudo-random variates is in fact a problem of considerable complexity and it affects directly the quality of the output of a simulation experiment. The generation of random numbers with a prescribed distribution is done generally in two steps. First we require a random number having a uniform distribution in $[0, 1)$. There are several ways to generate uniform random numbers. The best known and simplest algorithms

are based on *linear congruential generators.* Such a generator delivers a sequence of non-negative integers $\{X_i\}$, where

$$X_{i+1} = (aX_i + c) \mod m \qquad i = 0, 1, \ldots \tag{5.26}$$

m being called the *modulus*, a prespecified positive integer, X_0, a and c being given non-negative integers, the *seed*, the *multiplier* and the *increment*, respectively. The numbers X_i have values in the range $[0, m-1]$. A uniform random number is then obtained simply by dividing through m, i.e., $u_i = X_i/m$ and u_i will be approximately uniform in $[0, 1)$.

There are many deep and nontrivial problems with such generators. For instance one is that of *periodicity.* This means that there is a smallest integer p, such that $X_p = X_0$. Of course, we want p to be very large. This period cannot exceed the modulus and it can be shown, that maximum period is obtained if and only if (1) c is relatively prime to m, (2) $(a-1)$ is a multiple of each prime factor of m and (3) $(a-1)$ is a multiple of 4, if m is. Many improvements and refinements of the above technique are known.

Once we have a uniform random number, random variates with a prescribed distribution other than uniform may be generated. The situation is particularly simple, if we have a closed form expression of the distribution function, as is the case, for example, for the exponential distribution. Since the distribution function of a random variable X, say $F(x)$, is monotonically increasing, there exists a uniquely determined inverse function F^{-1}. The generation of exponential variates via an inverse transform has been dealt with in reference to modelling in Section 1.4. We prefer to repeat it here so as to unify different features in simulation.

Now it is known and not difficult to show that the random variable $F(X)$ has a uniform distribution in $[0, 1]$. Thus if u is a random number uniform in $[0, 1]$, then $x = F^{-1}(u)$ will be a random number from one with distribution function F. In the case of an exponential distribution we have

$$F(X) = 1 - e^{-\lambda X} = u,$$

and from this it follows that for each u uniform in $[0, 1]$,

$$x = -\frac{1}{\lambda} \ln(1-u) \text{ or } x = -\frac{1}{\lambda} \ln u, \tag{5.27}$$

will be exponentially distributed with mean $1/\lambda$, since u and $1-u$ have the same distribution and $1-u$ may be replaced by u. For distributions other than the exponential, the above procedure of inverse transform may be difficult to apply, if no closed form formula is known for $F(x)$, as it is the case, e.g., for the Gamma distribution. In such cases there are other methods available.

We illustrate the generation of Erlangian variates with mean $1/\mu$ and k stages. It is known that the random variable is the sum of k i.i.d. exponential variables each with mean $1/k\mu$. Therefore, the Erlangian variate is

$$x = \frac{1}{k\mu} \sum_{i=1}^{k} \ln u_i, \tag{5.28}$$

where $u_1, \ldots, u_k$ are independent uniform variates.

No closed form expression is available for the inverse of the distribution function of a normal distribution. However, by the central limit theorem we know that the sum of a large number of uniform variables becomes a normal variable. In practice it is often sufficient to add six uniform variates. Another way of generating normal variates is given in Exercise 10.

Generating random variates from discrete distributions is of sufficient interest. Let X be the discrete random variable such that $P(X = x_i) = p_i$, $i = 1, 2, \ldots, m$. Generate a uniform $[0, 1)$ random variate u and let it be such that

$$\sum_{i=1}^{j-1} p_i < u \leq \sum_{i=1}^{j} p_i. \tag{5.29}$$

Then x generated by (5.27) from u is a x_j variate from the discrete distribution. As an example in queues, suppose we are interested in whether the next arrival occurs before the service completion. Let $X = 1$ if this happens, otherwise $X = 0$. In the $M/M/1$ model, it can be proved that $P(X = 1) = \lambda/(\lambda + \mu)$ (see Exercise 7, Chapter 1). Therefore, if the generated u satisfies

$$0 < u \leq \lambda/(\lambda + \mu)$$

then the discrete variate is 1, otherwise it is zero.

3. Bookkeeping of the generated data. From a formal point of view, discrete event simulation is the generation of the sample paths of two vector valued stochastic processes $\{\mathbf{T}(t), t \geq 0\}$ and $\{\mathbf{Z}(t), t \geq 0\}$ in a finite time. $\mathbf{T}(t)$ represents times at which an event occurs and $\mathbf{Z}(t)$ the types of events occuring at each of those time epochs. Suppose the time epochs are $t_1, t_2, \ldots$ such that $0 < t_1 < t_2 < \cdots$. At event time t_n, let the vector $\mathbf{T}_n = \mathbf{T}(t_n)$ be a chronologically ordered list of the times of all future events that has been scheduled through the completion of the nth event. Thus the first component of $\mathbf{T}_n$ gives us the time of the $(n + 1)$th event. However, the time of the $(n + 2)$th event may not be known because it may be generated at time t_{n+1}. Let the vector $\mathbf{Z}_n = \mathbf{Z}(t_n)$ be similarly constructed for the state variables corresponding to $\mathbf{T}_n$.

The sample paths of the processes $\mathbf{T}(t)$ and $\mathbf{Z}(t)$ are generated as follows: at the beginning $\mathbf{T}(t)$ and $\mathbf{Z}(t)$ are set to their initial values $\mathbf{T}_0$ and $\mathbf{Z}_0$ and the system clock is set to zero. The time of the first event is obtained from the first component in $\mathbf{T}_0$, its event type is specified by $\mathbf{Z}_0$. Now the clock is set to time t_1, the first component of $\mathbf{T}_0$, and the first event is executed. Upon execution of this event two new vectors $\mathbf{T}_1$ and $\mathbf{Z}_1$ are generated based on the observation of $\mathbf{T}_0$ and $\mathbf{Z}_0$. The time of the next event is obtained from $\mathbf{T}_1$, its type from $\mathbf{Z}_1$. The clock is advanced to time t_2 and this procedure is repeated until a stopping condition is satisfied.

Observe the special way the events are triggered and the clock is set by this method. It is also called the *time to the next event* technique and is very efficient and effective, since the system is observed at event times only and no information is lost.

An implementation of this method requires, in addition to an appropriate data structure representing $\mathbf{T}_n$ and $\mathbf{Z}_n$, a facility called *scheduler* and a system clock. The scheduler does all the bookkeeping that is required:

- Retrieve the next event from the event list.
- Process that event and adjust system state accordingly.
- Schedule new events.

Let us have a look at a simple example, a $G/G/1$ system. We will denote the components of the bivariate vector $(\mathbf{T}_n, \mathbf{Z}_n)$ by $[t, x]$, t is the time when the event will occur, the label x signifies the type of event, A means arrival, S completion of a service. The ordered pair $[t, x]$ is also called *event data structure.* Assume that time is measured in minutes, the system starts empty at time zero. See Table 5.2. This procedure continues until the clock exceeds a certain preassigned stopping time T, which means the scheduler encounters an event scheduled for a time $\tau > T$. Observe that in our simple example the event list will never hold more than two events. In a more sophisticated experiment, there may be many different types of events the scheduler has to take care of, e.g., when customers are allowed to be of different type. In practice we also want the scheduler to report about the system state at regular time instances. This task can be accomplished very easily by defining a special event $[t, C]$, the label C signifying the scheduler that it has to collect system statistics at time t.

4. Obtaining reliable estimates for the system characteristics of interest. This is a point of utmost importance. Since we are creating observations in a random way, we need good statistical procedures for obtaining satisfactory conclusions. Besides obtaining point estimates of the system characteristics which is normally done, it is necessary to have reliable estimates of their variances as well as the

TABLE 5.2. Bookkeeping

Clock	Next Event	$Q(t)$	Action of Scheduler	Event List
0	[ ]	0	Generate next (=first) arrival time	$\{[5, A]\}$
5	$[5, A]$	1	Add customer to system	
			Generate time of first service completion	$\{[10, S]\}$
			Generate next arrival time	$\{[10, S], [8, A]\}$
8	$[8, A]$	2	Add customer to system	
			Generate next arrival time	$\{[10, S], [17, A]\}$
10	$[10, S]$	1	Customer leaves since system is not empty	
			Generate time of next service completion	$\{[17, A], [22, S]\}$
17	$[17, A]$	2	Add customer to system	
			Generate next arrival time	$\{[22, S], [31, A]\}$
...	...		...	...

confidence interval estimates of the characteristics. They are indicators of how good the estimates are (large values indicate less precision of the estimates). Another problem arises due to the assumption of independence of observations in the classical statistical analysis. This need not be the case, if our interest is in the steady-state behaviour of the system. In such cases, samples are usually collected from a single but sufficiently long sample path. Suppose, we are studying the steady-state mean waiting time, for which we sample waiting times from a single sample path and treat the observations to be independent. Then the estimated variance will underestimate the true variance. It will be so because waiting times from such a sampling procedure are often positively correlated. To fix this needs more complicated analysis. However, this problem may be avoided in the transient case since collection of independent observations can possibly be done without much complications. Thus, transient and steady-state characteristics are treated separately. Observe that this is quite in contrast to the mathematical theory in queues, where steady-state problems are generally much easier to analyse than problems concerned with transient characteristics.

5. Determining the length of simulation run and the number of replications. The length is the time to continue for one sample and the number of replications refers to the sample size or the number of samples. In studying the transient characteristics, the length of a run is fixed in advance by the very nature of the problem or the system, say studying a characteristic at time t fixes the length to be t or say a system runs only for a specific duration for which the

length cannot exceed this duration. In principle, both lengths of runs and number of replications should be determined by the prescribed accuracy of the point estimates.

The last two points will be elaborated in the next two subsections.

5.3.2. Simulation of transient characteristics. It turns out that assessing the transient behaviour is considerably simpler than getting information about the steady-state behaviour. To be more precise, let $Q(t)$ denote a queueing process with state space $\mathcal{S}$. Suppose we want to estimate the quantity

$$r(t) = E(f(Q(t))) \tag{5.30}$$

for some function f (we are omitting certain conditions needed for rigorous mathematical treatment but are satisfied in our specific cases). We assume that this expectation is finite and also that

$$E(f^2(Q(t))) < \infty.$$

At first sight it may seem that confining to the estimation of expectations is a serious restriction, because we may be interested also in the estimation of probabilities. This is no problem, because probabilities are also expectations. In particular, the probability of an event A is given by $P(A) = E(1(A))$, where $1(A)$ is the indicator function of A as defined by

$$1(A) = \begin{cases} 1 & \text{for } a \in A, \ a \in S, \\ 0 & \text{otherwise.} \end{cases}$$

Here are some interesting candidates for the function f:

1. $$f(Q(t)) = Q(t),$$
   in which case (5.30) simply reduces to the expected queue length at time t.

2. $$f(Q(t)) = \begin{cases} 1 & \text{if } Q(t) = k, k \in \mathcal{S}, \\ \text{otherwise.} & \end{cases}$$
   Now, of course, $E(f(Q(t))) = P(Q(t) = k)$.

3. $$f(Q(t)) = \max_{0 \le s \le t} Q(s),$$
   the maximum attained by the process $Q(t)$ during the time interval $(0, t)$. (5.30) then equals the expected maximum queue length in $(0, t)$.

4. $$f(Q(t)) = \int_0^t 1(Q(s) \in A) \, \mathrm{d}s,$$
   the sojourn time of $Q(t)$ in the subset of state $A \in \mathcal{S}$. For instance, we may choose $A = \{k + 1, k + 2, \dots\}$, for some positive integer k.

Then $r(t)$ equals the expected total time the system is loaded with more than k customers.

These are only some examples, and the reader will have no difficulty to find other interesting instances for the function f.

Estimation of (5.30) may be carried out as follows:

Perform N simulation runs. For each run sample, $f(Q(t))$ yields a sequence $f_1, f_2, \ldots, f_N$, of realizations of the i.i.d. random variables $f_1(Q(t))$, $f_2(Q(t))\ldots$, the values of the function $f(Q(t))$ in the first, the second run, etc. Then by the law of large numbers, the arithmetic mean

$$\hat{r}(t) = \frac{1}{N}\sum_{i=1}^{N} f_i, \tag{5.31}$$

approaches $r(t)$ with probability one as $N \to \infty$. Therefore, $\hat{r}(t)$ has the consistency property (see Section A.9 in Appendix A) and is taken as the estimate of $r(t)$. Furthermore, let $s^2(t) = \text{var}(f(Q(t)))$, the variance of the random variable $f(Q(t))$. A *consistent estimator* of $s^2(t)$ is given by

$$\hat{s}^2(t) = \frac{1}{N-1}\sum_{i=1}^{N}(f_i - \hat{r}(t))^2. \tag{5.32}$$

To construct a confidence interval for $r(t)$, we use the consistency and invoke *the central limit theorem* (see Section A.4), to get

$$\frac{\hat{r}(t) - r(t)}{\hat{s}(t)}\sqrt{N} \to N(0,1), \qquad \text{as } N \to \infty, \tag{5.33}$$

where $N(0,1)$ denotes the standard normal distribution. Therefore an approximate $100(1-\alpha)$-percent *confidence interval estimate* for $r(t)$ is given by (see Section A.9)

$$\left[\hat{r}(t) - z_{\frac{\alpha}{2}}\frac{\hat{s}(t)}{\sqrt{N}}, \hat{r}(t) + z_{\frac{\alpha}{2}}\frac{\hat{s}(t)}{\sqrt{N}}\right], \tag{5.34}$$

where z_p is such that

$$\int_{z_p}^{\infty} \frac{1}{\sqrt{2\pi}} e^{-\frac{x^2}{2}}\, dx = p.$$

See Table B-1 in Appendix B to determine z_p.

Observe that this formula allows us also to determine, at least approximately, the number of simulation runs required such that the estimate $\hat{r}(t)$ has a prescribed precision. This precision may be measured by the width of the confidence interval (5.34). If we denote it by δ, then the number of runs N has to be determined such that

$$2z_{\frac{\alpha}{2}}\frac{\hat{s}(t)}{\sqrt{N}} \le \delta,$$

from which it follows, that

$$N \ge \left(2z_{\frac{\alpha}{2}}\frac{\hat{s}(t)}{\delta}\right)^2. \tag{5.35}$$

Using simulation, we may study now the following interesting design problem: How does the system performance change, if we split the capacity of one strong server to two or more parallel servers which have less capacity? In particular we are interested in two quantities: (a) the mean time all servers are busy B_t, which is a measure of utilization, and (b) the mean maximum queue length M_t.

Two experiments have been performed. In the first one, interarrival times and service-times have been both exponential, in the second experiment we assumed deterministic service-times. Here are the results together with 95% confidence intervals. In each case we had $t = 50$ and the number of runs N was 10000. The traffic intensity was in all cases equal to 2/3.

For exponential service-times:

System	λ	Mean Service T.	B_t	M_t
$M/M/3$	2	1	21.4428 ± 0.1273	6.2167 ± 0.0546
$M/M/2$	2	2/3	25.9752 ± 0.1126	6.5739 ± 0.0532
$M/M/1$	2	1/3	32.1189 ± 0.0924	7.2023 ± 0.0546

For deterministic service-times:

System	λ	Mean Service T.	B_t	M_t
$M/D/3$	2	1	21.1620 ± 0.0942	4.5223 ± 0.0310
$M/D/2$	2	2/3	25.9673 ± 0.0856	4.6825 ± 0.0316
$M/D/1$	2	1/3	32.4269 ± 0.0650	4.8116 ± 0.0308

The results suggest the following conclusion: splitting the server capacity is generally not a good idea because it significantly reduces server utilization. However, there is also an (statistically significant) increase in the mean maximum queue length.

It is interesting also to see that the mean time all servers are busy is approximately the same for exponential and deterministic interarrival times; however, for the $M/D/c$ systems the mean maximum queue length is lower, and the observed differences are again significant and cannot be explained by random fluctuations only.

A practical problem examining the same question is dealt with in Section 5.4. There the queue length is of importance and the actual data are collected.

5.3.3. Simulation of steady-state characteristics. The procedure outlined above for simulation of the transient behaviour cannot be adopted when our interest is in steady-state characteristics of a queueing system. The reason is simply that we cannot generate infinitely long sample paths of a stochastic system. To overcome this problem, it is common usage to generate one single long sample path and to argue that after a certain time period, the so-called *initial transient*, the system will be approximately in steady state. After discarding the initial portion of the sample path, one observes the system for an additional period of time and hopes that the system is close to steady state. From this steady-state phase we collect our statistics. The problems with this idea are that systems may approach steady state very slowly. So what is needed is a criterion to decide when the transient phase is over. Furthermore, the samples collected in the steady-state phase are generally not independent, thus usually only point estimates can be given. The construction of confidence intervals is rather difficult in this case because of the lack of independence.

The dependence problem may be avoided in a trivial way, if we adopt the sampling procedure of 5.3.2, that is, replicate the runs N times, each time starting with a different random number seed. Unlike the previous situation where the length of the run is fixed in advance, the length has to be determined in the present case. It should be long enough so that any queueing characteristic of interest is observable. Usually, its length is taken to be 500 or until a certain number of events occur, after the initial transient period is over. The decision on initial length and run length is more an art than a science and becomes mature with more experience. The point and interval estimates are obtained in an analogous manner as that of the transient situation.

Consider the estimation of P_k, the probability of having k customers in the system. Let $f_{k,i}$ be the proportion of time of having k customers in the system during the ith run. Then

$$\hat{P}_k = \frac{1}{N}\sum_{i=1}^{N} f_{k,i}, \qquad k = 0, 1, \dots, \tag{5.36}$$

is an estimate of P_k. The estimate of its variance is given by

$$\hat{s}_k^2 = \frac{1}{N-1}\sum_{i=1}^{N}(f_{k,i} - \hat{P}_k)^2, \quad k = 0, 1, \dots . \tag{5.37}$$

A $100(1-\alpha)$ percent confidence interval estimate of P_k has the expression

$$\left[\hat{P}_k - z_{\frac{\alpha}{2}}\frac{\hat{s}_k}{\sqrt{N}}, \hat{P}_k + z_{\frac{\alpha}{2}}\frac{\hat{s}_k}{\sqrt{N}}\right], \quad k = 0, 1, \dots . \tag{5.38}$$

Remark: The expressions in (5.36), (5.37) and (5.38) are the same as those of (5.31), (5.32) and (5.34) with the substitution of $f_{k,i}$ for f_i and with the omission of "t". In the transient case, the characteristic is examined only at a fixed time instant "t" and leading to $f = f(Q(t))$ to be a function of t. Consider the system length as an illustration: $f_i = 1$ if $Q(t) = k$ and $f_i = 0$ otherwise. On the other hand, for the steady-state case we have to look for the sample representative of the characteristic in the entire run length. In our illustration, the representative is $f_{k,i}$ for P_k in the ith run.

Let us consider a particularly simple simulation experiment: We want to estimate the transient probability of an $M/M/1$ queue being empty at time $t = 10$.

First we fix our input parameters: Let arrival rate and service rate be $\lambda = 1$ and $\mu = 1.1$ and set the time parameter to $t = 10$.

A single replication of the experiment will consist of the following steps.

1. Set the system clock to $T = 0$.

2. Set $Q(0) = 0$.
3. Generate the sequence of arrivals and service completions as described above.
4. When the system time T of the next event to be processed exceeds 10, stop and return 1 if the queue is empty and zero otherwise.

The output of a single experiment is therefore a binary random variable, say $I(t)$. If we repeat that experiment n times we obtain an estimate of the probability $P_{00}(t)$ by

$$\hat{p} = \frac{1}{n}\sum_{i=1}^{n} I_i(t),$$

with estimated standard deviation

$$s = \sqrt{\frac{\hat{p}(1-\hat{p})}{n}}.$$

Therefore an approximate $100(1-\alpha)$ percent confidence interval is given by

$$\hat{p} \pm z_{(1-\frac{\alpha}{2})}\sqrt{\frac{\hat{p}(1-\hat{p})}{n}},$$

where z_p denotes the percentiles of the standard normal distribution.

Suppose we want to estimate $P_{00}(t)$ with a precision of $\pm\epsilon$. Let us put $\alpha = 0.05$. From Table B-1 in Appendix B, $z_{\frac{\alpha}{2}} = 1.96$. Then it is easy to calculate a lower bound for the number of replications n as:

$$1.96\sqrt{\frac{\hat{p}(1-\hat{p})}{n}} \leq \epsilon \Rightarrow n \geq \frac{1.96^2\hat{p}(1-\hat{p})}{\epsilon^2}.$$

Since $\hat{p}(1-\hat{p}) \geq 0.25$, a lower bound is given by

$$n \geq \frac{1.96^2 \cdot 0.25}{\epsilon^2} \approx \frac{1}{\epsilon^2}.$$

If we choose $\epsilon = 0.005$, this results in a minimum number of replications of $n \approx 40000$.

Performing the required number of replications we obtained the following ($\alpha = 0.05$) confidence interval

$$\hat{p} = 0.2166 \pm 0.004.$$

To check the validity of the result, we compare this estimate with the exact value. The exact value obtained by (2.50) is

$$P_{00}(t) = \frac{e^{-(\lambda+\mu)t}}{t\mu}\sum_{k\geq 1} k\rho^{-k/2} I_k(2\sqrt{\lambda\mu}\, t)$$

which equals

$$P_{00}(10) = 0.21457 \qquad \lambda = 1,\ \mu = 1.1.$$

Thus our estimate is pretty close to the exact value.

5.4. A Practical Application

In an automobile workshop, vehicles in the accident repair section have to wait for a considerable period of time to get repaired. The workshop has to bear the cost for waiting vehicles. Thus the objective is to structure the service system so as to minimize the total stay period of a vehicle. The ensuing benefits would be increased customer goodwill, reduced loss of customers and saving on waiting cost.

The existing service system consists of 6 working stations with a waiting space for 11 vehicles. Increase in the number of stations will reduce the waiting times, but is associated with increase in the idle time of workers. In view of this, the management may be willing to increase the number of stations so as to balance stay period of a vehicle against idle time of a worker.

The situation may be described by a queueing system having 6 servers with finite system capacity of 17 (11 + 6), where the queue discipline is the usual FCFS. Now, we are to have information on arrival process and service-time distribution. For the first, the past records of vehicle arrivals are used to model the input process. But the service-times are not directly available and are estimated from the labour cost estimates. In both these cases we refer to Section 5.2.

The record for vehicles requesting service is examined over 225 days divided into 75 consecutive groups of 3 days each. The data are presented in columns 1 and 2 of Table 5.3, as frequencies of number of vehicles arriving in a 3-day unit. The mean and variance are calculated and are found to be close (given at the bottom of the table), suggesting to consider Poisson distribution to fit the data. In order to do that, we apply the chi-square goodness-of-fit test (see Section 5.2) and compute χ^2 statistic as given in (5.20). Column 2 gives o_i, $i = 0, 1, \ldots, 10$. To compute e_i we note that F_0 is the distribution function of the assumed Poisson distribution which is completely determined by its mean. Since this is unknown, we estimate it by the sample mean 3.45 (see Table 5.3). Thus according to the assumed Poisson distribution with $\lambda = 3.45$, the probability of the random variable taking on value i is given by $e^{-3.45}(3.45)^i/i!$ and the expected frequency e_i of value i is $75(e^{-3.45}(3.45)^i/i!)$, $i = 0, 1, \ldots, 10$. For computational purposes, we may use $e_i = \frac{3.45}{i} e_{i-1}$, $i = 1, \ldots, 10$. Values e_i are recorded in column 3. Finally, column 4 lists $(o_i - e_i)^2/e_i$ for each i, the sum of which over all i gives the value of χ^2 statistic equal to 6.48. Strictly speaking $i = 11, 12, \ldots$ should have been considered, even if $o_i = 0$ for those i. However, corresponding e_i is negligible not changing χ^2 value significantly. That is why such i's are omitted. Observe that the last five values are pooled together in order to make the expected frequency to be at least 5 (see chi-square test). The degrees of freedom k for the chi-square distribution is $k = 7 - 1 - 1 = 5$, 7 being the number of classes and the last 1 being the number of estimated parameter. The tabulated value for $k = 5$ and $\alpha = 0.05$ is 11.07 which is greater than calculated χ^2 statistic 6.48. Therefore, the hypothesis that the number of

TABLE 5.3. Goodness-of-fit test for arrivals

Number of Arriving Vehicles/3 Days i	Observed Frequency o_i		Expected Frequency e_i		$\frac{(o_i-e_i)^2}{e_i}$
0	6		2.38		5.506
1	7		8.21		0.179
2	13		14.17		0.096
3	14		16.30		0.325
4	13		14.05		0.078
5	11		9.70		0.174
6	8		5.58		
7	0		2.75		
8	2	11	1.19	10.13	0.126
9	0		0.45		
10	1		0.16		
Total	75		74.9		6.48

Sample mean = 3.45
Sample variance = 4.105
χ^2 statistic = 6.48
$\chi^2 = 11.07$ is tabulated for 5 degrees of freedom and $\alpha = 0.05$

arriving vehicles has Poisson distribution with mean 1.15 vehicles per day is not rejected. (Note that the mean 3.45 in the table is based on observations per 3-day period.) It indicates that we may assume the arrival process to be Poisson with $\lambda = 1.15$. We believe that other conditions of a Poisson process seem to be reasonably satisfied.

The graphs of observed and expected frequencies in Table 5.3 are drawn in Figure 5.2 to observe the closeness of the fit.

Regarding the service-time distribution, the service-time in days per vehicle is estimated from the labour cost. Table 5.4 provides the data and the necessary calculations for the chi-square goodness-of-fit test.

Note that unlike the previous case the frequencies are recorded for class intervals of width 4. Also, the present fitting is on a continuous distribution, whereas the last one is on a discrete distribution for arrivals, namely, Poisson distribution. For calculation of sample mean and variance, the mid-point in each class interval (say, in the class interval 0–4, the mid-point is 2) is used. We find the sample mean and standard deviation to be 7.69 and 7.23 days, respectively. Since in an exponential distribution, the mean and standard deviation are equal and since their sample estimates are close, we may therefore hypothesize (i.e., take the null hypothesis) that the service-time distribution is exponential with its mean as 7.60 days (the sample value as

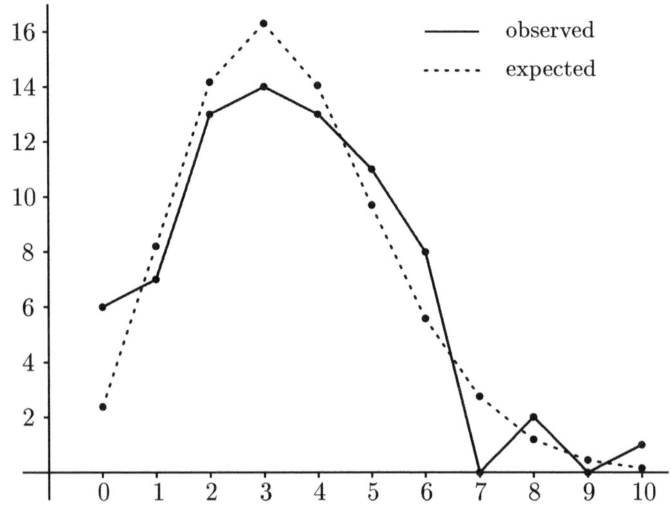

FIGURE 5.2. Observed and expected frequencies of Table 5.3

TABLE 5.4. Goodness-of-fit test for service-time distribution

Class Interval (service-time in days) i	Observed Frequency o_i		Expected Frequency e_i		$\frac{(o_i-e_i)^2}{e_i}$
0–4	38		44.20		0.870
4–8	35		26.28		2.893
8–12	16		15.62		0.009
12–16	5		9.29		1.981
16–20	6		5.52		0.042
20–24	5		3.28		
24–28	4	9	1.95	8.08	1.444
28–∞	0		2.85		
Total	109		108.99		7.239

Sample mean $= 7.69$
Sample deviation $= 7.23$
χ^2 statistic $= 7.239$
$\chi^2 = 9.488$ is tabulated for 4 degrees of freedom and $\alpha = 0.05$

an estimate is used for the mean of the distribution). Under this hypothesis the expected frequency e_i of the ith class is computed by the formula

$109 \int_{a_i}^{a_{i+1}} \frac{1}{7.69} e^{-t/7.69} dt$ where a_i and a_{i+1} are the lower and upper boundaries of the ith class, $i = 1, \ldots, 6$. The last three classes are pooled together to get corresponding $e_i > 5$. The χ^2 statistic turns out to be 7.239 which is less than the tabulated value of 9.488 for degrees of freedom 4 at $\alpha = 0.05$. Thus, we may assume the service-time distribution to be exponential with mean 7.69 days or with service rate $\mu = 0.13$ per day for each server.

To see the closeness of the fit of the exponential model, the histogram of observed class frequencies o_i along with the curve to represent expected frequencies are drawn in Figure 5.3. Note that in drawing the curve, the class width 4 is multiplied in the expression.

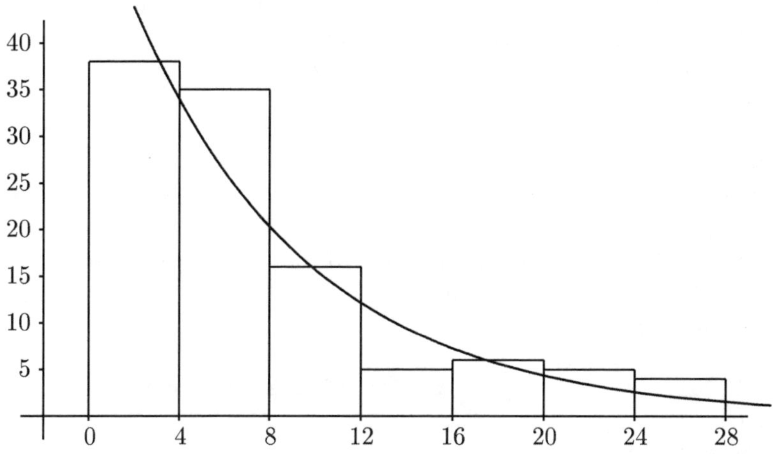

FIGURE 5.3. Observed frequencies of Table 5.4 and $109e^{-t/7.69}dt/7.69$ with $dt = 4$

On combining inferences on arrival process and service-time distribution, we are led to construct the model as $M/M/6$ with $\lambda = 1.15$ and $\mu = 0.13$ having finite system capacity $N = 17$ and FCFS queue discipline. Exercise 9 in Chapter 2 deals with this model and provides expressions for different measures of queueing characteristics. We compute $P_0, L_q, \lambda_{\text{eff}}$ and L for $c = 6$ from (b), (c), (d) and (e), respectively, as

$$P_0 = 0.00001, \;\; L_q = 8.95354, \;\; \lambda_{\text{eff}} = 0.77866 \text{ and } L = 14.94327.$$

These represent the probability of the system being empty, expected number of vehicles in queue, expected number of vehicles accepted per day and expected number of vehicles in the system, respectively. The use of Little's formula (the first one in (2.67)) gives $E(W_q)$, the expected stay period in queue as $E(W_q) = L_q/\lambda_{\text{eff}}$ since $\bar{\lambda} = \lambda_{\text{eff}}$ in our model. Similarly the expected stay period in system $E(W)$ is obtained from (2.66) and Exercise 9(e) in Chapter 2 as $E(W) = E(W_q) + 1/\mu$.

At the outset, we stated that our interest is to reduce waiting time by adding more stations. But the reduction in waiting time should be balanced against idle time of a station. In addition to earlier characteristics, we consider the percent of a day a station remains idle and percent of vehicles refused. Their formulas are given as $\frac{100}{c}\sum_{n=0}^{c}(c-n)P_n$ and $100P_{17}$ where P_n is computed by using Exercise 9(a) in Chapter 2.

Values of various characteristics for working stations $c = 6, 7, \ldots, 17$ are presented in Table 5.5. The associated computer program by using the MATLAB 6 package is given below, where $C = c$, $W_q = E(W_q)$ and $W = E(W)$.

```
lamda= input('Enter the value of lamda :'); %rate of arrivals
mu= input('Enter the value of mu :');  %rate of service
N= input('Enter the value of N :');  % system capacity
rho=lamda/mu;

for C=6:17 % No. of working bays
    a1=(rho^C*(1-(rho/C)^(N-C+1)))/(factorial(C)*(1-(rho/C)));
    sum=0;
    for n=0:C-1 %'n' is the number of units in the system
        b1=rho^n/factorial(n);
        sum=sum+b1;
    end
    P_0=(a1+sum)^-1; % probability of the system being empty
    sum1=0;
    sum2=0;
    for n=0:N
        if n>=0 & n<=C
            P_n= (rho^n/factorial(n))*P_0;
        else
            P_n= (rho^n/(factorial(C)*C^(n-C)))*P_0;
        end
        sum1=sum1+P_n;
        if n<=C
            idl=(C-n)*P_n;
        else
            idl=0;
        end
        sum2=sum2+idl;
    end
    Pc_idle=100*sum2/C; % Idle time of servers(%)
    Pc_vehi_refused= 100*P_n; %percent of vehicles refused
    c1=(1-(rho/C)^(N-C)-(N-C)*(rho/C)^(N-C)*(1-((rho/C))));
    c2=P_0*rho^(C+1)/(factorial(C-1)*(C-rho)^2);
    lamda_eff=lamda*(1-P_n); % No. of vehicles accepted per day
```

```
    L_q=c1*c2;  %expected no. of vehicles in the queue
    L=L_q+(lamda_eff/mu); %expected no. of vehicles in system
    W_q=L_q/lamda_eff; %expected stay period in the queue
    W=W_q+(1/mu);  %expected stay period in the system
    fprintf('C=%d\n',C);
    fprintf('P_0=%1.5f,  Pc_idle=%1.5f,
        lamda_eff=%1.5f \n',P_0,Pc_idle,lamda_eff);
    fprintf('L_q=%1.5f,  L=%1.5f,
        W_q=%1.5f,  W=%1.5f\n',L_q,L,W_q,W);
    fprintf('Pc_vehi_refused=%1.2f\n',Pc_vehi_refused);
    fprintf('\n%s\n');
    x=C;
    y=Pc_idle;
    plot(x,y,'k*');
    if C ==6
        hold on;
    end
    X=C;
    y=W;
    plot(x,y,'k.');
    if C ==6
        hold on;
    end

end
xlabel('No. of Bays');
axis([6 20 0 50]);
ylabel('Days');
```

In order to balance $E(W)$ and percent of idle time, these are graphed against the number of stations, as shown in Figure 5.4. It can be seen that the two curves cross at a point against the number of stations close to 9. Thus the decision is to increase the number of working stations from 6 to 9 for which the average stay period of a vehicle comes down from 19 to 10 days, whereas the idle time increases to 8.8 percent.

5.5. Discussion

Every theory that has been developed is based on some assumed models. In our case, it is, for instance, on the input process and on the distribution of service-times. The derived results and related computational techniques are as good as the assumed models behind them. To make sure they are good enough, we employ statistical tools for estimating the parameters of the assumed distributions and test whether any assumed distribution confirms the samples collected from such a queueing system. This has been the subject matter of Section 5.2. To begin with, it is always advisable to plot the observed

TABLE 5.5. Values of characteristics

No. of Working Stations (c)	No. of Vehicles Accepted per Day (λ_{eff})	Probability of the System Being Empty (P_0)	Expected No. of Vehicles in the System (L)	Expected No. of Vehicles in the Queue (L_q)	Expected Waiting Time in System (days) ($E(W)$)	Expected Waiting in Queue (days) ($E(W_q)$)	Idle Time at a Station (X)	Percent of Vehicles Refused
6	0.77866	0.00001	14.94327	8.95354	19.19089	11.49858	0.17115	32.29
7	0.90014	0.00002	13.76757	6.84343	15.29496	7.60266	1.08380	21.73
8	0.99994	0.00006	12.29011	4.59826	12.29085	4.59854	3.85194	13.05
9	1.06670	0.00009	10.92166	2.71628	10.23874	2.54643	8.82905	7.24
10	1.10417	0.00012	9.95332	1.45970	9.01430	1.32199	15.06385	3.99
11	1.12348	0.00013	9.37831	0.73613	8.34753	0.65522	21.43474	2.31
12	1.13336	0.00014	9.07016	0.35202	8.00291	0.31060	27.34885	1.45
13	1.13856	0.00014	8.91633	0.15817	7.83123	0.13892	32.62950	0.99
14	1.14140	0.00014	8.84484	0.06481	7.74908	0.05678	37.28549	0.75
15	1.14298	0.00014	8.81472	0.02255	7.71204	0.01973	41.38554	0.61
16	1.14382	0.00014	8.80401	0.00537	7.69700	0.00470	45.00850	0.54
17	1.14418	0.00014	8.80142	0.00000	7.69231	0.00000	48.22694	0.51

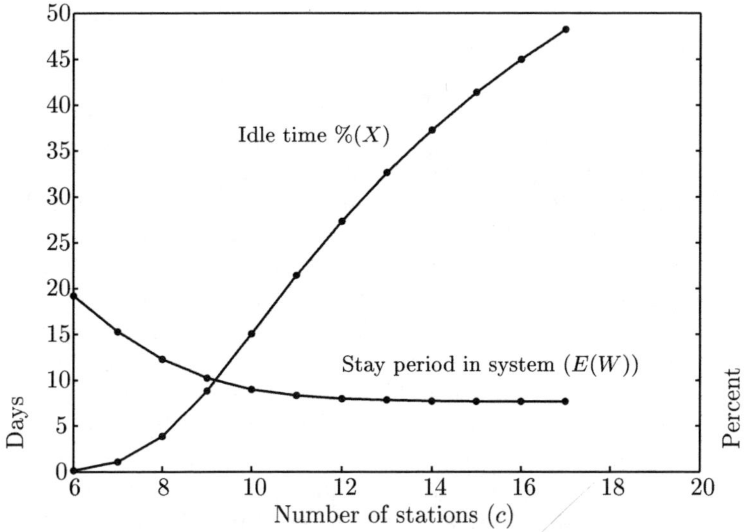

FIGURE 5.4. Stay period in system $E(W)$ and idle time (X) vs. number of stations (c)

values against their expected values under the assumed hypothesis. This will give an idea of the closeness between the two. Applying further the statistical goodness-of-fit tests will assist in the decision making process. A good reference on these tests is D'Agostino and Stephens (1986). However, for testing exponentiality, one may refer to Shapiro (1995).

What we have touched, in this chapter, on testing problems is whether the assumption of a given model is acceptable or not. In reality, we are observing processes over which we superimpose some models. These models have some assumptions for instance, time homogeneity which implies the process in the interval $(t, t+h)$ has the same behaviour as in $(u, u+h)$. For examinations of such questions one may refer to other books such as Basawa and Prakasa Rao (1980).

Structurally, statistical tests are developed to reject the null hypothesis, which in our case is a specified distribution. Therefore, one may be cautioned that not rejecting it only tells us lack of enough evidence in the sample to reject the assumed model. Thus when we do not reject the assumed distribution, we have only a guarded acceptance. Regarding estimation, it is simply not enough to get an estimate of a parameter, but to know how good it is by learning about its variance. A small variance provides a better estimate. Another desirable property of an estimate is its closeness to the parameter when sample size becomes large. This property is called consistency. Quite often, we look for this property in our context.

An important point for consideration is that drawing inferences is based on the method of sampling and the type of data collected, such as waiting time and length of the queue. A reference for MLEs from waiting time data is Basawa, Bhat and Lund (1996).

A little review of what has been done so far would reveal that under some model assumptions, the distributions of different characteristics are derived, sometimes conveniently computable and sometimes not, in which case we have attempted to provide numerical solutions. However, there is another approach for consideration, particularly when the distributions of the desired characteristics are not available, mostly due to lack of appropriate mathematical tools. In such situations, the method of simulation may be used in order to gain enough information on these characteristics. This method has been explained in Section 5.3. Basically, simulation is a mechanism to generate a very large number of realizations of the prescribed queueing process without actually observing them in real life and it is done with the aid of the computer, hardly using much resource and time. The number could be large enough to virtually cover the universe of the prescribed process. Here again, because we are dealing with samples, we are in the world of statistics. The desired characteristics are estimated from this large number of realizations and therefore the estimates are expected to be close to the actual value due to the property of consistency of estimates.

It seems we have done it without much sweat! Why bother about the theory and others. Not quite so. First of all, generating samples involves quite a few mathematical challenges. Second, the sampling procedure or what we call the selection of runs, could be tricky, at least that is the way it looks when dealing with steady-state characteristics. Here we refer to Gunther and Wolff (1980) and Heidelberger and Welch (1983). This gives rise to challenges in statistical methodology. While these reflect on the difficulties arising in the design aspect of simulation set-up, we also have to realize the limitations on the output of simulation experiment. One important limitation is that the information on characteristics is very much based on estimated values of parameters, which makes them tied to those specific values, whereas a theoretical result is broad-based, true for any value of parameters. For instance, consider the $M/M/1$ model. Using simulation, we generate sample realizations with the specific estimated values of λ and μ. On the basis of these realizations we estimate characteristics. If λ and μ are changed to another set of values, this method has to start all over again. In contrast, a theoretical result like the steady-state distribution of the number in the system has a symbolic expression which is true as long as $\lambda < \mu$. The second limitation is that simulation method only provides an estimate whereas the theoretical result is exact. Thus whenever there is a theoretical result available, by all means use it. Yet the unlimited world of models will entice anyone to tinker with simulation. When the exact solution is not in a computable form, such an adventure is justifiable. A natural consequence of increased use of simulation has generated an equal interest in looking into appropriate statistical tools. Ironic enough, in the past queueing theorists never wished to be bothered by statistical issues.

In Section 5.4, an attempt has been made to demonstrate the applicability of some concepts and results at our disposal. It is a simple application and throws some light as to how a queueing problem could be handled.

For statistical inference in Section 5.2, we refer to Bhat and Miller (2002), Bhat, Miller and Rao (1997), Shapiro (1995) and Gross and Harris (1998). Simulation study being of recent interest in queueing theory has an extensive literature, but for our purpose we list a few which are Cheng (1977), Fishman (1978), Lewis, Goodman and Miller (1969), Gordon (1978), Law and Kelton (1982) and Shedler (1987).

Time to pause! At the end of this chapter and after finishing five chapters, we have at least exposed the reader to the basic developments in queues: some theoretical results, computational methods and statistical techniques. It may be said that these disciplines form the triumvirate, taking charge of the responsibility to enrich the subject – the theory of queues. We mortals (and that ever-present inanimate computer) are merely serving their cause.

5.6. Exercises

1. Prove (5.6).

2. The p.d.f. of the Erlangian distribution may be written as (see (2.22))

$$f(t) = \frac{\theta^k}{(k-1)!} t^{k-1} e^{-\theta t}, \quad t \geq 0, \theta > 0.$$

   (a) Find the MLE of θ.
   (b) Find the estimates of θ and k by the method of moments.

3. Establish (5.12) and (5.13).

4. Prove (5.16) and show that the MLE $\hat{\rho}$ of ρ is the sample mean of the number of arrivals during first n service-times.

5. For the $M/D/1$ queue with service-time of length one, the probability that the number served during a busy period is k, is shown to be

$$\frac{e^{-\lambda k}(\lambda k)^{k-1}}{k!}, \qquad k = 1, 2, \ldots$$

   in Exercise 31, Chapter 3. This becomes a proper probability function with mean $(1-\lambda)^{-1}$ if $\lambda < 1$.

   (a) Show that the MLE $\hat{\lambda}$ of λ is given by

$$\hat{\lambda} = \frac{\bar{k} - 1}{\bar{k}}$$

   where $\bar{k}$ is the sample mean of numbers served during n random samples of busy period.

   (b) Show that the MLE and the estimate by the method of moments are equal.

6. If F_0 is the distribution function of the exponential(μ) distribution, estimate μ by using the technique of $Q - Q$ plot fitting similar to the one in (5.18).

7. Test whether the data in Table 5.1 comes from exponential (3.7).

8. By using the randomization method (Chapter 4), find the transient distribution $\{P_n(5)\}$ of the $M/M/2$ system with finite capacity $N = 10$, for $i = 5$, $\lambda = 6$ and $\mu = 3$. Obtain it by using the simulation procedure with 100 replications. Compute a 95% confidence interval for $P_{10}(5)$.

9. Suppose the queueing model is $M/E_2/1$ with $i = 1$, $\lambda = 4$ and $\mu = 5$. Use simulation for this exercise.
   (a) Estimate the expected queue length at time $t = 7$ by taking the number of replications $N = 100$.
   (b) How large N should be to achieve the length of a 95% confidence interval equal to 2?
   (c) Estimate the average waiting time of a customer if he arrives at $t = 7$.

10. Let X and Y be two independent random variables each having standard normal distribution (i.e., mean = 0 and variance = 1). Let these be transformed to R and θ by the transformation $x = R\cos\theta$ and $y = R\sin\theta$. Show that R and θ are independent and θ is uniform over $(0, 2\pi)$. Also show that $V = \frac{R^2}{2}$ has exponential distribution with mean 1. Hence generate two sets of independent standard normal variates.

References

Basawa, I. V. and Prakasa Rao, B. L. S. (1980). *Statistical Inference for Stochastic Processes*, Academic Press, New York.

Basawa, I. V., Bhat, U. N. and Lund, R. (1996). Maximum likelihood estimation of single server queues from waiting time data, *Queueing Syst.*, **24**, 155–167.

Bhat, U. N. and Miller, G. K. (2002). *Elements of Applied Stochastic Processes*, 3rd Ed., John Wiley & Sons, New York.

Bhat, U. N., Miller, G. K. and Rao, S. B. (1970). Statistical analysis of queueing systems, *Frontiers in Queueing*, Chapter 13, Edited by J. H. Dshalalow, CRC Press, New York.

Cheng, R. C. H. (1977). The generation of gamma variable with non-negative shape parameter, *Appl. Statist.*, **26**, 71–75.

D'Agostino, R. B. and Stephens, M. A. (Eds.) (1986). *Goodness-of-fit Techniques*, Marcel Dekker, New York.

Fishman, G. S. (1978). *Principle of Discrete Event Simulation*, John Wiley & Sons, New York.

Gordon, G. (1978), *System Simulation*, 2nd Ed., Englewood Cliffs-Prentice-Hall, New York.

Gross, D. and Harris, C. M. (1998). *Fundamentals of Queueing Theory*, 3rd Ed., John Wiley & Sons, New York.

Gunther, F. L. and Wolff, R. W. (1980). The almost regenerative method for stochastic system simulation, *Oper. Res.*, **28**, 375–386.

Heidelberger, P. and Welch, P. D. (1983). Simulation run length control in the presence of an initial transient, *Oper. Res.*, **31**, 1109–1144.

Law, A. M. and Kelton, W. D. (1982). *Simulation Modeling and Analysis*, McGraw-Hill, New York.

Lewis, P., Goodman, A. S. and Miller, J.M. (1969). Pseudorandom number generator for the system 360, *IBM System J.*, **8**, 136–146.

Shapiro, S. S. (1995). Goodness of fit tests, *The Exponential Distribution*, Chapter 13, pp. 205–220, N. Balakrishnan and A.P. Basu, Eds., Gordon and Breach Publishers, Amsterdam.

Shedler, G. S. (1987). *Regeneration and Networks of Queues*, Springer, New York.

Part 2

CHAPTER 6

Regenerative Non-Markovian Queues – II

6.1. Introduction

In Chapter 3, regenerative non-Markovian models, more specifically models $M/G/1$ and $G/M/1$, are introduced and studied through the imbedded Markov chain technique. Not-withstanding the fact that the chapter, by and large, has covered the distributions of various characteristics of interest, it is recognized at the end that certain related topics are neither included (such as transient solution) nor thoroughly dealt with (such as functional relations). The purpose of the present Chapter is to pick up the line of investigation from the point where it was left out in Chapter 3. However, this chapter differs in character in the sense that it is conceptually and theoretically heavy. A new feature, namely, "combinatorial methods," is added as Section 6.3. In spite of the overall theoretical overplay in most of the topics of this chapter, it is seen that their usefulness cannot be ignored and in fact is noticeably substantial in certain situations.

6.2. Non-Markovian Queues: Transient Solution

The study of the transient behavior of the $M/M/1$ model (Section 2.5 in Chapter 2) which is characterized by the process $\{X(t)\}$ was facilitated through the differential-difference equations, because $\{X(t)\}$ happens to be a Markov process. As observed in Chapter 3, $\{X(t)\}$ is not so as soon as one of input and service distributions is non-Markovian. This motivated us to induce the imbedded Markov chain at some suitable regeneration points, thus suggesting that the study of $X(t)$ alone at any arbitrary point is not possible through the Markovian property. However, the situation can be rectified by the *supplementary variable technique* which simply allows us to include a sufficient number of variables so as to retain the Markovian nature of the process.

6.2.1. The $M/G/1$ model. Let us consider the $M/G/1$ model. In Section 3.2 we noticed that the departure points serve as regeneration points since at that time the elapsed service-time of an entering unit is zero. Therefore it suggests that in addition to $X(t)$, the elapsed service-time of the unit getting the service at time t, considered to be the supplementary variable, will capture the Markovian property in the sense that the future is determined by the present.

Let

$$P_n(t,x) = \frac{\partial}{\partial x} P(X(t) = n \text{ and the elapsed service-time of the unit already in service at time } t \text{ is } \le x),$$

$$P(\text{a service is completed in } (t, t+\Delta) | \text{it started at time } t-x) = \eta(x)\Delta + o(\Delta),$$

and

$$P(z,t,x) = \sum_{n=1}^{\infty} P_n(x,t) z^n. \tag{6.1}$$

Observe the following:

1. $P_n(t,x)$ represents the joint p.f. of $X(t) = n$ and the elapsed service-time of the unit already in service equals x;
2. $\eta(x)$ is similar to the service rate μ except that it depends on the elapsed service-time;
3. In (6.1) $n = 0$ is not included and therefore $P(z,t,x)$ is not a p.g.f.

Because of the Markovian property, we have for $n \ge 2$,

$$P_n(t+\Delta, x+\Delta) = P_n(t,x)[(1-\lambda\Delta)(1-\eta(x)\Delta)] + P_{n-1}(t,x)\lambda\Delta + o(\Delta),$$
$$P_1(t+\Delta, x+\Delta) = P_1(t,x)[(1-\lambda\Delta)(1-\eta(x)\Delta)] + o(\Delta)$$

and

$$P_0(t+\Delta) = P_0(t)(1-\lambda\Delta) + \int_0^\infty P_1(t,x)\eta(x)\Delta\, dx + o(\Delta),$$

by the usual total probability argument as was done in Section 2.2 in Chapter 2 and by assuming the increments Δt in t and Δx in x are of the same order and are equal to Δ. In the last equation, notice that $x + \Delta x$ does not arise in the left side as the system is empty and that the integral in the right side is necessary whereas it does not exist in the first two equations. We can express the above equations as

$$\frac{\partial P_n(t,x)}{\partial t} + \frac{\partial P_n(t,x)}{\partial x} = \lambda P_{n-1}(t,x) - (\lambda + \eta(x))P_n(t,x), \qquad n \ge 2, \tag{6.2}$$

$$\frac{\partial P_1(t,x)}{\partial t} + \frac{\partial P_1(t,x)}{\partial x} = -(\lambda + \eta(x))P_1(t,x), \tag{6.3}$$

and

$$\frac{dP_0(t)}{dt} + \lambda P_0(t) = \int_0^\infty P_1(t,x)\eta(x)\, dx. \tag{6.4}$$

Suppose the system is initially empty. Then

$$P_0(0) = 1 \qquad \text{and} \qquad P_n(0,x) = 0, \quad n \ge 1. \tag{6.5}$$

Furthermore, the other boundary conditions are

$$P_n(t,0) = \int_0^\infty P_{n+1}(t,x)\eta(x)\, dx, \qquad n \ge 1, \tag{6.6}$$

and

$$P_1(t,0) = \int_0^\infty P_2(t,x)\eta(x)\,dx + \lambda\ P_0(t). \tag{6.7}$$

Let us follow Method 1 of Section 2.5 in the $M/M/1$ model. Making use of (6.1) on (6.2) and (6.3) we obtain

$$\frac{\partial}{\partial t}P(z,t,x) + \frac{\partial}{\partial x}P(z,t,x) = (\lambda z - \lambda - \eta(x))P(z,t,x).$$

Its L.T. becomes

$$\frac{\partial}{\partial x}P^*(z,\theta,x) + \theta P^*(z,\theta,x) = (\lambda z - \lambda - \eta(x))P^*(z,\theta,x), \tag{6.8}$$

with the help of (6.5). The L.T. of (6.4) is

$$(\lambda+\theta)P_0^*(\theta) - 1 = \int_0^\infty P_1^*(\theta,x)\eta(x)\,dx. \tag{6.9}$$

Likewise, combination of (6.6) and (6.7) leads to

$$P^*(z,\theta,0) = \frac{1}{z}\int_0^\infty P^*(z,\theta,x)\eta(x)\,dx - \int_0^\infty P_1^*(\theta,x)\eta(x)\,dx + z\lambda P_0^*(\theta). \tag{6.10}$$

At this point, further simplification in the line of the $M/M/1$ case is not possible because of the presence of the partial derivative in (6.8). However, the solution of the differential equation (6.8) is

$$P^*(z,\theta,x) = C(z,\theta)\exp\left(-(\lambda - \lambda z + \theta)x\right)\exp\left(-\int_0^x \eta(y)\,dy\right), \tag{6.11}$$

where $C(z,\theta)$ is to be determined. Using (6.11) in (6.10), we have

$$\begin{aligned}C(z,\theta) =& \frac{1}{z}c(z,\theta)\int_0^\infty \exp\left(-(\lambda - \lambda z + \theta)x\right)\eta(x)\exp\left(-\int_0^x \eta(y)\,dy\right)dx\\ &+ z\ \lambda P_0^*(\theta) - (\lambda+\theta)P_0^*(\theta) + 1.\end{aligned}$$

It can be checked that

$$f_S(x) = \eta(x)\exp\left(-\int_0^x \eta(y)\,dy\right).$$

Thus

$$C(z,\theta) = \frac{z(\lambda(1-z)+\theta)P_0^*(\theta) - z}{f_S^*(\theta+\lambda(1-z)) - z}. \tag{6.12}$$

By Rouché's Theorem (Section A.5) there is only one zero of the denominator in $|z| < 1$, for which the numerator must vanish so as to make $C(z,\theta)$ finite. Therefore $P_0^*(\theta)$ may be expressed as

$$P_0^*(\theta) = \frac{1}{\theta + \lambda(1 - z(\theta))}, \tag{6.13}$$

where $z(\theta)$ is the zero of

$$z = f_S^*(\theta + \lambda(1 - z)) \tag{6.14}$$

with $|z| < 1$ (observe that (6.14) is the same as (3.84)). Thus the transient solution is implicitly given through (6.11) where $C(z, \theta)$ is determined by (6.12), (6.13) and (6.14).

Remarks:

1. The solution of (6.14) is crucial for the final expression (6.11). For specific service-time distributions, one may attempt to find such solutions if not directly but by numerical techniques (see Section 4.2 in Chapter 4).

2. The result (6.11) seems to have more theoretical interest than practical value. Moreover, it does not give the transient solution $P_n(t)$ or $P_n(t, x)$ but gives something related to the p.g.f. of $P_n(t, x)$.

6.2.2. The $G/M/1$ model. In the $G/M/1$ model, the knowledge of the elapsed time since the last arrival epoch along with the number $X(t)$ will make the system Markovian. As the input is not necessarily Markovian, it is convenient to assume that $X(0-) = 0$ and $X(0) = 1$ (i.e., there is an arrival at time $t = 0$). In this model, we would be able to achieve more than the previous one and the development although looks similar is somewhat different.

Let

$$p_n(t, x) = \frac{\partial}{\partial x} P(X(t) = n \text{ and the elapsed time since the last arrival epoch is } \leq x)$$

which represents the joint p.f. of $X(t) = n$ and the elapsed time after the last arrival epoch equals to x. Clearly $p_n(t, 0)$ is the transition p.f. that the state of the system changes from $n - 1$ to n at time t. Furthermore, let $b_m(t)$ be the probability that m units complete service during an interval of length t assuming that there are more than m units at the beginning of the interval. Then

$$b_m(t) = e^{-\mu t} \frac{(\mu t)^m}{m!}. \tag{6.15}$$

On the other hand, if there are exactly m units at the beginning of the interval, the probability of their completion of service during the interval is

$$1 - \sum_{j=0}^{m-1} b_j(t)$$

which for simplicity may be denoted by $\bar{b}_m(t)$.

Now we are ready to express $p_n(t, 0)$ as a set of integro-difference equations (similar to differential-difference equations in Section 2.2 in Chapter 2). These are

$$p_1(t,0) = f_T(t)\bar{b}_1(t) + \int_0^t f_T(u) \sum_{m=0}^{\infty} p_{m+1}(t-u,0)\bar{b}_{m+1}(u)\,du,$$

$$p_2(t,0) = f_T(t)b_0(t) + \int_0^t f_T(u) \sum_{m=0}^{\infty} p_{m+1}(t-u,0)b_m(u)\,du,$$

$$p_n(t,0) = \int_0^t f_T(u) \sum_{m=0}^{\infty} p_{m+n-1}(t-u,0)b_m(u)\,du, \qquad n \geq 3. \tag{6.16}$$

In the right side of the first equation, the first term expresses the fact that the initial unit completes service before t with no arrival in between and the first arrival after the initial one occurs at t. The second term corresponds to the case when besides the occurrence of an arrival at t there are arrivals during the interval $(0,t)$ so that the last one occurs at epoch $t-u$ creating state $m+1$ and all these $m+1$ units finish being served during $(t-u,u)$. In the second equation, the first term arises from the fact that there is an arrival at t and the service of the initial unit is not completed during the interval. Its second term consists of the case with an arrival at t and arrivals during $(0,t)$ so that the last one occurs at $t-u$ creating state $m+1$, m of which finish being served during $(t-u,t)$. The explanation for the third one is similar except that the first term does not arise here.

Taking the L.T. of the last equation in (6.16), we get

$$p_n^*(\theta,0) = \sum_{m=0}^{\infty} \lambda_m(\theta) p_{m+n-1}^*(\theta,0), \quad n \geq 3 \tag{6.17}$$

where

$$\lambda_m(\theta) = \int_0^{\infty} e^{-\theta u} e^{-\mu u} \frac{(\mu u)^m}{m!} f_T(u)\,du.$$

Note that $\lambda_m(\theta)$ is the coefficient of z^m in the power series expansion of

$$f_T^*(\theta + \mu(1-z)).$$

In order to solve the set of homogeneous difference equations (6.17), substitute z^n for $p_n^*(\theta,0)$ in (6.17) and obtain the characteristic equation as

$$z = f_T^*(\theta + \mu(1-z)) \tag{6.18}$$

which is similar to (6.14). For the final solution of $p_n^*(\theta,0)$, not all roots of (6.18) are relevant which will be seen from the ensued discussion.

Let $p(t,0) = \sum_{n=1}^{\infty} p_n(t,0)$ which denotes the p.f. of an arrival at epoch t. It can be expressed as

$$p(t,0) = f_T(t) + \int_0^t p(t-u,0) f_T(u)\,du$$

by using an argument similar to that in (6.16), which leads to

$$p^*(\theta,0)=\sum_{n=1}^{\infty}p_n^*(\theta,0)=\frac{f_T^*(\theta)}{1-f_T^*(\theta)} \tag{6.19}$$

which should converge at least for $\Re(\theta)>0$. Thus, when $\Re(\theta)>0$, we consider for the solution of $p_n^*(\theta,0)$ only those roots of (6.18) with modulus less than unity so as to ensure the finiteness of $p^*(\theta,0)$. However, Rouché's Theorem shows that there is only one root with modulus less than unity. This line of argument is repeated in Section 2.2 in Chapter 2 and confinement to one root through Rouché's Theorem is done in Section 2.5 in Chapter 2 and in this section for the $M/G/1$ model.

Let the unique root of (6.18) be ξ. Then

$$p_n^*(\theta,0)=A\xi^n, \qquad n\geq 3 \tag{6.20}$$

where A is a constant to be determined. Interestingly, (6.20) is also true for $n=2$ which can be checked by considering the second equation in (6.16). If we take the first one in (6.16), we would get (see Exercise 4)

$$p_1^*(\theta,0)=A\xi-1. \tag{6.21}$$

Substitution of these expressions in (6.19) results in

$$\frac{A\xi}{1-\xi}-1=\frac{f_T^*(\theta)}{1-f_T^*(\theta)}$$

which yields

$$A=\frac{1-\xi}{\xi\,(1-f_T^*(\theta))}.$$

Hence

$$p_1^*(\theta,0)=\frac{f_T^*(\theta)-\xi}{1-f_T^*(\theta)},$$

and

$$p_n^*(\theta,0)=\frac{(1-\xi)\xi^{n-1}}{1-f_T^*(\theta)}, \qquad n\geq 2. \tag{6.22}$$

Following the argument with respect to $p_n(t,0)$, we may write the integro-difference equations for $P_n(t)$ as follows:

$$\begin{aligned}
P_0(t)&=\bar{F}_T(t)\bar{b}_1(t)+\int_0^t\bar{F}_T(u)\sum_{m=0}^{\infty}p_{m+1}(t-u,0)\bar{b}_{m+1}(u)\,du,\\
P_1(t)&=\bar{F}_T(t)b_0(t)+\int_0^t\bar{F}_T(u)\sum_{m=0}^{\infty}p_{m+1}(t-u,0)b_m(u)\,du,\\
P_n(t)&=\int_0^t\bar{F}_T(u)\sum_{m=0}^{\infty}p_{m+n}(t-u,0)b_m(u)\,du, \qquad n\geq 2.
\end{aligned} \tag{6.23}$$

Taking the L.T. of the last equation in (6.23) we obtain

$$P_n^*(\theta) = \sum_{m=0}^{\infty} p_{m+n}^*(\theta, 0) \int_0^{\infty} e^{-\theta u} b_m(u) \bar{F}_T(u)\, du.$$

Upon substitution of values of $p_{m+n}^*(\theta, u)$ from (6.22) and of $b_m(u)$ from (6.15), and after an elementary simplification, one gets

$$P_n^*(\theta) = \frac{(1-\xi)\xi^{n-1}}{1 - f_T^*(\theta)} \bar{F}_T^*(\theta + \mu(1-\xi)).$$

It is easy to verify that

$$\bar{F}_T^*(\theta) = \frac{1 - f_T^*(\theta)}{\theta}.$$

Thus

$$P_n^*(\theta) = \frac{(1-\xi)\xi^{n-1}}{1 - f_T^*(\theta)} \frac{1 - f_T^*(\theta + \mu(1-\xi))}{\theta + \mu(1-\xi)}.$$

Remembering that ξ satisfies (6.18), we get the final expression of $P_n^*(\theta)$ for $n \geq 2$ which incidentally is also valid for $n = 1$. The expression is

$$P_n^*(\theta) = \frac{(1-\xi)^2 \xi^{n-1}}{(1 - f_T^*(\theta))(\theta + \mu(1-\xi))}, \qquad n \geq 1. \tag{6.24}$$

Similarly the first equation of (6.23) leads to (see Exercise 5)

$$P_0^*(\theta) = \frac{1}{\theta} - \frac{1-\xi}{(1 - f_T^*(\theta))(\theta + \mu(1-\xi))}. \tag{6.25}$$

As a check for consistency, if we add up (6.24) and (6.25) for all values of n, we find

$$\sum_{n=0}^{\infty} P_n^*(\theta) = \frac{1}{\theta}$$

which is known to be true. What we have obtained is the transient solution in terms of the L.T.

In this model because of (6.24) and (6.25) we are in a position to evaluate the steady-state distribution $\{P_n\}$. Using the facts that (see Section A.5 in Appendix A)

$$\lim_{t\to\infty} P_n(t) = \lim_{\theta\to 0} \theta\ P_n^*(\theta)$$

and

$$\frac{1}{\lambda} = \int_0^{\infty} t\, dF_T(t)$$

where λ is the arrival rate, and that

$$f_T^*(\theta) = 1 - \frac{1}{\lambda}\theta + o(\theta)$$

and

$$\xi(\theta) = \xi_0 + \xi_1 \theta + o(\theta)$$

as the Taylor series expansions at $\theta = 0$ where $\xi_0 = \xi(0)$ and $\xi_1 = \xi'(0)$, it is left as an exercise to derive

$$P_n = \frac{\lambda}{\mu}(1 - \xi_0)\xi_0^{n-1}, \quad n \geq 1 \tag{6.26}$$

and

$$P_0 = 1 - \frac{\lambda}{\mu}.$$

6.3. Combinatorial Methods

In Section 3.6.2 in Chapter 3, the busy period distribution has been obtained by consideration of an urn problem which in essence is of combinatorial nature. A special version of the urn problem is the well-known ballot problem. Its relation to random walks and application to queueing models will be discussed first before coming to the urn problem.

6.3.1. The ballot problem, random walk and applications. The *ballot problem* is stated as follows:

> If in a ballot, candidate A scores a votes and candidate B, b votes, $a > b$, then what is the probability that A leads B throughout counting?
>
> The *ballot theorem* says: This probability is $(a - b)/(a + b)$.

The solution is usually derived by an elegant and simple combinatorial approach. Represent a vote for A by a horizontal unit and for B by a vertical unit. Then a sequence of voting record is represented by a *lattice path* from the origin to the point (a, b) and each path is equally likely. The representation of a sequence of votes for $(a, b) = (6, 4)$ is given in Figure 6.1.

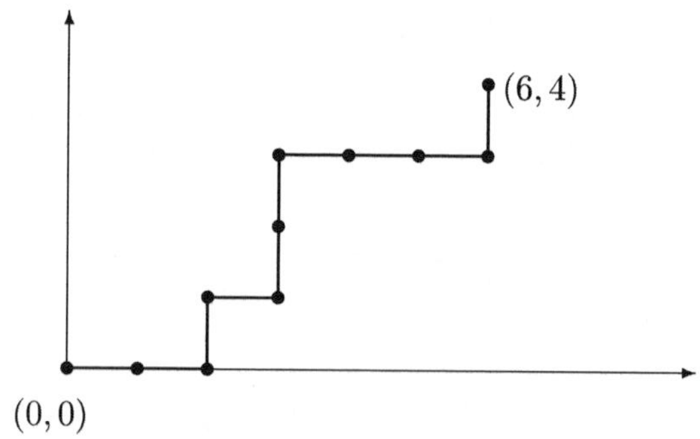

FIGURE 6.1. Lattice path for a sequence in $(6, 4)$

In terms of lattice paths, the solution of the ballot problem is the ratio of the number of paths that do not touch the line $x = y$ except at the origin to the total number of paths. the ratio can be written as

$$1 - \frac{N}{\binom{a+b}{b}}$$

where N is the number of paths that touch or cross the line $y = x$ at some point.

Consider counting the paths from $(0,0)$ to (a,b) which touch or cross the line $y = x + c$ $(c > 0)$. Denote the number by N_c. A typical path is shown in Figure 6.2.

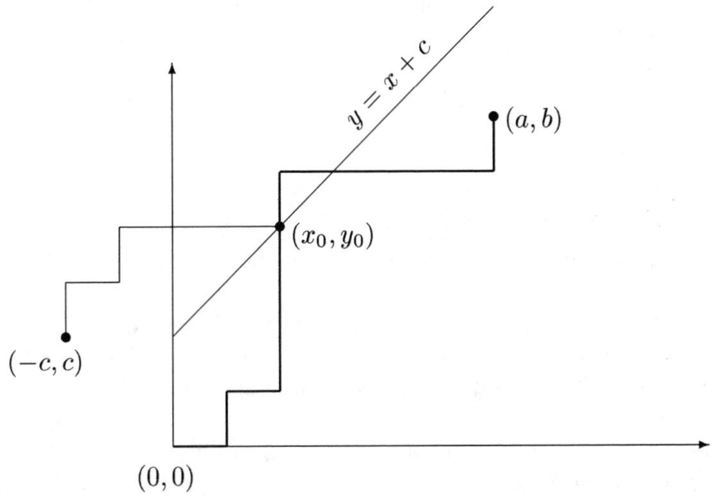

FIGURE 6.2. A path from $(0,0)$ to (a,b) crossing the line $y = x + c$

This path reaches the line $y = x + c$ for the first time say at (x_0, y_0). Reflect the portion of the path from the origin to (x_0, y_0) about the line $y = x + c$, as shown in the diagram and attach the remaining portion of the path to it. The new path is one from $(-c, c)$ to (a, b). It can be seen that to each path which reaches (touches or crosses) the line $y = x + c$, there corresponds a distinct path from $(-c, c)$ to (a, b) and vice versa. Therefore N_c is equal to the number of paths from $(-c, c)$ to (a, b) so that

$$N_c = \binom{a+b}{b-c}. \tag{6.27}$$

This combinatorial approach is known as the so-called *reflection principle.*

In order to evaluate N, we observe that N is the sum of the number of paths from $(1,0)$ to (a,b) that touch or cross the line $x = y$ and the number of paths from (0,1) to (a,b)(the second set always crosses the line at the origin). The first set of paths is equivalent to the paths from $(0,0)$ to $(a-1,b)$ that touch or cross the line $y = x + 1$ and thus its number is $\binom{a+b-1}{b-1}$ by (6.27). This leads to

$$N = 2\binom{a+b-1}{b-1}$$

and the desired probability of the ballot problem is given by

$$1 - \frac{N}{\binom{a+b}{b}} = 1 - \frac{2b}{a+b} = \frac{a-b}{a+b}\,.$$

In the urn problem (see (3.105)), suppose there are a cards marked with 0 and b cards marked with 2, implying that $n = a + b$ and $k = 2b$. Also $k < n$ implies $b < a$. In the context of the ballot problem, a vote for A or for B corresponds to 0 or 2, respectively. At the rth stage of counting let there be α votes for A and β for B. Then the event $\{\gamma_1 + \cdots + \gamma_r < r\}$ in (3.105) is equivalent to $\{2\beta < \alpha + \beta\}$, which is the same as saying that the number of votes for A is larger than that of B at any stage of counting. The probability by (3.105) is

$$1 - \frac{k}{n} = 1 - \frac{2b}{a+b} = \frac{a-b}{a+b}\,.$$

Thus we have shown that the classical ballot problem is a special case of the urn problem.

Before dealing with the urn problem and its applications, we rederive the transient solutions of the $M/M/1$ model (see (2.50)) by a random walk approach in which the newly developed combinatorial formula (6.27) will be useful.

Under the model assumptions (Section 2.2 in Chapter 2), there are two independent Poisson processes, one for arrivals with rate λ and the other for potential service completions with rate μ. An event in the second process is an actual service completion if the system is not empty. Let Y_t and Z_t, respectively, be the number of arrivals and the number of potential service completions during the time $(0, t]$. Then

$$P(Y_t = a) = e^{-\lambda t}\frac{(\lambda t)^a}{a!}, \quad a = 0, 1, \ldots \tag{6.28}$$

and

$$P(Z_t = b) = e^{-\mu t}\frac{(\mu t)^b}{b!}, \quad b = 0, 1, \ldots\,. \tag{6.29}$$

In the combined process, each event is either an arrival or a potential service completion with probabilities $\frac{\lambda}{\lambda+\mu}$ and $\frac{\mu}{\lambda+\mu}$, respectively. Let η_j be the number in the system after j events, $j = 0, 1, \ldots$ in the combined process. Then $\{\eta_j\}$ is an unrestricted random walk with transition probabilities

$$P(\eta_{j+1} = x + 1 \mid \eta_j = x) = \frac{\lambda}{\lambda + \mu}$$

and

$$P(\eta_{j+1} = x - 1 \mid \eta_j = x) = \frac{\mu}{\lambda + \mu}\,.$$

Here η_j can be negative.

Assume $Y_t = a$ and $Z_t = b$. We establish a one-to-one correspondence between the unrestricted random walk and the queueing process $M/M/1$ through lattice path representation.

Represent an arrival and a potential service completion by a horizontal and a vertical unit, respectively. In the queueing process, let there be i units initially. For $a = 8$, $b = 6$ and $i = 1$, the lattice paths corresponding to a given realization of the random walk and that of its associated queueing process are given in Figures 6.3 and 6.4.

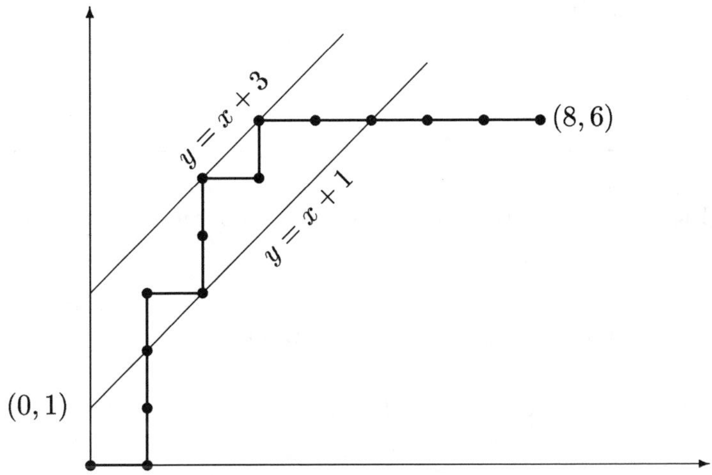

FIGURE 6.3. Lattice path representation of random walk

Observe that the line $y = x + 1$, in general $y = x + i$ acts as a boundary for the queueing process. The system becomes empty as soon as it hits the boundary. Therefore, any step corresponding to a potential service completion which is not actual is represented by a diagonal step on the line $y = x + i$. Moreover, it is obvious that if there are $k(\geq 0)$ diagonal steps then the corresponding random walk path touches but does not cross the line $y = x + k + i$ and vice versa. On the other hand, if the queueing process is not empty at any time then its path stays below the line $y = x + i$. In that case, the corresponding random walk path is the same. This establishes a one-to-one correspondence between the random walk and the queueing process.

Coming back to (2.50), we have

$$P_n(t) = \sum_a \sum_b P(X(t) = n \mid Y_t = a,\ Z_t = b) P(Y_t = a) P(Z_t = b), \quad (6.30)$$

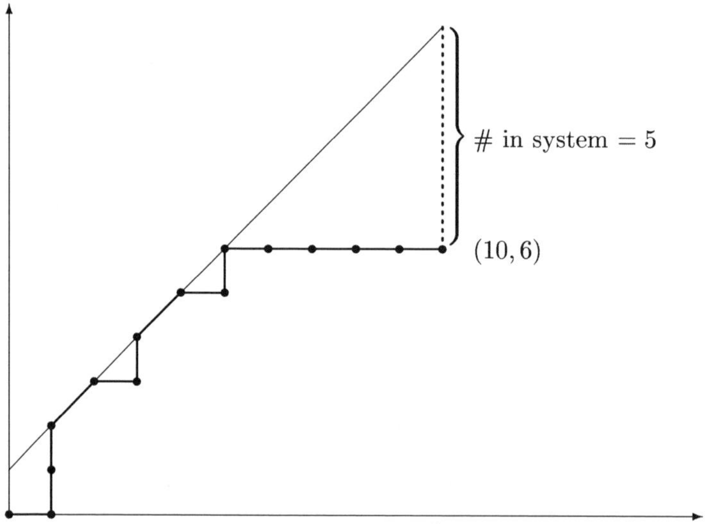

FIGURE 6.4. Lattice path representation of queueing process

where a, b and n are related, to be discussed later. Given $Y_t = a$ and $Z_t = b$, any realization of the random walk has the probability

$$\left(\frac{\lambda}{\lambda+\mu}\right)^a \left(\frac{\mu}{\lambda+\mu}\right)^b,$$

and therefore

$$P(X(t) = n \mid Y_t = a,\ Z_t = b) = \frac{N^* \left(\frac{\lambda}{\lambda+\mu}\right)^a \left(\frac{\mu}{\lambda+\mu}\right)^b}{\binom{a+b}{b} \left(\frac{\lambda}{\lambda+\mu}\right)^a \left(\frac{\mu}{\lambda+\mu}\right)^b} = \frac{N^*}{\binom{a+b}{b}} \tag{6.31}$$

where N^* is the number of random walk paths which correspond to the event $X(t) = n$.

For a given a and b, lattice paths corresponding to random walk paths contributing to N^*, start from $(0, 0)$ and end at (a, b). These paths are divided into two classes, (i) those that do not touch $y = x + i$ and (ii) those that touch or cross $y = x + i$. Consider (i) in which case $n = a + i - b$ and the number of paths is the total number of paths minus the number of paths that touch or cross the line $y = x + i$. This gives us the corresponding N^* as

$$N^* = \binom{a+b}{b} - \binom{a+b}{b-i} \tag{6.32}$$

by using (6.27), where a and b are constrained by $n = a + i - b$. Note that if $i = 1$, then (6.32) becomes $\frac{a+1-b}{a+b+1}\binom{a+b+1}{b}$ which is the number of lattice paths from the origin to $(a + 1, b)$ that do not touch the line $x = y$.

When we deal with case (ii), we find that $n = a+i-b+k$, $k = 0, 1, \ldots, b-i$, where k is determined by the line $y = x+i+k$ which is touched but not crossed by the random walk paths (k is also the number of diagonal steps on $y = x+i$ in queueing process paths). This prompts us to determine the number of paths that touch but do not cross the line $y = x + i + k$, $k = 0, 1, \ldots, b - i$. Again, an application of (6.27) yields the number to be

$$\binom{a+b}{b-i-k} - \binom{a+b}{b-i-k-1}, \quad k = 0, 1, \ldots, b-i. \tag{6.33}$$

Now, putting (6.28), (6.29), (6.31), (6.32) and (6.33) in (6.30) we have

$$\begin{aligned} P_n(t) \;=\; & e^{-(\lambda+\mu)t}\Bigg[\sum_a \sum_b \left(\frac{1}{a!b!} - \frac{1}{(a+i)!(b-i)!}\right)\lambda^a \mu^b t^{a+b} \\ & + \sum_a \sum_b \sum_{k=0}^{b-i}\Bigg(\frac{1}{(a+i+k)!(b-i-k)!} \\ & - \frac{1}{(a+i+k+1)!(b-i-k-1)!}\Bigg)\lambda^a \mu^b t^{a+b}\Bigg] \end{aligned} \tag{6.34}$$

subject to $n = a + i - b$ and $n = a + i - b + k$, $k = 0, 1, \ldots, b - i$ in the first and second sum, respectively. Incorporating these stipulations, we get

$$\begin{aligned} P_n(t) &= \\ &= e^{-(\lambda+\mu)t}\Bigg[\left(\sum_{b=0}^{\infty} \frac{1}{b!(b+n-i)!} - \sum_{b=i}^{\infty} \frac{1}{(b-i)!(b+n)!}\right)\lambda^{b+n-i}\mu^b t^{2b+n-i} \\ & + \sum_{b=i}^{\infty} \sum_{k=0}^{b-i}\Bigg(\frac{1}{(b-i-k)!(b+n)!} \\ & - \frac{1}{(b-i-k-1)!(b+n+1)!}\Bigg)\lambda^{b+n-i-k}\mu^b t^{2b+n-i-k}\Bigg] \end{aligned} \tag{6.35}$$

which is a combinatorial transient solution of the $M/M/1$ model.

Let us show that (6.35) agrees with (2.50). Recalling the expression for the modified Bessel function $I_n(z)$ in (2.48), we find that the first two sums in (6.35) become

$$e^{-(\lambda+\mu)t}\rho^{\frac{n-i}{2}}\left(I_{n-i}(2\sqrt{\lambda\mu}t) - I_{n+i}(2\sqrt{\lambda\mu}t)\right).$$

Put $r = b - i - k$, $c = n + i + k$ and $r = b - i - k - 1$, $c = n + i + k + 2$ in the first and second term, respectively, in the last sum of (6.35). It simplifies to

$$e^{-(\lambda+\mu)t}\left(\sum_{c=n+i}^{\infty} \rho^{n-\frac{c}{2}} I_c(2\sqrt{\lambda\mu}t) - \sum_{c=n+i+2}^{\infty} \rho^{n+1-\frac{c}{2}} I_c(2\sqrt{\lambda\mu}t)\right).$$

Thus, another expression for $P_n(t)$ is

$$\begin{aligned}P_n(t) &= e^{-(\lambda+\mu)t}\rho^{\frac{n-i}{2}}\left(I_{n-i}(2\sqrt{\lambda\mu}t) - I_{n+i}(2\sqrt{\lambda\mu}t)\right) \\ &\quad + e^{-(\lambda+\mu)t}\left(\sum_{c=n+i}^{\infty}\rho^{n-\frac{c}{2}}I_c(2\sqrt{\lambda\mu}t) - \sum_{c=n+i+2}^{\infty}\rho^{n+i-\frac{c}{2}}I_c(2\sqrt{\lambda\mu}t)\right)\end{aligned} \tag{6.36}$$

which by an elementary simplification, checks with (2.50). Compare (6.36) with (2.61) and note that the two terms in (6.36) and (2.61) (see Exercise 19, Chapter 2) arise out of division of radom walk paths into two mutually exclusive classes (i) and (ii).

Instead of fixing a and b, we may fix $m = a + b$ and b. Although one can use this substitution in (6.35), it is of interest to proceed directly with m, b and n. We start with

$$P_n(t) = \sum_{m=|n-i|}^{\infty} P(X(t) = n \mid Y_t + Z_t = m)P(Y_t + Z_t = m).$$

It is known that $Y_t + Z_t$ has Poisson distribution with parameter $(\lambda + \mu)t$ and thus

$$P(Y_t + Z_t = m) = e^{-(\lambda+\mu)t}\frac{(\lambda+\mu)^m t^m}{m!}\ .$$

(It is so because

$$\begin{aligned}P(Y_t + Z_t = m) &= \sum_{b=0}^{m} e^{-(\lambda+\mu)t}\frac{(\lambda t)^{m-b}}{(m-k)!}\frac{(\mu t)^b}{b!} \quad \text{by (6.28) and (6.29)} \\ &= e^{-(\lambda+\mu)t}\frac{(\lambda+\mu)^m t^m}{m!}\ .)\end{aligned}$$

The first factor can be written as

$$\begin{aligned}&P(X(t) = n \mid Y_t + Z_t = m) = \\ &= \sum_b P(X(t) = n \mid Y_t + Z_t = m, Z_t = b)P(Z_t = b \mid Y_t + Z_t = m) \\ &= \sum_b N^* \frac{\lambda^{m-b}\mu^b}{(\lambda+\mu)^m} \quad \text{(see (6.31))}.\end{aligned}$$

Here N^* will be evaluated a little differently.

For computing N^*, we fix $b - k = r$ (say), where r is the number of vertical steps in paths representing the queueing process or the number of actual service completions. If m is such that $|n - i| \le m \le n + i$, then m events consist entirely of arrivals and actual service completions. In that case $k = 0$ (remember k is the number of diagonal steps on $y = x + i$ in queueing process paths). The corresponding queueing paths are from $(0, 0)$ to $(m - r, r)$ without any restriction. Moreover, from $m = a + b$ and $n - i = a - b$, it follows that $b = r = \frac{1}{2}(m-n+i) = \ell$ (say). Therefore, the contribution to $P_n(t)$ is

$$\sum_{m=|n-i|}^{n+i} \binom{m}{b} \lambda^{m-b}\mu^b e^{-(\lambda+\mu)t}\frac{t^m}{m!} = \sum_{m=|n-i|}^{n+i} \binom{m}{\ell} \lambda^{n-i+\ell}\mu^\ell e^{-(\lambda+\mu)t}\frac{t^m}{m!}. \tag{6.37}$$

Note $\binom{m}{r} = 0$, if r is not a non-negative integer.

Suppose $m \geq n+i+1$. Since the number of vertical steps is r, there are $n-i+r$ horizontal steps and $k = m-n+i-2r$ diagonal steps. If $k=0$, then $r = \ell$. The corresponding queueing paths are from (0,0) to $(m-r, r)$ such that they do not cross the line $y = x+i$ without having any diagonal steps. The number of such paths is $\binom{m}{\ell} - \binom{m}{\ell-i-1}$ by (6.27). The contribution to $P_n(t)$ from this source is

$$\sum_{m=n+i+1}^{\infty} \left\{\binom{m}{\ell} - \binom{m}{\ell-i-1}\right\} \lambda^{n-i+\ell}\mu^\ell e^{-(\lambda+\mu)t}\frac{t^m}{m!}. \tag{6.38}$$

In case, $k > 0$, i.e., $k = 1, \ldots, b-i$ which is equivalent to $r = i, \ldots, \left[\frac{2\ell-1}{2}\right]$ (when $k = b-i$, one gets $r = i$ from $b-k = r$ and when $k=1$, we have $r = \left[\frac{m-n+i-1}{2}\right]$), the queueing paths will have k diagonal steps on the line $y = x+i$. Noting the discussion just before (6.33) and using (6.33), we have the contribution as

$$\sum_{m=n+i+1}^{\infty} \sum_{r=i}^{\left[\frac{2\ell-1}{2}\right]} \left\{\binom{m}{r-i} - \binom{m}{r-i-1}\right\} \lambda^{n-i+r}\mu^{m-n+i-r} e^{-(\lambda+\mu)t}\frac{t^m}{m!}. \tag{6.39}$$

When (6.37), (6.38) and (6.39) are combined, we obtain

$$\begin{aligned} P_n(t) &= e^{-(\lambda+\mu)t}\left\{\sum_{m=|n-i|}^{n+i} \binom{m}{\ell} \lambda^{n-i+\ell}\mu^\ell \frac{t^m}{m!}\right. \\ &+ \sum_{m=n+i+1}^{\infty} \left(\binom{m}{\ell} - \binom{m}{\ell-i-1}\right) \lambda^{n-i+\ell}\mu^\ell \frac{t^m}{m!} \\ &+ \left.\sum_{m=n+i+1}^{\infty} \sum_{r=i}^{\left[\frac{2\ell-1}{2}\right]} \left(\binom{m}{r-i} - \binom{m}{r-i-1}\right) \lambda^{n-i+r}\mu^{m-n+i-r} \frac{t^m}{m!}\right\}. \end{aligned} \tag{6.40}$$

Inverse Processes

We may yet take another route and still use the lattice path combinatorics to handle the same problem. The alternative route is primarily through *inverse process*. Consider the birth-death process and its flow diagram in Section 2.3 in Chapter 2, which is repeated in Figure 6.5.

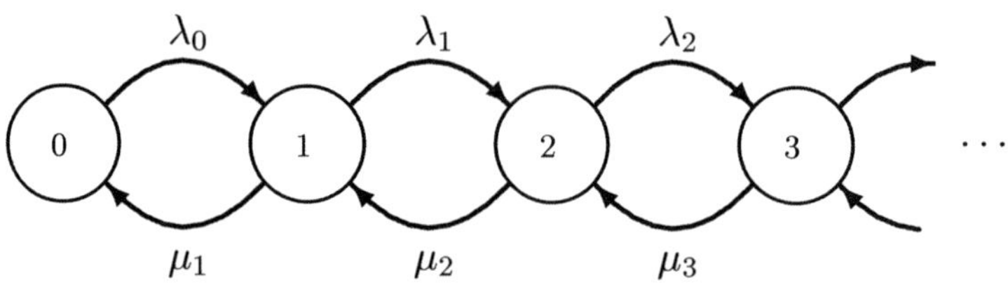

FIGURE 6.5. Flow diagram for birth-death process

The process having the flow diagram in Figure 6.6 is called its inverse process.

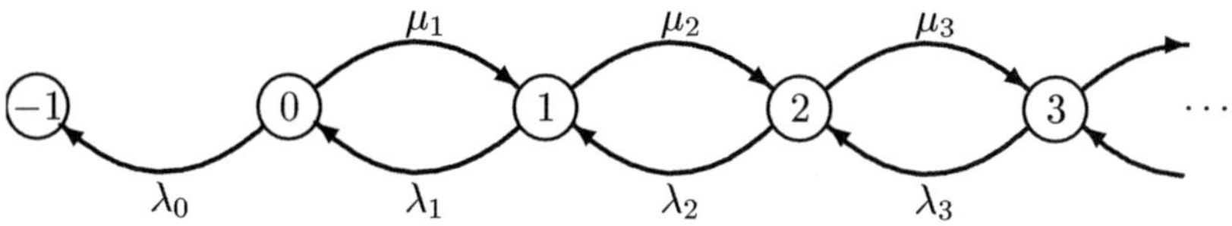

FIGURE 6.6. Flow diagram of the inverse process

Next we state the basic inversion result connecting these two processes.

Let $P_{i,j}(t)$ be the one-step transition probability of moving from state i at time 0 to state j at time t in the birth-death process and let $P^*_{i,j}(t)$ be the same for the inverse process. Then

$$P_{i,j}(t) = \sum_{k=i}^{\infty} [P^*_{j,k}(t) - P^*_{j-1,k}(t)]$$

and

$$P^*_{i,j}(t) = \sum_{k=0}^{i} [P_{j,k}(t) - P_{j+1,k}(t)] \tag{6.41}$$

for all states $i, j = 0, 1, 2, \ldots$ with the convention $P^*_{-1,k}(t) = 0$ if $k > -1$.

We may observe that a direct substitution of the second expression for different values of $P^*_{r,s}(t)$'s in the right side of the first checks it, and similarly the second one. Thus the two relations in (6.41) are merely inverse relations of each other and this is the rationale for calling the second process the inverse of the first. (In the literature it is referred to as the dual process. But our dual processes are different in which the roles of arrivals and service completions are interchanged. See Section 3.5 last remark, the last paragraph of Section 3.6 in Chapter 3 and Section 7.7 in Chapter 7.)

The construction of the inverse process still remains a mystery for the reader. One way to unfold it, is to traverse through the randomization method of Section 4.4.4 in Chapter 4. Invoking (4.77) and (4.75), we realize that

$$P_{i,j}(t) = e^{-qt} \sum_{n=0}^{\infty} \frac{(qt)^n}{n!} P_{i,j}^{[n]} \tag{6.42}$$

where $P_{i,j}^{[n]}$ is the n-step transition probability from i to j in the Markov chain whose transition matrix $\mathbf{P}$ is determined by

$$\mathbf{P} = \frac{1}{q}\mathbf{Q} + \mathbf{I}, \tag{6.43}$$

$\mathbf{Q}$ being the generator matrix of the birth-death process and $q \geq \sup_i |q_{ii}|$. (Realize that for the $M/M/1$ model $P_{i,j}^{[n]} = P(X(t) = j | Y_t + Z_t = n)$.) A similar statement holds good for the inverse process. Notations for this process are marked with superscript *.

These are

$$P_{i,j}^*(t) = e^{-qt} \sum_{n=0}^{\infty} \frac{(qt)^n}{n!} P_{i,j}^{*[n]} \tag{6.44}$$

and

$$\mathbf{P}^* = \frac{1}{q}\mathbf{Q}^* + \mathbf{I}. \tag{6.45}$$

We have used q instead of q^* since we can take q any number such that

$$q \geq \max(\sup |q_{ii}|, \sup |q_{ii}^*|).$$

The inversion result connecting the birth-death chains corresponding to the two processes is similar to (6.41) and is stated below.

$$P_{r,s}^{[n]} = \sum_{k=r}^{\infty} [P_{s,k}^{*[n]} - P_{s-1,k}^{*[n]}] \text{ and } P_{r,s}^{*[n]} = \sum_{k=0}^{r} [P_{s,k}^{[n]} - P_{s+1,k}^{[n]}] \tag{6.46}$$

for $n \geq 1$ and $r, s = 0, 1, 2, \ldots$ with the convention $P_{-1,k}^{[n]} = 0$ if $k > -1$.

Now, given $\mathbf{Q}$ we generate $\mathbf{P}$ by (6.43) and then $\mathbf{P}^*$ by (6.46) and finally come to $\mathbf{Q}^*$ through (6.45). This explains Figure 6.6. Thus

$$\mathbf{P} = \begin{bmatrix} 1 - \frac{\lambda_0}{q} & \frac{\lambda_0}{q} & & & \\ \frac{\mu_1}{q} & 1 - \frac{\lambda_1 + \mu_1}{q} & \frac{\lambda_1}{q} & & \\ & \frac{\mu_2}{q} & 1 - \frac{\lambda_2 + \mu_2}{q} & \frac{\lambda_2}{q} & \\ & & & \ddots & \end{bmatrix}, \tag{6.47}$$

where row and column indices $i, j = 0, 1, 2, \ldots$. The corresponding matrix $\mathbf{P}^*$ for the inverse process is

$$\mathbf{P}^* = \begin{bmatrix} 1 & & & & \\ \frac{\lambda_0}{q} & 1-\frac{\lambda_0+\mu_1}{q} & \frac{\mu_1}{q} & & \\ & \frac{\lambda_1}{q} & 1-\frac{\lambda_1+\mu_2}{q} & \frac{\mu_2}{q} & \\ & & & & \ddots \end{bmatrix} \tag{6.48}$$

where row and column indices $i, j = -1, 0, 1, \ldots$. Observe that both the inverse process and the associated Markov chain with transition matrix $\mathbf{P}^*$ have state -1 as the absorbing barrier.

Let us turn to the $M/M/1$ system in which $\lambda_i = \lambda$ and $\mu_i = \mu$ for all i. Take $q = \lambda + \mu$. In this case $\mathbf{P}^*$ defines a random walk with an absorbing barrier at state -1 having an arrival with probability $\frac{\mu}{\lambda+\mu}$ and a service completion with probability $\frac{\lambda}{\lambda+\mu}$. In order to evaluate $P^{*[n]}_{s,k}$, the n-step probability of the random walk moving from s to k without crossing state 0, we refer back to the lattice path representation in Figure 6.3 of a random walk. Realize that an arrival or a service completion is represented, respectively, by a horizontal or a vertical step in the lattice path. Consider paths having h horizontal steps and v vertical steps. Each such path has probability $\left(\frac{\mu}{\lambda+\mu}\right)^h \left(\frac{\lambda}{\lambda+\mu}\right)^v$. The number of paths equals the number of those from (0,0) to (h, v) without touching the line $y = x + s + 1$, which is

$$\binom{h+v}{v} - \binom{h+v}{v-s-1} \tag{6.49}$$

by (6.27). But $h + v = n$ and $h - v = k - s$ yielding

$$h = \frac{n+k-s}{2} \text{ and } v = \frac{n-k+s}{2}$$

so that

$$P^{*[n]}_{s,k} = \left[\binom{n}{\frac{n-k+s}{2}} - \binom{n}{\frac{n-k-s-2}{2}}\right]\left(\frac{\mu}{\lambda+\mu}\right)^{\frac{n+k-s}{2}}\left(\frac{\lambda}{\lambda+\mu}\right)^{\frac{n-k+s}{2}} \tag{6.50}$$

with the convention that $\binom{n}{r} = 0$ if $r < 0$, r is not an integer or $n < r$. Tracing back our steps, we can get $P^*_{s,k}(t)$ from (6.44) and finally $P_{i,j}(t)$ from (6.41). Thus

$$\begin{aligned} P_{i,j}(t) &= e^{-(\lambda+\mu)t}\sum_{n=0}^{\infty}\frac{t^n}{n!}\sum_{k=i}^{\infty}\left[\left(\binom{n}{\frac{n-k+j}{2}} - \binom{n}{\frac{n-k-j-2}{2}}\right)\mu^{\frac{n-k-j}{2}}\lambda^{\frac{n-k+j}{2}}\right. \\ &\left. - \left(\binom{n}{\frac{n-k+j-1}{2}} - \binom{n}{\frac{n-k-j-1}{2}}\right)\mu^{\frac{n+k-j+1}{2}}\lambda^{\frac{n-k+j-1}{2}}\right]. \end{aligned} \tag{6.51}$$

Remarks:

1. It seems odd to derive $P_{i,j}(t)$ in a round about manner via inverse process. The basic idea of introducing the inverse process is that it, or its associated Markov chain, might have an absorbing barrier and that counting lattice paths is relatively easier under this situation as can be seen in the $M/M/1$ model in the original treatment. This can be further evidenced if we consider for example the same model with finite system capacity N. We will discuss this model very shortly.
2. Whether there is an absorbing barrier can be checked by creating the inverse process through $\mathbf{Q}$ to $\mathbf{P}$ by (6.43), to $\mathbf{P}^*$ by (6.46) and to $\mathbf{Q}^*$ by (6.45).
3. On an examination of (6.40) and (6.51), one recognizes that both are series expansion in powers of t and the coefficient of $\frac{t^n}{n!}e^{-(\lambda+\mu)t}$ is $P_{i,j}^{[n]}$. This derivation is arrived at by a probabilistic argument in one case and by an algebraic one leading to the randomization method in the other case.

We want to apply the inverse process technique to $M/M/1$ model having finite system capacity N. Here the inverse process is the same as in $M/M/1$ with an additional absorbing barrier at N. Proceeding similarly, we have only to change (6.49) to the appropriate number of lattice paths. By using the usual inclusion-exclusion method and repeated reflection principle it can be proved that the number of lattice paths from the origin to (m, n) that do not touch the lines $x = y + t$ and $x = y - s$, $t > 0$, $s > 0$ is

$$\sum_{h=-\infty}^{\infty} \left[\binom{m+n}{m - h(t+s)} - \binom{m+n}{m - h(t+s) - t} \right]. \tag{6.52}$$

Thus, denoting by $P_{s,k}^{*[n]} \mid N$ the respective probability under finite capacity N, we can see that

$$\begin{aligned} P_{s,k}^{*[n]} \mid N &= \sum_{h=-\infty}^{\infty} \left[\binom{n}{\frac{n-k+s}{2} - h(N+1)} - \binom{n}{\frac{n+k+s}{2} + h(N+1) + 1} \right] \\ &\quad \times \left(\frac{\mu}{\lambda+\mu} \right)^{\frac{n+k-s}{2}} \left(\frac{\lambda}{\lambda+\mu} \right)^{\frac{n-k+s}{2}}. \end{aligned} \tag{6.53}$$

The same problem and that of $M/M/1$ model when $N \to \infty$ have non-combinatorial solutions in Section 4.5 in Chapter 4 ((4.86), (4.87), (4.88) and (4.89)), which are obtained by an algebraic approach.

6.3.2. The urn problem and applications. Now we give the solution of the ballot problem that the probability of A leading B throughout counting equals $(a - b)/(a + b)$, by another technique called the method of penetrating analysis which will be used in the urn problem. For illustration, let us take $a = 7$, $b = 4$ for which a typical path is PQ in Figure 6.7.

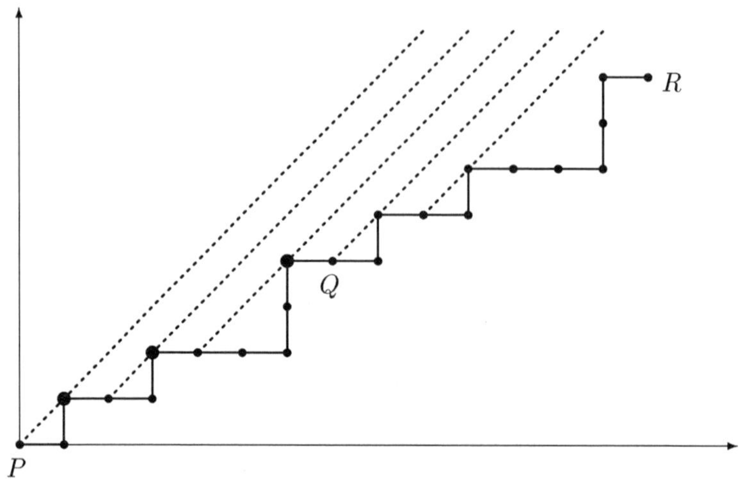

FIGURE 6.7. Penetrating analysis

In the figure QR is the repetition of PQ. Any path starting from a lattice point in PQ excluding Q and ending after 11 (i.e., $7+4$) units is said to be a cyclic permutation of the path PQ. Thus there are 11 cyclic permutations of PQ.

Imagine PQR to be an opaque screen which is hit by an infinitely distant light source in the positive direction of the line $x = y$ (see the figure). It can be seen that any cyclic permutation which only starts from one of the points in the light (not in the shade) will not touch the line $x = y$ and any path starting with a vertical unit will touch the line $x = y$. Among points in PQ excluding Q, only 3 points (each marked by a big dot in the diagram) are in the light. Observe that every vertical unit blocks a horizontal unit. Thus among 7 points starting with a horizontal unit, only 3 ($= 7 - 4$) are in the light. This analysis shows that among 11 (in general $a+b$) cyclic permutations of PQ, only 3 (in general $a-b$) will not touch the line $x = y$. A little reflection shows that this is true if we begin with any path. Therefore, the probability that A leads B, being independent of any selection of path, is $3/11$ and in general is $(a-b)/(a+b)$.

The same effect can be achieved by considering a fixed sequence of a A's and b B's (A for A's vote and B for B's vote) put on an oriented circle and deleting all adjacent pairs AB in the chosen direction. The left-over $a-b$ A's are the starting points of those sequences among $a+b$ sequences for which A leads B and this yields our result.

Next let us turn to the urn problem (3.105). Here also consider a particular sequence of numbers on the cards. To each card in the sequence, associate a horizontal unit followed by vertical units the number of which is equal to the number on the card. The sequence is thus represented by a lattice path from $(0,0)$ to (n,k). By repeating the path we find that any cyclic arrangement of the sequence is represented by a path which is a segment of the two adjoining paths having the same length. Use the procedure of penetrating analysis (i.e., light, opaque screen and shade) as in the diagram. The event $\{\nu_1+\ldots+\nu_r<r, r=1,\ldots,n\}$ corresponds to those cyclic paths, the starting point of each of which is in the light. The number of such paths is again the number of horizontal units minus the number of vertical units (i.e., $n-k$) and therefore (3.105) is established.

Let us go back to (3.103) or (3.104). Consider a sequence $(\nu_1,\ldots,\nu_n)$ of *cyclically interchangeable* (for definition see Section A.2 in Appendix A) r.v.'s taking on non-negative integer values, i.e.,

$$P(\nu_1=x_1,\ldots,\nu_n=x_n)=P(\nu_{i_1}=x_1,\ldots,\nu_{i_n}=x_n)$$

for every $(i_1,\ldots,i_n)$, a cyclic permutation of $(1,\ldots,n)$, where $x_1,\ldots,x_n$ are non-negative integers. Then the urn problem leads to the following result:

Let $\nu_1,\ldots,\nu_n$ be non-negative integer-valued cyclically interchangeable r.v.'s. Then

$$P\left\{\sum_{i=1}^{r}\nu_i<r, 1\le r\le n\,\middle|\,\sum_{i=1}^{n}\nu_i=k\right\}=\begin{cases}1-\dfrac{k}{n} & \text{if } 0\le k\le n,\\ 0 & \text{otherwise.}\end{cases} \tag{6.54}$$

A formal proof of the result along the lines discussed above is presented below.

Proof: Let $k_1,\ldots,k_n$ be non-negative integers such that $k_1+\ldots+k_n=k$, $0\le k\le n$ and let $k_{n+r}=k_r$, $r=1,\ldots,n$. Denote by δ_r for $r=0,1,\ldots,n$ the following:

$$\delta_r=\begin{cases}1 & \text{if } k_{r+1}+\ldots+k_i<i-r \text{ for } r<i\le r+n,\\ 0 & \text{otherwise.}\end{cases}$$

We can easily see that

$$\sum_{j=1}^{n}\delta_j=n-k. \tag{6.55}$$

If we replace $k_1,\ldots,k_n$ by $\nu_1,\ldots,\nu_n$ then δ_j's are r.v.'s, and the identity (6.55) becomes

$$\sum_{j=1}^{n}\delta_j=\left(n-\sum_{j=1}^{n}\nu_j\right)^+ \tag{6.56}$$

with probability 1 where $(x)^+=\max\{0,x\}$. Therefore

$$P\left(\sum_{i=1}^{r}\nu_i<r \text{ for } 1\le r\le n\,\middle|\,\sum_{i=1}^{n}\nu_i=k\right)=E\left(\delta_0\,\middle|\,\sum_{i=1}^{n}\nu_i=k\right).$$

Because $\nu_1, \ldots, \nu_n$ are cyclically interchangeable, δ_i's have the same distribution and hence

$$\begin{aligned} E\left(\delta_0 \middle| \sum_{i=1}^{n} \nu_i = k\right) &= \frac{1}{n} E\left(\sum_{i=1}^{n} \delta_i \middle| \sum_{i=1}^{n} \nu_i = k\right) \\ &= \frac{1}{n}\left((n - \sum_{i=1}^{n} \nu_i)^+ \middle| \sum_{i=1}^{n} \nu_i = k\right) \quad \text{by (6.56)} \\ &= \frac{n-k}{n} . \end{aligned}$$

This completes the proof.

Remark: Since i.i.d. variables are *interchangeable* (see Section A.2 in Appendix A for definition) which in turn are cyclically interchangeable, (6.54) is true for interchangeable and i.i.d. variables. Thus we have not only provided a direct combinatorial proof for (3.104) (therefore for (3.103)), but also have a result which is true for a larger class of variables than i.i.d. variables.

We have seen in Section 6.3.1 that the solution to the ballot problem which forms the combinatorial nucleus of transient analysis of $M/M/1$ model, is a special case of (6.54). In fact, more can be squeezed out of (6.54). Suppose there are a ν_i's equal to zero and the remaining equal to $m+1$, m being a non-negative integer. If among the first r ν_i's there are α_r zeroes and β_r $(m+1)$'s, then the event $\{\sum_{i=1}^{r} \nu_i < r, 1 \leq r \leq n\}$ is equivalent to $\{m\beta_r < \alpha_r, 1 \leq r \leq n\}$ and $\sum_{i=1}^{n} \nu_i = (m+1)(n-a)$. According to (6.54),

$$P\{m\beta_r < \alpha_r, 1 \leq r \leq n\} = \frac{a - m(n-a)}{n} = \frac{a - mb}{a+b}, \quad b = n - a.$$

Refer back to the ballot problem in Section 6.3.1. The left side is the probability that A's votes are larger than m times B's votes at every stage of counting.

If all sequences of voting records are equally likely such as the situation in the denominator of (6.31), then the total number of sequences in the event is

$$\frac{a - mb}{a+b}\binom{a+b}{b}. \tag{6.57}$$

In terms of lattice paths, it is the number of paths from the origin to (a, b) that do not touch the line $x = my$ except at the origin. Translating it to queueing models, one may think of $M/M/1$ model with batch arrivals of size m and is interested in the busy period analysis.

Suppose the busy period starts when a batch of m customers arrives. Denote by $G(n;t)$ the probability that the length of the busy period is $\leq t$ and the number of arriving batches during the busy period is $n-1$. Note that $G(n;t) = G(n,t)$ (see (3.101)) for $m = 1$. In order to derive $G(n;t)$, we use a probabilistic conditioning argument as it was done in Section 3.6.2 in Chapter 3. However, we take advantage of the special properties of $M/M/1$ model similar to the one in Section 6.3.1 for $P_n(t)$ which is derived by conditioning $Y_t + Z_t = m$. In our case, given that the length of the busy period is u during which $nm + n - 1$ ($n-1$ arrivals and nm service completions) events occur from the combined Poisson process $(\lambda + \mu)$ and the last one occurs during $(u, u + \Delta u)$, we derive the conditional probability of the busy period of interest. Under this condition, the probability of any sequence of events having $n-1$ arrivals and nm service completions is

$$\left(\frac{\lambda}{\lambda+\mu}\right)^{n-1} \left(\frac{\mu}{\lambda+\mu}\right)^{nm}.$$

To find the number of sequences, we represent a service completion by a horizontal unit and an arrival by a vertical unit. Then a sequence in the busy period corresponds to a lattice path from the origin to $(nm, n-1)$ that does not touch the line $x = my + m$ except at the end. If we shift the origin to $(nm, n-1)$ and view the path in the reverse order, then the path becomes one from the origin to $(nm, n-1)$ which does not touch the line $x = my$ except at the origin. According to (6.57), the number of such paths is

$$\frac{m}{n(m+1)-1}\binom{n(m+1)-1}{n-1}.$$

Taking into account the above three factors, we get

$$\begin{aligned} G(n;t) &= \int_0^t \frac{m}{n(m+1)-1}\binom{n(m+1)-1}{n-1}\left(\frac{\lambda}{\lambda+\mu}\right)^{n-1}\left(\frac{\mu}{\lambda+\mu}\right)^{nm} \\ &\quad \times e^{-(\lambda+\mu)u}\frac{((\lambda+\mu)u)^{n(m+1)-2}}{(n(m+1)-2)!}(\lambda+\mu)du \\ &= \frac{\lambda^{n-1}\mu^{nm}}{n!(nm-1)!}\int_0^t e^{-(\lambda+\mu)u}u^{n(m+1)-2}du. \end{aligned} \tag{6.58}$$

When $m = 1$, it checks with (3.101). Again coming back to (6.54), let us suppose $\xi_i = m\nu_i$ for all i, m being a positive integer. Clearly ξ_i's are non-negative integer-valued cyclically interchangeable r.v.'s. Therefore, by (6.54) we have

$$P\left(\sum_{i=1}^{r}\xi_i < r, 1 \le r \le n \,\middle|\, \sum_{i=1}^{n}\xi_i = mk\right) = \begin{cases} 1 - \dfrac{mk}{n} & \text{if } 0 \le mk \le n, \\ 0 & \text{otherwise,} \end{cases}$$

which leads to

$$P\left(\sum_{i=1}^{r}\nu_i < \frac{r}{m}, 1 \le r \le n \,\middle|\, \sum_{i=1}^{n}\nu_i = k\right) = \begin{cases} 1 - \dfrac{mk}{n} & \text{if } 0 \le mk \le n, \\ 0 & \text{otherwise.} \end{cases} \tag{6.59}$$

Suppose ν_i's are i.i.d. r.v.'s. Then $\nu_1, \ldots, \nu_r$ may be replaced by $\nu_{n-r+1}, \ldots, \nu_n$. In that case, (6.59) becomes

$$P\left(\sum_{i=1}^{r} \nu_i > k - \frac{n-r}{m}, 1 \le r \le n \middle| \sum_{i=1}^{n} \nu_i = k\right) = \begin{cases} 1 - \frac{mk}{n} & \text{if } 0 \le mk \le n, \\ 0 & \text{otherwise.} \end{cases} \tag{6.60}$$

When $m = 1$, they check in (3.103) and (3.104).

As an application, consider the $M/G/1$ model of Section 3.2 in Chapter 3 except that the arrivals are in batches of size m. We are analyzing the busy period which starts when a batch arrives and finds the service idle. Denote by $G(n, m; t)$ the probability that the length of the busy period is $\le x$ and it consists of serving nm customers. Let $\psi_1, \ldots, \psi_{nm}$ be the length of successive service-times and $\nu_1, \ldots, \nu_{nm}$ the number of batches arriving during the 1st, 2nd, ..., (nm)th service-time, respectively. Obviously $n - 1$ batches have arrived during the busy period, i.e., $\nu_1 + \cdots + \nu_{nm} = n - 1$. In order that the busy period continues until the (nm)th customer is being served, $\nu_1 + \cdots + \nu_n \ge [\frac{r}{m}]$, $r = 1, \ldots, nm - 1$. Now

$$\begin{aligned} G(n, m; t) &= \\ &= P\left\{\sum_{i=1}^{nm} \psi_i \le t, \sum_{i=1}^{r} \nu_i \ge [\frac{r}{m}], r = 1, \ldots, nm - 1, \sum_{i=1}^{nm} \nu_i = n - 1\right\}. \end{aligned} \tag{6.61}$$

But the event

$$\left\{\sum_{i=1}^{r} \nu_i \ge [\frac{r}{m}], r = 1, \ldots, nm - 1\right\}$$

is equivalent to

$$\left\{\sum_{i=1}^{r} \nu_i > \frac{r}{m} - 1, r = 1, \ldots, nm - 1\right\}.$$

Moreover, $\nu_1, \ldots, \nu_{nm}$ are i.i.d. r.v.'s. Therefore by (6.60) we obtain for $u > 0$,

$$\begin{aligned} P&\left\{\sum_{i=1}^{r} \nu_i \ge \left[\frac{r}{m}\right], 1 \le r \le nm \middle| \sum_{i=1}^{nm} \nu_i = n - 1, \sum_{i=1}^{nm} \psi_i = u\right\} \\ &= 1 - \frac{m(n-1)}{nm} = \frac{1}{n}. \end{aligned}$$

Also

$$P\left\{\sum_{i=1}^{nm} \nu_i = n - 1 \middle| \sum_{i=1}^{nm} \psi_i = u\right\} = e^{-\lambda u} \frac{(\lambda u)^{n-1}}{(n-1)!}.$$

Hence

$$G(n, m; t) = \frac{\lambda^{n-1}}{n!} \int_0^t e^{-\lambda u} u^{n-1} dF_S^{(nm)}(u), \tag{6.62}$$

which checks with (3.101) for $m = 1$. Furthermore, if the service-times are i.i.d. exponential(μ) then it checks with (6.58).

Observe that the derivation of (6.58) follows the joint probability structure in (6.61) and the conditioning argument although the presentation is wordy.

A continuous version of (6.54) (but somewhat different) leads to the following:

Let $\nu_1, \ldots, \nu_n$ be non-negative interchangeable r.v.'s and let $\tau_1, \ldots, \tau_n$ (note that as an exception we are miserly in these notations for good reasons, since we will see soon that our usual τ's are consistent with these notations) be order statistics based on n i.i.d. uniform r.v.'s over $(0, t)$. Suppose ν_i's and τ_i's are independent. Then

$$P\left(\sum_{i=1}^{r} \nu_i < \tau_r, 1 \le r \le n \middle| \sum_{i=1}^{n} \nu_i = y\right) = \begin{cases} 1 - \dfrac{y}{t} & \text{if } 0 \le y \le t, \\ 0 & \text{otherwise.} \end{cases} \tag{6.63}$$

Proof: For $n = 1$,

$$P(\nu_1 < \tau_1 | \nu_1 = y) = P(\tau_1 \ge y) = 1 - \frac{y}{t}, \qquad \text{if } 0 \le y \le t$$

which checks with (6.63). Assume it holds for $n - 1$. Let $\nu_1 + \ldots + \nu_{n-1} = z$, $0 \le z \le y$ and $\tau_n = u$, $0 \le u \le t$. All calculations are done under the condition $\sum_{i=1}^{n} \nu_i = y$ and for brevity this condition is not written. If $y \le u \le t$, then

$$\begin{aligned} &P\left(\sum_{i=1}^{r} \nu_i < \tau_r \ , 1 \le r \le n \middle| \sum_{i=1}^{n-1} \nu_i = z, \tau_n = u\right) \\ &\qquad = P\left(\sum_{i=1}^{r} \nu_i < \tau_r, 1 \le r \le n-1 \middle| \sum_{i=1}^{n-1} \nu_i = z, \tau_n = u\right) \\ &\qquad = \begin{cases} 1 - \dfrac{z}{u} & \text{when } 0 \le z \le u, \\ 0 & \text{when } z \ge u, \end{cases} \end{aligned}$$

by the induction hypothesis, since $(\tau_1, \ldots, \tau_{n-1})$ under the condition that $\tau_n = u$ are order statistics based on $n - 1$ i.i.d. uniform variables over $[0, u]$. Hence for $y \le u \le t$,

$$\begin{aligned} &P(\sum_{i=1}^{r} \nu_i < \tau_r, 1 \le r \le n | \tau_n = u) \\ &\quad = \int_0^y \left(1 - \frac{z}{u}\right) f(z) dz, \qquad f \text{ for simplicity being the p.f. of } \sum_{i=1}^{n-1} \nu_i, \\ &\quad = 1 - \frac{1}{u} E\left(\sum_{i=1}^{n-1} \nu_i\right) \\ &\quad = 1 - \frac{1}{u} \frac{n-1}{n} y, \end{aligned} \tag{6.64}$$

because $\nu_1, \ldots, \nu_{n-1}$ are interchangeable given $\sum_{i=1}^{n} \nu_i = y$ and thus $E(\nu_i) = \frac{y}{n}$. On the other hand, if $u < y$, the above probability equals zero.

Now, τ_n being the largest among n variables from the uniform distribution over $(0, t)$, its p.f. is

$$\frac{n}{t}\left(\frac{u}{t}\right)^{n-1}, \qquad 0 \le u \le t. \tag{6.65}$$

Finally, from (6.64) and (6.65) we obtain for $0 \le y \le t$,

$$\begin{aligned}
P&\left(\sum_{i=1}^{r} \nu_i < \tau_r, 1 \le r \le n \middle| \sum_{i=1}^{n} \nu_i = y\right) = \\
&= \int_0^t \left(1 - \frac{1}{u}\frac{n-1}{n}y\right)\frac{n}{t}\left(\frac{u}{t}\right)^{n-1} du \\
&= 1 - \frac{y}{t}.
\end{aligned}$$

The last part of (6.63) being obvious, the proof is complete.

In Section 3.6, we have derived $G(n, t; x)$ for the $M/G/1$ model by two different analytic methods (see (3.100) and Remark 1). Now it is derived by using (6.63) and the procedure is similar to the derivation of (3.106). If $n = 0$, then $t = x$ and the probability of no arrivals during $(0, x)$ is $e^{-\lambda x}$. Therefore it checks with (3.100).

Consider $n \ge 1$. S_n being the service-time of the customer C_n,

$$\tau_j \le S_1 + \ldots + S_{j-1} + x, \qquad j = 1, \ldots, n$$

must be satisfied in the initial busy period. Thus we can express that

$$\begin{aligned}
G(n, t; x) &= P\left(\sum_{i=1}^{j-1} S_i + x \ge \tau_j, 1 \le j \le n, N(x) = n, \sum_{i=1}^{n} S_i \le t - x\right) \\
&= \int_0^{t-x} P\left(\sum_{i=1}^{j-1} S_i + x \ge \tau_j, 1 \le j \le n \middle| N(x) = n, \sum_{i=1}^{n} S_i = u\right) \\
&\quad \times P\left(N(x) = n \middle| \sum_{i=1}^{n} S_i = u\right) dF_S^{(n)}(u).
\end{aligned} \tag{6.66}$$

Obviously,

$$P\left(N(x) = n \middle| \sum_{i=1}^{n} S_i = u\right) = e^{-\lambda(x+u)}\frac{(x+u)^n}{n!}\lambda^n. \tag{6.67}$$

In order to compute the first probability under the integral sign in (6.66), we recall from the property of Poisson processes that $\tau_1, \ldots, \tau_n$ form a set of order statistics of n i.i.d. uniform variables over $(0,\ x+u)$. Also $S_1, \ldots, S_n$ being i.i.d. are interchangeable. Furthermore, if the inequalities

$$\sum_{i=1}^{j-1} S_i + x \ge \tau_j, \qquad 1 \le j \le n$$

are subtracted from $\sum_{i=1}^{n} S_i + x = u + x$, we have

$$S_n + S_{n-1} + \ldots + S_j \leq \tau_{n+1-j}^*, \qquad 1 \leq j \leq n$$

by putting $\tau_{n+1-j}^* = x + u - \tau_j$. Note that $\tau_1^*, \ldots, \tau_n^*$ also form a set of order statistics of n i.i.d. uniform variables over $(0, x+u)$ and $S_n, S_{n-1}, \ldots, S_j$ have the same distribution as that of $S_1, \ldots, S_{n-j+1}$. Therefore,

$$\begin{aligned} P&\left(\sum_{i=1}^{j-1} S_i + x \geq \tau_j, 1 \leq j \leq n \middle| N(x) = n_1, \sum_{i=1}^{n} S_i = u\right) \\ &= P\left(\sum_{i=1}^{j} S_i \leq \tau_j^*, 1 \leq j \leq n \middle| N(x) = n, \sum_{i=1}^{n} S_i = u\right) \\ &= 1 - \frac{u}{x+u} \end{aligned} \tag{6.68}$$

with the aid of (6.63). On combining (6.66), (6.67) and (6.68), we finally get

$$G(n, t; x) = \frac{x\lambda^n}{n!} \int_0^{t-x} e^{-\lambda(x+u)}(x+u)^{n-1}\, dF_S^{(n)}(u) \tag{6.69}$$

which checks with (3.100).

Remarks:

1. $G(n, t; x)$ was derived in Section 3.6 by an ad hoc analytic method, whereas it is obtained here by a set procedure which is laid out in Section 3.6 under combinatorial methods as an illustration for the $G/M/1$ model. This procedure unifies the treatment of the busy period distributions and works out well for obtaining both $G(n, t; x)$ and $G(n, t : j)$.

2. The combinatorial approach is powerful enough to be applicable in queueing models involving batches (i.e., bulk queues) with suitable change in the formulation of (6.54).

3. Some of the distributions can be derived from others through the concept of duality to be developed in Section 7.7 in Chapter 7.

Although the combinatorial method is mainly used for the study of the busy period, it is not too surprising that the same method provides a way to find the transient solution (see Chapter 2) at regeneration points. As an example, let us consider the $M/G/1$ model in which departure epochs are known to be regeneration points.

At first we may evaluate $P(X_n = 0|X_0 = j)$. This event occurs if and only if

$$\sum_{i=1}^{n} \gamma_i \leq n - j \quad \text{and} \quad \sum_{i=r}^{n} \gamma_i < n - r + 1, \qquad r = j+1, \ldots, n.$$

If we fix $\sum_{i=1}^{n}\gamma_i = k(\leq n-j)$, then the second set of inequalities can be rewritten as

$$\sum_{i=1}^{r}\gamma_i > r-n+k, \qquad r=n-k,\ldots,n-1.$$

Therefore, by applying (3.103), we obtain

$$\begin{aligned}P(X_n=0|X_0=j) &= \sum_{k=0}^{n-j}\left(1-\frac{k}{n}\right)P\left(\sum_{i=1}^{n}\gamma_i=k\right)\\ &= \sum_{k=0}^{n-j}\left(1-\frac{k}{n}\right)\int_0^\infty e^{-\lambda u}\frac{(\lambda u)^k}{k!}\,dF_S^{(n)}(u). \qquad (6.70)\end{aligned}$$

Let us proceed to evaluate $P\left(X_n\leq r|X_0=j\right)$. As before, this event is equivalent to

$$\left\{\sum_{i=1}^{n}\gamma_i\leq n+r-j,\sum_{i=k}^{n}\gamma_i<n+r-k+1, k=j+1,\ldots,n\right\}.$$

Thus

$$\begin{aligned}P(X_n\leq r|X_0=j) &= P\Big(\sum_{i=1}^{n}\gamma_i\leq n+r-j\Big)\\ &\quad -P\Big(\sum_{i=1}^{n}\gamma_i\leq n+r-j,\sum_{i=k}^{n}\gamma_i\geq n+r-k+1\\ &\qquad\qquad \text{for some } k=j+1,\ldots,n\Big). \qquad (6.71)\end{aligned}$$

Let the smallest k for which $\sum_{i=k}^{n}\gamma_i\geq n+r-k+1$ holds be $\ell+1$, $\ell=j,\ldots,n-1$. Then the second expression on the right side of (6.71) is given by

$$\sum_{\ell=j}^{n-1} P\left(\sum_{i=1}^{n}\gamma_i \le n+r-j, \sum_{i=k}^{n}\gamma_i < n+r-k+1, j+1 \le k \le \ell, \right.$$
$$\left. \sum_{i=\ell+1}^{n} \gamma_i = n+r-\ell\right) =$$
$$= \sum_{\ell=j}^{n-1} P\left(\sum_{i=k}^{\ell}\gamma_i < \ell-k+1, j+1 \le k \le \ell, \sum_{i=1}^{\ell}\gamma_i \le \ell - j, \right.$$
$$\left. \sum_{i=\ell+1}^{n} \gamma_i = n+r-\ell\right)$$
$$= \sum_{\ell=j}^{n-1} P\left(\sum_{i=\ell+1}^{n} \gamma_i = n+r-\ell\right)$$
$$\times P\left(\sum_{i=k}^{\ell}\gamma_i < \ell-k+1, j+1 \le k \le \ell, \sum_{i=1}^{\ell}\gamma_i \le \ell - j\right). \tag{6.72}$$

But because of (3.104), the last factor in (6.72) equals

$$\sum_{m=0}^{\ell-j}\left(1-\frac{m}{\ell}\right)P\left(\sum_{i=1}^{\ell}\gamma_i = m\right). \tag{6.73}$$

Finally, substituting (6.72) and (6.73) appropriately in (6.71) and simplifying it further we get

$$P(X_n \le r | X_0 = j) = \sum_{k=0}^{n+r-j}\int_0^\infty e^{-\lambda x}\frac{(\lambda x)^k}{k!}\,dF_S^{(n)}(x)$$
$$-\sum_{\ell=j}^{n-1}\sum_{m=0}^{\ell-j}\left(1-\frac{m}{\ell}\right)\left(\int_0^\infty e^{-\lambda x}\frac{(\lambda x)^m}{m!}\,dF_S^{(\ell)}(x)\right)$$
$$\times \int_0^\infty e^{-\lambda x}\frac{(\lambda x)^{n+r-\ell}}{(n+r-\ell)!}\,dF_S^{(n-\ell)}(x). \tag{6.74}$$

Indeed (6.74) is an explicit expression but not a very nice one for computation.

We may find the distribution of Y_n, the number of customers in the system just before the arrival of C_n, from the distribution of X_n. The event $\{X_{n+j} \le k | X_0 = j\}$ implies that there are at most k customers in the system just before the $(n+k+1)$st arrival epoch (i.e., $\{Y_{n+k+1} \le k | X_0 = j\}$) which follows the $(n+j)$th departure epoch. Conversely, consider the event $\{Y_{n+k+1} \le k | X_0 = j\}$, which implies that the $(n+j)$th departure epoch must precede the $(n+k+1)$st arrival epoch and there can be at most k customers just after the $(n+j)$th departure epoch. Thus we have established the identity

$$P\left(Y_{n+k+1} \leq k | X_0 = j\right) = P\left(X_{n+j} \leq k | X_0 = j\right), \tag{6.75}$$

which immediately provides one distribution from the other.

In the $G/M/1$ model, we may evaluate $P\left(Y_n \geq k\right)$ when initially there are no customers. The event $\{Y_n \geq k\}$ occurs if and only if there is a j such that

$$j - \sum_{i=1}^{j} \gamma_i^* = k, \qquad k \leq j \leq n. \tag{6.76}$$

If ℓ is the smallest index for which (6.76) is satisfied, then we can show that

$$P\left(Y_n \geq k\right) = \sum_{\ell=k}^{n} P\left(\sum_{i=1}^{\ell} \gamma_i^* = \ell - k, \sum_{i=r+1}^{\ell} \gamma_i^* < \ell - r, 1 \leq r \leq \ell - 1\right), \tag{6.77}$$

which ultimately leads to

$$P\left(Y_n \geq k\right) = \sum_{\ell=k}^{n} \frac{k}{\ell} \int_0^\infty e^{-\mu x} \frac{(\mu x)^{\ell-k}}{(\ell-k)!} \, dF_T^{(j)}(x) \tag{6.78}$$

for $1 \leq k \leq n$, with the help of (3.104).

6.4. Functional Relations

What has been developed heuristically in Section 3.5 in Chapter 3, is done more thoroughly here. In addition, a few more relations will be brought forth.

We start with the $M/G^k/1$ model and establish (3.77). Let $Q_j(n) = P(Y_n = j)$. Since units are served in batches of size k, it is clear that for $r = 0, 1, \ldots$

$$Q_j(n) \begin{cases} = 0 & \text{if } n \neq rk + j + 1, \\ \neq 0 & \text{if } n = rk + j + 1, \end{cases}$$

and hence

$$Q_j = \frac{1}{k} \lim_{r \to \infty} Q_j(rk + j + 1). \tag{6.79}$$

provided that the limit exists.

We observe that when $j < k$, the events $\{Y_{nk+j+1} = j\}$ and $\{X_n \leq j\}$ are equivalent. Note that X_n refers to the number of units in the system at the departure of the nth batch. Therefore,

$$Q_j(nk + j + 1) = \sum_{i=0}^{j} \Pi_i(n), \tag{6.80}$$

where $\Pi_i(n) = P(X_n = i)$. If $j \geq k$ and $j = sk+r$ for some $s > 0$, $r < k$, then the events $\{Y_{nk+j+1} = ik + r, i = 0, 1, \ldots, s\}$ and $\{X_n \leq j\}$ are equivalent and thus

$$\sum_{i=0}^{s} Q_{ik+r}(nk + j + 1) = \sum_{i=0}^{j} \Pi_i(n). \tag{6.81}$$

Suppose $\rho < k$. Then $\Pi_i(n) \to \Pi_i$ as $n \to \infty$. In view of (6.79), (6.80) in the limit becomes

$$Q_j = \frac{1}{k}\left(\Pi_0 + \ldots + \Pi_j\right), \qquad j < k. \tag{6.82}$$

Likewise, the limit of (6.81) yields

$$\sum_{i=0}^{s} Q_{ik+r} = \frac{1}{k}\sum_{i=0}^{j} \Pi_i, \qquad j \geq k, j = sk + r$$

which can be written as

$$Q_j + Q_{j-k} + \ldots + Q_r = \frac{1}{k}\sum_{i=0}^{j} \Pi_i. \tag{6.83}$$

Replacing j by $j - k$ and subtracting from (6.83), we get

$$Q_j = \frac{1}{k}\left(\Pi_{j-k+1} + \ldots + \Pi_j\right), \qquad j \geq k. \tag{6.84}$$

Note that (6.82) and (6.84) coincide with (3.76). Finally, taking the generating function (see Exercise 28 in Chapter 3), we obtain

$$Q(z) = \frac{1}{k}\frac{1 - z^k}{1 - z}\Pi(z), \tag{6.85}$$

which checks with the second part of (3.77). Moreover, (6.85) is valid for the $G/G^k/1$ model, since no use of Markovian input is made.

Although the heuristic argument provides a one-line proof for $P(z) = Q(z)$ in Section 3.5 in Chapter 3, we give a direct proof of the relation between $P(z)$ and $\Pi(z)$ so that in conjunction with (6.85) the relation among $P(z)$, $Q(z)$ and $\Pi(z)$ will be complete.

Consider an arbitrary period T, the expected number of arrivals and services during which are λT an $\lambda T/k$, respectively. As the expected length of each service is $1/\mu$, the expected time of the server being busy equals $(\lambda T/k)(1/\mu) = \rho T/k$. Hence the probability that at an arbitrary instant of time the server is busy in ρ/k and that the server is idle is $1 - \rho/k$. This simply means

$$\sum_{j=0}^{k-1} P_j = 1 - \frac{\rho}{k}. \tag{6.86}$$

We get an expression for P_j in the idle period (i.e., when $j < k$). In such a situation note that

$$P_j = \sum_{i=0}^{k} P(\text{the selected random instant falls in an idle period which begins with } i \text{ units, } 0 \le i \le j) P(\text{at the random instant there are } j \text{ units given the previous event}).$$

The probability of the first event is evidently proportional to Π_i and the expected length of such an idle period which is indeed $(k-i)/\lambda$. The conditional probability of the second that there are j units ($j = i, \ldots, k-1$) at the random instant is $1/(k-i)$. Thus

$$\begin{aligned} P_j &= \sum_{i=0}^{j} c\Pi_i (k-i) \frac{1}{k-i} \\ &= \sum_{i=0}^{j} c\Pi_i, \qquad j = 0, 1, \ldots, k-1, \end{aligned} \tag{6.87}$$

where c is the normalizing constant to be determined from (6.86). Substituting P_j in (6.86) and simplifying we have

$$c \sum_{j=0}^{k-1} (k-j)\Pi_j = 1 - \frac{\rho}{k}.$$

However, from (3.44) it can be derived (see Exercise 16) that

$$\sum_{j=0}^{k-1} (k-j)\Pi_j = k - \rho. \tag{6.88}$$

Therefore $c = 1/k$ and from (6.87) we get

$$P_j = \frac{1}{k} \sum_{i=0}^{j} \Pi_i, \qquad j = 0, 1, \ldots, k-1. \tag{6.89}$$

Next we consider P_j in the busy period. Denote by V the expired part of the service-time that is going on at the random instant considered. Let ξ, ξ^+ and ζ, respectively, denote the queue size at an arbitrary instant, the queue size just after the immediately preceding service completion instant and the number of arrivals during V. Then it can be seen that

$$\xi = \max(\xi^+, k) + \zeta, \qquad \xi \ge k. \tag{6.90}$$

Unlike the previous case, we will derive the generating function of $\{P_j, j \ge k\}$ for which we need the p.g.f. of $P(\zeta = j)$ to be immediately determined.

As a consequence of a known result in renewal theory (see Section A.6), we get

$$F_V(x) = \int_0^x \frac{\rho}{k}\mu \bar{F}_S(u)\,du, \tag{6.91}$$

provided F_S is not a lattice distribution. Note that ρ/k represents the fact that the system is busy. Let $m_j = P(\zeta = j)$ and $M(z)$ be the p.g.f. of $\{m_j\}$. Then

$$m_j = \frac{\lambda}{k}\int_0^\infty e^{-\lambda x}\frac{(\lambda x)^j}{j!}\bar{F}_S(x)\,dx$$

due to (6.91), and

$$\begin{aligned} M(z) &= \frac{\lambda}{k}\int_0^\infty e^{-\lambda(1-z)x}\bar{F}_S(x)\,dx \\ &= \frac{1}{k}\frac{1-\alpha(z)}{1-z} \end{aligned} \tag{6.92}$$

which is obtained by integration by parts and (3.13).

From (6.90) it follows that

$$P_j = \sum_{i=k}^{j}\Pi_i m_{j-i} + \sum_{i=0}^{k-1}\Pi_i m_{j-k}.$$

Therefore,

$$\sum_{j=k}^{\infty} P_j z^j = \left(\Pi(z) - \sum_{j=0}^{k-1}\Pi_j z^j\right) M(z) + \sum_{j=0}^{k-1}\Pi_j z^k M(z). \tag{6.93}$$

Realizing that

$$\sum_{j=k}^{\infty} P_j z^j = P(z) - \sum_{j=0}^{k-1} P_j z^j$$

and utilizing (6.89) for P_j in the right side, from (6.93) we obtain

$$\begin{aligned} P(z) &= \frac{1}{k}\sum_{j=0}^{k-1} z^j \sum_{i=0}^{j}\Pi_i + \Pi(z)M(z) + M(z)\sum_{j=0}^{k-1}\Pi_j(z^k - z^j) \\ &= \frac{1}{k}\sum_{j=0}^{k-1}\Pi_j\frac{z^k - z^j}{z-1} + \Pi(z)M(z) + M(z)\sum_{j=0}^{k-1}\Pi_j(z^k - z^j) \\ &= \frac{\Pi(z)}{k(z-1)}\left(\frac{z^k}{\alpha(z)} - 1\right) + \frac{z^k}{\alpha(z)}M(z)\Pi(z), \end{aligned} \tag{6.94}$$

the last line of simplification being carried out by substituting the value $\sum_{j=0}^{k-1}\Pi_j(z^k - z^j)$ obtained from (3.44). When (6.92) is used to replace $M(z)$ in (6.94), $P(z)$ has the final form

$$P(z) = \frac{1}{k}\frac{1-z^k}{1-z}\Pi(z), \tag{6.95}$$

provided $\rho < k$ and F_S is not a lattice distribution. This conforms with (3.78).

Clearly when $k = 1$,

$$P(z) = Q(z) = \Pi(z), \rho < 1$$

which was done in (3.74). At the end, we remind ourselves regarding the usefulness of the functional relations that $P(z)$ cannot be directly derived by the induced Markov chain technique but is done so through the functional relations.

An analogous treatment for the $G^k/M/1$ model leads us to

$$\Pi_j = \frac{1}{k} \lim_{r\to\infty} \Pi_j(rk - j), \tag{6.96}$$

$$\Pi_j = \frac{1}{k} (Q_0 + \ldots + Q_j), \qquad j < k \tag{6.97}$$

$$\Pi_j = \frac{1}{k} (Q_{j-k+1} + \ldots + Q_j), \quad j \geq k \tag{6.98}$$

and ultimately to the final relation as

$$\Pi(z) = \frac{1}{k} \frac{1 - z^k}{1 - z} Q(z). \tag{6.99}$$

Observe that these in the respective order correspond to (6.79), (6.82), (6.84) and (6.85), respectively. The proof is left as an exercise (see Exercise 15).

So far we have examined the relationship between three important stationary distributions with regard to the number in the system in a given model. In the strict sense, the purpose of this section is to dwell upon those types of relationships. Nevertheless, it is tempting to explore on relationships between characteristics arising in two different models, more particularly so, due to the correspondence, say, between $M/E_k/1$ and $M^k/M/1$ (M^k referring to Markovian input in batches of size k), as observed in Remark 2 following (2.28). Indeed, an exploitation of the same correspondence enables us to establish a relation between $\Pi(z)$ of $M/G^k/1$ and $E_k/G/1$.

Evidently, if $\tau_1, \tau_2, \ldots$ are the arrival instants in $M/G^k/1$, the arrival instants in $E_k/G/1$ are $\tau_k, \tau_{2k}, \ldots$, since an arrival unit in Erlangian input consists of k stages of arrivals. However, the departure instants are the same for both models. The well-known notation $X(t)$ is used in $M/G^k/1$, whereas the corresponding notation in $E_k/G/1$ is $X^*(t)$. Then, clearly

$$X^*(t) = \left[\frac{X(t)}{k}\right], \tag{6.100}$$

where $[x]$ is the integral part of x. The nonintegral part is the proportion of stages completed by an arrival in the Erlangian input. Similar to the $M/G^*/1$ model, denote by $\Pi_j^*(n)$, Π_j^*, P_j^* and Q_j^* the corresponding notations for the $E_*/G/1$ model. From (6.100) it follows that

$$\Pi_j^*(n) = \sum_{i=jk}^{jk+k-1} \Pi_i(n)$$

and in the limit as $n \to \infty$,

$$\Pi_j^* = \sum_{i=jk}^{jk+k-1} \Pi_i, \qquad \rho < k. \tag{6.101}$$

Clearly,

$$\sum_{j=0}^{\infty} z^j \sum_{i=0}^{j} \Pi_i^* = \frac{\Pi^*(z)}{1-z} \tag{6.102}$$

where $\Pi^*(z)$ is the p.g.f. of $\{\Pi_j^*\}$. But

$$\begin{aligned}\sum_{j=0}^{\infty} z^j \sum_{i=0}^{j} \Pi_i^* &= \sum_{j=0}^{\infty} z^j \sum_{i=0}^{jk+k-1} \Pi_i \qquad \text{by (6.101)}\\ &= \sum_{j=0}^{\infty} \frac{z^j}{2\pi i} \oint_C \frac{\Pi(v)}{(1-v)v^{jk+k}}\, dv\end{aligned}$$

(see Section A.5 in Appendix A) where C is a contour around the origin excluding the poles of $\Pi(z)/(1-z)$. Therefore, from (6.102) we get

$$\begin{aligned}\Pi^*(z) &= \frac{1-z}{2\pi i} \oint_C \sum_{j=0}^{\infty} v^{-k}(zv^{-k})^j \frac{\Pi(v)}{1-v}\, dv\\ &= \frac{1-z}{2\pi i} \oint_C \frac{\Pi(v)}{(1-v)(v^k - z)}\, dv. \end{aligned} \tag{6.103}$$

Because no new argument will be advanced, a relation between $P^*(z)$, the p.g.f. of $\{P^*\}$, and $P(z)$ is exactly the same as (6.103) except that $\Pi^*(z)$ and $\Pi(v)$ are replaced by $P^*(z)$ and $P(v)$, respectively. Furthermore, since $E_k/G/1$ belongs to $G/G/1$, we know from either (3.78) or (3.79) that

$$Q^*(z) = \Pi(z), \tag{6.104}$$

$Q^*(z)$ being the p.g.f. of $\{Q_j^*\}$.

A parallel set of relations can be derived for $G^k/M/1$ and $G/E_k/1$ models (see Exercise 16).

Remark: Observe that $E_k/G/1$ and $G/E_k/1$ are models with neither Markovian input nor Markovian service-time. Yet, we are able to get information on them through their correspondence with models that can be analyzed. This justifies the inclusion of these relations in this section.

6.5. Discussion

Right at the beginning it is mentioned that this chapter is a continuation of Chapter 3 on regenerative non-Markovian models. Most of the materials are of theoretical interest but reveal some of the finer aspects in the theory of queues, such as application of combinatorics and functional relations.

In Section 6.2, we have studied the transient behaviour of the number of units in the system with the aid of supplementary variables, which is yet another method added to the list of methods (viz., the method of stages in Section 2.4 and the imbedded Markov chain technique in Chapter 3), developed in order to retrieve the Markovian property for the analysis. Learning from the experience of the same study for $M/M/1$ in Chapter 2, we never expected to make any substantial achievement from our endeavour except to gain an insight into the richness of the theory. That is precisely what has happened to us without any element of surprise. The busy period study which is essentially equivalent to the inquiry into the first passage time problem has been dealt with extensively in Chapter 3 through analytical methods.

Section 6.3 dealing with combinatorial methods is novel in the sense that these methods are more recent in the study of transient analysis and have not been included in most textbooks so far. In Section 3.6.2 in Chapter 3, the concept of a combinatorial method was alluded. It is in this chapter that these methods have taken two streams, namely, ballot problem and the urn problem. The first one is restricted to Markovian models and mainly is dealing with $M/M/1$ models. The combinatorial nature of the models is exposed once a correspondence to random walks and to lattice paths is established. Amazingly a simple construction called the reflection principle has done the trick in attaining elegant looking expressions, when the structure of the random walk process is unraveled. This was noticed by Champernowne (1956) who essentially derived (6.35) when the transform method was dominating the scene. An interesting feature is added by introducing the inverse process which makes the basic counting simpler. Its impact is more significant when $M/M/1$ with finite capacity is considered. Remarkably this technique is even applicable to models with catastrophes in which there is a catastrophe rate of going to state 0 from any other state (see Krinik et al. (2005)). In the case of urn problem, the combinatorial nature is revealed through the invariance property of cyclic permutations of a given non-negative sequence of integers. The property is so powerful that it immediately simplifies some complex analytical arguments. We have explored for possible ramifications so as to be applicable in busy period analysis of non-Markovian queues.

Unlike the analytical transform method which puts any process immediately into a black box of transforms without much revelation of its special features, the combinatorial approach starts with examining the process for the existence of some structure that could lead to a solution and combines different component solutions into one. Invariably the approach emanates from probabilistic arguments based on mutual exclusivity and conditioning and sometimes ends up having elegant solutions. Remember our problems as formulated are probabilistic in nature. When the structure of the system is examined, we may end up having solutions with meaningful probabilistic interpretation. One example is (6.36) or (2.61), in which the first term refers to *zero-avoiding* (i.e., not reaching state 0) probability. We will come across this in Chapter 9 when dealing with discrete-time queues. In this respect, one may term the latter as the bottom-to-top approach in contrast to top-to-bottom approach of the transform method. Incidentally we remark that besides these two approaches one may recall the algebraic approach in Sections 4.3 and 4.4 in Chapter 4 and satisfy oneself that each has its own advantages and merits to be learnt.

Section 6.4 on functional relations provides proofs of functional relations on $P(z)$, $Q(z)$ and $\Pi(z)$ that are given in Section 3.5 in Chapter 3. However, we must reinforce the fact that $P(z)$ for these models cannot be derived directly due to the non-Markovian nature of $\{P_n\}$. The inclusion of new relations in $M/G^k/1$ and $E_k/G/1$ or in $G^k/M/1$ and $G/E_k/1$ is significant in the sense that models $E_k/G/1$ and $G/E_k/1$ cannot otherwise be handled by the conventional method because neither the input nor the service mechanism is Markovian.

Retrospecting over the imbedded Markov chain technique, what we realize is that the technique simply considers a non-Markovian process of interest such as $\{X(t)\}$ at regeneration points say at departure instants in $M/G/1$, so as to make it Markovian. Because $X(t)$ is our focus of attention, there arises the need of imbedding. In contrast, the processes of interest might themselves be Markov processes. A case in point is the process $\{V(t)\}$, the unfinished work in the system at time t in the $M/G/1$ model. It is of interest as a queueing characteristic when it is interpreted to represent the waiting time in the queue of a customer arriving at time t and thus is called the virtual waiting time at t.

The process $\{V(t)\}$ decreases linearly with slope -1 except that it jumps at $\tau_n = t_n$ with the magnitude of the jump equal to $S_n = s_n$, $n = 1, 2, \dots$ and when it hits zero it remains there until a jump occurs. This is illustrated in Figure 6.8.

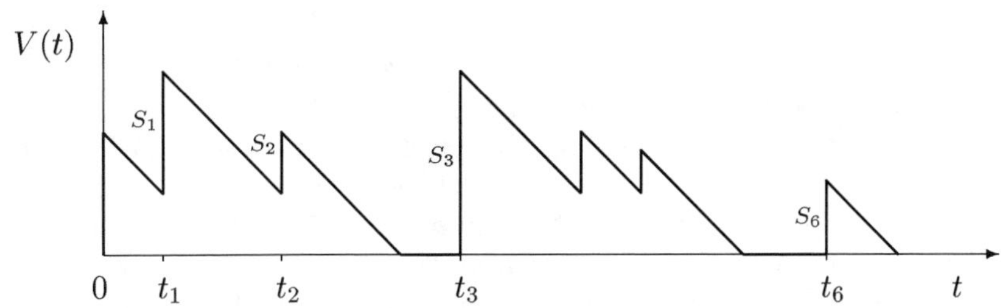

FIGURE 6.8. Unfinished work $V(t)$

Clearly, the process $\{V(t)\}$ is a continuous parameter Markov process with changes of states occurring in jumps as well as continuously in time.

Let $W(x,t) = P(V(t) \leq x)$. Considering the interval $(t, t + \Delta t)$ we see that $V(t + \Delta t)$ is at most x if either there are no arrivals in $(t, t + \Delta t)$ and $V(t) < x + \Delta t$ or an arrival occurs in $(t, t + \Delta t)$ and $V(t)$ jumps from $y(0 \leq y \leq x)$, to at most x. The case of more than one arrivals is not considered since it occurs with probability $o(\Delta t)$. Thus,

$$W(x, t + \Delta t) = (1 - \lambda\Delta t)W(x + \Delta t, t) + \lambda\Delta t \int_0^x W(y, t)F_S(x - y)\, dy. \tag{6.105}$$

This confirms that $\{V(t)\}$ is a Markov process. Using the expansion

$$W(x + \Delta t, t) = W(x, t) + \Delta t \frac{\partial}{\partial x} W(x, t) + o(\Delta t)$$

and taking the limit as $\Delta t \to 0$, we have

$$\frac{\partial}{\partial t} W(x, t) = \frac{\partial}{\partial x} W(x, t) - \lambda \int_0^x W(y, t)\bar{F}_S(x - y)\, dy, \tag{6.106}$$

which is known as the *Takács integro-differential equation.* It can be shown that except a jump at $x = 0$, $W(x, t)$ is a continuous function of x, thus justifying the above expansion. Note that Equation (6.106) remains valid even if λ is replaced by $\lambda(t)$.

On taking the L.T. of (6.106), we get

$$\begin{aligned}\frac{\partial}{\partial t} W^*(\theta, t) &= \theta W^*(\theta, t) - W(0, t) - \lambda \left(\theta W^*(\theta, t) - W(0, t)\right) \bar{F}_S^*(\theta) \\ &\quad - \lambda\, W(0, t)\bar{F}_s^*(\theta) \\ &= \theta W^*(\theta, t)\left(1 - \lambda \bar{F}_S^*(\theta)\right) - W(0, t).\end{aligned}$$

The stationary solution is obtained by suppressing t and setting the left side equal to zero. This yields

$$\theta\, W^*(\theta) = \frac{W(0)}{1 - \lambda \bar{F}_S^*(\theta)}.$$

When $\theta \to 0$, we get $W(0) = 1 - \rho$ (note that $W(0) = P_0$). Hence

$$\theta\, W^*(\theta) = \frac{1 - \rho}{1 - \lambda \bar{F}_S^*(\theta)} = \frac{1 - \rho}{1 - \frac{\lambda}{\theta}\left(1 - f_S^*(\theta)\right)}, \tag{6.107}$$

which is seen to be the $P - K$ formula as given in (3.19).

Returning to the discussion on the Markovian nature of $\{V(t)\}$, it is important to realize that when the model moves from $M/G/1$ to $G/G/1$ (Chapter 7), $\{V(t)\}$ is no longer a Markov process, since the arrival in $(t, t + \Delta t)$ depends on how far t is removed from the arrival time just preceding t. A Markov process will be imbedded at suitable regeneration points just as it was done for $\{X(t)\}$ when we moved from $M/M/1$ to $M/G/1$. With this observation, we conclude this section and lead the way to Chapter 7.

Because the present chapter is an extension of Chapter 3, some of its references are mentioned in that chapter. The supplementary variable technique in Section 6.2 is taken from the papers by Cox (1955), Conolly (1958) and Takács (1955). Most materials in Section 6.3 are from Champernowne (1956), Karlin and Taylor (1981), Jain and Mohanty (1981), Mohanty (1972, 1979), Mohanty and Jain (1970) and Takács (1962, 1967b), but on inverse process, one may refer to Anderson (1991) and Krinik et al. (2005). For Section 6.4

see Foster (1961, 1964) and Foster and Nyunt (1961) and Foster and Perera (1964, 1965).

6.6. Exercises

1. Using the supplementary variable technique, obtain the steady-state distribution for $M/G/1$ model.

2. Find the solution for $M/M/1$ as a special case of $M/G/1$.

3. For the queueing model $G/M/1$, results of interest are expressed in terms of least positive root of the functional equation (6.18). Discuss the computational procedure to obtain this root numerically for the models $E_K/M/1$ and $D/M/1$.

4. Verify (6.21).

5. Derive (6.25).

6. Verify (6.36) checks with (2.50).

7. (a) Let $\nu_1, \ldots, \nu_n$ be non-negative interchangeable r.v.'s such that $\nu_1 + \cdots \nu_n = t$ and let $\tau_1, \ldots, \tau_{n-1}$ be order statistics based on $n-1$ i.i.d. uniform r.v.'s over $(0, t)$. Suppose these two sequences are independent of each other. By using (6.66), prove

$$P\left(\sum_{i=1}^{r} \nu_i < \tau_r, \quad 1 \le r \le n-1\right) = \frac{1}{n} .$$

   (b) Hence derive (3.101), the busy period distribution of $M/G/1$ model.

8. Consider the $M/M/1$ model where customers are served in batches of size m. The server is idle if there are fewer than m customers in the system. Let $G_n(t)$ be the probability that a busy period consists of serving n batches and has length $\le t$. Show that

$$\begin{aligned} G_n(t) &= \frac{\mu^n}{(n-1)!} \sum_{j=0}^{m-1} \frac{(j+1)\lambda^{(n-1)m+j}}{((n-1)m+j+1)!} \\ &\quad \times \int_0^t e^{-(\lambda+\mu)u} u^{(n-1)(m+1)+j} du. \end{aligned}$$

   Hint: See Takács (1962).

9. Consider the $G/M/1$ model (see Section 3.3) where the customers are served in batches of size m. The server is idle if there are fewer than m customers in the system. Let $G_{n,j}(t)$ be the probability that a busy period consists of serving n batches leaving j $(< m)$

customers in the system and has length $\leq t$. Show that

$$G_{n,j}(t) = \\ = \begin{cases} \mu \int_0^t e^{-\mu v}(1 - F_T(v))dv & \text{when } n = 1,\, m = 1 \text{ and } j = 0, \\ \frac{\mu^n}{(n-1)!} \sum_{k=0}^{n} \binom{n-1}{k-1} \frac{m(k-1)+j}{m(n-1)+j} \\ \quad \int\int_R e^{-\mu(u+v)} u^{n-k} v^{k-1} (1 - F_T(v)) dF_T^{(m(n-1)+j)}(u) dv, \\ \text{where } R = \{(u,v) : u + v \leq x, u \geq 0, v \geq 0\}, & \text{otherwise.} \end{cases}$$

Hint: See Mohanty and Jain (1970).

10. Consider the queueing model with heterogeneous arrivals in which customers arrive in batches of size u_i in accordance with the Poisson process (λ_i), $i = 1, \ldots, k$, from k independent sources and are served one by one with service-times independent of arrival times, distributed as i.i.d. exponential(λ_0). Suppose, there are initially $j \geq 1$ customers when service begins. Let $G(j; a_1, \ldots, a_k; t)$ denote the probability that a busy period consists of serving a_i batches from source i, $i = 1, \ldots, k$ and has length $\leq t$. Show that

$$G(j; a_1, \ldots, a_k; t) = \\ = g(j; a_1, \ldots, a_k) \int_0^t e^{-\lambda y} \frac{(\lambda y)^{\sum_{i=1}^{k}(u_i+1)a_i + j - 1}}{(\sum_{i=1}^{k}(u_i + 1)a_i + j - 1)!} \lambda dy$$

where

$$\lambda = \lambda_0 + \lambda_1 + \cdots + \lambda_k$$

and

$$g(j; a_1, \ldots, a_k) = \\ = \frac{j}{j + \sum_{i=1}^{k}(u_i + 1)a_i} \binom{j + \sum_{i=1}^{k}(u_i + 1)a_i}{a_1, \ldots, a_k} \\ \times \left(\frac{\lambda_0}{\lambda}\right)^{\sum_{i=1}^{k} u_i a_i + j} \prod_{i=1}^{k} \left(\frac{\lambda_i}{\lambda}\right)^{a_i}.$$

Hint: See Mohanty (1972).

11. In the $M/G/1$ model with i initial customers, show that the nth customer finding the server idle is given by

$$\sum_{j=0}^{n-i} \left(1 - \frac{j}{n}\right) \int_0^\infty e^{-\lambda x} \frac{(\lambda x)^j}{j!} dF_S^{(n)}(x).$$

See Takács (1962).

12. In the $M/G/1$ model, let $P_i(t|y)$ be the probability that the server is idle at time t given that the initial queue size is i and the initial

occupation time of the server is y. Show that

$$P_i(t|y) = \sum_{n=1}^{\infty} e^{-\lambda t} \frac{(\lambda t)^{n-i}}{(n-i)!} \int_0^{t-y} \left(1 - \left(1 - \frac{i}{n}\right) \frac{u}{t}\right) dF_S^{(n)}(u).$$

See Takács (1967a).

13. For the $G/M/1$ model, derive a formula similar to (6.77).
14. For the $G^k/M/1$ model show that

$$\Pi(z) = \frac{1}{k} \frac{1-z^k}{1-z} Q(z).$$

15. Derive relations similar to (6.96), (6.97), (6.98) and (6.99) for the $G^k/M/1$ and $G/E_k/1$ models.
16. Derive (6.88).

References

Anderson, W. J. (1991). *Continuous-Time Markov Chains, An Application-Oriented Approach*, Springer-Verlag, New York.

Champernowne, D. G. (1956). An elementary method of solution of the queueing problem with a single server and constant parameters, *J. Roy. Statist. Soc., Series B*, **18**, 125–128.

Conolly, B. W. (1958). A difference equation technique applied to the simple queue with arbitrary arrival interval distribution, *J. Roy. Statist. Soc., Series B*, **20**, 167–175.

Cox, D. R. (1955). The analysis of non-Markovian stochastic processes by the inclusion of supplementary variables, *Proc. Camb. Phil. Soc.*, **51**, 433–441.

Foster, F. G. (1961). Queue with batch arrivals I, *Acta Math. Acad. Sci. Hung.*, **12**, 1–10.

Foster, F. G. (1964). Batched queueing processes, *Oper. Res.*, **12**, 441–449.

Foster, F. G. and Nyunt, K. M. (1961). Queues with batch departure – I, *Ann. of Math. Statist.*, **32**, 1324–1332.

Foster, F. G. and Perera, A. G. A. D. (1964). Queues with batch departure – II, *Ann. of Math. Statist.*, **35**, 1147–1156.

Foster, F. G. and Perera, A. G. A. D. (1965). Queues with batch arrivals – II, *Acta. math. Acad. Sci. Hung.*, **16**, 275–287.

Jain, J. L. and Mohanty, S. G. (1981). Busy period distributions for two heterogeneous queueing models involving batches, *INFOR.*, **19**, 133–139.

Karlin, S. and Taylor, H. M. (1981). *A Second Course in Stochastic Processes*, Academic Press, New York.

Krinik, A., Rubino, G., Marcus, D., Swift, R. J., Kasfy, H. and Lam, H. (2005). Dual processes to solve single service systems, *Journ. Stat. Planning and Inference*, **135**, 121–147.

Mohanty, S. G. (1972). On queues involving batches, *J. Appl. Prob.*, **9**, 430–435.

Mohanty, S. G. (1979). *Lattice Path Counting and Applications*, Academic Press, New York.

Mohanty, S. G. and Jain, J. L. (1970). On two types of queueing process involving batches, *Can. Oper. Res. Soc. J.*, **8**, 38–43.

Takács, L. (1955). Investigation of waiting time problems by reduction to Markov processes, *Acta Math. Acad. Sci. Hung.*, **6**, 101–129.

Takács, L. (1962). Generalization of the ballot problem and its application in the theory of queues, *JASA*, **57**, 327–337.

Takács, L. (1967a). Combinatorial methods in the theory of queues. *Rev. Inst. Internat. Statist.*, **34**, 207–219.

Takács, L. (1967b). *Combinatorial Methods in the Theory of Stochastic Processes*, John Wiley & Sons, New York.

CHAPTER 7

General Queues

7.1. Introduction

Starting from the well-structured Markovian models such as $M/M/1$ or in general the models arising out of birth-death processes in Chapter 2, we move away to partially Markovian models such as $M/G/1$ and $G/M/1$ in Chapters 3 and 5, and finally intend to consider in this chapter a general model $G/G/1$ which covers all earlier models. While in the process of transition, the Markovian nature in the structure of the models is dropped off gradually, the study of the characteristics is done through the imbedded Markov chain technique in which, paradoxically enough, we cling to the Markovian property that is being exhibited at regeneration points.

In the past, the number of units $X(t)$ in the system was the main characteristic of study. It is a Markov process in $M/M/1$, but not so in $M/G/1$ or $G/M/1$. Therefore, in these models $X(t)$ was considered at regeneration points in order to retain the Markovian property. Now removed one step further to $G/G/1$, we are unable to find regeneration points for $X(t)$.

Recall that the unfinished work $V(t)$ at time t (see Section 6.5) is a Markov process for $M/G/1$. However, it is not so in $G/G/1$. Thus, analogous to $X(t)$ the Markovian property of $V(t)$ is retained in $G/G/1$ when it is considered at suitable regeneration points. Therefore, instead of $V(t)$, we study $V(t)$ at an arrival instant which has the Markovian structure and is equivalent to the waiting time of a customer. In this chapter, its distribution is examined through the spectrum factorization technique.

Sometimes it is advantageous to find some suitable approximations or bounds for the exact probabilistic nature of the characteristics of interest. First of all, even though we are considering a very general model, the model assumptions such as independence and stationarity are seldom true and therefore an exact result based on these assumptions may be far removed from the correct one. Second, some of the exact expressions are complex enough so that no useful results can be derived from those. Thus simpler approximations or bounds may satisfy our need and become better alternatives to the exact results. With this view in mind, bounds for the mean waiting time and some approximations of the waiting time distribution are discussed. Finally, the idea of duality which has been whispered out at the end of Sections 3.5 and 3.6, finds a place here for a general model.

7.2. Waiting Time and Idle Time: An Analytic Method

Let us consider a single-server general model, which covers all earlier models. The model is called the $G/G/1$ model with the following properties:

(i) The interarrival times are i.i.d. with a distribution function F_T having mean $1/\lambda$;

(ii) The service-times independent of arrival times are i.i.d. with a distribution function F_S having mean $1/\mu$ and the service is provided by one server;

(iii) Infinite waiting space is available;

(iv) The queue discipline is FCFS.

Notation:

W_n : waiting time in the queue of C_n

As $V(t+\Delta t)$ depends on $V(t)$ as well as on the elapsed time from the immediately preceding arrival instant to t, it can be made Markovian if the regeneration points are chosen to be the arrival instants. Property 4 is crucial for our development since in that case $V(t)$ representing the virtual waiting time becomes W_n at the arrival instant of $C_n (n = 1, 2, \ldots)$, assuming C_1 arrives at $t = 0$. It is easy to see that W_n is defined by the following recurrence relation:

$$W_1 = V(0)$$

and for $n \geq 1$

$$W_{n+1} = \begin{cases} W_n + U_n & \text{if } W_n + U_n > 0, \\ 0 & \text{otherwise,} \end{cases} \tag{7.1}$$

where $U_n = S_n - T_{n+1}$. Relation (7.1) which can be alternatively expressed as

$$W_{n+1} = \max\{0, W_n + U_n\} \tag{7.2}$$

confirms that $\{W_n, n = 0, 1, \ldots\}$ is a Markov process. It has discrete time parameter and continuous state space. Its increments being i.i.d., it is represented by a single variable U.

Now let us examine the stationary distribution of W_n. Before trying to derive it, it is natural that as usual the question of its existence should haunt us. We may satisfy ourselves by accepting that it exists when $E(U) < 0$ which can be seen to be equal to the condition $\rho < 1$. Then the stationary variable is $\lim_{n\to\infty} W_n = W_q$, the waiting time in the queue of an arbitrary customer, which is independent of the initial state.

From (7.1) or (7.2) it follows that

$$F_{W_{n+1}}(y) = \begin{cases} \displaystyle\int_{-\infty}^{y} F_{W_n}(y-u)\, dF_{U_n}(u) & \text{if } y \geq 0, \\ 0 & \text{if } y < 0, \end{cases}$$

which in the limit becomes

$$F_{W_q}(y) = \begin{cases} \displaystyle\int_{-\infty}^{y} F_{W_q}(y-u)\, dF_U(u) & \text{if } y \geq 0, \\ 0 & \text{if } y < 0. \end{cases} \tag{7.3}$$

It can be easily checked that Equation (7.3) has the following alternative form:

$$F_{W_q}(y) = \begin{cases} \displaystyle\int_{0}^{y} F_U(y-u)\, dF_{W_q}(u) & \text{if } y \geq 0, \\ 0 & \text{if } y < 0. \end{cases} \tag{7.4}$$

This is known as *Lindley's integral equation*, and it is of *Wiener-Hopf* type.

We describe an analytic method of solving (7.3) by the use of spectrum factorization to be explained later. In Section 7.6 a probabilistic method of solution will be described. Had the integral in (7.3) not been truncated, relation (7.3) would have been an ordinary convolution. In order to get rid of the truncated part, we introduce a d.f. on the negative side having the same expression as on the positive side, namely,

$$F_-(y) = \begin{cases} 0 & \text{if } y \geq 0, \\ \displaystyle\int_{-\infty}^{y} F_{W_q}(y-u)\, dF_U(u) & \text{if } y < 0. \end{cases} \tag{7.5}$$

For brevity, let us write $F_{W_q} = F_+$. Together (7.3) and (7.5) can be expressed as

$$F_+(y) + F_-(y) = \int_{-\infty}^{y} F_+(y-u)\, dF_U(u) \tag{7.6}$$

for all y. Introduce notations $F_+^*(\theta)$ and $F_-^*(\theta)$ as given by

$$F_+^*(\theta) = \int_0^{\infty} e^{-\theta y} F_+(y)\, dy, \tag{7.7}$$

and

$$F_-^*(\theta) = \int_{-\infty}^{0} e^{-\theta y} F_-(y)\, dy. \tag{7.8}$$

Since

$$\int_{-\infty}^{\infty} e^{-\theta u}\, dF_U(u) = \Phi_S(\theta)\Phi_T(-\theta), \tag{7.9}$$

we have from (7.6)

$$\int_{-\infty}^{\infty} \Big(F_+(y) + F_-(y)\Big) e^{-\theta y}\, dy = F_+^*(\theta)\Phi_S(\theta)\Phi_T(-\theta)$$

which simplifies to

$$F_{-}^{*}(\theta) = F_{+}^{*}(\theta)\Big(\Phi_S(\theta)\Phi_T(-\theta) - 1\Big). \tag{7.10}$$

Before we proceed to introduce the spectrum factorization, let us establish some properties of $F_{+}^{*}(\theta)$ and $F_{-}^{*}(\theta)$. Clearly, $F_{+}^{*}(\theta)$ is analytic for $\Re(\theta) > 0$, due to the fact that it is the L.T. of a d.f. Next, consider the analyticity of $F_{-}^{*}(\theta)$. Assuming that the p.f. of T is $O(e^{-Dt})$ as $t \to \infty$ for some real positive D, we realize that the behaviour of $F_U(u)$ will be dominated by that of T when $u \to -\infty$ and thus $F_U(u)$ is $O(e^{Du})$ as $u \to -\infty$. This in conjunction with (7.5) gives rise to the fact that $F_{-}(y)$ is $O(e^{Dy})$ as $y \to -\infty$, which in turn implies that $F_{-}^{*}(\theta)$ is analytic in the region $\Re(\theta) < D$.

What is of interest to us in an expression of $F_{W_q}(y)$ in (7.3) or $F_{+}^{*}(\theta)$ in (7.7). The method of spectrum factorization suggests us to decompose $\Phi_S(\theta)\Phi_T(-\theta) - 1$ in (7.10) into two factors with appropriate properties so that by using some complex variable methodology we are able to determine $F_{+}^{*}(\theta)$. Suppose the spectrum factorization is

$$\Phi_S(\theta)\Phi_T(-\theta) - 1 = \frac{\psi_{+}(\theta)}{\psi_{-}(\theta)} \tag{7.11}$$

with the following properties:

(i) $\psi_{+}(\theta)$ is analytic with no zeroes in $\Re(\theta) > 0$,

(ii) $\psi_{-}(\theta)$ is analytic with no zeroes in $\Re(\theta) < D$,

(iii) $\lim_{|\theta|\to\infty} \frac{\psi_{+}(\theta)}{\theta} = 1$ in $\Re(\theta) > 0$,

(iv) $\lim_{|\theta|\to\infty} \frac{\psi_{-}(\theta)}{\theta} = -1$ $\Re(\theta) < D$.

To determine the factors with the above stated properties is the difficult part of the present method of solution. Once so obtained, we are able to write (7.10) as

$$F_{-}^{*}(\theta)\psi_{-}(\theta) = F_{+}^{*}(\theta)\psi_{+}(\theta). \tag{7.12}$$

From the region of analyticity of each function that is either discussed or stated above, we find the common region for both sides to be $0 < \Re(\theta) < D$. By analytic continuation either side is analytic and bounded for all finite values of θ and thus when we apply Liouville's theorem (See Section A.5 in Appendix A), we get

$$F_{-}^{*}(\theta)\psi_{-}(\theta) = F_{+}^{*}(\theta)\psi_{+}(\theta) = K \tag{7.13}$$

where K is a constant. This leads to

$$F_{+}^{*}(\theta) = \frac{K}{\psi_{+}(\theta)}. \tag{7.14}$$

In order to determine K, note from (7.7) that

$$\int_0^\infty e^{-\theta y}\, dF_{W_q}(y) = \theta\ F_{+}^{*}(\theta). \tag{7.15}$$

But

$$\lim_{\theta \to 0} \int_0^\infty e^{-\theta y}\, dF_{W_q}(y) = 1,$$

which from (7.14) and (7.15) yields

$$\lim_{\theta \to 0} \theta F_+^*(\theta) = \lim_{\theta \to 0} \frac{\theta K}{\psi_+(\theta)} = 1. \tag{7.16}$$

Therefore combining (7.14) and (7.16) we get

$$F_+^*(\theta) = \frac{1}{\psi_+(\theta)} \lim_{\theta \to 0} \frac{\psi_+(\theta)}{\theta}. \tag{7.17}$$

which gives an expression for L.T. of W_q, Moreover, it can be shown that

$$K = P(W_q = 0) \tag{7.18}$$

(see Exercise 1). One may note that $P(W_q = 0)$ is not necessarily equal to $1 - \rho$ (see (3.41)).

The good old $M/M/1$ may serve as an example to understand the method of spectrum factorization. In this case, we have

$$\begin{aligned} \Phi_S(\theta)\Phi_T(-\theta) - 1 &= \frac{\lambda}{\lambda - \theta}\frac{\mu}{\mu + \theta} - 1 \\ &= \frac{\theta(\theta + \mu - \lambda)}{(\lambda - \theta)(\mu + \theta)}. \end{aligned} \tag{7.19}$$

In (7.11) if we take

$$\psi_+(\theta) = \frac{\theta(\theta + \mu - \lambda)}{\mu + \theta} \tag{7.20}$$

and

$$\psi_-(\theta) = \lambda - \theta$$

then all properties (1)–(4) are satisfied. Thus from (7.17)

$$F_+^*(\theta) = \frac{(\mu + \theta)(1 - \rho)}{\theta(\theta + \mu - \lambda)} = \frac{1}{\theta} - \frac{\rho}{\theta + \mu - \lambda}. \tag{7.21}$$

Its inverse gives

$$F_{W_q}(y) = 1 - \rho \exp\{-(\mu - \lambda)y\}, \qquad y \geq 0, \tag{7.22}$$

which checks with the well-known result (2.68). This completes our derivation of the distribution of the waiting time of a customer with the aid of a specific method, namely the spectrum factorization method. However, for some models we may be able to solve the integral equation (7.3) without using the factorization (see Exercise 3).

Returning to (7.2), suppose we consider a variable

$$\widetilde{I}_n = -\min\{0, W_n + U_n\} \tag{7.23}$$

a natural counterpart to be thought of, relating to the length of the idle period. In fact if $\widetilde{I}_n > 0$ (i.e., $W_{n+1} = 0$), then it represents the length of the idle period to be terminated at the arrival of C_{n+1} and thus is a suitable notation. Also it is obvious from (7.2) and (7.23) that

$$W_{n+1}\widetilde{I}_n = 0. \tag{7.24}$$

In order to avoid ambiguity, we assume that the idle period exists only when $\widetilde{I}_n > 0$. The purpose of introducing $\widetilde{I}_n$ is not only necessarily due to its theoretical naturalness as a random variable, but also because its distribution essentially depends on the factor $\psi_-(\theta)$ which has yet played no role in the distribution of the waiting time.

To show this, we begin with the observation that

$$W_{n+1} - \widetilde{I}_n = W_n + U_n, \tag{7.25}$$

the L.T. of which is

$$E\left(e^{-\theta(W_{n+1}-\widetilde{I}_n)}\right) = E\left(e^{-\theta\ W_n}\right) E\left(e^{-\theta\ U_n}\right), \tag{7.26}$$

due to the independence of W_n and U_n. Let us evaluate the left side which by the use of (7.24) can be expressed as

$$E\left(e^{-\theta W_{n+1}}\middle|\widetilde{I}_n = 0\right) P\left(\widetilde{I}_n = 0\right) + E\left(e^{\theta \widetilde{I}_n}\middle|\widetilde{I}_n > 0\right) P\left(\widetilde{I}_n > 0\right).$$

The first term is replaced by

$$E\left(e^{-\theta W_{n+1}}\right) - E\left(e^{\theta W_{n+1}}\middle|\widetilde{I}_n > 0\right) P\left(\widetilde{I}_n > 0\right)$$

which is equal to

$$E\left(e^{-\theta W_{n+1}}\right) - P\left(\widetilde{I}_n > 0\right)$$

again by (7.24). When these substitutions are made and the limit is taken as $n \to \infty$, the left side in (7.26) ends up being

$$\Phi_{W_q}(\theta) - a_0 + a_0\Phi_I(-\theta), \tag{7.27}$$

where

$$\begin{aligned} a_0 = \lim_{n\to\infty} P\left(\widetilde{I}_n > 0\right) &= P(\text{an arrival finds the system being empty}) \\ &= P(W_q = 0). \end{aligned} \tag{7.28}$$

Equating (7.27) with the limit of the right side in (7.26), we have

$$\Phi_{W_q}(\theta) - a_0\left(1 - \Phi_I(-\theta)\right) = \Phi_{W_q}(\theta)\Phi_U(\theta),$$

leading to

$$\Phi_{W_q}(\theta) = \frac{a_0\left(\Phi_I(-\theta) - 1\right)}{\Phi_U(\theta) - 1}, \tag{7.29}$$

which with the help of (7.9) and (7.11) simplifies to

$$\Phi_{W_q}(\theta) = a_0\left(\Phi_I(-\theta) - 1\right)\frac{\psi_-(\theta)}{\psi_+(\theta)}. \tag{7.30}$$

Note that

$$\begin{aligned}\Phi_{W_q}(\theta) &= \int_0^\infty e^{-\theta y}\, dF_{W_q}(y) \\ &= \theta F_+^*(\theta) \quad \text{by (7.15)} \\ &= \frac{\theta a_0}{\psi_+(\theta)} \qquad \text{by (7.14), (7.18) and (7.28).}\end{aligned} \tag{7.31}$$

Combining (7.30) and (7.31) and changing $\theta \to -\theta$, we get

$$1 - \Phi_I(\theta) = \frac{\theta}{\psi_-(-\theta)}. \tag{7.32}$$

As a check for consistency of (7.32), we may again refer back to the $M/M/1$ model and use (7.20) to obtain

$$\Phi_I(\theta) = \frac{\lambda}{\lambda + \theta}$$

which implies

$$f_I(t) = \lambda e^{-\lambda t}, \qquad t > 0$$

agreeing with (2.74).

Other examples on the application of the general formula are given as exercises (Exercises 4 and 5).

7.3. Bounds for the Average Waiting Time

In Section 7.2 we have obtained the distribution of W_q from which a direct evaluation of average waiting time is not straightforward. Here we obtain an expression for the same from the first principle. This in turn leads to a simple upper bound. From now on we assume σ_S^2 and σ_T^2 to exist.

On taking the expectation on both sides of (7.25) and then taking the limit as $n \to \infty$, we have

$$E(\tilde{I}) = -E(U) \tag{7.33}$$

where $\lim_{n\to\infty} \widetilde{I_n} = \tilde{I}$. Remember that $I = \tilde{I}$ subject to the condition $\tilde{I} > 0$. Furthermore, if the limit of the expected value of the square of (7.25) is taken then it becomes

$$E(\tilde{I}^2) = 2E(W_q)E(U) + E(U^2), \tag{7.34}$$

by remembering (7.24) and that W and U are independent. This leads to

$$E(W_q) = -\frac{E(U^2)}{2E(U)} - \frac{E(\tilde{I}^2)}{2E(\tilde{I})}, \tag{7.35}$$

with the help of (7.33). But

$$\begin{aligned}E(\tilde{I}) &= E(\tilde{I}|\tilde{I} > 0)P(\tilde{I} > 0) \\ &= a_0 E(I) \qquad \text{by (7.28).}\end{aligned} \tag{7.36}$$

Similarly

$$E(\tilde{I}^2) = a_0 E(I^2). \tag{7.37}$$

Besides (7.36) and (7.37) if we use the fact

$$E(U) = -\frac{1}{\lambda}(1-\rho)$$

and

$$E(U^2) = \sigma_S^2 + \sigma_T^2 + \frac{1}{\lambda^2}(1-\rho)^2$$

in (7.35), we get

$$E(W_q) = \frac{\lambda(\sigma_S^2 + \sigma_T^2)}{2(1-\rho)} + \frac{1}{2}\left(\frac{1-\rho}{\lambda} - \mu_I\right) - \frac{\sigma_I^2}{2\mu_I}. \tag{7.38}$$

Moreover, from (7.36) and (7.33) it is obvious that

$$\mu_I = -\frac{E(U)}{a_0} = \frac{1-\rho}{\lambda a_0} \geq \frac{1-\rho}{\lambda}$$

due to $0 \leq a_0 \leq 1$. Therefore, (7.38) yields an upper bound for $E(W_q)$ as

$$E(W_q) \leq \frac{\lambda(\sigma_S^2 + \sigma_T^2)}{2(1-\rho)}. \tag{7.39}$$

Remarks:

(i) The significance of the bound (7.39) is realized when we note that it only depends on the means and variances of the T and S (but not on F_T and F_S), whereas the exact value in (7.38) needs μ_I and σ_I^2 which require the knowledge of the complete distributions F_T and F_S.

(ii) In order that the upper bound coincides with the exact value in the $M/G/1$ model, a new bound

$$E(W_q) \leq \frac{1 + (C_S)^2}{(1/\rho)^2 + (C_S)^2} \frac{\sigma_S^2 + \sigma_T^2}{2(1-\rho)} \tag{7.40}$$

is suggested where $C_S(= \sigma_S/\mu_S)$ is the service-time coefficient of variation. The numerical computation shows that (7.40) is reasonably good for $G/M/1$ but is fair for $G/G/1$ so far as percentage of error is concerned.

For a lower bound of $E(W_q)$, a different approach is put forth. Let

$$(X)^+ = \max\{0, X\} \qquad \text{and} \qquad (X)^- = -\min\{0, X\}.$$

Then

$$X = (X)^+ - (X)^- \qquad \text{and} \qquad (X)^+(X)^- = 0.$$

Hence

$$X^2 = \left((X)^+\right)^2 + \left((X)^-\right)^2, \tag{7.41}$$

$$E(X) = E\left((X)^+\right) - E\left((X)^-\right), \tag{7.42}$$

and

$$Var(X) = Var\left((X)^+\right) + Var\left((X)^-\right) + 2E\left((X)^+\right)E\left((X)^-\right). \tag{7.43}$$

Substituting $X = W_n + U_n$ in (7.43) and taking the limit as $n \to \infty$, we get

$$Var(W_q) + Var(U) = Var(W_q) + Var(\tilde{I}) + 2E(W_q)E(\tilde{I}), \tag{7.44}$$

since W_n and U_n are independent and $(X)^+ = W_{n+1}$ and $(X)^- = \widetilde{I_n}$ which follow from (7.2) and (7.23), respectively. From (7.44) another expression for $E(W_q)$ is obtained as

$$E(W_q) = \frac{Var(U)}{2E(\tilde{I})} - \frac{Var(\tilde{I})}{2E(\tilde{I})}, \tag{7.45}$$

but

$$Var(\tilde{I}) = E(\tilde{I}^2) - (E(U))^2 \tag{7.46}$$

due to (7.33). Again since $W_n \geq 0$, we have $W_n + U_n \geq U_n$ and hence $\widetilde{I_n} = (W_n + U_n)^- \leq (U_n)^-$. Therefore

$$\begin{aligned} E(\tilde{I}^2) &\leq E((U)^-)^2 \\ &= E(U^2) - E((U)^+)^2 \qquad \text{by (7.41).} \end{aligned} \tag{7.47}$$

Utilizing (7.46) and (7.47) in (7.45) and recalling (7.33), we end up having a lower bound for $E(W_q)$ as

$$E(W_q) \geq \frac{E((U)^+)^2}{2\lambda(1-\rho)}. \tag{7.48}$$

Observe that unlike the upper bound the above lower bound depends on more than just the first two moments of T and S.

7.4. A Heavy Traffic Approximation

The heavy-traffic condition in the $G/G/1$ model is attained when ρ is approximately equal to one, written as $\rho \cong 1$. Since we would be dealing with stable queues, it is necessary to assume $\rho < 1$. Under this situation, it is possible to provide a simple approximation for the distribution of W_q.

In Section 7.2, we have seen that $F_+^*(\theta)$ is the L.T. of the d.f. of W_q and $\theta F_+^*(\theta)$ is the L.T. of the p.f. of W_q when the p.f. exists (see (7.15)). Second, the final value theorem of the L.T. (See Section A.5) implies that the behaviour of the p.f. of W_q in the tail is governed by values of θ which are close to zero. In addition, if θ_0 is a pole of $F_+^*(\theta)$ then the corresponding exponential component in the d.f. of W_q has the mean equal to $-1/\theta_0$ (obviously θ_0 has to be negative). All these suggest that if $F_+^*(\theta)$ is approximated by considering its poles closer to zero then its inversion should be a good approximation of the distribution of W_q at least near the tail.

With the above objective in mind, we start searching for poles of $F_+^*(\theta)$ near zero. It is, however, clear from (7.17) that we may look for zeroes of

$\Phi_S(\theta)\Phi_T(-\theta) - 1$ in the vicinity of $\theta = 0$ and for that purpose examine the following Taylor series expansion:

$$\begin{aligned}
&\Phi_S(\theta)\Phi_T(-\theta) - 1 \\
&= \left(1 - \theta\mu_S + \frac{\theta^2}{2}E(S^2)\right)\left(1 + \theta\mu_T + \frac{\theta^2}{2}E(T^2)\right) - 1 + o(\theta^2) \\
&= \theta\left(\mu_T - \mu_S + \theta\left(\frac{E(S^2) + E(T)^2}{2} - \mu_T\mu_S\right)\right) + o(\theta^2), \qquad (7.49)
\end{aligned}$$

when higher order terms of θ are neglected (which is justified because of our interest near zero), obviously $\theta = 0$ is a zero. Another zero may be obtained from the second factor. Note that

$$\begin{aligned}
\mu_T - \mu_S + \theta&\left(\frac{E(S^2) + E(T^2)}{2} - \mu_T\mu_S\right) \\
&= \mu_T - \mu_S + \theta\left(\frac{\sigma_S^2 + \sigma_T^2}{2} + \frac{(\mu_S - \mu_T)^2}{2}\right) \\
&\cong \mu_T - \mu_S + \theta\frac{\sigma_S^2 + \sigma_T^2}{2}, \qquad \text{since } \rho \cong 1.
\end{aligned}$$

This leads to the second zero which is approximated as

$$\theta_0 \cong -\frac{2\mu_T(1-\rho)}{\sigma_S^2 + \sigma_T^2}. \tag{7.50}$$

Clearly θ_0 is negative and close to zero. Thus we can express

$$\Phi_S(\theta)\Phi_T(-\theta) - 1 \cong \theta(\theta - \theta_0)\frac{\sigma_S^2 + \sigma_T^2}{2}, \tag{7.51}$$

as an approximation in the neighbourhood of zero. From (7.11) and (7.51), it follows that

$$\psi_+(\theta) \cong C\theta(\theta - \theta_0) \tag{7.52}$$

where

$$C = \psi_-(0)\frac{\sigma_S^2 + \sigma_T^2}{2}.$$

Finally, using (7.17) and (7.52) we obtain for $\rho \cong 1$,

$$\begin{aligned}
F_+^*(\theta) &\cong -\frac{\theta_0}{\theta(1 - \theta_0)} \\
&= \frac{1}{\theta} - \frac{1}{\theta - \theta_0} \qquad (7.53)
\end{aligned}$$

which when inverted gives rise to

$$F_{W_q}(y) \cong 1 - \exp\left(-\frac{2\mu_T(1-\rho)}{\sigma_S^2 + \sigma_T^2}y\right) \tag{7.54}$$

by remembering (7.50). Evidently the distribution is exponential with mean

$$E(W_q) \cong \frac{\sigma_S^2 + \sigma_T^2}{2\mu_T(1-\rho)}. \tag{7.55}$$

Remarks:

(i) Results (7.54) and (7.55) having the same flavour as the central limit theorem in probability theory may be considered as the same type of theorem in queueing theory. The exponential distribution of W_q in heavy traffic situation plays a similar role just as the normal distribution does for sums of i.i.d. r.v.'s.

(ii) The approximation is extremely robust in the sense that it is insensitive to the detailed form of the arrival and service-time distributions.

(iii) The approximation (7.55) for the average waiting time is nothing but the upper bound (7.39) for the same and does not depend on the distributions of S and T except the first two moments. It being inversely proportional to $1-\rho$ shoots up fast to infinity as ρ comes closer to unity from the left.

Manufacturing—a just-in-time production system. An assembly station consists of a single robot which is capable of working on a variety of parts. Raw material required for production is delivered by a supply handling system at a rate of one lot per u hours. However, deliveries are never exactly in time, the times between deliveries vary with a standard deviation σ_T. The processing time of a lot is constant and equals v hours. As is typical of a just-in-time system, u is close to v but still $v < u$, so that $\rho = v/u < 1$ but $\rho \cong 1$. An application of the approximation formula (7.55) yields for the mean waiting time of a lot

$$E(W_q) \cong \frac{\sigma_T^2}{2\mu_T(1-\rho)} = \frac{\sigma_T^2}{2(u-v)},$$

and for the *mean lead time* of a lot:

$$E(W) \cong v + \frac{\sigma_T^2}{2(u-v)}.$$

Using these formulas, interesting design problems can be solved, for instance, we can find out how much the mean lead time of a lot will be reduced, if we are able to improve the reliability of the supply handling system. To be concrete, let us assume that currently the system is working with processing time $v = 9$ hours, mean time between deliveries $u = 10$ hours and $\sigma_T = 4$ hours. Then $E(W) \cong 17$ hours. Now suppose that the standard deviation of the time between deliveries can be reduced to $\sigma_T = 3$ hours. How much will be the reduction of the mean lead time? This time $E(W) = 13.5$ hours, which is an improvement of about 20%.

7.5. Diffusion Approximation

In the previous section, an approximation to the solution in a given model has been derived. The present section deals with an approach of finding an exact solution to an approximation of the given model.

The main idea of approximation is to replace the usual discrete-state and continuous-time queueing process (assumed to be an exact model) by a continuous-state and continuous-time process (an approximation). This is justified by an application of the central limit theorem in probability theory in which we assume the knowledge of only the first two moments of a distribution. To begin with, let us bring back notations Y_t and Z_t from Section 6.3 which represent the number of arrivals and potential departures during $(0, t)$, respectively. Let

$$\xi(t) = Y_t - Z_t. \tag{7.56}$$

Observe that the events $\{Y_t \geq n\}$ and $\{\tau_n < t\}$ are equivalent events. This leads to a remarkable and useful identity

$$P(Y_t \geq n) = P(\tau_n < t) \tag{7.57}$$

connecting a p.f. of a discrete r.v. with a p.f. of a continuous r.v. But

$$\tau_n = T_1 + \ldots + T_n,$$

T_i's being i.i.d. interarrival times. By applying the central limit theorem (see Section A.4 in Appendix A), we get for large n

$$P(\tau_n < t) \cong \Phi\left(\frac{\lambda t - n}{\sqrt{n}\,\lambda\sigma_T}\right) \tag{7.58}$$

where

$$\Phi(y) = \frac{1}{\sqrt{2\pi}}\int_{-\infty}^{y} e^{-u^2/2}\,du.$$

As n becomes large, we expect the number of arrivals during $(0, t)$ is approximately λt. It is reasonable to replace n in the denominator by λt, for large n. Therefore from (7.57) we obtain the following by (7.58):

$$P\left(Y_t < n\right) \cong \Phi\left(\frac{n - \lambda t}{\sqrt{\lambda^3 t \sigma_T^2}}\right). \tag{7.59}$$

In other words, Y_t is approximately distributed as a normal r.v. with mean λt and variance $\lambda^3 t \sigma_T^2$.

By considering the unrestricted departure process Z_t, not constrained by the absence of customers in the system, a similar approximation shows that

$$P(Z_t < n) \cong \Phi\left(\frac{n - \mu t}{\sqrt{\mu^3 t \sigma_S^2}}\right). \tag{7.60}$$

Hence from (7.56) it follows that the distribution of the unconstrained $\xi(t)$ is approximated by the normal distribution with mean $(\lambda - \mu)t$ and variance $\left(\lambda^3\sigma_T^2 + \mu^3\sigma_S^2\right)t$. Moreover, as a process it has stationary and independent increments. These suggest that $\{\xi(t),\, t \geq 0\}$ is a *Wiener-Levy process* or a *Brownian motion* with a drift (see Section A.6 in Appendix A), subject to the condition $\xi(t) \geq 0$. The approximation is of second order since the whole process is summarized by the first two moments only.

Denoting by $F(x_0, 0; x, t) = F$ for brevity, the conditional d.f. of $\xi(t)$ at x given $\xi(0) = x_0$ and using the above approximation, we find that

$$\begin{aligned} F = P(\xi(t) \leq x \mid \xi(0) = x_0) &= \int_{-\infty}^{x} \frac{1}{\sqrt{2\pi\beta t}} \exp\left(-\frac{(y - x_0 - \alpha t)^2}{2\beta t}\right) dy \\ &= \int_{-\infty}^{\frac{x - x_0 - \alpha t}{\sqrt{\beta t}}} \frac{1}{\sqrt{2\pi}}\, e^{-\frac{u^2}{2}} du, \end{aligned} \tag{7.61}$$

where $\alpha = \lambda - \mu$ and $\beta = \lambda^3\sigma_T^2 + \mu^3\sigma_S^2$. Observe that

$$\begin{aligned} \frac{\partial F}{\partial t} &= \left[\frac{1}{\sqrt{2\pi}} \exp\left(-\frac{(x - x_0 - \alpha t)^2}{2\beta t}\right)\right]\left[-\frac{\alpha}{\sqrt{\beta t}} - \frac{x - x_0 - \alpha t}{2\sqrt{\beta t}t}\right] \\ &= -\alpha f - \frac{1}{2}\,\frac{x - x_0 - \alpha t}{t}\, f \end{aligned}$$

where $f = \frac{\partial F}{\partial x}$. This leads to the fact that F satisfies

$$\frac{\partial F}{\partial t} = -\alpha \frac{\partial F}{\partial x} + \frac{\beta}{2}\,\frac{\partial^2 F}{\partial x^2} \tag{7.62}$$

which is the *forward Kolmogorov equation* (called the *Fokker-Plank equation* in physics) or simply the *diffusion equation*. Clearly in addition to (7.62), F satisfies the initial condition

$$F(x_0, 0; x, 0) = \begin{cases} 0 & \text{for } x < x_0, \\ 1 & \text{for } x \geq x_0, \end{cases} \tag{7.63}$$

but not the boundary condition

$$F(x_0, 0; 0, t) = 0, \qquad \text{for } x_0 > 0 \text{ and } t > 0, \tag{7.64}$$

which is equivalent to $\xi(t) \geq 0$, $t > 0$. Condition (7.64) obviously suggests that $x = 0$ is a reflecting barrier for $\xi(t)$.

Because of two constraints, we find another particular solution of (7.62) which happens to be $e^{2\alpha x/\beta} F(x_0, 0; -x, t)$. Then consider the solution to be of the form $A\, F(x_0, 0; x, t) + B\, e^{2\alpha x/\beta} F(x_0, 0; -x, t)$. Using (7.63) and (7.64), we obtain $A = 1$ and $B = -1$. Thus, the transient solution of approximated model is given by the d.f.

$$F(x_0, 0; x, t) = \int_{-\infty}^{\frac{x - x_0 - \alpha t}{\sqrt{\beta t}}} \frac{1}{\sqrt{2\pi}} e^{-\frac{u^2}{2}} du - e^{2\alpha x/\beta} \int_{-\infty}^{-\frac{x + x_0 + \alpha t}{\sqrt{\beta t}}} \frac{1}{\sqrt{2\pi}} e^{-\frac{u^2}{2}} du, \tag{7.65}$$

for $x \geq 0$.

The result (7.65) is surprisingly simple and covers all kinds of situations. It may be necessary to comment that the selection of the second solution is based on some rational analysis. For our purpose we content ourselves by finding it and that satisfies our need.

The steady-state solution is as follows:
For $\rho < 1$,

$$\lim_{t \to \infty} F(x_0, 0; x, t) = 1 - e^{2\alpha x/\beta}, \qquad \text{if } x \geq 0. \tag{7.66}$$

On the other hand, when $\rho \geq 1$, this limit is zero suggesting that no proper distribution exists.

7.6. Waiting Time and Idle Time: A Probabilistic Method

In Section 7.2 we have derived F_{W_q} and F_I by an analytic method, namely, by using spectrum factorization. Here we apply a probabilistic approach for the same purpose.

Referring to $U_n = S_n - T_{n+1}$ in (7.1), let $U'_r = \sum_{j=0}^{r} U_j$, $r \geq 0$ with $U'_0 = 0$. Since U_n's are i.i.d., the sequence $\{U'_r,\ r \geq 0\}$ is a random walk, generated by $\{U_n,\ n \geq 0\}$.

For any random walk $\{R_n,\ n \geq 0\}, R_0 = 0$, generated by $\{X_n,\ n \geq 0\}$, define the *ladder points* recursively as follows: M_k is the kth *ascending ladder point* if

$$M_0 = 0 \text{ and } M_k = \min\{n :\ R_n > R_{M_{k-1}}\},\ k \geq 1 \tag{7.67}$$

and $\bar{M}_k$ is the kth *descending ladder point* if

$$\bar{M}_0 = 0 \text{ and } \bar{M}_k = \min\{n :\ R_n \leq R_{\bar{M}_{k-1}}\},\ k \geq 1. \tag{7.68}$$

R_{M_k} and $R_{\bar{M}_k}$ are their respective *ladder heights.*

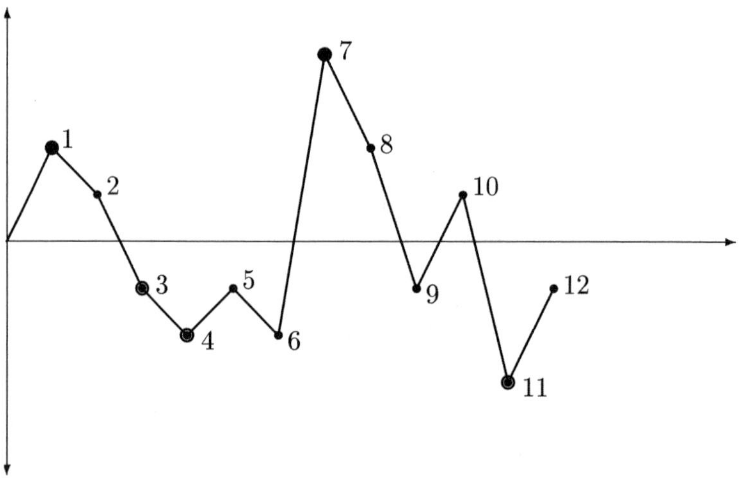

FIGURE 7.1. Ladder indices

In Figure 7.1, 1 and 7 are the first and second ascending ladder points and 3, 4 and 11 are the first, second and third descending ladder points.

The pivotal result that is used, establishes again a factorization through probabilistic arguments, which is stated as follows without proof.

For $0 < z < 1$, w real and $i = \sqrt{-1}$, we have the factorization

$$1 - zE(e^{iwX_n}) = \left[1 - E(z^M e^{iwR_M})\right]\left[1 - E(z^{\bar{M}} e^{iwR_{\bar{M}}})\right] \tag{7.69}$$

where $M = M_1$ and $\bar{M} = \bar{M}_1$.

In the theory of probability $E(e^{iwX})$ is known to be the *characteristic function* of a random variable X. The above factorization is called *Wiener-Hopf factorization.*

Another result of importance is given below without proof:

For $x \geq 0$,

$$\lim_{n\to\infty} P(\max_{0\leq j\leq n}\{R_j\} \leq x) = \\ = (1-p)(1 + E(\text{number of ascending ladder points in the interval } (0,x])) \tag{7.70}$$

is a proper distribution if and only if $p = P(M < \infty) < 1$, and

$$\begin{aligned}\lim_{n\to\infty} P(\min_{0\leq j\leq n}\{R_j\} \geq -x) &= \\ = \quad (1-\bar{p})(1 + E(&\text{number of descending ladder points} \\ &\text{in the interval } [-x, 0)),\end{aligned} \tag{7.71}$$

is a proper distribution if and only if $\bar{p} = P(\bar{M} < \infty) < 1$.

It is intuitively reasonable that for a limiting distribution of the maximum to exist the random walk should drift to $-\infty$, which is equivalent to saying $p < 1$ and a similar explanation for the minimum.

Let us apply the second result to the model $G/G/1$. Assume $W_1 = V(0) = 0$. From (7.2), it follows that

$$W_n = \max\{0, U_{n-1}, U_{n-1} + U_{n-2}, \ldots, U_{n-1} + \cdots + U_1\}. \tag{7.72}$$

Since $\{U_j\}$ is a sequence of i.i.d. r.v.'s, W_n has the same distribution as

$$\max\{0, U_1, U_1 + U_2, \ldots, U_1 + \cdots + U_{n-1}\} = \max_{0\leq r\leq n-1}\{U'_r\}$$

where U_j is replaced by U_{n-j}. We may write this as

$$W_n = \max_{0\leq r\leq n-1}\{U'_r\}. \tag{7.73}$$

Since $\{U'_r,\ r \geq 0\}$ is a random walk, F_{W_q} is obtained from (7.70).

Again from

$$X = \max\{0, X\} + \min\{0, X\},$$

we get $-\min\{0, X\} = \max\{0, X\} - X$. Thus $\tilde{I}_n$ from (7.23) becomes $W_{n+1} - W_n - U_n$. Then

$$\sum_{j=1}^{n} \tilde{I}_j = W_{n+1} - U'_n.$$

Also from (7.73),

$$\begin{aligned} W_{n+1} &= \max\{U'_n - U'_n, U'_n - U'_{n-1}, \ldots, U'_n - U'_1, U'_n\} \\ &= U'_n + \max\{-U'_n, -U'_{n-1}, \ldots, -U'_1, 0\} \\ &= U'_n - \min\{0, U'_1, U'_2, \ldots, U'_n\}. \end{aligned}$$

Therefore, $\sum_{j=1}^{n} \tilde{I}_j = -\min\{0, U_1', U_2', \ldots, U_n'\}$, the limiting distribution of which is obtained by using (7.71).

Consider the model $M/M/1$ to illustrate the use of the previous ideas. To begin with, we find the p.f. of $U_n = U$ (say)

$$f_U(u) = \begin{cases} \frac{\lambda\mu}{\lambda+\mu} e^{\lambda u} & \text{for } u \leq 0, \\ \frac{\lambda\mu}{\lambda+\mu} e^{-\mu u} & \text{for } u \geq 0. \end{cases}$$

Its characteristic function is

$$E(e^{iwU}) = \int_{-\infty}^{\infty} e^{iwu} f_U(u) du = \frac{\lambda\mu}{(\lambda + iw)(\mu - iw)}.$$

Now we apply the factorization theorem (7.69). The left side of (7.69) becomes

$$1 - zE(e^{iwU}) = \frac{(a - iw)(iw - b)}{(\mu - iw)(\lambda + iw)} \tag{7.74}$$

where

$$a = \frac{\mu - \lambda + \sqrt{(\mu - \lambda)^2 + 4\lambda\mu(1 - z)}}{2}$$

and

$$b = \frac{\mu - \lambda - \sqrt{(\mu - \lambda)^2 + 4\lambda\mu(1 - z)}}{2}$$

are the roots of the quadratic equation

$$(iw)^2 - iw(\mu - \lambda) - \lambda\mu(1 - z) = 0.$$

Next, it can be checked that (7.74) can be factorized as

$$\frac{(a - iw)(iw - b)}{(\mu - iw)(\lambda + iw)} = \left(1 - \frac{\mu - a}{\mu - iw}\right)\left(1 - \frac{\lambda + b}{\lambda + iw}\right). \tag{7.75}$$

Letting

$$h(z) = \frac{\lambda + \mu - \sqrt{(\lambda + \mu)^2 - 4\lambda\mu z}}{2\lambda} \tag{7.76}$$

we see that

$$h(z) = 1 + \frac{b}{\lambda} \text{ and } \frac{\lambda}{\mu} h(z) = 1 - \frac{a}{\mu}.$$

When a and b are replaced by the above expressions, the right side of (7.75) simplifies to

$$\left(1 - \frac{\lambda h(z)}{\mu - iw}\right)\left(1 - \frac{\lambda h(z)}{\lambda + iw}\right).$$

From the factorization (7.69), we may identify the factors as

$$E(z^M e^{iwR_M}) = \frac{\lambda h(z)}{\mu - iw} \text{ and } E(z^{\bar{M}} e^{iwR_{\bar{M}}}) = \frac{\lambda h(z)}{\lambda + iw}. \tag{7.77}$$

As $z \to 1$, the first one gives an expression for $E(e^{iwR_M})$ from which the distribution of R_M can be obtained. This distribution will be helpful in evaluating the right side of (7.70) and hence F_{W_q} (see Exercise 17).

Let us analyze the second relation in (7.77). First, noting that $\tilde{I} = -R_{\bar{M}}$, we have

$$E(z^{\bar{M}} e^{iw\tilde{I}}) = \frac{\lambda h(z)}{\lambda - iw} \,.$$

As $z \to 1$, the above leads to

$$E(e^{iw\tilde{I}}) = \rho^{-1} \frac{\lambda}{\lambda - iw} \tag{7.78}$$

when $\rho > 1$, since under that condition

$$\lim_{z \to 0} h(z) = \frac{\mu}{\lambda} = \rho^{-1}.$$

The inversion of (7.78) yields

$$f_{\tilde{I}}(x) = \rho^{-1} \lambda e^{-\lambda x} \qquad x > 0,$$

leading to $P(\tilde{I} > 0) = \rho^{-1}$ and therefore to $P(\tilde{I} = 0) = 1 - \rho^{-1}$. This is consistent with the fact that the conditional p.f. of $\tilde{I}$ given $\tilde{I} > 0$ or of I (see Section 7.2) is given by the last relation in Section 7.2. Condition $\rho > 1$ is vital. This assures the distribution of $\tilde{I}$ to exist.

7.7. Duality

The concept of duality has been mentioned in an informal way without elaboration in the last remark of Section 3.5 and at the end of Section 3.6. If in a queueing model the interarrival and service-time distributions are interchanged, the new model so formed is called the dual of the original and conversely. For example $M/E_k/1$ and $E_k/M/1$ are dual of each other. The reason for introducing duality is that the distributions of certain random variables in a system may be derived from the knowledge of distributions of variables in the dual. As an instance, in the busy period study we can obtain the distribution $P(N = n)$ for $G/M/1$ directly from $P(N(x) = n)$ in $M/G/1$ through (3.100)–(3.101) which is otherwise obtained from (3.110) as $G(n, \infty)$. This is so because of the duality relation between the two models.

Consider only single server queues which have the usual assumptions for both original and dual models. We reserve the standard notations $\{T_n\}$, $\{S_n\}$, T and S for the original model and use correspondingly $\{\hat{T}_n\}$, $\{\hat{S}_n\}$, $\hat{T}$ and $\hat{S}$ for the dual model. In general, any notation with a hat is the corresponding notation in the dual model. For convenience, denote by A and B the distribution functions of T and S, respectively, and hence by $G_A/G_B/1$ the original model. Then, by definition, the distribution functions of $\hat{T}$ and $\hat{S}$ are B and A, respectively, and the dual model is $G_B/G_A/1$. Let $V = V(0)$, independent of T and S, represent the initial workload in the original model.

In the original model, define U, a random variable given by

$$U = \begin{cases} V + \sum_{i=1}^{N(V)} S_i - \sum_{i=1}^{N(V)} T_i & \text{if } N(V) < \infty \\ \infty & \text{otherwise} \end{cases} \tag{7.79}$$

Clearly, the last arrival during the busy period is at the instant $\Delta(V) - U$ and

$$\Delta(V) = V + \sum_{i=1}^{N(V)} S_i = U + \sum_{i=1}^{N(V)} T_i. \tag{7.80}$$

For fixed $V = v$ and $U = u$, the busy period consists of serving n customers, if and only if

$$\sum_{i=1}^{j} T_i < v + \sum_{i=1}^{j-1} S_i, \qquad j = 1, \dots, n$$

and

$$u + \sum_{i=1}^{n} T_i = v + \sum_{i=1}^{n} S_i. \tag{7.81}$$

Let the probability of this conditional event be denoted by $\zeta(n; v, u)$. Then

$$P\left(N(v) = n, U \le u\right) = \int_0^y \zeta(n; v, u)(1 - A(u))\, du. \tag{7.82}$$

Suppose E is a random event associated with the busy period and denote by $\zeta_E(n; v, u)$ the conditional probability that E occurs and n customers are served during the busy period given $V = v$ and $U = u$. It is easy to see that

$$P(E, U \le y \mid V = v) = \int_0^\infty \sum_{n=1}^{\infty} \zeta_E(n; v, u)(1 - A(u))\, du. \tag{7.83}$$

Now consider the dual model $G_B/G_A/1$, for which $\hat{\zeta}(n; u, v)$ denotes the conditional probability that the busy period consists of serving n customers given $\hat{V} = u$ and $\hat{U} = v$. The event is equivalent to (see (7.81))

$$\sum_{i=1}^{j} \hat{T}_i < u + \sum_{i=1}^{j-1} \hat{S}_i, \qquad j = 1, \dots, n \tag{7.84}$$

and

$$v + \sum_{i=1}^{n} \hat{T}_i = u + \sum_{i=1}^{n} \hat{S}_i.$$

Because of duality, the transformations

$$\hat{T}_i = S_{n+1-i}, \quad \hat{S}_i = T_{n+1-i}, \qquad i = 1, \dots, n \tag{7.85}$$

are justified for the purpose of evaluating the probability; remarkably the transformations lead to

$$\sum_{i=1}^{j} T_i < v + \sum_{i=1}^{j-1} S_i, \qquad j = 1, \ldots, n$$

and

$$u + \sum_{i=1}^{n} T_i = v + \sum_{i=1}^{n} S_i. \tag{7.86}$$

Thus $\hat{\zeta}(n; u, v)$ is the probability of (7.86). But, when we compare (7.86) with (7.81), we observe that

$$\hat{\zeta}(n; u, v) = \zeta(n; v, u) \tag{7.87}$$

and note that the roles of V and U are interchanged.

The above discussion suggests that the dual process starts at the end of the original process and moves backward to reach the beginning. Clearly, the service-times and interarrival times are interchanged but in a reverse manner as seen in (7.85),

As in (7.82), we have

$$P\left(\hat{N}(v) = n, \hat{U} \leq y\right) = \int_0^y \hat{\zeta}(n; u, v)(1 - B(v))\, dv. \tag{7.88}$$

If $\hat{E}$ is the transformation of E under (7.85) and $\hat{\zeta}_{\hat{E}}(n; u, v)$ corresponds to $\zeta_E(n; v, u)$ according to (7.87), then similar to (7.83) we have

$$P\left(\hat{E} \middle| \hat{V} = u\right) = \int_0^\infty \sum_{n=1}^{\infty} \hat{\zeta}_{\hat{E}}(n; u, v)(1 - B(v))dv. \tag{7.89}$$

From (7.83), (7.87) and (7.89), it follows that

$$\int_0^\infty P\left(E \middle| V = v\right)(1 - B(v))\, dv = \int_0^\infty P\left(\hat{E} \middle| \hat{V} = u\right)(1 - A(u))\, du, \tag{7.90}$$

without depending on the distributions of V and $\hat{V}$. Expression (7.90) is an important duality relation for a busy period.

Suppose $A(x) = 1 - e^{-\lambda x}$. Then the original and dual models are $M/G/1$ and $G/M/1$, respectively. Using that value of $1 - A(x) = e^{-\lambda x}$ in the right side of (7.90), we have

$$\lambda \int_0^\infty P\left(E \middle| V = v\right)(1 - B(v))\, dv = \int_0^\infty P\left(\hat{E} \middle| \hat{V} = u\right) \lambda e^{-\lambda u}\, du. \tag{7.91}$$

If the busy period in the dual model starts when a customer arrives, then $F_{\hat{V}}(x) = A(x)$ and hence (7.91) becomes

$$P(\hat{E}) = \lambda \int_0^\infty P\left(E \middle| V = v\right)(1 - B(v))\, dv \tag{7.92}$$

relating the unconditional probability $\hat{E}$ in $G/M/1$ with conditional probability of E in $M/G/1$. Set $E = \{N = n-1\}$ which implies $\{E|V = v\} = \{N(v) = n-1\}$ and $\hat{E} = \{\hat{N} = n\}$. Therefore,

$$\begin{aligned}
P(\hat{N} = n) &= \lambda \int_0^\infty P\left(N(v) = n-1\right)\left(1 - B(v)\right) dv \\
&= \lambda \int_0^\infty \left[\int_v^\infty dG(n-1, t; v)\right] \left(1 - B(v)\right) dv \\
&\qquad \text{(see (3.95) for notation)} \\
&= \lambda \int_0^\infty \left[\int_v^\infty e^{-\lambda t} \frac{(\lambda t)^{n-2}}{(n-1)!} \lambda v \, dB^{(n-1)}(t-v)\right] \left(1 - B(v)\right) dv \\
&\qquad \text{(by using (3.100))} \\
&= \frac{\lambda^n}{(n-1)!} \int_0^\infty t^{n-2} e^{-\lambda t} \int_0^t v\left(1 - B(v)\right) dB^{(n-1)}(t-v) dv.
\end{aligned} \tag{7.93}$$

The same expression could be obtained from (3.110) by integrating over t and remembering that μ and $F_T(v)$ are to be replaced by λ and $B(v)$. Thus we have confirmed our observation which is made at the end of the first paragraph.

We develop a second duality relation. In the original model, let

$$P(k|i) = P\left(\sup_{0 \le t \le \Delta_i} X(t) \le k | X(0) = i\right)$$

which is the probability that the maximum number in the system during the initial busy period is $\le k$, given that the initial queue size is i. Evidently, $\hat{P}(k|i)$ is the notation for the dual model. The duality relation is that if

$$P\left(\sup_{1 \le n < \infty} \left|\sum_{j=1}^n S_j - \sum_{j=1}^n T_j\right| = \infty\right) = 1, \tag{7.94}$$

then for $0 < i \le k$,

$$\hat{P}(k|k-i) = 1 - P(k|i). \tag{7.95}$$

Condition (7.94) implies that the busy period must end for either model. For the proof of (7.95), consider a process $\{\eta(t),\, t \ge 0\}$ such that $\eta(0) = 0$ and there is a jump of $+1$ at T_i and of -1 at $\hat{T}_i$ for every i. Then $P(k|i)$ is the probability that the process reaches $-i$ first before reaching $k+1-i$. Similarly $\hat{P}(k|k-i)$ is the probability that the process reaches $k+1-i$ first before reaching $-i$. Because of (7.94) $P(k|i) + \hat{P}(k|k-i) = 1$ which completes the proof.

As earlier, we will utilize (7.95) in finding $\hat{P}(k|i)$ in $G/M/1$ from $P(k|i)$ in $M/G/1$. Observe that the process $\{\eta(t),\, t \ge 0\}$ which reaches $-i$ first before reaching $k+1-i$ can be decomposed into two independent processes, one which reaches $-(i-j)$ first before reaching $k+1-i$ followed by another which starting from $-(i-j)$ reaches $-i$ first before reaching $k+1-i$, where $j \leq i$. Therefore, we get

$$P(j|i) = P(k-j|i-j)P(k|j)$$

the solution of which takes the form

$$P(k|i) = \frac{a_{k-i}}{a_k}, \qquad 0 \leq i \leq k. \tag{7.96}$$

If we take into consideration that during the first service-time the number of arrivals may be j, $j = 0, 1, \ldots$ and note from Section 3.2 in Chapter 3 that α_j represents the probability of j arrivals during a service-time of a customer, then

$$P(k+i|i) = \sum_{j=0}^{k} \alpha_j P(k+i|i+j-1),$$

for $i \geq 0$ and $k \geq 0$, which by using (7.96) leads to the recurrence relation

$$a_k = \sum_{j=0}^{k} \alpha_j a_{k+1-j}, \qquad k = 0, 1, \ldots \tag{7.97}$$

Introducing the generating function $a(z)$ of $\{a_k\}$ in (7.97), we have

$$\begin{aligned} a(z) &= \sum_{k=0}^{\infty} \sum_{j=0}^{k} \alpha_j a_{k+1-j} z^k \\ &= \sum_{j=0}^{\infty} \alpha_j z^{j-1} \sum_{k=j}^{\infty} a_{k+1-j} z^{k+1-j} \\ &= \frac{\alpha(z)}{z} \left(a(z) - a_0\right). \end{aligned}$$

This yields

$$a(z) = \frac{a_0 \alpha(z)}{\alpha(z) - z} \tag{7.98}$$

which is convergent for $|z| < \delta$, δ being the smallest non-negative real root of $\alpha(z) = z$. Thus $P(k|i)$ can be evaluated from (7.96) and (7.98) when an expression for $\alpha(z)$ is known from a specified service-time distribution. For instance, if it is exponential (i.e., the model is $M/M/1$), we can show that

$$P(k|i) = \frac{1-\rho^k}{1-\rho^{k+1}}, \tag{7.99}$$

(see Exercise 20).

Denote by $Q(x|c)$ the probability that the maximal waiting time during the initial busy period is $\leq x$, given that the initial waiting time is c. A parallel argument yields the following:

For $0 \leq c \leq x$,

$$(a) \qquad \hat{Q}(x|x-c) = 1 - Q(x|c); \tag{7.100}$$

$$(b) \qquad Q(x|c) = \frac{b(x-c)}{b(x)} \tag{7.101}$$

where

$$b(x) = P(W_q \leq x). \tag{7.102}$$

The proof is left as an exercise (see Exercise 21).

Let us turn our attention to duality relations involving the waiting time. Unlike the previous case, the dual process considered hereafter is the reflection of the original process so that S_i and T_i become $\hat{T}_i$ and $\hat{S}_i$, respectively. We have noted that W_n and $\max_{0 \leq r \leq n-1}\{U_r'\}$ have the same distribution. When $n \to \infty$, this implies that for any $x \geq 0$,

$$\begin{aligned}
F_{W_q}(x) &= P(U_r^{'} \leq x, r \geq 0) \\
&= P\left(\sum_{i=1}^{r}(S_i - T_{i+1}) \leq x, r \geq 1\right) \\
&= P\left(\sum_{i=1}^{r}\left(\hat{T}_i - \hat{S}_{i+1}\right) \leq x, r \geq 1\right) \\
&= P\left(\sum_{i=1}^{r}\left(\hat{T}_{i+1} - \hat{S}_i\right) \leq x, r \geq 1\right) \quad \text{since } \hat{T}_i's \text{ and } \hat{S}_i's \text{ are i.i.d.} \\
&= P\left(x + \hat{U}_r \geq 0, r \geq 1\right), \text{ where } \hat{U}_r = \sum_{i=1}^{r}(\hat{S}_i - \hat{T}_{i+1}), \\
&= P\left(\hat{\Delta}(x) = \infty\right),
\end{aligned} \tag{7.103}$$

an interesting dual relation.

Observe that in (7.103), we have also shown that

$$P\left(U_r^{'} \leq x\right) = P\left(-\hat{U}_r^{'} \leq x\right) \qquad \text{for } x \geq 0 \quad \text{and} \quad r \geq 1,$$

i.e.,

$$U_r^{'} = -\hat{U}_r, \qquad r \geq 1. \tag{7.104}$$

At the end, we establish a relation between the waiting time in the original model and the idle time in the dual model.

It is recognized that the process $\left\{U_n^{'}, n \geq 0\right\}$ is a random walk (see Section 7.6). In Section 7.2 the condition $E(U) < 0$ is stated to be necessary for the existence stationary solution of W_n. Thus, one can intuitively conclude that $\lim_{n\to\infty} U_n^{'} = -\infty$ which implies that there exists a largest ascending ladder point, say M_K (see Section 7.6 for ladder points). That W_n and

$\max_{0 \le r \le n-1}\{U'_r\}$ have the same distribution implies

$$W_q = U'_{M_K}. \tag{7.105}$$

Consider the dual process $\left\{\hat{U}'_n, n \ge 0\right\}$. Because of (7.104) and that U''s are continuous, it is clear that the ascending and descending ladder points as well as the respective ladder heights are interchanged for the original and the dual models, i.e.,

$$M_r = \hat{\bar{M}}_r \quad \text{and} \quad \hat{M}_r = \bar{M}_r \quad \text{for every } r. \tag{7.106}$$

Moreover, observe that any idle period is stopped by an arrival only at the time corresponding to a descending ladder index. Thus if $\hat{I}_r$ represents the length of the rth idle period in the dual model, then it is given by

$$\begin{aligned}\hat{I}_r &= \sum_{i=\hat{\bar{m}}_{r-1}+1}^{\hat{\bar{m}}_r} \left(\hat{T}_{i+1} - \hat{S}_i\right) \\ &= U'_{m_r} - U'_{m_{r-1}} \end{aligned} \tag{7.107}$$

by using (7.104) and (7.106). It is obvious that $\hat{I}_r$'s have the same distribution. Therefore,

$$\begin{aligned}\Phi_{W_q}(\theta) &= E\left(E(e^{-\theta U'_{M_K}} \middle| M_K)\right) \qquad \text{by (7.105)} \\ &= E\left(E\left(e^{-\theta(\hat{I}_1+\ldots+\hat{I}_K)} \middle| M_K\right)\right) \\ &= E\left(\Phi_{\hat{I}}(\theta)\right)^{M_K}. \end{aligned} \tag{7.108}$$

Let $1 - \zeta = P\left(U'_n \le U'_{m_k} \text{ for all } n > m_k\right)$. Then

$$P(M_K = m) = (1-\zeta)\zeta^m. \tag{7.109}$$

Combining (7.108) and (7.109) we obtain the duality relation

$$\Phi_{W_q}(\theta) = \frac{1-\zeta}{1-\zeta\Phi_{\hat{I}}(\theta)}, \tag{7.110}$$

which is also another expression for $\Phi_{W_q}(\theta)$ (see (7.29) and (7.30)).

7.8. Discussion

In earlier chapters, the assumption of exponentiality either for interarrival times T or for service-times S or both is crucial for the theoretical development. However, in reality this may not be true. Without precise knowledge of the nature of distributions that govern T and S, it is but natural to search for results which could hold good in a general setup. In this chapter, we have considered a wide class of distributions for which only the mean and in some cases variance are given.

In spite of the consideration of generality as far as the distribution is concerned, namely, the model $G/G/1$, the overall technical tool for dealing with the model has been again the utilization of the Markovian property. Even in the general situation, it happens that the waiting times $\{W_n\}$ of customers form a Markovian process and therefore their study has become possible. This has become the central focus of the present chapter.

First of all, the waiting time is seen to be a characteristic of a process which is constrained at zero (see (7.2)). Had there been no constraint, (7.3) would be a convolution and the analysis would be carried out by standard procedures. The challenge arises because of the constraint. We recall a similar constraint at zero is placed in the random walk process arising in the queueing process in Section 6.3. As in there, the procedure in Section 7.2 is to consider the unconstrained process and obtain the necessary adjustment for the constraint. This is achieved by the spectrum factorization (7.11), which eventually leads to solution (7.17). Two points to ponder: (i) the solution is in terms of the L.T. and therefore needs the inverse transformation (see Section 7.2); (ii) the factorization (7.11) along with the conditions may not work. However by using the factorization a solution to the idle period distributions is derived in (7.32). Note that what we have dealt with are the limiting distributions of waiting and idle times.

In Section 7.6, a probabilistic approach to solve the same problem is described. Here the waiting time is shown to be a maximum of a random walk sequence constrained at zero (see (7.73)). The distribution of a maximum is related to ladder points and ladder heights of a random walk, and that seems to be natural from the definition. The Wiener-Hopf factorization ultimately facilitates us to determine the actual distributions (not their L.T.s) of limiting waiting and idle times, namely F_{W_q} and F_I. In both sections, only the knowledge of means of T and S are assumed. As a side remark we may say that the topic of distribution of maxima in a process is a subject in the fluctuation theory which has been developed extensively.

Although (7.17) does not in general tell us much about F_{W_q}, we can have an approximation under heavy traffic condition, that is when $\rho \simeq 1$. Given the first two moments of T and S, it is found that an approximate distribution of W_q happens to be exponential as given in (7.54), which is derived with the help of the Taylor series expansion.

Sections 7.5 and 7.7 somewhat stand apart from the rest as far as the theme of waiting and idle times are concerned. Nevertheless, the theme of Section 7.5 is on $G/G/1$, only from a different perspective. The main idea is to approximate the discrete state space of the system by a continuous state space. It is achieved by the diffusion approximation, as provided by the central limit theorem (see Section A.4 in Appendix A) which essentially makes the sum of large number of any i.i.d. random variables with given first two moments to become a normal variable. Thus the unknown attributes of G are bypassed in the limit by reaching the normal distribution, the one which depends only on its first two moments and which is very thoroughly investigated. Now from the approximated model, we can study any characteristic of the queue. As an example, we have considered the queue length, the distribution of which is given in (7.66).

In Section 7.7, the concept of duality for any general model is developed and illustrated in quite a few cases. It is natural to think of a process called the dual process in which T ad S of the original process are interchanged. The change could be visualized by looking either at the process backward or at its reflection. This transform leads to some interesting relations between the characteristics of the original process and those of the dual process.

Regarding the references, the Lindley integral equation in Section 7.2 was introduced in Lindley (1952) and solved by the analytic method of spectrum factorization in Smith (1953). On the other hand, its treatment by a probabilistic method is done in Prabhu (1980). For bounds and heavy traffic approximation (Sections 7.3 and 7.4), see Kingman (1961), (1962) and (1970) and Marshall (1968b). The relation between W_q and I is taken from Marshall (1968c). For diffusion approximation (Section 7.5), we refer to Kleinrock (1976), and for duality (Section 7.7) to Greenberg (1969), Prabhu (1965) and Takács (1969). References for applications in manufacturing are Suri (1988), and Suri, Sanders and Kamath (1993).

7.9. Exercises

1. Prove (7.18).

2. for $M/D/1$, with a as the constant service-time, verify that

$$F_U(u) = \begin{cases} 1 & a < u \\ e^{-\lambda(a-u)} & a \geq u \end{cases}$$

3. Show that for $M/D/1$ model, with a being the constant service-time, Lindley's integral equation reduces to the following integral equation:

$$F_{W_q}(x) = \int_0^\infty F_{W_q}(x+u-a)\lambda e^{-\lambda u} du.$$

   Prove that

$$F_{W_q}(x) = (1-\lambda a)e^{\lambda x} \sum_{k=0}^{n} \frac{\lambda^k}{k!} e^{-k\lambda a}(k-x)^k \quad na < x \leq (n+1)a$$

   for $n = 0, 1, 2, \ldots$, is the solution of the integral equation.

4. Show that for $D/M/1$ queue with b being the fixed length of the interarrival time,

$$F_{W_q}(x) = 1 - (1-p_0)e^{-\mu p_0 x}$$

   where p_0 satisfies

$$e^{-\mu p_0 b} = 1 - p_0.$$

5. For the $M/G/1$ system,

$$\Phi_U(\theta) - 1 = \frac{\lambda\Phi_S(\theta) - \lambda + \theta}{\lambda - \theta}.$$

Thus

$$\psi_+(\theta) = \lambda\Phi_S(\theta) - \lambda + \theta$$

and

$$\psi_-(\theta) = \lambda - \theta \qquad D < \lambda$$

satisfy the requirement for spectrum factorization as in (7.11). Show the

$$\Phi_{W_q}(\theta) = \frac{(1-\rho)}{1 - \frac{\lambda}{\theta}[1 - \Phi_S(\theta)]}.$$

6. Find the distribution of W_q for the $HE_2/M/1$ system.

7. For the $G/G/1$ model show that

$$\Phi_{W_q}(\theta) = \frac{a_0[1 - \Phi_I(-\theta)]}{1 - \Phi_U(\theta)}$$

(see Marshall (1968c)).

8. Recalling from (7.11) that

$$1 - \Phi_S(\theta)\Phi_T(-\theta) = -\frac{\psi_+(\theta)}{\psi_-(\theta)}$$

is the spectrum factorization, and

$$\Phi_{W_q}(\theta) = \frac{a_0\theta}{\psi_+(\theta)},$$

verify that

$$\bar{F}_I^*(\theta) = \frac{1}{\psi_-(\theta)}$$

where $\bar{F}_I^*(\theta)$ is the L.T. of the $P(I > x)$.

9. Check that for the $M/M/1$ model I (idle time) follows an exponential distribution, by using the spectrum factorization method.

10. Using the spectrum factorization procedure for the queueing model $M/D/1$, show that

$$\begin{aligned}\Phi_{W_q}(\theta) &= \frac{(1-\rho)}{(1-\frac{\lambda}{\theta}) + \frac{\lambda}{\theta}e^{-\frac{1}{\mu}\theta}} \\ \bar{F}_I(t) &= e^{-\lambda t}.\end{aligned}$$

11. Using the fact that

$$a_0 E(I) = -E(U),$$

check that

$$E(I) \geq \frac{1}{\lambda} - \frac{1}{\mu}$$

and for the $D/D/1$ system

$$E(I) = \frac{1}{\lambda} - \frac{1}{\mu}.$$

12. Show that the L.S.T. of the distribution function of D (the inter-departure time) for $G/G/1$ system is given by

$$\Phi_D(\theta) = (1 - a_0)\Phi_S(\theta) + a_0\Phi_T(\theta)\Phi_S(\theta)$$

where $a_0 = P$(an arrival finds the system empty), and for $M/G/1$ model

$$\Phi_D(\theta) = \frac{\lambda}{\mu}\left(\frac{\mu+\theta}{\lambda+\theta}\right)\Phi_S(\theta),$$

and check that D has an exponential distribution with parameter λ, if and only if the service-time is exponentially distributed with parameter μ.
Show that $\mathrm{Var}(D) = \sigma_T^2 + 2\sigma_S^2 - \frac{2}{\lambda}(1-\rho)E(W_q)$ for $G/G/1$ system and

$$\mathrm{Var}(D) = \frac{1}{\lambda^2} - \frac{1}{\mu^2} + \sigma_S^2$$

for $M/G/1$ system.
(see Marshall (1968a)).

13. Obtain the L.T. of the probability density function of D and I for the queueing model $G/M/1$.

14. For the $G/G/1$ model

$$E[W_q] = -\frac{E[U^2]}{2E[U]} - \frac{E[I^2]}{2E[I]},$$

where $U = S - T$, and I is the idle period. Check that

$$E[W_q] = \frac{\sigma_T^2 + \sigma_S^2 + [E(T)]^2(1-\rho)^2}{2E(T)(1-\rho)} - \frac{E[I^2]}{2E[I]}.$$

Show that for $M/G/1$ queue

$$E[I] = \frac{1-\rho}{\lambda}, \qquad \mathrm{Var}[I] = \frac{1-\rho^2}{\lambda^2}.$$

15. Show that the upper bound on $E(W_q)$ as given by (7.39) for $G/G/1$ model is equal to the exact value of $E(W_q)$ for the special case $M/G/1$.

16. For $G/M/1$ model (as dual of $M/G/1$), show that

$$\Phi_{\hat{I}}(\theta) = \mu[1 - \Phi_S(\theta)]/\theta.$$

where $E(S) = 1/\mu$.

17. Using (7.70), show that for $M/M/1$ system

$$F_{W_q}(x) = 1 - \rho e^{-(\mu-\lambda)x}$$

(see Prabhu (1980)).

18. For the $M/G/1$ system, we have seen that

$$P\{\Delta(v) \le t\} = \int_{x=v}^{t} \frac{v}{x} \sum_{j=0}^{\infty} e^{-\lambda x} \frac{(\lambda x)^j}{j!} dB^{(j)}(x-v)$$

where $B(\cdot)$ is the distribution function of service-time.

Using the duality relation on busy period given in (7.90), show that for a $G/M/1$ system (as dual of $M/G/1$)

$$P\{\Delta(v) < t\} = \int_{v=0}^{t} \int_{x=v}^{t} \lambda \frac{v}{x} \sum_{j=0}^{\infty} e^{-\lambda x} \frac{(\lambda x)^j}{j!} B^{(j)}(x-v)\bar{B}(v)dv\, dx,$$

where λ is the service rate and $B(\cdot)$ is the distribution function of interarrival time in $G/M/1$.

19. For the $M/G/1$ queueing system, assuming $B(\cdot)$ as the distribution function of service-times, it is seen that if $\frac{\lambda}{\mu} < 1$, then $P\{\Delta(v) < \infty\} = 1$ for all values of v. Check that for the $G/M/1$ model as a dual of $M/G/1$,

$$P\{\hat{\Delta} < \infty\} = \int_{0}^{\infty} \lambda \bar{B}(v)dv.$$

Here λ is the rate of service and $B(\cdot)$ is the distribution function of interarrival time.

20. Check (7.99).

21. Complete the derivation of (7.100) and (7.101).

References

Greenberg, I. (1969). Some duality results in the theory of queues, *J. Appl. Prob.*, **6**, 99–121.

Kingman, J. F. C. (1961). The single server queue in heavy traffic, *Proc. Camb. Phil. Soc.*, **57**, 902–904.

Kingman, J. F. C. (1962). Some inequalities for $GI/G/1$, *Biometrica* **49**, 315–324.

Kingman, J. F. C. (1970). Inequalities in the theory of queues, *J. Royal Stat. Soc. Ser. B*, **32**, 102–110.

Kleinrock, L. (1975). *Queueing Systems, Volume I*, John Wiley, New York.

Kleinrock, L. (1976). *Queueing Systems, Volume II*, John Wiley, New York.

Lindley, D. V. (1952). The theory of queues with a single server, *Proc. Camb. Phil. Soc.*, **48**, 277–289.

Marshall, K. T. (1968a). Some inequalities in queueing, *Oper. Res.* **16**, 651–665.

Marshall, K. T. (1968b). Bounds for some generalization for $GI/G/1$ queue, *Ope. Res.*, **16**, 841–848.

Marshall, K. T. (1968c). Some relationships between the distributions of waiting time, idle time and inter output time in the $GI/G/1$ queue, *Siam J. Appl. Math.* **16**, 324–327.

Prabhu, N. U. (1965). *Queues and Inventories*, John Wiley, New York.

Prabhu, N. U. (1980). *Stochastic Storage Processes Queues, Insurance Risk and Dams*, Springer-Verlag, New York.

Smith, W. L. (1953). On the distribution of queueing times, *Camb. Phil. Soc.* **49**, 449–461.

Suri, R. (1988). RMT puts production at the helm. *Manuf. Engrg.*, 100 (2), 41–44.

Suri, R., Sanders, J. L., Kamath, M. (1993), Performance Evaluation of Production Networks, in *Handbooks in Operations Reseach and Management Science*, S. C. Graces et al. Eds., Amsterdam.

Takács, L. (1969). On inverse queueing process, *Z. Matematyki*, **10**, 213–224.

CHAPTER 8

Computational Methods – II

8.1. Introduction

Repeatedly we are haunted with the question: Theory or not, can we lay our hands on numerically computable queueing characteristics? Of course, theory has helped and guided us in our pursuit, sometimes directly and sometimes indirectly. Nevertheless, our quest for computable solutions has led us to separately consider this problem in Chapter 4 and as a consequence, quite a few novel and interesting ideas and procedures are revealed to us, most of which have no direct resemblance with materials of Chapters 2 and 3. Continuing in the same vein, the present chapter deals with more computational methods, in some way as extensions of earlier concepts. But remarkable enough, they cover models for which theoretical treatments are not easily available.

Unlike Chapter 4, the primary focus in this chapter is on the computability of steady-state solutions. In Sections 8.2 and 8.3, the main idea is to consider models in which the corresponding $\mathbf{P}$ or $\mathbf{Q}$ matrix has the familiar structure that we are acquainted with in Chapters 2 and 3 except that the entries are matrices instead of scalars. In Section 8.2, we give a procedure which mimics the $M/M/1$ model in a very limited sense, in order to obtain a geometric-type solution in matrices. Here, the concept of phase-type distribution is introduced which particularly leads models to confirm to the new structure. Section 8.3 is an extension of the state reduction method in Section 4.3 to the new situation with matrix or block entries.

In Sections 8.2 and 8.3, the approaches are recursive and algebraic. Thus there is a lurking allurement to think of the transform methods which we are so used to in most of our theoretical development. The challenge is to find the inversion of a transformed solution. Root finding techniques answer this problem when the roots of a polynomial occurring in the transformed function are required. In Section 8.4, we propose numerical techniques of inverting a probability generating function and Laplace transform by the Fourier series method.

8.2. The Matrix-Geometric Solution

In Chapter 4, we have discussed some methods for obtaining stationary solutions. We again consider the same problem but for more general models and thus in a way extend earlier techniques in solving new models. In order to generalize models, let us for example look at $M/M/1$ (Chapter 2) and $G/M/1$ (Chapter 3), particularly so because the stationary distribution in each case is attractively simple and is nothing but the geometric distribution. This may prompt us to ask the question: Are there other models we still can have the stationary distribution to be geometric in nature (for example, $P_n = (1-\rho)\rho^n$ in the $M/M/1$ model)?

8.2.1. Examples. In order to understand the geometric nature of the solution, we consider a slightly modified version of the $M/M/1$ model (in fact a particular case of the birth-death model), in which the arrival rate is changed from λ to λ' when the system is empty. In this case the generator $\mathbf{Q}$ is given by

$$\mathbf{Q} = \begin{bmatrix} -\lambda' & \lambda' & & & & \\ \mu & -(\lambda+\mu) & \lambda & & & \\ & \mu & -(\lambda+\mu) & \lambda & & \\ & & \mu & -(\lambda+\mu) & \lambda & \\ & & & & & \ddots \end{bmatrix}. \tag{8.1}$$

Without any ambiguity, we use the earlier notation $\{P_n\}$ for stationary probabilities. The balance equations become

$$\begin{aligned} &P_0\lambda' - P_1\mu = 0 \\ &P_0\lambda' - P_1(\lambda+\mu) + P_2\mu = 0, \\ &P_{j-1}\lambda - P_j(\lambda+\mu) + P_{j+1}\mu = 0, \qquad j \geq 2. \end{aligned} \tag{8.2}$$

Assume that

$$P_j = P_1 c^{j-1}, \qquad j \geq 2,$$

for some constant c. This assumption is consistent with the structure and solution in the $M/M/1$ model. On substitution, we get the quadratic equation

$$\lambda - (\lambda+\mu)c + \mu c^2 = 0, \tag{8.3}$$

which has two roots, namely $c = 1$ and $c = \rho = \lambda/\mu$. The first solution $c = 1$ does not agree with the normalizing equation $\sum_{j=0}^{\infty} P_j = 1$ and therefore is discarded. Note that the present analysis resembles the technique of characteristic equation in $M/M/1$ model. Using boundary equations with $P_2 = \rho P_1$ and the normalizing equation, it is easy to establish

$$\begin{aligned} P_0 &= \frac{1}{1-\rho'/(1-\rho)}, \qquad \rho' = \lambda'/\mu \\ P_1 &= \frac{\rho'}{1-\rho'/(1-\rho)}, \end{aligned} \tag{8.4}$$

and

$$P_j = \rho^{j-1} P_1, \qquad j \geq 2.$$

Observe that the geometric nature of the solution originates from the shifted repetitive structure of the entries in $\mathbf{Q}$ and from the ability to derive a root of (8.3) having value less than 1.

For a possibility of generalization, let us consider a modified version of a $M/E_2/1$ model in which the arrival rate is λ' and λ according as the queue is empty or not, and the mean service-times for first and second phase are $1/\mu_1$ and $1/\mu_2$, respectively. The state space of the induced Markov process consists of

$$\{(0,0) \cup (i,j), i = 0,1,\ldots; j = 1,2\}$$

where the first entry represents the queue size and the second one the phase of service except in $(0,0)$, in which the second zero is irrelevant.

Arrange the elements of the state space in lexicographic order. The generator of the process is given by the *block structured matrix*

$$\mathbf{Q} = \begin{bmatrix} \mathbf{B}_0 & \mathbf{C}_0 & & & \\ \mathbf{B}_1 & \mathbf{A}_1 & \mathbf{A}_0 & & \\ & \mathbf{A}_2 & \mathbf{A}_1 & \mathbf{A}_0 & \\ & & \mathbf{A}_2 & \mathbf{A}_1 & \mathbf{A}_0 \\ & & & & \ddots \end{bmatrix} \tag{8.5}$$

where the blocks $\mathbf{B}_0, \mathbf{B}_1, \mathbf{C}_0, \mathbf{A}_0, \mathbf{A}_1$ and $\mathbf{A}_2$ are given by the matrices

$$\mathbf{B}_0 = \begin{bmatrix} -\lambda' & \lambda' & 0 \\ 0 & -(\lambda+\mu_1) & \mu_1 \\ \mu_2 & 0 & -(\lambda+\mu_2) \end{bmatrix}, \quad \mathbf{B}_1 = \begin{bmatrix} 0 & 0 & 0 \\ 0 & \mu_2 & 0 \end{bmatrix},$$

$$\mathbf{C}_0 = \begin{bmatrix} 0 & 0 \\ \lambda & 0 \\ 0 & \lambda \end{bmatrix}, \quad \mathbf{A}_0 = \begin{bmatrix} \lambda & 0 \\ 0 & \lambda \end{bmatrix},$$

$$\mathbf{A}_1 = \begin{bmatrix} -(\lambda+\mu_1) & \mu_1 \\ 0 & -(\lambda+\mu_2) \end{bmatrix}, \mathbf{A}_2 = \begin{bmatrix} 0 & 0 \\ \mu_2 & 0 \end{bmatrix}.$$

Notice the similarity between the shifted repetitive structure in (8.1) and (8.5), except that in (8.5) the entries are matrices.

Letting

$$\mathbf{P}_0 = (P(0,0), P(0,1), P(0,2))$$

and

$$\mathbf{P}_i = (P(i,1), P(i,2)), \qquad i \geq 1,$$

we can write the balance equations for the repeating structure part of the system as

$$\mathbf{P}_{j-1}\mathbf{A}_0 + \mathbf{P}_j\mathbf{A}_1 + \mathbf{P}_{j+1}\mathbf{A}_2 = \mathbf{0}, \qquad j \geq 2,$$

which is very similar to the last equation in (8.2) except that the scalars are replaced by matrices. As in the previous case, let us assume that

$$\mathbf{P}_j = \mathbf{P}_1\mathbf{R}^{j-1}, \qquad j \geq 2,$$

$\mathbf{R}$ being a 2×2 matrix. This leads to the quadratic equation

$$\mathbf{A}_0 + \mathbf{R}\mathbf{A}_1 + \mathbf{R}^2\mathbf{A}_2 = \mathbf{0}. \tag{8.6}$$

At this point, the concept of spectral radius of a matrix is introduced. The spectral radius of a matrix $\mathbf{X}$ of order $m \times m$, denoted by $\text{sp}(\mathbf{X})$ is defined as

$$\text{sp}(\mathbf{X}) = \max\{|\lambda_i|, 1 \leq i \leq m\},$$

where λ_i's are the eigenvalues of $\mathbf{X}$. For (8.6), we pick up a solution $\mathbf{R}^*$, such that $\text{sp}(\mathbf{R}^*) < 1$. Its existence and computation will be discussed under a general setting in Section 8.2.2. Once $\mathbf{R}^*$ is determined, the geometric nature of the solution

$$\mathbf{P}_j = \mathbf{P}_1\mathbf{R}^{*j-1}, \qquad j \geq 2 \tag{8.7}$$

is established. However, $\mathbf{P}_0$ and $\mathbf{P}_1$ are obtained from the boundary and normalizing equations

$$\begin{cases} \mathbf{P}_0\mathbf{B}_0 + \mathbf{P}_1\mathbf{B}_1 = \mathbf{0} \\ \mathbf{P}_0\mathbf{e}' + \mathbf{P}_1\sum_{j=1}^{\infty}\mathbf{R}^{*j-1}\mathbf{e}' = \mathbf{P}_0 + \mathbf{P}_1(\mathbf{I} - \mathbf{R}^*)^{-1}\mathbf{e}' = \mathbf{e}', \end{cases} \tag{8.8}$$

where $\mathbf{e}$ is a row vector of 1's. In this chapter vectors are always understood to be *row vectors* for notational convenience.

The elements in $\mathbf{Q}$ are matrices and the solution is of geometric nature – thus arises the term *matrix-geometric solution.*

Recall that the generator matrix of a birth-death process (see Chapter 2) turns out to be tridiagonal. Hence a model for which the generator matrix is tridiagonal with matrix elements is called *quasi-birth-death* process. Its generator matrix has the form (8.5), in which the matrices have any order. Clearly the modified $M/E_2/1$ model discussed above is a quasi-birth-death process.

Just as birth-death processes cover many interesting models (see Chapter 2), it is natural that quite a few models like $M/E_k/1$ model are cases of quasi-birth-death processes. For stationary solution of such a process, one proceeds similarly as in $M/E_2/1$ model.

8.2.2. Quasi-birth-death processes. Let us consider a Markov process with bivariate state space $\{(0,0) \cup (i,j)\ i = 0,1,2,\ldots,\ 1 \leq j \leq m\}$, where i represents the level and j the phase of the ith level. In the $M/E_2/1$ model, the level represents the number of customers in the queue and the phase is the stage of the service so that in this example $j = 1, 2$. Essentially, phases are playing the role to make the model Markovian. The number of phases m may be infinite although for our purpose we take m to be finite.

The Markov process is called a QBD (quasi-birth-death) process if one-step transitions are restricted to the same level or to the two adjacent levels. Thus it is only possible to move from (n,j) to (m,k) in one step if $m = n, n+1,$

or $n-1$, for $n \geq 1$ and if $m = 0, 1$, for $n = 0$. The generator matrix $\mathbf{Q}$ is tridiagonal and has the following general form:

$$\mathbf{Q} = \begin{bmatrix} \mathbf{B}_0 & \mathbf{C}_0 & & & \\ \mathbf{B}_1 & \mathbf{A}_1 & \mathbf{A}_0 & & \\ & \mathbf{A}_2 & \mathbf{A}_1 & \mathbf{A}_0 & \\ & & \mathbf{A}_2 & \mathbf{A}_1 & \mathbf{A}_0 \\ & & & & \ddots \end{bmatrix} \tag{8.9}$$

where the matrices $\mathbf{C}_0, \mathbf{B}_1, \mathbf{A}_0$ and $\mathbf{A}_2$ are non-negative and the matrices $\mathbf{B}_0$ and $\mathbf{A}_1$ have non-negative off diagonal elements but strictly negative diagonal elements.

The row sums of $\mathbf{Q}$ are necessarily equal to zero, which implies that

$$\begin{aligned} \mathbf{B}_0\mathbf{e}' + \mathbf{C}_0\mathbf{e}' &= 0, \\ \mathbf{B}_1\mathbf{e}' + \mathbf{A}_1\mathbf{e}' + \mathbf{A}_0\mathbf{e}' &= 0, \end{aligned}$$

and

$$\mathbf{A}_0\mathbf{e}' + \mathbf{A}_1\mathbf{e}' + \mathbf{A}_2\mathbf{e}' = 0.$$

Although the order of $\mathbf{e}$ may not be the same in each case (the order of $\mathbf{e}$ in the third equation is obviously m), we have taken the liberty of using the same notation in every case without causing any ambiguity.

Assume that the process is irreducible and ergodic and therefore has a unique stationary solution. Denote by $\mathbf{P}$ its stationary probability vector. Let us partition $\mathbf{P}$ into subvectors $\mathbf{P}_0, \mathbf{P}_1, \ldots$, where $\mathbf{P}_n$, $n \geq 0$ represents the set of stationary solutions of the nth level. Evidently the dimension of $\mathbf{P}_n$ is m for $n \geq 1$. Thus we can write $\mathbf{P}$ as $\mathbf{P} = [\mathbf{P}_0, \mathbf{P}_1, \ldots]$. The stationary equations can be seen to be as follows:

$$\begin{aligned} &\mathbf{P}_0\mathbf{B}_0 + \mathbf{P}_1\mathbf{B}_1 = \mathbf{0} \\ &\mathbf{P}_0\mathbf{C}_0 + \mathbf{P}_1\mathbf{A}_1 + \mathbf{P}_2\mathbf{A}_2 = \mathbf{0} \end{aligned} \tag{8.10}$$

and

$$\mathbf{P}_{n-1}\mathbf{A}_0 + \mathbf{P}_n\mathbf{A}_1 + \mathbf{P}_{n+1}\mathbf{A}_2 = 0, \quad n = 2, 3, \ldots, \tag{8.11}$$

where (8.10) are boundary equations. The normalizing equation is given by

$$\sum_{n \geq 0} \mathbf{P}_n\mathbf{e}' = 1. \tag{8.12}$$

For the existence of the stationary solution, the properties of irreducibility and ergodicity (positive recurrence) of the QBD process are needed. A simple necessary and sufficient condition exists for finite m. Let $\mathbf{A} = \mathbf{A}_0 + \mathbf{A}_1 + \mathbf{A}_2$. Matrix $\mathbf{A}$ is obviously the generator of a finite Markov process. If it is irreducible, then its unique stationary solution denoted by $\boldsymbol{\nu}$ exists, which is determined by

$$\boldsymbol{\nu}\mathbf{A} = \mathbf{0} \quad \text{and} \quad \boldsymbol{\nu}\mathbf{e}' = 1. \tag{8.13}$$

Assuming $\mathbf{Q}$ to be irreducible and $\mathbf{A}$ to be finite and irreducible, a necessary and sufficient condition for the stationary solution of the QBD process to exist is

$$\boldsymbol{\nu}\mathbf{A}_2\mathbf{e}' > \boldsymbol{\nu}\mathbf{A}_0\mathbf{e}'. \tag{8.14}$$

Condition (8.14) is an easily verifiable property. Moreover, it can be shown that there exists a non-negative matrix $\mathbf{R}$ of order m such that

$$\mathbf{P}_n = \mathbf{P}_1\mathbf{R}^{n-1}, \quad n \geq 2. \tag{8.15}$$

Upon substitution in (8.11), we see that $\mathbf{R}$ satisfies

$$\mathbf{A}_0 + \mathbf{R}\mathbf{A}_1 + \mathbf{R}^2\mathbf{A}_2 = \mathbf{0}. \tag{8.16}$$

It is also known that for the stationary distribution $\{\mathbf{P}_n\}$ to exist, $\mathbf{R}^*$ as a solution of (8.16) must satisfy $\text{sp}(\mathbf{R}^*) < 1$. Analogous to the condition $\rho < 1$ in the $M/M/1$ model, this condition makes $\sum_{n\geq 0} \mathbf{P}_n$ convergent. $\mathbf{P}_1$ and $\mathbf{P}_0$ are determined by (8.10) and (8.12).

Remark:

We can define a discrete-time QBD chain analogously and continue the analysis in which (8.15) will be the same but (8.16) will be different (Exercise 1).

On the basis of (8.16), the following recursive procedure provides a computational algorithm for evaluating $\mathbf{R}^*$:

$$\begin{aligned} &\mathbf{R}(0) = 0 \\ &\mathbf{R}(n+1) = \mathbf{A}_0(-\mathbf{A}_1)^{-1} + \mathbf{R}^2(n)\mathbf{A}_2(-\mathbf{A}_1)^{-1}, \; n = 0, 1, 2, \ldots \end{aligned} \tag{8.17}$$

until $\mathbf{R}(n+1)$ is close to $\mathbf{R}(n)$. It can be proved that

$$\mathbf{R}^* = \lim_{n\to\infty} \mathbf{R}(n) \tag{8.18}$$

and this limiting values is called the *rate matrix* of the matrix-geometric solution.

Using the fact that the sum of each row of $\mathbf{Q}$ is zero, we get

$$\mathbf{A}_0\mathbf{e}' + \mathbf{A}_1\mathbf{e}' + \mathbf{A}_2\mathbf{e}' = 0,$$

from which we have

$$\mathbf{A}_1\mathbf{e}' = -(\mathbf{A}_0 + \mathbf{A}_2)\mathbf{e}'. \tag{8.19}$$

But

$$(\mathbf{A}_0 + \mathbf{R}\mathbf{A}_1 + \mathbf{R}^2\mathbf{A}_2)\mathbf{e}' = \mathbf{0},$$

which with the help of (8.19) simplifies to

$$\mathbf{A}_0\mathbf{e}' - \mathbf{R}(\mathbf{A}_0 + \mathbf{A}_2)\mathbf{e}' + \mathbf{R}^2\mathbf{A}_2\mathbf{e}' = \mathbf{0},$$

or to

$$(\mathbf{I} - \mathbf{R})(\mathbf{A}_0\mathbf{e}' - \mathbf{R}\mathbf{A}_2\mathbf{e}') = \mathbf{0}.$$

Therefore

$$\mathbf{R}\mathbf{A}_2\mathbf{e}' = \mathbf{A}_0\mathbf{e}'. \tag{8.20}$$

Thus $\mathbf{R}^*$ must satisfy (8.20).

Interestingly, $\mathbf{R}^*$ can be obtained explicitely if

$$\mathbf{A}_2 = \mathbf{v}'\mathbf{w} \tag{8.21}$$

for some row vectors $\mathbf{v}$ and $\mathbf{w}$, such that $\mathbf{w}\mathbf{e}' = 1$. In that case (8.20) becomes

$$\mathbf{R}\mathbf{v}' = \mathbf{A}_0\mathbf{e}',$$

which gives

$$\mathbf{R}\mathbf{A}_2 = \mathbf{R}\mathbf{v}'\mathbf{w} = \mathbf{A}_0\mathbf{e}'\mathbf{w}. \tag{8.22}$$

Using this in the quadratic matrix Equation (8.16), we get

$$\mathbf{A}_0 + \mathbf{R}\mathbf{A}_1 + \mathbf{R}\mathbf{A}_0\mathbf{e}'\mathbf{w} = \mathbf{0},$$

from which we obtain

$$\mathbf{R}^* = -\mathbf{A}_0(\mathbf{A}_1 + \mathbf{A}_0\mathbf{e}'\mathbf{w})^{-1}. \tag{8.23}$$

The non-singularity of $\mathbf{A}_1 + \mathbf{A}_0\mathbf{e}'\mathbf{w}$ can be verified.

$M/M/1$ **queue in a Markovian random environment.** A simple example of QBD process is to consider the $M/M/1$ queue in a Markovian random environment, which is nothing but an $M/M/1$ queue with arrival and service rates varying over time. In addition to the arrival and service process, one introduces a so-called environmental process $\{Y(t),\ t \geq 0\}$, a Markov process on a finite state space $\{1, 2, \ldots, m\}$ with $\mathbf{A}$ as its generator (i.e., with instantaneous transition rates a_{ij}, $i \neq j$, $1 \leq i \leq m$, $1 \leq j \leq m$). We assume that the Markov process is irreducible.

The environment controls the arrival and service processes as follows. Suppose at time t, $Y(t) = j$, $j = 1, \ldots, m$, then the arrival rate is λ_j and the service rate is μ_j provided that the server is busy at time t.

Thus the $M/M/1$ queue in a Markovian random environment is a bivariate Markov process $\{X(t), Y(t);\ t \geq 0\}$ on the state space $\{(i, j);\ i = 0, 1, 2, \ldots,\ 1 \leq j \leq m\}$, where $X(t)$ is the number of units in the system and $Y(t)$ is the state of the environment at time t. Here various states of the random environment act like phases in the original set-up.

Changes of bivariate states occur either when the environment changes, or when a new customer arrives, or when a service is completed. The possible transitions, and the corresponding instantaneous rates, are given in the table below:

From	To	Rate		
(n, j)	$(n-1, j)$	μ_j	for $n \geq 1$	(8.24)
(n, i)	(n, j)	a_{ij}	for $n \geq 0$ $i \neq j$	
(n, j)	$(n+1, j)$	λ_j	for $n \geq 0$.	

For general value of the possible environment state m and the generator $\mathbf{A}$ of $\{Y(t),\ t \geq 0\}$, the generator $\mathbf{Q}$ of the $M/M/1$ queue under Markovian environment is given by:

$$\mathbf{Q} = \begin{bmatrix} \mathbf{A}_1 + \mathbf{A}_2 & \mathbf{A}_0 & & & \\ \mathbf{A}_2 & \mathbf{A}_1 & \mathbf{A}_0 & & \\ & \mathbf{A}_2 & \mathbf{A}_1 & \mathbf{A}_0 & \\ & & & & \ddots \end{bmatrix} \tag{8.25}$$

where

$$\begin{aligned} \mathbf{A}_0 &= \operatorname{diag}(\lambda_1, \lambda_2, \ldots, \lambda_m), \\ \mathbf{A}_1 &= \mathbf{A} - \operatorname{diag}(\lambda_1 + \mu_1, \lambda_2 + \mu_2, \ldots, \lambda_m + \mu_m), \end{aligned}$$

and

$$\mathbf{A}_2 = \operatorname{diag}(\mu_1, \mu_2, \ldots, \mu_m).$$

Thus the $M/M/1$ queue in a Markovian environment is nothing but a QBD process.

8.2.3. Phase type (PH) distributions. We have seen in 8.2.1 that Erlangian distribution E_k's through the phases stipulate matrix elements of a special nature. The question arises: are there other distributions besides E_k's for which this structure holds?

Interestingly enough, a close look at E_k suggests a generalization leading to the so-called *phase type (PH) distributions.* The Erlangian distribution E_k can be seen to be the distribution of the time till absorption of a particle moving sequentially from phase 1 to phase k and finally getting absorbed in phase $k+1$, the movement being from phase i to phase $i+1$, $i = 1, \ldots, k$. If we allow the movements from any phase i to any phase j, then the distribution of the time until absorption in phase or state $k+1$ is called the PH-type distribution and is denoted by PH_k, or simply PH.

In this set-up, the generator matrix of the Markov process leading to E_k with rate λ is

$$\left[\begin{array}{ccccc|c} -\lambda & \lambda & 0 & \cdots & 0 & 0 \\ 0 & -\lambda & \lambda & & 0 & 0 \\ & & \ddots & & & \vdots \\ 0 & 0 & 0 & \cdots & -\lambda & \lambda \\ \hline 0 & 0 & 0 & \cdots & 0 & 0 \end{array}\right]_{(k+1)\times(k+1)} . \tag{8.26}$$

Notice that the generator matrix can be partitioned into four matrices as shown above. Thus the generator matrix of a Markov process leading to a PH_k distribution with $k+1$ phases, $(k+1)$th phase being the absorbing one, can be represented by

$$\mathbf{Q} = \begin{pmatrix} \mathbf{T} & \mathbf{T}^0 \\ \mathbf{0} & 0 \end{pmatrix}, \tag{8.27}$$

where $\mathbf{T}, \mathbf{T}^0$ and $\mathbf{0}$ are of orders $k \times k$, $k \times 1$ and $1 \times k$, respectively. They satisfy the following

$$\begin{aligned} T_{ii} &< 0, && 1 \le i \le k, \\ T_{ij} &\ge 0, && i \ne j, 1 \le i \le k, 1 \le j \le k, \\ T_i^0 &\ge 0, && \text{and } T_i^0 > 0 \text{ for at least one } i,\ 1 \le i \le k, \end{aligned} \tag{8.28}$$

and

$$\mathbf{T}\mathbf{e}' + \mathbf{T}^0 = \mathbf{0}.$$

Let the initial distribution of the Markov process be given by $(\boldsymbol{\alpha}, \alpha_{k+1})$, $\boldsymbol{\alpha}$ being of order $1 \times k$. Then

$$\boldsymbol{\alpha}\mathbf{e}' + \alpha_{k+1} = 1. \tag{8.29}$$

We assume that the absorption into state $k+1$ is certain, which is so, if and only if $\mathbf{T}$ is non-singular. Therefore the time to absorption is a random variable, to be denoted by τ. In view of relations (8.28) and (8.29), the random variable or equivalently PH_k distribution may be represented by $(\boldsymbol{\alpha}, \mathbf{T})$. Note that this representation is not unique.

A PH_k distribution represented by $(\boldsymbol{\alpha}, \mathbf{T})$ has the following properties:

(i) The distribution function is given by

$$F_\tau(x) = 1 - \boldsymbol{\alpha}\exp(\mathbf{T}x)\mathbf{e}', \qquad x \ge 0.$$

It has a jump of height α_{k+1} at $x = 0$ and its p.d.f. is

$$f_\tau(x) = \boldsymbol{\alpha}\exp(\mathbf{T}x)\mathbf{T}^0, \qquad x \ge 0.$$

(ii) The Laplace-Stieltjes transform $\Phi_\tau(\theta)$ is given by

$$\Phi_\tau(\theta) = \alpha_{k+1} + \boldsymbol{\alpha}(\theta\mathbf{I} - \mathbf{T})^{-1}\mathbf{T}^0.$$

(iii) The non-central moments μ'_r are all finite and are given by

$$\mu'_r = (-1)^r r!(\boldsymbol{\alpha}\mathbf{T}^{-r}\mathbf{e}'), \qquad r \ge 0.$$

Remarks:

1. The class of PH distributions can encompass a very large class of distributions by changing k, α and $\mathbf{T}$. It includes a generalized Erlangian distribution with parameters $\lambda_1, \ldots, \lambda_k$ by setting $\alpha = (1, 0, \ldots, 0)$ and

$$\mathbf{T} = \begin{bmatrix} -\lambda_1 & \lambda_1 & & \\ & -\lambda_2 & \lambda_2 & \\ & & \ddots & \\ & & & \lambda_k \end{bmatrix}. \tag{8.30}$$

The PH distribution possesses a rich structure which is mathematically tractable as well as computationally amenable.

2. A discrete version of PH distributions can be defined in which the generator $\mathbf{Q}$ in (8.26) is replaced by the transition matrix

$$\mathbf{P} = \begin{pmatrix} \mathbf{T} & \mathbf{T}^0 \\ \mathbf{0} & 1 \end{pmatrix}. \tag{8.31}$$

As earlier, the distribution is represented by $(\boldsymbol{\alpha}, \mathbf{T})$.

8.2.4. Bivariate Markov chains and processes. After extending the geometric solution arising in a $M/M/1$ model to matrix-geometric solution in a QBD model, we may examine the geometric stationary solution in a $G/M/1$ model for a similar extension. Denoting by $\mathbf{P}$ the transition matrix (P_{ij}), P_{ij}'s being given by (3.25), we can write

$$\mathbf{P} = \begin{bmatrix} b_0 & \alpha_0 & & & \\ b_1 & \alpha_1 & \alpha_0 & & \\ b_2 & \alpha_2 & \alpha_1 & \alpha_0 & \\ & & & & \ddots \end{bmatrix}$$

where $\alpha_i = \beta_i$ and $b_i = 1 - \sum_{j=0}^{i} \alpha_j$. Again, if we generalize the $G/M/1$ model to the $G/E_k/1$ model, the corresponding transition matrix will look like $\mathbf{P}$ except that its elements will be matrices (see Section 8.2.1). This is so because E_k introduces a bivariate state space $\{(i,j)\}$ where i refers to the number of customers in the queue and j to the stage or phase.

In general, let us consider a Markov chain with bivariate state space $\{(i,j),\ i \geq 0,\ 1 \leq j \leq k\}$, where i may represent the level and j the phase of the chain. Its transition probability matrix $\mathbf{P}$ when partitioned for different values of i may look like the following

$$\mathbf{P} = \begin{bmatrix} \mathbf{B}_0 & \mathbf{A}_0 & & & \\ \mathbf{B}_1 & \mathbf{A}_1 & \mathbf{A}_0 & & \\ \mathbf{B}_2 & \mathbf{A}_2 & \mathbf{A}_1 & \mathbf{A}_0 & \\ & & & & \ddots \end{bmatrix} \tag{8.32}$$

(see the matrix following (3.25)) where matrices $\mathbf{A}_r$ and $\mathbf{B}_r, r = 0, 1, \ldots$ are $k \times k$ non-negative matrices satisfying

$$\sum_{r=0}^{n} \mathbf{A}_r \mathbf{e}' + \mathbf{B}_n \mathbf{e}' = \mathbf{e}', \qquad n = 0, 1, \ldots. \tag{8.33}$$

Such a model is called $G/M/1$ *type* model whose transition probability matrix is of the form (8.32). In $\mathbf{P}$, the (m,n)th entry represents the matrix of transition probabilities from states $\{(m,j) : 1 \leq j \leq k\}$ to $\{(n,\nu) : 1 \leq \nu \leq k\}$. The form (8.32) is called the canonical form because the basic result follows from its structure. What is of interest to us is to find stationary probabilities of the states, denoted by a row vector $\mathbf{q}$.

First we should check the criteria of ergodicity of $\mathbf{P}$ such that the stationary solution exists (see Section A.8). Assume $\mathbf{P}$ to be irreducible. Also assume

$\mathbf{A} = \sum_{j=0}^{\infty} \mathbf{A}_j$ to be stochastic which is usually so in practical applications. $\mathbf{A}$ represents the marginal transition probabilities of phases and happens to be irreducible since $\mathbf{P}$ is so. As $\mathbf{A}$ is irreducible and finite (therefore ergodic), the corresponding Markov chain possesses a unique stationary solution to be denoted by $\boldsymbol{\kappa}$ (a row vector). Letting

$$\boldsymbol{\gamma}' = \sum_{r=1}^{\infty} r\mathbf{A}_r\mathbf{e}' \tag{8.34}$$

(in most cases $\boldsymbol{\gamma}' \neq \mathbf{0}$), it can be proved that $\boldsymbol{\kappa}\boldsymbol{\gamma}' > 1$ is a necessary and sufficient condition for the original Markov chain corresponding to $\mathbf{P}$ to be ergodic.

Once the condition for the existence of the stationary solutions are met, we are now to solve the stationary equations

$$\mathbf{qP} = \mathbf{q} \qquad \text{and} \qquad \mathbf{qe}' = 1, \tag{8.35}$$

where $\mathbf{q}$ is the stationary solution vector. Partitioning the vector $\mathbf{q}$ in a natural way into vectors (each of dimension k) $\mathbf{q}_0, \mathbf{q}_1, \ldots$, the above equations may be rewritten as

$$\sum_{r=0}^{\infty} \mathbf{q}_r\mathbf{B}_r = \mathbf{q}_0, \tag{8.36}$$

$$\sum_{r=0}^{\infty} \mathbf{q}_{n+r-1}\mathbf{A}_r = \mathbf{q}_n, \qquad n = 1, 2, \ldots \tag{8.37}$$

and

$$\sum_{r=0}^{\infty} \mathbf{q}_r\mathbf{e}' = 1. \tag{8.38}$$

Equations (8.36), (8.37) and (8.38) are, respectively, the boundary equations, the queueing equations and the normalizing equation. Remembering that the stationary solution of the $G/M/1$ model is a geometric distribution, we may attempt to check whether (8.37) has a matrix-geometric solution, say of the form

$$\mathbf{q}_n = \mathbf{q}_0\mathbf{R}^n, \qquad n = 0, 1, \ldots \tag{8.39}$$

where $\mathbf{R}$ is a matrix to be determined.

Define a formal power series in matrix $\mathbf{X}$ as

$$\mathbf{A}[\mathbf{X}] = \sum_{r=0}^{\infty} \mathbf{X}^r\mathbf{A}_r \tag{8.40}$$

which is in a way a matrix generating function. $\mathbf{A}[\mathbf{X}]$ converges (convergence of a matrix is element-wise convergence), if $\text{sp}(\mathbf{X}) < 1$.

On substitution of (8.39) in (8.37), we obtain

$$\mathbf{q}_0\mathbf{R}^{n-1}(\mathbf{R} - \mathbf{A}[\mathbf{R}]) = \mathbf{0}, \qquad n = 1, 2, \ldots$$

which implies

$$\mathbf{R} = \mathbf{A}[\mathbf{R}], \tag{8.41}$$

an expression similar to (3.29).

We have to find a solution of (8.41) such that its spectral radius is less than 1. It turns out that if $\boldsymbol{\kappa}\gamma' > 1$, i.e., if $\mathbf{P}$ is ergodic, then (8.41) has a unique solution $\mathbf{R}^* > \mathbf{0}$ with $\text{sp}(\mathbf{R}^*) < 1$. There are situations when $\mathbf{R}^*$ can be analytically determined. Otherwise, the following result provides a numerical technique for its determination:

Once $\mathbf{R}^*$ is determined, we use it in (8.39) and (8.36) to get

$$\mathbf{q}_0 = \mathbf{q}_0 \sum_{r=0}^{\infty} \mathbf{R}^{*r}\mathbf{B}_r = \mathbf{q}_0\mathbf{B}[\mathbf{R}^*]. \tag{8.42}$$

One can check that the matrix $\mathbf{B}[\mathbf{R}^*]$ is stochastic and therefore has an eigenvalue equal to 1. Hence $\mathbf{q}_0$ is a left eigenvector of $\mathbf{B}[\mathbf{R}^*]$ corresponding to the eigenvalue 1. Finally the constant multiplier of the eigenvector is evaluated from (8.38) which becomes

$$\mathbf{q}_0(\mathbf{I} - \mathbf{R}^*)^{-1}\mathbf{e}' = 1. \tag{8.43}$$

In summary, the solution of $G/M/1$ type model is derived from (8.39), (8.41), (8.42) and (8.43), provided $\mathbf{R}^*$ is computed. $\mathbf{R}^*$ is uniquely determined by (8.41) and $\mathbf{q}_0$ uniquely by (8.42) and (8.43). Then $\mathbf{q}_n$, $n = 1, 2, \ldots$ is obtained from (8.39).

Similar to the algorithm developed in Section 8.2.2 (see (8.17)), we have the following computational procedure for evaluating $\mathbf{R}^*$:

$$\mathbf{R}(0) = \mathbf{0},$$

and

$$\mathbf{R}(n+1) = \mathbf{A}[\mathbf{R}(n)], \qquad n \geq 0, \tag{8.44}$$

until $\mathbf{R}(n+1)$ is close to $\mathbf{R}(n)$. It can be proved that

$$\mathbf{R}^* = \lim_{n\to\infty} \mathbf{R}(n).$$

This is known as the *natural algorithm.* Another one called the *traditional algorithm* seems to reduce the number of iterations. Since $\mathbf{I} - \mathbf{A}_1$ is never singular, we may write (8.41) as

$$\mathbf{R}(\mathbf{I} - \mathbf{A}_1) = \mathbf{A}_0 + \sum_{r=2}^{\infty} \mathbf{R}^r\mathbf{A}_r.$$

Thus the algorithm is

$$\mathbf{R}(0) = \mathbf{0}$$

and

$$\mathbf{R}(n+1) = \left(\mathbf{A}_0 + \sum_{r=2}^{\infty} \mathbf{R}^r(n)\mathbf{A}_r\right)(\mathbf{I} - \mathbf{A}_1)^{-1}. \tag{8.45}$$

Yet a different algorithm is proposed which seems to be better than the other two. Again starting from (8.41), we can write

$$\mathbf{R} = \mathbf{A}_0 + \mathbf{R}\mathbf{U}$$

where

$$\mathbf{U} = \sum_{r=0}^{\infty} \mathbf{R}^r \mathbf{A}_{r+1}, \tag{8.46}$$

which implies

$$\mathbf{R} = \mathbf{A}_0(\mathbf{I} - \mathbf{U})^{-1}, \tag{8.47}$$

thus the new scheme is to first find an approximation of $\mathbf{R}$, which is used to approximate $\mathbf{U}$ from (8.46) by

$$\mathbf{U}(n) = \sum_{r=0}^{\infty} \mathbf{R}^r(n)\mathbf{A}_{n+1}. \tag{8.48}$$

Then we get a new approximation for $\mathbf{R}$ from (8.47) as

$$\mathbf{R}(n+1) = \mathbf{A}_0(\mathbf{I} - \mathbf{U}(n))^{-1}. \tag{8.49}$$

There is an entirely analogous theory for a bivariate Markov process, whose generator $\mathbf{Q}$ may be partitioned, similar to the matrix $\mathbf{P}$ as in (8.32), into the following canonical form:

$$\mathbf{Q} = \begin{bmatrix} \mathbf{B}_0 & \mathbf{A}_0 & & \\ \mathbf{B}_1 & \mathbf{A}_1 & \mathbf{A}_0 & \\ \mathbf{B}_2 & \mathbf{A}_2 & \mathbf{A}_1 & \mathbf{A}_0 \\ & & & \ddots \end{bmatrix} \tag{8.50}$$

where the off-diagonal elements of $\mathbf{Q}$ are non-negative and diagonal elements are negative such that

$$\sum_{r=0}^{n} \mathbf{A}_r\mathbf{e}' + \mathbf{B}_n\mathbf{e}' = \mathbf{0}, \qquad n = 0, 1, \ldots. \tag{8.51}$$

We assume that $\mathbf{Q}$ is irreducible, a necessary condition for that being $\mathbf{B}_0$ and $\mathbf{A}_1$ are non-singular. As before let $\mathbf{A} = \sum_{r=0}^{\infty} \mathbf{A}_r$ and we limit ourselves to the case $\mathbf{A}\mathbf{e}' = \mathbf{0}$, so that $\mathbf{A}$ is a generator. $\mathbf{A}$ has a unique stationary solution, denoted by $\boldsymbol{\kappa}$. Under the condition

$$\boldsymbol{\kappa}\mathbf{A}_0\mathbf{e}' < \sum_{r=2}^{\infty} (r-1)\boldsymbol{\kappa}\mathbf{A}_r\mathbf{e}' \tag{8.52}$$

an irreducible Markov process with generator $\mathbf{Q}$ possesses a unique stationary solution vector $\mathbf{p} = (\mathbf{p}_0, \mathbf{p}_1, \ldots)$ satisfying the following equations:

$$\sum_{r=0}^{\infty} \mathbf{p}_r \mathbf{B}_r = \mathbf{0}, \tag{8.53}$$

$$\sum_{r=0}^{\infty} \mathbf{p}_{n+r-1} \mathbf{A}_r = \mathbf{0}, \qquad r = 1, 2, \ldots \tag{8.54}$$

and

$$\sum_{r=0}^{\infty} \mathbf{p}_r \mathbf{e}' = 1. \tag{8.55}$$

These are analogous to (8.36), (8.37) and (8.38), respectively. The stationary distribution is then given by the matrix-geometric form

$$\mathbf{p}_n = \mathbf{p}_0 \mathbf{R}^{*n}, \qquad n = 0, 1, \ldots \tag{8.56}$$

where $\mathbf{R}^*$ is the unique solution of the matrix equation

$$\mathbf{A}[\mathbf{R}] = \mathbf{0}, \tag{8.57}$$

with $\mathrm{sp}(\mathbf{R}^*) < 1$. Now $\mathbf{p}_0$ is determined by the initial boundary equation which leads to

$$\mathbf{p}_0 \mathbf{B}[\mathbf{R}] = \mathbf{0} \tag{8.58}$$

and the normalizing equation which simplifies to

$$\mathbf{p}_0 (\mathbf{I} - \mathbf{R})^{-1} \mathbf{e}' = 1. \tag{8.59}$$

For computational purposes we use the fact (which can be proved)

$$\mathbf{R} = \lim_{n \to \infty} \mathbf{R}(n),$$

where

$$\mathbf{R}(0) = \mathbf{0},$$

and

$$\mathbf{R}(n+1) = -\mathbf{A}_0 \mathbf{A}_1^{-1} - \mathbf{R}^2(n) \mathbf{A}_2 \mathbf{A}_1^{-1} - \mathbf{R}^3(n) \mathbf{A}_3 \mathbf{A}_1^{-1} - \ldots, \qquad n \geq 0. \tag{8.60}$$

The other two algorithms can be developed analogously.

Some models involving PH distributions are discussed below.

1. Model I: $G/PH/1$

We consider a class of $G/PH/1$ queueing system in which the service-time distribution is of phase type (PH) with a representation $(\boldsymbol{\beta}, \tilde{\mathbf{S}})$ of order k. An algorithmic solution to this model can be obtained by embedding a Markov chain at epochs of arrivals.

Let $k \times k$ matrices $\mathbf{P}(n, t)$, $n \geq 0$, $t \geq 0$ be such that $P_{ij}(n, t)$ is the conditional probability that at time $t+$, the Markov process with generator

$\mathbf{Q}^* = \tilde{\mathbf{S}} + \mathbf{S}^0\boldsymbol{\beta}$ is in state j and that n renewals have occurred in $(0, t]$, given that the initial phase is i. The Kolmogorov equations for $\mathbf{P}(n, t)$ can be seen to be the following:

$$\begin{aligned} \mathbf{P}'(0,t) &= \mathbf{P}(0,t)\tilde{\mathbf{S}} \\ \mathbf{P}'(n,t) &= \mathbf{P}(n,t)\tilde{\mathbf{S}} + \mathbf{P}(n-1,t)\mathbf{S}^0\boldsymbol{\beta}, \qquad \text{for } n \geq 1, \end{aligned}$$

with $\mathbf{P}(0, 0) = I$, $\mathbf{P}(n, 0) = \mathbf{0}$, for $n \geq 1$.

Let Y_n be the number of units in the system just before the arrival time of the nth unit and J_n be the phase of the service immediately after the nth arrival. The embedded bivariate Markov chain $\{(Y_n, J_n),\ n \geq 0\}$ has the transition probability matrix $\mathbf{P}$ of the type given by (8.32) where $\mathbf{A}_r$ and $\mathbf{B}_r$, $r = 0, 1, \ldots$, are $k \times k$ non-negative matrices given by

$$\mathbf{A}_r = \int_0^\infty \mathbf{P}(r,t)dF_T(t), \quad r = 0, 1, \ldots \tag{8.61}$$

$$\mathbf{B}_r = \sum_{n=r+1}^{\infty} \int_0^\infty \mathbf{P}(n,t)dF_T(t)\mathbf{B}^{00}, \quad r = 0, 1, \ldots \tag{8.62}$$

and the matrix $\mathbf{B}^{00}$ is given by

$$\mathbf{B}_{ij}^{00} = \beta_j, \qquad \text{for} \quad 1 \leq i, j \leq k.$$

It can be checked that $\boldsymbol{\kappa}\gamma' = 1/\rho$. Therefore the present Markov chain is ergodic if and only if the classical equilibrium condition $\rho < 1$ holds good.

For the matrix-geometric solution (8.39), we have to solve

$$\mathbf{R} = \mathbf{A}[\mathbf{R}]$$

where the form $\mathbf{A}_r$ is given in (8.61).

One can check that

$$\mathbf{q}_0 = c\boldsymbol{\beta}$$

(given as an exercise). The use of normalizing condition (8.38) yields

$$c = \left[\boldsymbol{\beta}(\mathbf{I} - \mathbf{R})^{-1}\mathbf{e}'\right]^{-1}.$$

The quantity c satisfies $0 < c < 1$. Thus letting q_n to represent the stationary probability of having n in the system, we obtain

$$q_n = c\boldsymbol{\beta}\mathbf{R}^n\mathbf{e}', \text{ for } n \geq 0.$$

What is left is the evaluation of $\mathbf{R}$. $\mathbf{R}$ cannot be obtained analytically and the different computational procedures to determine $\mathbf{R}$ will be discussed in Section 8.5.

2. Model II: $M/PH/1$

In this case, since the number in the system can be studied at any instant, it is obvious to start with a bivariate Markov process with the state space $E = \{0, (i, j);\ i \geq 1,\ 1 \leq j \leq k\}$. The state 0 corresponds to the empty system and the state (i, j) corresponds to having $i \geq 1$ customers in the system and the service process in the jth phase, $1 \le j \le k$. The generator is given by

$$\mathbf{Q} = \begin{bmatrix} -\lambda & \lambda\boldsymbol{\beta} & \mathbf{0} & \mathbf{0} & \cdots \\ \mathbf{S}^0 & \mathbf{S}-\lambda\mathbf{I} & \lambda\mathbf{I} & \mathbf{0} & \cdots \\ \mathbf{0} & \mathbf{S}^0\boldsymbol{\beta} & \mathbf{S}-\lambda\mathbf{I} & \lambda\mathbf{I} & \cdots \\ \mathbf{0} & \mathbf{0} & \mathbf{S}^0\boldsymbol{\beta} & \mathbf{S}-\lambda\mathbf{I} & \cdots \\ \mathbf{0} & \mathbf{0} & \mathbf{0} & \mathbf{S}^0\boldsymbol{\beta} & \cdots \\ \mathbf{0} & \mathbf{0} & \mathbf{0} & \mathbf{0} & \cdots \\ \vdots & \vdots & \vdots & \vdots & \vdots \end{bmatrix}. \tag{8.63}$$

Let $\mathbf{p} = [p_0, \mathbf{p}_1, \mathbf{p}_2, \ldots,]$ be the stationary distribution of the queueing system. The steady-state equations are as follows:

$$-\lambda p_0 + \mathbf{p}_1\mathbf{S}^0 = 0 \tag{8.64}$$

$$\lambda p_0\boldsymbol{\beta} + \mathbf{p}_1(\mathbf{S}-\lambda\mathbf{I}) + \mathbf{p}_2\mathbf{S}^0\boldsymbol{\beta} = \mathbf{0} \tag{8.65}$$

$$\lambda\mathbf{p}_{i-1} + \mathbf{p}_i(\mathbf{S}-\lambda\mathbf{I}) + \mathbf{p}_{i+1}\mathbf{S}^0\boldsymbol{\beta} = \mathbf{0}, \text{ for } i \ge 2. \tag{8.66}$$

The normalizing equation is

$$p_0 + \mathbf{p}_1\mathbf{e}' + bp_2\mathbf{e}'_2 + \cdots = 1. \tag{8.67}$$

Multiplying (8.65) and (8.66) on the right by the column vector $\mathbf{e}'$ and using (8.64), we get

$$\begin{aligned} \mathbf{p}_1\mathbf{S}^0 &= \lambda p_0, \\ \mathbf{p}_{i+1}\mathbf{S}^0 &= \lambda\mathbf{p}_i\mathbf{e}', \qquad \text{for } i \ge 1. \end{aligned} \tag{8.68}$$

Again multiplying on the right by $\boldsymbol{\beta}$ in (8.68) we have

$$\mathbf{p}_{i+1}\mathbf{S}^0\boldsymbol{\beta} = \lambda\mathbf{p}_i\mathbf{e}'\boldsymbol{\beta}, \qquad \text{for } i \ge 1. \tag{8.69}$$

From (8.66) and (8.67), we find

$$\mathbf{p}_i(\lambda\mathbf{I} - \lambda\mathbf{e}'\boldsymbol{\beta} - \mathbf{S}) = \lambda\mathbf{p}_{i-1}, \qquad \text{for } i \ge 2, \tag{8.70}$$

and similarly

$$\mathbf{p}_1(\lambda\mathbf{I} - \lambda\mathbf{e}'\boldsymbol{\beta} - \mathbf{S}) = \lambda p_0\boldsymbol{\beta}. \tag{8.71}$$

It can be easily checked that $(\lambda\mathbf{I} - \lambda\mathbf{e}'\boldsymbol{\beta} - \mathbf{S})$ is a nonsingular matrix. Thus

$$\mathbf{p}_i = \lambda p_0\boldsymbol{\beta}[(\lambda\mathbf{I} - \lambda\mathbf{e}'\boldsymbol{\beta} - \mathbf{S})^{-1}]^i, \quad i = 1, 2.$$

Using the normalizing condition (8.67), we get

$$p_0 = 1 - \rho.$$

8.3. The Block Elimination Method

Now we would like to extend the state reduction method (Section 4.3) to compute the stationary probabilities for bivariate Markov models. In the study of bivariate Markov models which frequently arise in queueing systems, the transition matrix will have the block structure, implying that the elements in the matrix are themselves matrices.

In the previous section, solutions of special types of bivariate Markov models are obtained by an algorithmic procedure. It involves the computation of a matrix R, leading to the so-called matrix-geometric solution. In this section, we deal with a class of bivariate Markov models which include those in the last section. The procedure of obtaining the solution is similar to that of the state reduction method. While the state reduction method is applied to a univariate Markov chain or Markov process, the block elimination method is to solve a bivariate Markov chain or Markov process, in which the steady-state system of equations involves vectors and matrices. Before we come to this method, we describe a general approach called *censoring*.

8.3.1. Censoring. Equation (8.42) may be viewed as a stationary equation of a Markov chain with only one level 0 having transition probability matrix $\mathbf{B}[\mathbf{R}^*]$. Essentially, levels other than 0 are censored. In general, one may start with a Markov chain with state space partitioned into two sets E and E'. In this situation, E consists of all states in level 0 and E' the remaining. Let $\{X_n\}$ be the original Markov chain. Define $\{Y_i,\ i > 0\}$ as $Y_i = X_{n_i}$ where n_i is the time of the ith visit of the chain to E, $i > 0$. This new process $\{Y_i\}$ which is a Markov chain is called the chain imbedded in E and the states in E' are said to be censored.

In the censored chain, there will be two parts of contribution to the transition probability matrix, one from the original one and the other from returning to E after departing to E'. Thus if $\mathbf{P}$ is the transition probability matrix of the original chain having the induced partition as

$$\mathbf{P} = \begin{bmatrix} \mathbf{T} & \mathbf{H} \\ \mathbf{L} & \mathbf{Q} \end{bmatrix} \tag{8.72}$$

based on E and E', then the transition probability matrix $\mathbf{P}^E$ of the Markov chain in E is given by

$$\mathbf{P}^E = \mathbf{T} + \mathbf{HNL} \tag{8.73}$$

where

$$\mathbf{N} = \sum_{r=0}^{\infty} \mathbf{Q}. \tag{8.74}$$

Clearly, $(\mathbf{HNL})_{ij}$ represents the probability of having a sojourn in E' which ends by entering state $j \in E$, given the sojourn in E' started from the state $i \in E$.

Assume all matrices to be finite. Let $\hat{\mathbf{\Pi}}$ and $\bar{\mathbf{\Pi}}$, respectively, be the vector of steady-state probabilities for E and E'. Then we can write

$$\hat{\mathbf{\Pi}} = \hat{\mathbf{\Pi}}\,\mathbf{T} + \bar{\mathbf{\Pi}}\,\mathbf{L}$$

and

$$\bar{\mathbf{\Pi}} = \hat{\mathbf{\Pi}}\,\mathbf{H} + \bar{\mathbf{\Pi}}\,\mathbf{Q}$$

which lead to

$$\begin{aligned} \bar{\mathbf{\Pi}} &= \hat{\mathbf{\Pi}}\,\mathbf{H}(\mathbf{I} - \mathbf{Q})^{-1} \\ \hat{\mathbf{\Pi}} &= \hat{\mathbf{\Pi}}(\mathbf{T} + \mathbf{H}(\mathbf{I} - \mathbf{Q})^{-1}\mathbf{L}). \end{aligned} \tag{8.75}$$

From (8.74), we have

$$\mathbf{N} = \mathbf{I} + \sum_{r=1}^{\infty} \mathbf{Q}^r = \mathbf{I} + \mathbf{QN}$$

implying

$$\mathbf{N} = (\mathbf{I} - \mathbf{Q})^{-1}. \tag{8.76}$$

Thus, on combining (8.73), (8.75) and (8.76) we get

$$\mathbf{P}^E = \mathbf{T} + \mathbf{H}(\mathbf{I} - \mathbf{Q})^{-1}\mathbf{L} \tag{8.77}$$

and

$$\hat{\mathbf{\Pi}} = \hat{\mathbf{\Pi}}\,\mathbf{P}^E$$

and hence $\hat{\mathbf{\Pi}}$ is the steady-state probabilities corresponding to $\mathbf{P}^E$. Note that while the expression for $\mathbf{P}^E$ in (8.73) is obtained by a probabilistic argument, expression (8.77) is derived by an algebraic operation, namely a block elimination (eliminating states in E').

This approach will be used as a repeated block elimination in a general bivariate Markov chain.

8.3.2. Block elimination as censoring. Given the state space of the Markov chain $\{(i,j);\ i = 0,1,\ldots,N,\ j = 1,2,\ldots,m\}$, denote by $p_{i,r;k,s}$ the probability of one step transition from (i,r) to (k,s). Let $\mathbf{P}_{ik}$ be a matrix of order $m \times m$ having $p_{i,r;k,s}$ as its elements. Clearly $\mathbf{P}_{ik}$ represents the transition from level i to level k. Therefore, the entire transition matrix of the process is one with its elements as $\mathbf{P}_{ik}$, $i,k = 1,2,\ldots,N$. Denote the matrix by $\mathbf{P}$.

Let the steady-state distribution be $\mathbf{\Pi}$. Partitioning the vector $\mathbf{\Pi}$ in a natural way into vectors (each of dimension m), $\mathbf{\Pi}_0, \mathbf{\Pi}_1, \ldots, \mathbf{\Pi}_N$, the steady-state equations are

$$\mathbf{\Pi}_j = \sum_{i=0}^{N} \mathbf{\Pi}_i \mathbf{P}_{ij}, \qquad j = 0,1,\ldots,N \tag{8.78}$$

subject to

$$\sum_{j=0}^{N} \mathbf{\Pi}_j \mathbf{e}' = 1.$$

For $j = N$, we get

$$\mathbf{\Pi}_N(\mathbf{I} - \mathbf{P}_{NN}) = \sum_{i=0}^{N-1} \mathbf{\Pi}_i \mathbf{P}_{iN},$$

which gives

$$\mathbf{\Pi}_N = \sum_{i=0}^{N-1} \mathbf{\Pi}_i \mathbf{R}_{iN}, \tag{8.79}$$

where

$$\mathbf{R}_{iN} = \mathbf{P}_{iN}(\mathbf{I} - \mathbf{P}_{NN})^{-1}, \quad i = 0, 1, \ldots, N-1. \tag{8.80}$$

Substitute (8.79) in the rest of $N-1$ equations in (8.78). As in the case of the state reduction method, one can form a new matrix

$$\mathbf{P}^{(N-1)} = \left(\mathbf{P}_{ij}^{(N-1)}\right),$$

where

$$\mathbf{P}_{ij}^{(N-1)} = \mathbf{P}_{ij} + \mathbf{P}_{iN}(\mathbf{I} - \mathbf{P}_N)^{-1}\mathbf{P}_{Nj}. \tag{8.81}$$

This is the matrix form of (4.27). It is also similar to (8.77) with E' as level N which is censored. In this notation $\mathbf{P}_{ij} = \mathbf{P}_{ij}^{(N)}$.

It can be shown that $\left(\mathbf{P}_{ij}^{(N-1)}\right)$ is a stochastic matrix with levels $0, 1, \ldots,$ $N-1$. Define recursively for $n = N, N-1, \ldots, 1$

$$\mathbf{R}_{in} = \mathbf{P}_{in}^{(n)}(\mathbf{I} - \mathbf{P}_{nn}^{(n)})^{-1}, \; i = 0, 1, \ldots, n-1, \tag{8.82}$$

and

$$\mathbf{P}_{ij}^{(n-1)} = \mathbf{P}_{ij}^{(n)} + \mathbf{R}_{in}\mathbf{P}_{nj}^{(n)}, \quad i, j = 0, 1, \ldots, n-1. \tag{8.83}$$

This recursive procedure is know as the *block elimination method*, and it finally yields $\mathbf{P}_{00}^{(0)}$. At this stage, we obtain a nonzero solution $\mathbf{\Pi}_0$ of

$$\mathbf{\Pi}_0 = \mathbf{\Pi}_0 \mathbf{P}_{00}^{(0)}, \qquad m > 1.$$

Note that when $m = 1$, the procedure is the state reduction method. In that case, the above equation becomes an identity without giving a solution. Therefore, we set $\mathbf{\Pi}_0 = 1$ and proceed.

Back substitution in (8.78) gives

$$\mathbf{\Pi}_n = \sum_{i=0}^{n-1} \mathbf{\Pi}_i \mathbf{R}_{in}, \qquad n = 1, 2, \ldots, N. \tag{8.84}$$

One can use the normalizing condition to find $\mathbf{\Pi}_0$.

These are the operational steps for solving (8.78) by applying the block elimination method. Its probabilistic interpretation comes from censoring.

The block elimination method can be extended to bivariate Markov chains with infinite levels, and this is being illustrated through the special case given below.

Discrete-time homogeneous QBD chain. We consider a discrete-time QBD model with state space $\{(i, j) : \ i \geq 0, \ 1 \leq j \leq m\}$. Let its transition matrix be

$$\mathbf{P} = \begin{bmatrix} \mathbf{B} & \mathbf{A}_0 & & & \\ \mathbf{A}_2 & \mathbf{A}_1 & \mathbf{A}_0 & & \\ & \mathbf{A}_2 & \mathbf{A}_1 & \mathbf{A}_0 & \\ & & & \ddots & \end{bmatrix}, \tag{8.85}$$

where we have merely used notations $\mathbf{A}_0, \mathbf{A}_1$ and $\mathbf{A}_2$, as in $\mathbf{Q}$ matrix in (8.9) without creating any confusion. These matrices are of order m and $\mathbf{B} = \mathbf{A}_1 + \mathbf{A}_2$. The transition probabilities (in block format) are assumed to be level independent. We will solve the problem using the block elimination method.

The block elimination method cannot be applied to a matrix of infinite order, because it starts at the last row of the matrix, and an infinite matrix does not have a last row. The natural way is to truncate the matrix $\mathbf{P}$ and augment the lower right-hand side block to $\mathbf{A}_1 + \mathbf{A}_0$ to make $\mathbf{P}$ a stochastic matrix. Solve the truncated system for the stationary distribution. Suppose $\mathbf{P}$ has been truncated at level N. Let the truncated matrix be $\mathbf{P}^*$. $\mathbf{P}^*$ so modified has $N = km$ states, where k is an arbitrarily large positive integer.

Let $\mathbf{\Pi} = (\mathbf{\Pi}_0, \mathbf{\Pi}_1, \ldots, \mathbf{\Pi}_N)$ be the stationary distribution of the QBD with stochastic matrix $\mathbf{P}^*$. $\mathbf{\Pi}$ will satisfy the following set of equations:

$$\mathbf{\Pi}_0 = \mathbf{\Pi}_0(\mathbf{A}_1 + \mathbf{A}_2) + \mathbf{\Pi}_1\mathbf{A}_2, \tag{8.86}$$

$$\mathbf{\Pi}_n = \mathbf{\Pi}_{n-1}\mathbf{A}_0 + \mathbf{\Pi}_n\mathbf{A}_1 + \mathbf{\Pi}_{n+1}\mathbf{A}_2, \ n = 1, 2, \ldots, N-1, \tag{8.87}$$

$$\mathbf{\Pi}_N = \mathbf{\Pi}_{N-1}\mathbf{A}_0 + \mathbf{\Pi}_N(\mathbf{A}_0 + \mathbf{A}_1). \tag{8.88}$$

The normalizing equation is

$$\sum_{n=0}^{N} \mathbf{\Pi}_n \mathbf{e}' = 1. \tag{8.89}$$

One can check that

$$\mathbf{\Pi}_{N-n} = \mathbf{\Pi}_{N-n-1}\mathbf{A}_0(\mathbf{I} - \mathbf{U}_n)^{-1}, \qquad n = 1, 2, \ldots, N-1, \tag{8.90}$$

and

$$\mathbf{\Pi}_N = \mathbf{\Pi}_{N-1}\mathbf{A}_0(\mathbf{I} - \mathbf{U}_0)^{-1}, \tag{8.91}$$

where $\mathbf{U}_n$ is given by

$$\mathbf{U}_0 = \mathbf{A}_0 + \mathbf{A}_1, \tag{8.92}$$

$$\mathbf{U}_n = \mathbf{A}_1 + \mathbf{A}_0(\mathbf{I} - \mathbf{U}_{n-1})^{-1}\mathbf{A}_2, \qquad n = 1, \ldots, N-1. \tag{8.93}$$

Once $\mathbf{U}$'s are computed, we can express $\mathbf{\Pi}_n$'s in terms of $\mathbf{\Pi}_0$ and then determine $\mathbf{\Pi}_0$ from (8.89) so that every $\mathbf{\Pi}_n$ is evaluated.

For large N, we may write from (8.90) and (8.91) that

$$\mathbf{\Pi}_n = \mathbf{\Pi}_{n-1}\mathbf{A}_0(\mathbf{I} - \mathbf{U})^{-1} = \mathbf{\Pi}_0(\mathbf{A}_0(\mathbf{I} - \mathbf{U})^{-1})^n \quad n = 1, 2, \ldots \tag{8.94}$$

where $\mathbf{U}$ is computed by the following algorithm:

1.
$$\mathbf{U}_0 = \mathbf{A}_0 + \mathbf{A}_1$$

2.
$$\mathbf{U}_{n+1} = \mathbf{A}_1 + \mathbf{A}_0(\mathbf{I} - \mathbf{U}_n)^{-1}\mathbf{A}_2, \quad n = 0, 1, 2, \ldots \tag{8.95}$$

3. Continue until

   $\mathbf{U}_{n+1}$ is close to $\mathbf{U}_n$ (i.e., n is as large as N)

   and take
$$\mathbf{U} = \lim_{n \to \infty} \mathbf{U}_n. \tag{8.96}$$

Indeed, we are assuming $\mathbf{U}_n$ converging for large n, which is true. Once $\mathbf{U}$ is evaluated, we can express $\mathbf{\Pi}_n$'s in terms of $\mathbf{\Pi}_0$ from (8.94) and determine $\mathbf{\Pi}_0$ from (8.89) so that every $\mathbf{\Pi}_n$ is computed.

Expression (8.94) is due to the existence of this limit. Using (8.96) in (8.95), we observe that $\mathbf{U}$ satisfies

$$\mathbf{U} = \mathbf{A}_1 + \mathbf{A}_0(\mathbf{I} - \mathbf{U})^{-1}\mathbf{A}_2. \tag{8.97}$$

On comparison between (8.15) – see the Remark following (8.15) – and (8.94), it can be seen that $\mathbf{U}$ and $\mathbf{R}$ are related by the following relation:

$$\mathbf{R} = \mathbf{A}_0(\mathbf{I} - \mathbf{U})^{-1}. \tag{8.98}$$

This establishes a connection between the two sections.

Remarks:

1. $\mathbf{R}$ and $\mathbf{U}$ have an interesting probabilistic interpretation. $\mathbf{R}$ records the expected number of visits to level $n+1$ starting from level n and avoiding levels $0, 1, \ldots, n$, whereas $\mathbf{U}$ records the probability of returning to level n, starting from level n and avoiding level $n-1$.
2. The Gaussian elimination procedure as used in the state reduction method (Section 4.3) or its extension in the block elimination method is merely an operational procedure to derive the steady-state distribution, whereas the concept of censoring provides a probabilistic interpretation.

8.4. The Fourier Series Method for Inverting Transforms

In this book we have encountered many examples where important results have been derived by means of transforms. Basically we worked with two types of transforms: (a) probability generating functions and (b) Laplace transforms. They are very powerful tools of analysis and offer many advantages. However, it is only fair to point out some typical problems which inevitably arise when working with transforms. Quite often the probabilistic structure of the problems to be solved is somewhat hidden behind the algebra of transforms, and more importantly the inversion of transforms to recover the original function is not easy. It is this question we are going to discuss in the present section.

From a purely mathematical point of view, the problem of inverting generating functions and Laplace transforms has been completely solved; moreover, not only one but several methods have been devised. Consider for instance the generating function

$$P(z) = \sum_{n\geq 0} p_n z^n.$$

As we already know, the numbers p_n can be determined by successive differentiation:

$$p_n = \frac{1}{n!}\frac{d^n}{dz^n}P(z)\mid_{z=0}.$$

It is quite remarkable (and not widely known) that a very similar formula holds also for Laplace transforms. Let

$$f(s) = \int_0^\infty e^{-st}F(t)dt$$

be the Laplace transform of a function $F(t)$. Then by the *Post-Widder formula*:

$$F(t) = \lim_{n\to\infty} \frac{(-1)^n}{n!}\left(\frac{n}{t}\right)^{n+1} \frac{d^n}{ds^n}f(s)\bigg|_{s=n/t}, \tag{8.99}$$

which holds at all points of continuity $t > 0$. The reader may observe the striking similarity of these formulas. Unfortunately, both methods are of rather limited value from a numerical point of view.

Thus we have to pursue another route: in complex analysis it is shown that both inversion problems can be reduced to the evaluation of a contour integral in the complex plane. Now the good news is: such integrals can be evaluated numerically and that task rests on very well-established methods in numerical mathematics. The bad news is: the integrands may behave in a rather unpleasant way, but fortunately that may usually happen only in case of Laplace transforms. Our discussion will show that these integrals involve *periodic functions.* Interestingly it turns out that in such a situation the best method to integrate numerically is the simple *trapezoidal rule* (see Section A.11 in Appendix A).

Suppose we want to calculate

$$F = \int_0^1 f(t)dt.$$

The trapezoidal rule leads to the approximation

$$F_N \approx \frac{1}{N}\left(\frac{1}{2}f(t_0) + f(t_1) + f(t_2) + \ldots + f(t_{N-1}) + \frac{1}{2}f(t_N)\right), \tag{8.100}$$

where $t_0 = 0$, $t_k = k/N$, and $k = 1, \ldots, N$. It is quite remarkable that this seemingly unsophisticated method works very well, if $f(t)$ is a periodic function. There is one more advantage that is offered by the trapezoidal rule: the sums (8.100) can be calculated very fast in a numerically stable manner by the powerful device of Fast Fourier Transforms (FFTs).

8.4.1. Inversion of probability generating functions. In most cases the probability generating functions we encounter in queueing theory appear in two forms: (a) as rational or meromorphic functions and (b) as algebraic functions. For rational and meromorphic functions we have the fairly standard procedure of partial fractions decomposition at our disposal. Although being standard, this procedure requires the numerical determination of all roots of polynomials, if we are interested in exact results. This nontrivial task has been discussed elsewhere. For algebraic functions the situation is far more delicate.

In this section we will demonstrate how generating functions can be inverted numerically by means of Fourier series. This approach combines two very nice features: it is fast and efficient, and it is a very accurate method.

Let $P(z) = \sum_{k\geq 0} p_k z^k$ be a probability generating function. We assume that $P(z)$ is convergent for $|z| < R$. Thus $P(z)$ is a function analytic in a neighborhood of 0 and as such, it has derivatives of any order at zero. Then in principle the probabilities p_k can be determined by successive differentiation, as we have already mentioned, but unfortunately this approach is feasible only for small values of k. Interestingly a much more promising method to invert $P(z)$ is based on doing exactly the opposite of differentiation: the coefficients p_k can be expressed as integrals. In fact, by Cauchy's integral formula (see Section A.5 in Appendix A):

$$p_k = \frac{1}{2\pi i}\oint_C \frac{P(z)}{z^{k+1}}dz, \tag{8.101}$$

where the contour of integration C is such that $P(z)$ has no singularities inside the region enclosed by C or on C. Choose C to be a circle centered at the origin and having radius $\rho < R$. This choice is crucial for in case $\rho \geq R$, C would enclose at least one singularity of $P(z)$ such as a pole, and in that case the coefficient formula (8.101) no longer holds, at least in this simple form.

To parameterize a circle of radius ρ in the complex plane we use the following substitution in (8.101)

$$z = \rho e^{2\pi i t} \quad \text{with differential } dz = 2\pi i\rho e^{2\pi i t}dt, \quad 0 \leq t \leq 1,$$

and this yields

$$p_k = \rho^{-k}\int_0^1 P(\rho e^{2\pi i t})e^{-2\pi i k t}dt. \tag{8.102}$$

In what follows it will be more convenient to work with $q_k = \rho^k p_k$, thus we have

$$q_k = \int_0^1 P(\rho e^{2\pi i t})e^{-2\pi i k t}dt. \tag{8.103}$$

We observe that q_k is a standard integral and having the powerful machinery of numerical integration at our disposal, the numerical evaluation of (8.103) is no longer a serious problem.

Unfortunately these optimistic expectations are not fully justified. In general we are interested in calculating the probabilities p_k not only for a single index k, but also over a more or less wide range of values of k. At this point computational efficiency becomes an issue which cannot be neglected. And there is a second problem: so far nothing has been said about an appropriate choice of ρ except that the contour must lie inside the region of convergence of $P(z)$.

Both issues, efficiency and accuracy can be properly handled by means of Fourier series. To see this, it is instructive to have a closer look at the integral (8.103).

Let $f(t)$ be a complex valued function which is periodic with period 1. Then under rather mild regularity conditions $f(t)$ can be represented by its Fourier series:

$$f(t) = \sum_{m=-\infty}^{+\infty} a_m e^{2\pi i m t}, \tag{8.104}$$

the numbers a_m being called the Fourier coefficients of $f(t)$. It is known that for the representation (8.104) to hold, it is required that $\sum_m |a_m| < \infty$. Furthermore these coefficients are given by the integrals

$$a_m = \int_0^1 f(t)e^{-2\pi i m t}\,dt. \tag{8.105}$$

Comparing (8.105) with (8.103) we find that q_k is just the k-th Fourier coefficient of the function $P(\rho e^{2\pi i t})$.

But how to evaluate the integral (8.105) numerically? The simplest approach is to use the trapezoidal rule of integration. To apply this rule, we split the unit interval into n slots each of length $1/n$ and approximate (8.105) by

$$a_m \approx \hat{a}_m = \frac{1}{n}\sum_{j=0}^{n-1} f\left(\frac{j}{n}\right) e^{-2\pi i j m/n}.$$

Putting

$$f_j = f\left(\frac{j}{n}\right) \quad \text{and } w = \exp\left(\frac{2\pi i}{n}\right),$$

we have

$$\hat{a}_m = \frac{1}{n}\sum_{j=0}^{n-1} f_j w^{-jm}. \tag{8.106}$$

These approximations $\hat{a}_m$ can be calculated very fast and numerically stable using any FFT algorithm.

Remarks:

1. Although having now a fast, stable and accurate method at our disposal, we are still left in the somewhat unpleasant situation that the error depends on two parameters, n and ρ. The following heuristic has been suggested: if we want p_i for $i = 0, 1, \ldots, k$ with an error of not more than 10^{-M}, M being a positive integer, then put $n = 2k$ and determine ρ, such that
$$\rho^n = \rho^{2k} = 10^{-M} \quad \text{implying} \quad \rho = 10^{-M/2k}. \tag{8.107}$$
At first sight it would be tempting to choose ρ as small as possible, but this is generally not a good idea, for very small values of ρ inevitably cause serious roundoff problems.
2. Fourier series and hence FFT should not be used to calculate p_k for very large values of k. In that situation asymptotic methods are a more appropriate tool of analysis.

We demonstrate the effectiveness of the method by means of the following example: consider the system $M/M/1$ with arrivals in batches of random size, see Chapter 2. The generating function of the steady-state distribution is given by (2.32):

$$P(z) = \frac{\mu(1 - \rho G'(1))(1 - z)}{\mu(1 - z) - \lambda z(1 - G(z))},$$

where $G(z) = \sum_{n \geq 0} \alpha_n z^n$ denotes the generating function of the batch size. To be precise, we assume that the batch sizes have a discrete uniform distribution over the integers $1, 2, \ldots, N$, i.e.,

$$G(z) = \frac{1}{N}(z + z^2 + \ldots + z^N) = \frac{z}{N}\frac{1 - z^N}{1 - z}.$$

The mean batch size is easily found to be $G'(1) = (N + 1)/2$, thus a steady state exists if $\rho < 2/(N + 1)$.

Let us put $N = 5, \lambda = 1$ and $\mu = 3.1$. If we want to compute p_n up to $k = 256$ with an error of 10^{-10}, we put $n = 512$ and determine the radius of the contour of integration by (8.107):

$$\rho = 10^{-10/512} = 0.9560239011\,.$$

The results are displayed in Table 8.1. Apparently the algorithm works remarkably good. It is not only fast but also very accurate.

8.4.2. Inversion of Laplace transforms. In this section we consider the problem to recover numerically a function $F(t)$ from its Laplace transform $f(s)$. In contrast to the numerical inversion of generating functions, the situation is now much more difficult. Although there is a general inversion formula, in a certain sense comparable to Cauchy's formula (8.101) for generating functions, this formula causes nontrivial numerical problems, as we shall see. As a matter of fact there is not only one technique to compute the inverse Laplace transform, many different procedures have been developed, and

TABLE 8.1

k	$\hat{p}_k$	$\lvert p_k - \hat{p}_k \rvert$
0	0.03225806452	$8.9 \cdot 10^{-19}$
1	0.01040582726	$2.4 \cdot 10^{-18}$
2	0.01168138028	$2.4 \cdot 10^{-18}$
3	0.01269705832	$1.0 \cdot 10^{-18}$
4	0.01328673700	$2.2 \cdot 10^{-18}$
5	0.01324746956	$3.0 \cdot 10^{-18}$
10	0.01190992289	$9.9 \cdot 10^{-19}$
100	0.00338179649	$2.3 \cdot 10^{-17}$
200	0.00083499683	$4.6 \cdot 10^{-14}$
300	0.00020616844	$6.1 \cdot 10^{-12}$
350	0.00010244493	$9.6 \cdot 10^{-11}$
400	0.00005090699	0.00000000209

as almost all of them require the numerical evaluation of integrals of periodic functions, FFT will prove again to be a very useful tool.

The Fourier series method. Among the techniques we are going to discuss this is the one of most general applicability. It is based on the *complex inversion formula*, which allows us to recover the original function of a Laplace transform $f(s)$ by the following contour integral:

The complex inversion formula. *Let $F(t)$ be an original function with $F(t) = 0$ for $t < 0$ and $f(s)$ its Laplace transform*

$$f(s) = \int_0^\infty e^{-st} F(t) dt,$$

convergent for $\Re(s) \geq s_0$, furthermore, $F(t)$ satisfies

$$\int_{-\infty}^{\infty} \left| e^{-s_0 t} F(t) \right| dt < \infty.$$

Then by the Fourier integral theorem:

$$F(t) = \frac{e^{\sigma t}}{2\pi} \int_{-\infty}^{\infty} f(\sigma + \omega i) e^{i\omega t} d\omega, \qquad \sigma \geq s_0, \tag{8.108}$$

where (8.108) has to be understood as principal value integral, i.e., the approach to the limits has to be symmetric.

We will not prove this theorem here and remark only that the contour of integration is the vertical straight line $s = \sigma + \omega i$, where ω runs from $-\infty$ to ∞.

Let us first rewrite (8.108) in light of the fact that $F(t)$ is a real valued function:

$$\begin{aligned}F(t) &= \frac{e^{\sigma t}}{2\pi}\int_{-\infty}^{\infty} f(\sigma+i\omega)\left[\cos\omega t + i\sin\omega t\right]d\omega \\ &= \frac{e^{\sigma t}}{2\pi}\int_{-\infty}^{\infty}\left[\Re\left(f(\sigma+i\omega)\right)\cos\omega t - \Im\left(f(\sigma+i\omega)\right)\sin\omega t\right]d\omega \\ &= \frac{2e^{\sigma t}}{\pi}\int_{0}^{\infty}\Re\left(f(\sigma+i\omega)\right)\cos\omega t\,d\omega \qquad (8.109)\end{aligned}$$

which is a much more convenient representation.

Now the integral (8.109) is approximated by the trapezoidal rule. For this purpose we truncate the interval of integration to $[0,T]$. Using a step size of $h = T/N$ for some big integer N, we obtain:

$$F_N(t) = \frac{he^{\sigma t}}{\pi}\Re\left(f(\sigma)\right) + 2\frac{he^{\sigma t}}{\pi}\sum_{k\geq 1}\Re\left(f(\sigma+ikh)\right)\cos(kht). \qquad (8.110)$$

Only a slight modification of the FFT algorithm is required to evaluate this summation (see Exercise 7).

Let us put $h = \pi/2t$, then

$$F_N(t) = \frac{e^{\sigma t}}{2t}\Re\left(f(\sigma)\right) + \frac{e^{\sigma t}}{t}\sum_{k\geq 1}\Re\left[f\left(\sigma+\frac{ik\pi}{2t}\right)\right]\cos\frac{k\pi}{2}.$$

Observe that in this sum all terms vanish for which k is an odd integer. Putting $\sigma = A/2t$ we get finally

$$F_N(t) = \frac{e^{A/2}}{2t}\Re\left[f\left(\frac{A}{2t}\right)\right] + \frac{e^{A/2}}{t}\sum_{k\geq 1}(-1)^k\Re\left[f\left(\frac{A+2k\pi i}{2t}\right)\right]. \qquad (8.111)$$

For practical purposes the summation in (8.111) has to be truncated at an index $k = N$, say, with the proviso that the last term included is multiplied by $1/2$, as is required by the trapezoidal rule.

The most important difficulty in using (8.111) is its slow convergence. A possible solution is to transform the series so that it converges faster. Since the series is (almost) alternating, Euler's transformation can be applied. This works as follows: write the sum in (8.111) in this way:

$$\sum_{k\geq 1}(-1)^k\Re\left[f\left(\frac{A+2k\pi i}{2t}\right)\right] = \sum_{k\geq 1}(-1)^k f_k.$$

Then it can be shown that

$$\sum_{k\geq 1}(-1)^k f_k = \sum_{p\geq 0}\frac{(-1)^p\Delta^p f_1}{2^{p+1}} \qquad (8.112)$$

where Δ^p denotes the $p-$th difference operator applied to the sequence $\{f_k\}$:

$$\Delta^p f_k = f_{k+p} - \binom{p}{1}f_{k+p-1} + \binom{p}{2}f_{k+p-2} - \ldots + (-1)^p f_k.$$

In many cases the new series will converge much more rapidly than the original series.

Inversion based on Pincherle's theorem. As we have remarked, the Fourier series method is the most general approach to the inversion problem of Laplace transforms, all we need is a formula for $f(s)$. Sometimes however, *a priori* information about the original function $F(t)$ is available. Then alternative methods of inversion may be more efficient.

Suppose that $F(t)$ is an entire function, which means that $F(t)$ is defined and analytic in the whole complex plane. It has a convergent Taylor series:

$$F(t) = \sum_{n \geq 0} a_n t^n, \qquad |t| < \infty.$$

Furthermore let $F(t)$ be of exponential type:

$$|F(t)| < Me^{\gamma t}$$

for some positive constant M. The real number $\gamma \geq 0$ is called the growth parameter of $F(t)$.

If all these requirements are met we are facing a particularly pleasing situation because due to Pincherle's theorem the contour of infinite length in (8.108) can be replaced by a closed curve of finite length. More precisely, let C be a positively oriented circle of radius $\rho > \gamma$, then Pincherle's theorem states that

$$F(t) = \frac{1}{2\pi i} \oint_C f(s) e^{st} ds = \rho \int_0^1 f\left(\rho e^{2\pi i z}\right) \exp\left[t\rho e^{2\pi i z} + 2\pi i z\right] dz. \quad (8.113)$$

It can be seen that the integrand is analytic and periodic with period 1; therefore, the trapezoidal rule of integration is the optimal method to evaluate this integral. Some fine tuning is required regarding the proper choice of the radius ρ.

Inversion based on Pincherle's theorem is a particularly interesting alternative, if the Laplace transform $f(s)$ is a rational function, i.e., it can be continued to the whole complex plane except for poles $s_1, s_2, \ldots, s_K$. A partial fraction decomposition may be given by means of calculus of residues and the inverse Laplace transform is identified as exponential polynomial

$$F(t) = \sum_{i=0}^{K} e^{s_i t} P_i(t),$$

where the $P_i(t)$ are polynomials of order $m_i - 1$, m_i being the order of the pole located at s_i. Clearly $F(t)$ is in this case also an entire function of exponential type. Thus inversion can be carried out by the method outlined above with the obvious advantage that it is not necessary to determine the poles by a root finding procedure. All we need to know is the modulus of the largest pole, which in turn gives the growth parameter; however, that task poses no real difficulty.

The Laguerre series method. This is one more alternative to the Fourier series approach. It is particularly efficient, if the coefficients of a certain power series associated to the Laplace transform $f(s)$ tend to zero sufficiently fast. It is also an interesting alternative to the method we discussed in the previous section: If the Laplace transform $f(s)$ cannot be continued analytically into the whole complex plane, with the possible exception of poles, then Pincherle's theorem cannot be applied. A classical example is the density of busy period in an $M/M/1$ system with Laplace transform (see Section 2.7 in Chapter 2)

$$f(s) = \frac{\lambda + \mu + s - \sqrt{(\lambda + \mu + s)^2 - 4\lambda\mu}}{2\lambda}. \tag{8.114}$$

This function has a branch point at $s = 0$ and therefore $f(s)$ is analytic only in the *cut plane*, the complex plane excluding the negative real axis. In such cases we may consider the Laguerre Method as an appropriate tool of inversion. Its requirements are:

(A) The Laplace transform $f(s)$ is analytic in the half plane $\Re(s) > 0$.
(B) $f(s) \to 0$ for $s \to \infty$ in this half plane.

If $f(s)$ happens to be analytic only for $\Re(s) > s_0$, $s_0 > 0$, then we apply the damping rule for Laplace transforms:

$$e^{-s_0 t} F(t) \quad \leftrightarrow \quad f(s + s_0).$$

For instance let $f(s) = 1/\sqrt{s^2 - 1}$, the Laplace transform of modified Bessel function $I_0(t)$. It has branch points at ± 1 and is therefore analytic only for $s > 1$. But

$$f(s+1) = \frac{1}{\sqrt{(s+1)^2 - 1}} = \frac{1}{\sqrt{s(s+2)}} \quad \leftrightarrow \quad e^{-t} I_0(t)$$

and now $f(s)$ has its branch points at 0 and -2, thus is analytic in $\Re(s) > 0$.

Consider now the mapping

$$z = \frac{s - \alpha}{s + \alpha}, \qquad \alpha > 0 \tag{8.115}$$

which maps the half plane $\Re(s) > 0$ into the unit circle $|z| < 1$. Any function analytic in $\Re(s) > 0$ can be expanded as a power series in z. This is just the idea of the Laguerre method: expand the Laplace transform $f(s)$ as a power series in z and perform inversion by taking the inverse Laplace transforms of

$$z^n = \left(\frac{s - \alpha}{s + \alpha}\right)^m.$$

Unfortunately this will not work, since condition (B) is violated, but this condition must be satisfied for *any* Laplace transform. However, there is a simple solution: consider instead the functions

$$l_n(s) = \frac{z^n}{s + \alpha} = \frac{(s - \alpha)^n}{(s + \alpha)^{n+1}}.$$

These functions are Laplace transforms and their inverse is well known:

$$l_n(s) \quad \leftrightarrow \quad e^{-\alpha t} L_n(2\alpha t),$$

where $L_n(t)$ denotes the Laguerre polynomial of order n. The latter are defined by the generating function

$$\sum_{n\geq 0} z^n L_n(t) = \frac{e^{-tz/(1-z)}}{1-z},$$

an explicit expression is given by

$$L_n(t) = \sum_{k=0}^{n} \binom{n}{k} \frac{(-t)^k}{k!},$$

but it is much more efficient to calculate these polynomials by the recurrence relation

$$\begin{aligned} L_0(t) &= 1 \\ L_1(t) &= 1 - t \\ L_n(t) &= \frac{1}{n}((2n-1-t)L_{n-1}(t) - (n-1)L_{n-2}(t)) \\ n &= 2, 3, \ldots, \end{aligned} \tag{8.116}$$

which turns out to be numerically stable. Hence the inversion of $f(s)$ reduces to the problem of determining the coefficients in the expansion

$$f(s) = \sum_{n\geq 0} q_n \frac{(s-\alpha)^n}{(s+\alpha)^{n+1}},$$

or equivalently, determine the coefficients in the power series

$$Q(z) = \sum_{n\geq 0} q_n z^n = (s+\alpha)f(s) = \frac{2\alpha}{1-z} f\left(\alpha \frac{1+z}{1-z}\right). \tag{8.117}$$

The reader may notice that this is just the problem we have solved in Section 8.4.1. Using Fourier series and FFT we may calculate the numbers q_n to any desired order with an error that tends to zero exponentially fast.

Once the coefficients q_n have been found, we may recover the original function $F(t)$ by

$$F(t) = e^{-\alpha t} \sum_{n\geq 0} q_n L_n(2\alpha t). \tag{8.118}$$

For practical purposes we use the truncated series as approximation:

$$F(t) \approx F_N(t) = e^{-\alpha t} \sum_{n=0}^{N} q_n L_n(2\alpha t). \tag{8.119}$$

8.5. Discussion

We have almost reached the end of the book except for the last two chapters, one dealing with discrete-time queues and the last one is on some relevant miscellaneous topics. We have certainly concluded our treatment on computational methods (not quite so; the method of recurrence relations which are effective in discrete-time queues will come in the next chapter) and by doing so, completed the major thrust of the subject, namely, the study of characteristics of interest in some well-structured models.

Whenever we think of a characteristic for study, it is invariably the distribution of the system length. Whatever models have been considered, they possess directly or indirectly the Markovian structure, with the state space consisting of numbers of units in the system. From the very nature of Markovian structure, it is known that the stationary distribution is the solution of

$$\mathbf{xP} = \mathbf{x} \quad \text{and} \quad \mathbf{xe}' = 1, \tag{8.120}$$

if we are dealing with a Markov chain, or

$$\mathbf{xQ} = \mathbf{0} \quad \text{and} \quad \mathbf{xe}' = 1, \tag{8.121}$$

if we are dealing with a Markov process, where $\mathbf{P}$ is the transition probability matrix arising in the Markov chain, $\mathbf{Q}$ is the generator matrix arising in the Markov process and $\mathbf{e}$ is the row vector of 1's (see for example (2.11), (2.12), (3.7) and (3.8)). There are of course, theoretical maneuvers to obtain explicit solutions with some success. Had $\mathbf{P}$ and $\mathbf{Q}$ been finite, the problem boils down to solve a finite system of linear equations. Even if they are not, the approach is to take the size of the matrix large enough so that the solution is close to the actual one. This is precisely the point of view that has been expressed in (2.111) and (2.112) and the discussion around it. A simple Cramer's rule could have been sufficient to obtain the exact solution, provided the matrix size is small. Otherwise, an algorithmic approach like the Gaussian elimination method looks a reasonable one, which gives rise to the state reduction method in Section 4.3.

The route that the theoretical development often takes is to apply the transform method, in the form of p.g.f. and L.T. and ends in solutions in terms of these transforms. Thus arises the need to invert back from a transformed solution to the original distribution. For this purpose, numerical inversion methods for general situations are presented in Section 8.4. However, when the transformed solutions are looked at closely, one finds that in some models particularly with bulk queues, the inversion depends on roots of certain polynomials. It demands us to learn some root finding techniques and these are described in Section 4.2.

Until now the discussion has been on models in which Markovian nature is exhibited only through one variable, namely, the number in the system which is also called the level of the system. But right in Chapter 2, we have recognized that in models involving Erlangian distribution, the Markovian property is achieved only when another variable, namely, phase within the level is taken in account. Such bivariate Markovian models are the subject of study in Sections 8.2 and 8.3 of this chapter and the objective is to obtain numerical stationary solutions. The main feature of their transition matrix or generator matrix when presented in terms of levels is that its elements are now matrices which enlist the contribution of phases. Thus **P** and **Q** will have matrix elements or block entries. Interestingly, bivariate Markovian models appear in many applications due to the possibility that any distribution either of interarrival times or service-times may be approximated by a PH distribution, introduced in Section 8.2.3.

The procedure of solution for deriving the stationary distribution in either section is an extension of something we are familiar with in earlier chapters. Keeping in mind the celebrated geometric distribution as a solution, our quest ends in finding matrix-geometric solution for models having some special structure like QBD or $G/M/1$ type in Section 8.2. This approach is theoretical indeed but ends in being algorithmic and numerical in computing **R**, the rate matrix (see (8.44)). In fact, the real feature in this chapter is the numerical part. Thus we may ask whether such a paradigm can be used in the other important $M/G/1$ type models which is nothing but $M/G/1$ with matrix elements (see **P** after (3.5) and change the elements to appropriate matrices).

Using matrix notations, the balance equations and the normalizing equation are (see (3.7) and (3.8))

$$\mathbf{\Pi}_i = \mathbf{\Pi}_0 \boldsymbol{\alpha}_j + \sum_{i=1}^{j+1} \mathbf{\Pi}_i \boldsymbol{\alpha}_{j-i+1}, \quad j = 0, 1, 2, \ldots \tag{8.122}$$

and

$$\sum_{j=0}^{\infty} \mathbf{\Pi}_j = 1. \tag{8.123}$$

What is needed is the evaluation of $\mathbf{\Pi}_0$. Once it is done, other $\mathbf{\Pi}$'s can be obtained, in principle, from (8.122).

Now, let **G** be the matrix with (i, j)th element as the probability of starting from a given level (above level 0) in phase i and entering for the first time the next lower level in phase j. Using the matrix notations, it can be proved that

$$\mathbf{\Pi}_0 = \mathbf{\Pi}_0 \sum_{r=0}^{\infty} \boldsymbol{\alpha}_r \mathbf{G}^r \tag{8.124}$$

and

$$\mathbf{G} = \sum_{r=0}^{\infty} \boldsymbol{\alpha}_r \mathbf{G}^r. \tag{8.125}$$

Similar to the computation of $\mathbf{R}$ as in (8.44), we compute $\mathbf{G}$ by the following algorithm:

$$\mathbf{G}(0) = \mathbf{0},$$

and

$$\mathbf{G}(n+1) = \sum_{r=0}^{\infty} \boldsymbol{\alpha}_r \mathbf{G}^r(n), \quad n \geq 0 \tag{8.126}$$

until $\mathbf{G}(n+1)$ is close to $\mathbf{G}(n)$. When $\mathbf{G}$ is computed, it can help determine $\boldsymbol{\Pi}_0$ from (8.123) and (8.124).

In Section 8.3, an extension of scalar elimination in the state reduction method to block elimination is presented. Both methods are algorithmic and in the QBD case the connection between them is shown in (8.98). Again, if the matrix has some other type of structure, special computational procedures are discussed, one in Lal and Bhat (1987, 1988) and the other in Kao and Lin (1989) and Kao (1991). These are based on the concept of censoring, which finds its roots as early as 1960 in Kemeny and Snell (1960).

What about the transient solution? Well, while the whole question of stationary solution is reduced to solve a system of linear equations and therefore the need of an algebraic treatment has become a plausible approach, it is no longer that simple for the transient case. The first attempt is made by transform method applied to $M/M/1$ in Section 2.5 and the solution appears nothing closer to the stationary solution. Other alternative exact solutions are also presented (see Section 2.5.3, Section 4.4.5 and Section 4.5). In some expressions, the amount of discrepancy from the stationary solution is separated out. A few numerical procedures are discussed in Section 4.4. The first two use transform and algebraic methods. Combinatorial techniques are also applied to obtain exact solutions in some special models (see Section 6.3). It seems there is no unified approach for finding the transient solution.

For matrix-geometric solution approach in Section 8.2, the references are Evans (1967), Neuts (1978a,b, 1981, 1982), Gillent and Latouche (1983), Latouche and Ramaswami (1989, 1999), Latouche and Neuts (1999) and Lucantoni and Ramaswami (1985). In Evan's paper, one may see the beginning of this type of solution. See Grassmann and Heyman (1990, 1993) and Grassmann and Stanford (2000) for Section 8.3. The references for Section 8.4 are Abate and Dubner (1968), Abate, Choudhury and Whitt (1996), Abate and Whitt (1992a,b), Doetsch (1974), Feller (1971), Henrici (1977), Henrici and Kenan (1986) and Jagerman (1982).

8.6. Exercises

1. What is the equation corresponding to (8.16) for a discrete-time QBD chain?

2. Suppose the interarrival time T is a hyper exponential variable with k phases. Its distribution function is given by

$$F_T(x) = \sum_{j=1}^{k} \alpha_j (1 - e^{-\lambda_j x}).$$

   Give a PH-distribution representation of this distribution.

3. PH/M/1 Queue: Consider a single server queue with exponential service-time of rate μ and arrival according to a PH-renewal process. The representation of the PH-distribution $F_{\mathbf{T}}(\cdot)$ of the interarrival time is $(\boldsymbol{\alpha}, \mathbf{T})$, where $\boldsymbol{\alpha}$ is a row vector of size m and $\mathbf{T}$ is a $m \times m$ matrix. (The PH-renewal process is simply related to the irreducible Markov chain with generator $\mathbf{Q}^* = \mathbf{T} + \mathbf{T}^0\boldsymbol{\alpha}$).

   Obtain the generator matrix of this queueing system. Discuss the steps for obtaining the stationary distribution of the number of units in the system.

4. For a discrete-time QBD process, the stationary probability $\mathbf{x}_0$ of the boundary states in level 0 is the unique solution of the system:

$$\mathbf{x}_0(\mathbf{A}_1 + \mathbf{A}_2 + \mathbf{R}\mathbf{A}_2) = \mathbf{x}_0$$
$$\mathbf{x}_0(\mathbf{I} - \mathbf{R})^{-1}\mathbf{e}' = 1.$$

   Show that $\mathbf{x}_0 = \boldsymbol{\alpha}(\mathbf{I} - \mathbf{R})$, where $\boldsymbol{\alpha}$ is the unique solution of the system:

$$\boldsymbol{\alpha}\mathbf{A} = \boldsymbol{\alpha}, \quad \boldsymbol{\alpha}\mathbf{e}' = 1,$$

   with $\mathbf{A} = \mathbf{A}_0 + \mathbf{A}_1 + \mathbf{A}_2$.

5. Finite QBD model: In many applications the level is the number of units in the system. Let $\ell(0)$ represent the states in level n. Consider a discrete-time QBD system, in which the state space is restricted to $\bigcup_{0 \le n \le N} \ell(n)$, where N is finite and representing the capacity of the system. Show that by using the block elimination method the stationary distribution is given by:

$$\boldsymbol{\Pi}_n = \boldsymbol{\Pi}_{n-1}\mathbf{R}_n, \qquad 1 \le n \le N$$

   where

$$\begin{aligned} \mathbf{R}_n &= \mathbf{A}_0(\mathbf{I} - \mathbf{U}_n)^{-1}, \qquad 1 \le n \le N \\ \mathbf{U}_n &= \mathbf{A}_1 + \mathbf{A}_0(\mathbf{I} - \mathbf{U}_{n+1})^{-1}\mathbf{A}_2, \qquad 1 \le n \le N-1 \\ \mathbf{U}_N &= \mathbf{A}_0(I - (\mathbf{A}_1 + \mathbf{A}_0))^{-1}. \end{aligned}$$

   In the finite case, the matrices $\mathbf{U}_i$'s depend on the level, since there is a finite upper boundary N.

6. Prove the representation (8.109).
7. Show how an FFT algorithm has to be applied (and modified, if necessary) to evaluate the trapezoidal sum in (8.110).
8. Consider the Laplace transform $f(s)$ of a function $F(t)$ with associated series $Q(z)$ which is supposed to have an algebraic singularity at $z = 1$, so that the corresponding Laguerre series converges slowly. Using

$$t^n F(t) \quad \leftrightarrow \quad (-1)^n f^{(n)}(s), \qquad n = 1, 2, \ldots,$$

where $f^{(n)}(s)$ denotes the nth derivative, show that differentiation of $f(s)$ improves the speed of convergence of the Laguerre series considerably in many cases. Perform numerical experiments and find out to what extent convergence can be accelerated. What problems have to be expected when n becomes large?
9. Prove (8.124) and (8.125).

References

Abate, J. and Dubner, H. (1968). A new method for generating power series expansions of functions, *SIAM J. Numer. Anal.*, **5**, 102–112.

Abate, J., Choudhury, G. L. and Whitt, W. (1996). On the Laguerre method for numerically inverting Laplace transforms, *INFORMS Journal on Computing*, **8**, 413–427.

Abate, J. and Whitt, W. (1992a), The Fourier-series method for inverting transforms of probability distributions, *Queueing Systems*, **10**, 5–88.

Abate, J. and Whitt, W. (1992b). Numerical inversion of probability generating functions, *Operations Research Letters*, **12**, 245–251.

Doetsch, G. (1974). *Introduction to the Theory and Application of Laplace Transforms*, Springer, New York.

Evans, R.V. (1967). Geometric distribution in some two-dimensional queueing systems, *Oper. Res.*, **15**, 830–845.

Feller, W. (1971). *An Introduction to Probability Theory and its Applications*, Vol. 2, John Wiley & Sons, New York.

Gillent, F. and Latouche, G. (1983). Semi-explicit solutions for M/PH/1 like queueing systems, *Europ. J. of Oper. Res.*, **13**, 151–160.

Grassmann, W. K. and Heyman, D. P. (1990). Equilibrium distribution of block-structured Markov chains with repeating rows, *J. Appl. Prob.*, **27**, 557–576.

Grassmann, W. K. and Heyman, D. P. (1993). Computation of steady-state probabilities for infinite-state Markov chains with repeating rows, *ORSA J. on Computing*, **5**, 282–303.

Grassmann, W. K. and Stanford, D. A. (2000). Matrix analytic methods, *Computational Probability*, Grassmann, W. K., Ed., Kluwer Academic Publisher, Boston, Chapter 6, 153–202.

Henrici, P. (1977). *Applied & Computational Complex Analysis: Special Functions, Integral Transforms, Asymptotics, Continued Fractions*, Vol. 2, John Wiley & Sons, New York.

Henrici, P. and Kenan W. R. (1986). *Applied & Computational Complex Analysis: Integration, Conformal Mapping, Location of Zeros*, Vol. 3, John Wiley & Sons, New York.

Jagerman, D. L. (1982). An inversion technique for Laplace transforms, *Bell System Tech. J.*, **61**, 1995–2002.

Kao, E. P. C. (1991). Using state reduction for computing steady state probabilities of queues of GI/PH/1 type, *ORSA J. on Computing*, **3**, 231–240.

Kao, E. P. C. and Lin, C. (1989). The M/M/1 queue with randomly varying arriving and service rates: A phase substitution solutions, *Management Science*, **35**, 561–570.

Kemeny, J. G. and Snell, J. L. (1960). *Finite Markov Chains*, Von Nostrand Co., Princeton, NJ, 114–115.

Lal, R. and Bhat, U. N. (1987). Reduced systems in Markov chains and their applications in queueing theory, *Queueing Systems*, **2**, 147–172.

Lal, R. and Bhat, U. N. (1988). Reduced systems algorithms for Markov chains, *Management Sci.*, **34**, 1202–1220.

Latouche, G. and Neuts, M. F. (1999). *Introduction to Matrix Analytic Methods in Stochastic Modeling*, SIAM-ASA, Philadelphia, PA.

Latouche, G. and Ramaswami, V. (1989). An experimental evaluation of the matrix-geometric method for the GI/PH/1 queue, *Commun. Statist. Stochastic Models*, **5**, 629–667.

Lucantoni, D. M. and Ramaswami, V. (1985). Efficient algorithms for solving the non-linear matrix equations arising in phase type queues, *Commun. Statist. Stochastic Models*, **1**, 29–52.

Neuts, M. F. (1978a). The M/M/1 queue with randomly varying arrival and service rates, *Opsearch*, **15**, 139–157.

Neuts, M. F. (1978b). Further results on the M/M/1 queue with randomly varying rates, *Opsearch*, **15**, 158–168.

Neuts, M. F. (1981). *Matrix-Geometric Solutions in Stochastic Models: An Algorithmic Approach*, The Johns Hopkins University Press, Baltimore, MD.

Neuts, M. F. (1982). Explicit steady-state solutions of some elementary queueing models, *Oper. Res.*, **30**, 480–489.

CHAPTER 9

Discrete-Time Queues: Transient Solutions

9.1. Introduction

Interest in discrete-time queues is a recent phenomenon. Its motivation has been provided in Section 3.7 in Chapter 3 wherein this topic has been introduced. There, we followed the classical methods and obtained analogous results on stationary behaviour of queues. The study of transient behaviour has been postponed to this chapter, until we gained some insight in the continuous time case.

It has been observed at several points, but mainly in Section 6.3, that combinatorial techniques can be fruitfully utilized to obtain transient results. Keeping this in mind and remembering that we are dealing with discreteness, it is but natural to think of such techniques as tools for finding solutions. In Section 9.2, we use lattice path combinatorics for the same purpose to deal with several models.

Past experience tells us that not every model leads to an explicit solution and as an alternative we look for some numerical approaches. Here again in Section 9.3, we exploit the discreteness of the process to apply the age old method of recurrence relations and provide numerical solutions in the study of transient behaviour of general models.

9.2. Combinatorial Methods: Lattice Path Approach

It is the objective of this section to give the reader some flavor of the impressing power of combinatorial methods based on the lattice path counting. The striking simplicity and elegance of this approach is especially revealed when applied to the analysis of the transient behaviour of Markovian systems in discrete time. By discrete time we mean, that time has been slotted into a sequence of contiguous subintervals of unit length

$$(0,1],(0,2],\ldots,(n-1,n],\ldots$$

which are henceforth referred to as a *slot*.

The route we will follow in this section starts with a discussion of the classical $M/M/1$ system in discrete time, known as the $Geo/Geo/1$ system (see Section 3.7). The ideas developed there are then generalized to systems which allow for customers to arrive in batches (and thus offers also a discussion of $M/E_k/1$ systems in discrete time, denoted by $Geo/Geo/1$) and to systems with several parallel servers. Then we show applications of path counting in higher dimensional spaces, in particular models with heterogeneous arrival streams and simple tandem queues. Finally we indicate briefly how to perform a limiting procedure which yields in a surprisingly simple way results for continuous time models.

9.2.1. The *Geo/Geo/1* model. In this section we consider the analogue of the classical $M/M/1$ system in discrete time.

We assume that the following holds:

(i) Customers arrive according to a Bernoulli process (see Section A.6 in Appendix A). The probability of an arrival in a slot equals α.
(ii) Customers are served one after the other, the service-times being i.i.d. random variables having a geometric distribution with parameter γ, which means, the probability that during a particular slot a customer leaves the system has probability γ.
(iii) Events in different slots are mutually independent and the probability of observing more than one event in a slot is zero.

Let Q_n denote the number of customers in the system at time n.

With the process Q_n we associate a random walk process S_n, which is defined by

$$S_n = \sum_{i=1}^{n} X_i, \qquad S_0 = 0,$$

having i.i.d. increments X_i which assume their values in the set $\{-1, 0, 1\}$ with probability function

$$P(X_i = 1) = \alpha, \quad P(X_i = -1) = \gamma, \quad P(X_i = 0) = \beta = 1 - \alpha - \gamma.$$

S_n is called the *basic process* associated with Q_n, because Q_n is obtained from S_n by imposing an impenetrable barrier at zero. This barrier acts as a reflecting boundary and as a result the sample paths of Q_n and S_n are related by the famous formula

$$Q_n = \max[m + S_n, S_n - \min_{0 \le i \le n} S_i], \quad m \ge 0. \tag{9.1}$$

Thus Q_n is a regulated random walk process and the boundary at zero is called a one-sided regulator.

Formula (9.1) shows that the joint distribution of S_n and of $\min_i S_i$ completely determines the distribution of Q_n. However, the determination of the joint distribution of S_n and $\min_i S_i$ essentially boils down to a discussion of the transition probabilities of S_n restricted by an absorbing boundary. Such problems can be conveniently dealt with by lattice path combinatorics.

In order to apply results from lattice path counting we first have to translate the sample paths of Q_n and S_n into lattice paths. This is done by representing an arrival by a $(1,0)$-step, a departure by a $(0,1)$- step and empty slots (i.e., slots where neither an arrival nor a departure occurs) by a $(1,1)$ step, the latter will be called also a *diagonal* step. We refer to Section 6.3 in Chapter 6 on lattice path representation. Joining these steps together we obtain a two-dimensional lattice path which starts at the origin and is subject to various restrictions. This way of looking at the sample paths of Q_n and S_n stresses one important point: it is the order in which arrivals and departures are arranged, which is essential, and only certain arrangements give rise to events which make sense in a queueing context. For instance we require the process Q_n to be non-negative which means for the corresponding lattice paths that they must stay below a certain line. Lattice path combinatorics provides us with extremely powerful tools to determine the number of such arrangements.

Our first task will be to determine the transition probabilities of Q_n during time intervals where the service station is continuously busy. For this purpose it is convenient to introduce a random stopping time T_m which is defined by

$$T_m = \inf[n : Q_n = 0, Q_0 = m], \qquad m > 0.$$

Let A_n and D_n denote the total number of arrivals and the total number of service completions in the time interval $(0, n]$ and consider first

$$\begin{aligned} &P(A_n = n_1, D_n = n_2, T_m > n | Q_0 = m) \\ &= \quad P(Q_n = m + n_1 - n_2, D_n = n_2, T_m > n | Q_0 = m), \quad m > 0. \end{aligned} \tag{9.2}$$

The lattice paths corresponding to (9.2) have the following properties:

(i) The paths start at the origin, have n_1 horizontal, n_2 vertical and $n_3 = n - n_1 - n_2$ diagonal steps.

(ii) The paths are not allowed to touch or cross the line $y = x + m$.

Let $N(n_1, n_2, n_3|m)$ denote the number of such paths. If $n_3 = 0$, i.e., there are no diagonal steps, then the classical reflection principle (see (6.27)) tells us that

$$N(n_1, n_2, 0|m) = \binom{n_1 + n_2}{n_1} - \binom{n_1 + n_2}{m + n_1}. \tag{9.3}$$

Diagonal steps are introduced quite easily. Any path counted by (9.3) passes through $n_1 + n_2 + 1$ lattice points and in each such point we may insert diagonal steps, say n_3 in number. Observe that this is equivalent to putting n_3 indistinguishable balls into $n_1 + n_2 + 1$ distinct boxes. The number of ways this can be done is

$$\binom{n_1 + n_2 + n_3}{n_3},$$

and therefore we have

$$N_1(n_1, n_2, n_3|m) = \left[\binom{n_1 + n_2}{n_1} - \binom{n_1 + n_2}{m + n_1}\right]\binom{n_1 + n_2 + n_3}{n_3}. \tag{9.4}$$

Now any such path has the same probability, viz., $\alpha^{n_1}\gamma^{n_2}\beta^{n_3}$ and upon multiplication with $N_1(n_1, n_2, n_3|m)$ we get for $m > 0$:

$$\begin{aligned} &P(Q_n = m + n_1 - n_2, D_n = n_2, T_m > n | Q_0 = m) \\ &\qquad = N_1(n_1, n_2, n - n_1 - n_2|m)\alpha^{n_1}\gamma^{n_2}\beta^{n-n_1-n_2}. \end{aligned} \tag{9.5}$$

If we sum over n_1 and n_2 such that $n_1 - n_2 = k - m$, $k > 0$, then we obtain at once the joint distribution of S_n and $\min S_i$, or equivalently the *zero-avoiding* transition probabilities of Q_n:

$$\begin{aligned} P(Q_n = k, T_m > n | Q_0 = m) &= P^{(0)}_{m,k}(n), \quad \text{(say)} \\ &= P(m + S_n = k, m + \min_i S_i > 0) \\ &= \sum_{\substack{n_1+n_2 \le n \\ n_1-n_2=k-m}} N_1(n_1, n_2, n - n_1 - n_2 | m)\alpha^{n_1}\gamma^{n_2}\beta^{n-n_1-n_2}. \end{aligned} \tag{9.6}$$

Using the above simple argument again we get without difficulties the joint distribution of the duration of a busy period and the number of customers served, i.e., $P(T_m = n, D_{T_m} = k)$, for $m > 0$. The lattice paths corresponding to the event $\{T_m = n, D_{T_m} = k\}$ must stay strictly below the line $y = x + m$, except at the end, where the path touches for the first time.

To determine the number of such paths we forget for the moment about diagonal steps. Assuming that there are n_1 horizontal and n_2 vertical steps, where $n_1 = n_2 - m$, we immediately get from (9.3) the required number:

$$\binom{2n_2 + 1 - m}{n_2} - \binom{2n_2 + 1 - m}{n_2 + 1} = \frac{m}{n_2 + 1}\binom{2n_2 + 1 - m}{n_2}. \tag{9.7}$$

Now insert diagonal steps, say n_3 in number. This is possible in all points on the path except for the last point, where it touches the line $y = x + m$. Thus there are

$$\binom{n_1 + n_2 + n_3 - 1}{n_3}$$

different ways this can be done, and therefore, recalling that $n = n_1 + n_2 + n_3$:

$$P(T_m = n, D_{T_m} = k) = \frac{m}{k}\binom{2k - 1 - m}{k - 1}\binom{n - 1}{2k - 1 - m}\alpha^{k-m}\gamma^k\beta^{n-2k+m}. \tag{9.8}$$

Our next task will be to find the general time-dependent distribution of Q_n. This poses no particular problem, since the joint distribution of S_n and $\min S_i$ is known now. Recalling that

$$Q_n = \max[m + S_n, S_n - \min_i S_i],$$

we immediately obtain

$$\begin{aligned} P(Q_n < k | Q_0 = m) &= P(m + S_n < k, S_n - \min_i S_i < k) \\ &= \sum_{\nu > m} P(\nu + S_n = k, S_n - \min_i S_i < k) \\ &= \sum_{\nu > m} P(\nu + S_n = k, \nu + \min_i S_i > 0) \\ &= \sum_{\nu > m} P^{(0)}_{\nu,k}(n). \end{aligned} \tag{9.9}$$

As a result we get the transient distribution of Q_n by a simple binomial summation:

$$P(Q_n < k|Q_0 = m) = \sum_{\nu > m} \sum_{\substack{n_1+n_2 \le n \\ n_1-n_2=k-\nu}} N_1(n_1, n_2, n-n_1-n_2|\nu)\alpha^{n_1}\gamma^{n_2}\beta^{n-n_1-n_2}. \tag{9.10}$$

One more interesting result is readily available using another famous formula from lattice path counting.

Let $N_2(n_1, n_2, n_3|a, b)$ denote the number of paths starting at the origin which stay strictly between the lines $y = x + a$ and $y = x - b$, $a, b > 0$. Setting first $n_3 = 0$, it can be proved by applying repeatedly the reflection principle (see Section 6.3 in Chapter 6) that

$$N_2(n_1, n_2, 0|a, b) = \sum_{\nu \in \mathbb{Z}} \left[\binom{n_1 + n_2}{n_1 - \nu(a+b)} - \binom{n_1 + n_2}{n_1 - \nu(a+b) + a} \right].$$

Inserting n_3 diagonal steps in the usual way we get

$$N_2(n_1, n_2, n_3|a, b) = \sum_{\nu \in \mathbb{Z}} \left[\binom{n_1 + n_2}{n_1 - \nu(a+b)} - \binom{n_1 + n_2}{n_1 - \nu(a+b) + a} \right] \binom{n_1 + n_2 + n_3}{n_3}. \tag{9.11}$$

Now let $M_n = \max_{0 \le i \le n} Q_i$, the maximum of Q_n. The lattice paths corresponding to the event

$$\{Q_n = k, M_n < h, T_m > n|Q_0 = m\}$$

are just those counted by (9.11). Thus we get almost without any efforts the joint distribution of Q_n and its maximum during a period where the server is continuously busy, more precisely

$$P(Q_n = k, M_n < h, T_m > n|Q_0 = m) = \sum_{\substack{n_1-n_2=k-m \\ n_1+n_2 \le n}} N_2(n_1, n_2, n-n_1-n_2|m, h-m)\alpha^{n_1}\gamma^{n_2}\beta^{n-n_1-n_2}. \tag{9.12}$$

9.2.2. The $Geo^R/Geo/1$ model. Let us assume that customers arrive in batches of fixed size $R \ge 1$ at a single service station. The probability that during a particular slot a batch enters the system is α. All the other assumptions remain unchanged.

Now we have to modify the basic process S_n associated with the queueing process Q_n. S_n has increments X_i which assume their values in the set

$\{R, 0, -1\}$ with probabilities

$$P(X_i = R) = \alpha, \quad P(X_i = -1) = \gamma, \quad P(X_i = 0) = \beta = 1 - \alpha - \gamma.$$

The lattice paths corresponding to Q_n and S_n are constructed as follows: the arrival of a batch of size R is represented by a $(1,0)$-step, a single departure by a $(0,1)$-step and empty slots are encoded by diagonal steps, which are now parallel to the line $y = Rx$ and have the form $(1, R)$.

The principal argument is the same as in the $Geo/Geo/1$ case: first determine the joint distribution of S_n and $\min_i S_i$, and from this we get the transient distribution of Q_n. Unfortunately the situation is now considerably more difficult, because we have to count lattice paths restricted by lines of the type $y = Rx + a$, R being an integer ≥ 1, and if $R \neq 1$, then the classical reflection principle cannot be applied. The reason is, that reflection of a path on the line $y = Rx + a$ will alter the step set of the path. So different methods are required.

The key result is the generalization of the classical ballot problem, which when restated in terms of lattice paths, reads like follows:

The number of lattice paths from the origin to the point (n_1, n_2) which do not touch or cross the line $Ry = x$ except at the beginning is given by (see (6.57))

$$\frac{n_1 - Rn_2}{n_1 + n_2}\binom{n_1 + n_2}{n_1}, \qquad (R \in \mathbb{N}). \tag{9.13}$$

By interchanging horizontal and vertical steps we immediately obtain: the number of paths from the origin to the point $(n_1, m + Rn_1)$ which stay strictly above the line $y = Rx + m, m \in \mathbb{N}$ except for the endpoint, is given by

$$\frac{m}{m + n_1(R+1)}\binom{m + n_1(R+1)}{n_1}. \tag{9.14}$$

This formula enables us to derive the distribution of the stopping time

$$T_m = \inf(n : Q_n = 0, Q_0 = m),$$

the duration of a busy period initiated by $m > 0$ customers. First we have to insert diagonal steps in the paths counted by (9.14). If there are n_3 such steps, these can be inserted in all points of the paths except for the endpoint, because the last step has to be a vertical one. This can be done in

$$\binom{n_1 + n_2 + n_3 - 1}{n_3} \qquad \text{(note that } n_2 = m + Rn_1\text{)}$$

different ways. Thus the number of lattice paths which stay above the line $y = Rx + m$ and touch at the end, say $N_0^R(n_1, n_2, n_3|m)$, is given by

$$N_0^R(n_1, n_2, n_3|m) = \frac{m}{m + n_1(R+1)}\binom{m + n_1(R+1)}{n_2}\binom{n-1}{n_3}, \tag{9.15}$$

where $n_2 = m + Rn_1$, and $n = n_1 + n_2 + n_3$. But n_2, the number of vertical steps equals D_{T_m}, the number of customers served in the busy period, therefore, noting that $n_3 = n - m - n_1(R+1)$, we obtain

$$\begin{aligned} P(T_m = n, D_{T_m} = m + Rn_1|Q_0 = m) &= \\ &= \frac{m}{m+n_1(R+1)}\binom{m+n_1(R+1)}{n_2}\binom{n-1}{m+n_1(R+1)-1} \\ &\quad \times \alpha^{n_1}\gamma^{m+Rn_1}\beta^{n-m-n_1(R+1)}. \end{aligned} \tag{9.16}$$

Summing on n_1 we get $f_m(n) = P(T_m = n)$, the probability function of T_m:

$$\begin{aligned} f_m(n) &= \sum_{n_1 \geq 0} \frac{m}{m+n_1(R+1)}\binom{m+n_1(R+1)}{n_1}\binom{n-1}{m+n_1(R+1)-1} \\ &\quad \times \alpha^{n_1}\gamma^{m+Rn_1}\beta^{n-m-n_1(R+1)}. \end{aligned} \tag{9.17}$$

Unfortunately this formula cannot be simplified further except for the case $R = 1$. However, with this formula we may now determine the probability that a path touches or crosses the line $y = Rx + m$ and from this we get the probability that a path stays strictly above $y = Rx + m$. The latter probability is none but the joint distribution of S_n and $\min_i S_i$. So let us define $g(n,k) = P(S_n = k)$, the transition function of the basic process S_n. It is given by

$$g(n,k) = \sum_{\substack{n_1+n_2+n_3=n \\ Rn_1-n_2=k}} \binom{n_1+n_2}{n_1}\binom{n}{n_3}\alpha^{n_1}\gamma^{n_2}\beta^{n_3}. \tag{9.18}$$

Using this formula and (9.17) we obtain

$$\begin{aligned} P_{m,k}^{(0)}(n) &= P(Q_n = k, T_m > n|Q_0 = m) \\ &= P(m + S_n = k, m + \min_i S_i > 0) \\ &= g(n, k-m) - f_m(n) * g(n,k), \end{aligned} \tag{9.19}$$

where $*$ denotes convolution with respect to n. The representation formula (9.1) is still valid, and by an almost verbal repetition of (9.9) we find the general transient distribution

$$P(Q_n < k|Q_0 = m) = \sum_{\nu > m} [g(n, k-\nu) - f_\nu(n) * g(n,k)]. \tag{9.20}$$

9.2.3. The *Geo/Geo/c* model. Let us now turn to discrete-time systems with many parallel servers. The transient analysis of such systems is notoriously difficult. Still lattice path combinatorics can contribute to a deeper understanding of this problem and can even provide solutions.

In particular it will turn out that standard counting techniques encountered above, together with the technique of putting balls into boxes which we have used so many times to handle diagonal steps, can be successfully applied to this apparently difficult problem.

The derivations below and the arguments used there are almost elementary and have interesting applications going far beyond simple $M/M/c$ systems. In fact they apply also to general birth-death processes in discrete time. To give the reader an impression of the basic combinatorial idea we will discuss here only the case of $M/M/2$. However, the reader will have no difficulties to extend this approach to systems with more than two parallel servers. Additionally we assume that the number of customers in the system at time zero m, and the number of customers in the system at time n, say k, both are ≥ 2. This assumption is for convenience only and can be relaxed easily.

First of all we need a suitable lattice path representation, and for this purpose we use the same coding we used when discussing simple $M/M/1$, i.e., arrivals are represented by horizontal steps, departures by vertical steps and empty slots are encoded by diagonal steps. Sample paths of Q_n are again represented by a lattice path in the plane which is restricted by a reflecting barrier $y = x + m$. Now let us delete all diagonal steps in the paths, and observe that we have three types of such steps:

1. Diagonal steps on the line $y = x + m$, which have probability $\beta_0 = 1 - \alpha$;
2. Diagonal steps on the line $y = x + m - 1$, which have probability $\beta_1 = 1 - \alpha - \gamma$;
3. Diagonal steps below the line $y = x + m - 1$, which have probability $\beta_2 = 1 - \alpha - 2\gamma$.

Next let us delete also all steps between the lines $y = x+m$ and $y = x+m-1$. The resulting paths now consist only of horizontal steps with probability α and vertical steps having probability 2γ. Furthermore these paths have a certain number of reflection points on the line $y = x + m - 1$, which may be zero also.

Thus our first task will be to count paths having a prescribed number of reflection points on a given line. Here a standard counting argument applies: the number of paths from the origin to the point (n_1, n_2), which have exactly $r \geq 0$ reflections on the line $y = x + a, a > 0$, is given by:

$$\begin{aligned} &\binom{n_1+n_2}{n_1} - \binom{n_1+n_2}{n_1+a} && \text{if } r = 0 \\ &\binom{n_1+n_2-r}{n_1+a-1} - \binom{n_1+n_2-r}{n_1+a} && \text{if } r > 0. \end{aligned} \tag{9.21}$$

The number of such paths will be denoted by $N_r(n_1, n_2|a)$. Now let us reintroduce the horizontal and vertical steps between the lines $y = x + m$ and $y = x+m-1$. Since $m, k \geq 2$, these steps form a sequence of *northeast turns.* Suppose there are n_3 such turns, these can be inserted in r reflection points and the number of ways this can be done equals

$$\binom{n_3+r-1}{n_3}. \tag{9.22}$$

Note that, if $r = 0$, then the above binomial coefficient will be equal to zero necessarily.

The next is to introduce diagonal steps below the line $y = x + m - 1$, which have probability β_2. Our path has $n_1 + n_2 - r + 1$ lattice points below $y = x + m - 1$, and therefore there are

$$\binom{n_1 + n_2 + n_4 - r}{n_4} \tag{9.23}$$

possibilities to insert n_4 such steps in these points.

Now consider diagonal steps on the line $y = x + m - 1$. Our path has $r + n_3$ points on this line. If we want to insert n_5 β_1-steps, then this can be done in

$$\binom{r + n_3 + n_5 - 1}{n_5} \tag{9.24}$$

different ways. It remains to insert diagonal steps on the line $y = x + m$. On this line we have n_3 points (equal to the number of northeast turns we have inserted), so the number of ways to put in total n_6 β_0-steps in these points equals

$$\binom{n_3 + n_6 - 1}{n_6}. \tag{9.25}$$

Putting everything together, we find that the total number of lattice paths corresponding to the event

$$\{Q_n = k | Q_0 = m\}, \qquad m, k \geq 2$$

is given by (except for a final summation):

$$\begin{aligned} N(\mathbf{n}, r|m, k) &\\ = \ N_r(n_1, n_2|m) &\binom{n_3 + r - 1}{n_3}\binom{n_1 + n_2 + n_4 - r}{n_4}\binom{r + n_3 + n_5 - 1}{n_5} \\ &\times \binom{n_3 + n_6 - 1}{n_6}, \end{aligned} \tag{9.26}$$

where $\mathbf{n} = (n_1, \ldots, n_6)$. Any such path has probability

$$\alpha^{n_1}(2\gamma)^{n_2}(\alpha\gamma)^{n_3}\beta_2^{n_4}\beta_1^{n_5}\beta_0^{n_6} = 2^{n_2}\alpha^{n_1+n_3}\gamma^{n_2+n_3}\beta_2^{n_4}\beta_1^{n_5}\beta_0^{n_6}$$

Thus we get finally

$$P(Q_n = k | Q_0 = m) = \sum_S N(\mathbf{n}, r|m, k) 2^{n_2}\alpha^{n_1+n_3}\gamma^{n_2+n_3}\beta_2^{n_4}\beta_1^{n_5}\beta_0^{n_6} \tag{9.27}$$

where the summation extends over the set

$$S = \{r + n_3 + \sum_{i=1}^{6} n_i = n, r, n_i \geq 0, n_1 - n_2 = k - m\}, \tag{9.28}$$

and note that since we have inserted n_3 northeast turns, they contribute $2n_3$ steps.

9.2.4. Queues with heterogeneous arrivals. Let us consider a discrete-time queueing system which satisfies the following assumptions:

1. There is a single service facility.
2. Customers arrive in batches of size b_i $(i = 1, \ldots, r)$ from $r \geq 1$ different sources.
3. The probability that during a particular time slot, say $(n-1, n]$ there is an arrival of a batch of size b_i from source i equals α_i, $(i = 1, \ldots, r)$.
4. Customers are served one after the other. The probability that a customer finishes service in $(n-1, n]$ equals γ provided that the queue is not empty and independent of the origin of the customers. The probability of a service completion is zero otherwise.
5. Events in different slots are mutually independent.
6. The probability of more than one arrival or service or the simultaneous occurrence of arrivals and services during a particular slot is zero.
7. The probability that there is neither an arrival nor a service completion during a particular slot is
$$\beta = 1 - \gamma - \sum_{i=1}^{r} \alpha_i,$$
if the queue is not empty and equals $1 - \sum_{i=1}^{r} \alpha_i$ otherwise.
8. Initially there are $m > 0$ customers waiting and the server is busy.

Thus we have a Markovian system in which customers arrive in batches, the arrivals, however, being heterogeneous insofar, as customers arrive from various sources in batches of various types. Such models have interesting applications the theory of data packet transmission, where we may think of the server as a communication channel which takes his input from various heterogeneous data streams.

The transient analysis of this queueing system is most conveniently done by representing the sequence of arrivals and service completions as well as the event {no change in system state} by a lattice path in $r+2$-dimensional space. For this purpose consider the lattice in $\mathbb{Z}_+^{r+2}$ and let us represent

- A service completion by a step of unit length parallel to the positive x_0 axis,
- An arrival of a batch of customers of size b_i from source i by a step of unit length parallel to the positive x_i-axis $(i = 1, \ldots, r)$,
- The event {no change in system state} by a step of unit length parallel to the positive x_{r+1}-axis.

Let $A_n(i)$ denote the number of batch arrivals from source i, $(i = 1, \ldots, r)$ in the time interval $(0, n]$ and define an r-dimensional random vector $\mathbf{A}_n$ with components $A_n(i)$. Furthermore, let D_n denote the total number of service completions in $(0,n]$ and let

$$A_n = \sum_{i=1}^{r} b_i A_n(i),$$

be the total number of customers arrived in $(0,n]$.

For convenience we introduce also the following notations:

$$\mathbf{n} = (n_1, \ldots, n_r), \quad n = \sum_{i=0}^{r+1} n_i, \quad n_i \geq 0, \quad i = 0, 1, \ldots, r+1.$$

Using lattice path counting in higher dimensional spaces we will derive some interesting formulas concerning the behaviour of this system during busy periods. In particular we will find the joint distribution of the random vector $\mathbf{A}_n$ and D_n during time intervals where the server is continuously busy.

For this purpose we need the following generalization of the classical ballot theorem:

Let E_m denote the hyperplane

$$E_m : x_0 = m + \sum_{i=1}^{r} b_i x_i.$$

The number of lattice paths from the origin to the point $(n_0, \mathbf{n}, n_{r+1})$ which do not touch or cross the hyperplane E_m except at the end, where the path terminates with a x_0-step on E_m, is given by

$$Q(m, n_0, \mathbf{n}, n_{r+1}) = \frac{m}{n}\binom{n}{\mathbf{n}, n_{r+1}}, \tag{9.29}$$

where

$$\binom{n}{a_1, \ldots, a_k} = \frac{n!}{\prod_{i=1}^{k}(a_i!)(n - \sum_{i=1}^{k} a_i)!}$$

denotes the usual multinomial coefficient, with

$$n_0 = m + \sum_{i=1}^{r} b_i n_i,$$

and

$$m > 0, \quad n_i \geq 0, \quad i = 1, \ldots, r+1.$$

Furthermore, let $R(m, n_0, \mathbf{n}, n_{r+1})$ denote the number of lattice paths from the origin to the point $(n_0, \mathbf{n}, n_{r+1})$, which do not touch or cross the hyperplane E_m. Then for $m > 0$ and $n_i \geq 0$, $i = 1, \ldots, r+1$:

$$R(m, n_0, \mathbf{n}, n_{r+1}) = \binom{n}{\mathbf{n}, n_{r+1}} - m\binom{n}{n_{r+1}} \sum_{S} \frac{(\sum_{i=0}^{r} c_i - 1)!(\sum_{i=0}^{r} a_i)!}{\prod_{i=0}^{r}(c_i!)\prod_{i=0}^{r}(a_i!)}, \tag{9.30}$$

where the summation is over the set S:

$$S = \{c_i + a_i = n_i; i = 0, \ldots, r\}$$

with a_i, c_i $(i = 0, \ldots, r)$ non-negative integers and

$$c_0 = m + \sum_{i=1}^{r} b_i c_i.$$

This formula may be proved by observing that the set of all paths from the origin to the point $(n_0, \mathbf{n}, n_{r+1})$ has

$$\binom{n}{\mathbf{n}, n_{r+1}} \tag{9.31}$$

elements and is the disjoint union of the set of paths which do not touch the hyperplane E_m and the set of paths which touch or cross this plane at least once. To find the number of paths in the latter set, observe, that any path which touches E_m will do so for a first time, say in a lattice point $(c_0, \mathbf{c}, c_{r+1})$. From that touching point onwards the rest of the path may be regarded as a new path without any restriction which, if we move the origin to the point $(c_0, \mathbf{c}, c_{r+1})$, moves to a point $(a_0, \mathbf{a}, a_{r+1})$. Thus the number of paths which touch E_m at least once is given by the convolution of (9.29) and (9.31). In this convolution we have to perform summation over the set of all a_i, c_i, $(i = 0, \ldots, r+1)$, such that

$$c_i + a_i = n_i; \quad i = 0, \ldots, r+1,$$

and

$$c_0 = m + \sum_{i=1}^{r} b_i c_i,$$

since at the point $(c_0, \mathbf{c}, c_{r+1})$ the path has to touch E_m. In particular we are left with the summation

$$\sum_S \sum_{c_{r+1}+a_{r+1}=n_{r+1}} \frac{m}{c} \binom{c}{\mathbf{c}, c_{r+1}} \binom{a}{\mathbf{a}, a_{r+1}}.$$

The inner summation can be done explicitly because c_0 does not depend on c_{r+1}:

$$\begin{aligned}
&\sum_S \sum_{c_{r+1}+a_{r+1}=n_{r+1}} \frac{m}{c} \binom{c}{\mathbf{c}, c_{r+1}} \binom{a}{\mathbf{a}, a_{r+1}} = \\
&= \sum_S \frac{(\sum_{i=0}^{r} c_i)!(\sum_{i=0}^{r} a_i)!}{\prod_{i=0}^{r}(c_i!)\prod_{i=0}^{r}(a_i!)} \sum_{c_{r+1}+a_{r+1}=n_{r+1}} \frac{m}{c} \binom{c}{c_{r+1}} \binom{a}{a_{r+1}} \\
&= m \sum_S \frac{(\sum_{i=0}^{r} c_i - 1)!(\sum_{i=0}^{r} a_i)!}{\prod_{i=0}^{r}(c_i!)\prod_{i=0}^{r}(a_i!)} \binom{n}{n_{r+1}}.
\end{aligned}$$

Consider now the event

$$\{\mathbf{A}_n = \mathbf{n}, D_n = n_0, T_m > n\}.$$

The lattice paths corresponding to this event must not touch or cross the hyperplane E_m. Any such path has the same probability which we find by assigning probability α_i to steps in the x_i-direction, $i = 1, 2, \ldots, r$, γ to x_0-steps and β to x_{r+1}-steps. Let

$$T_m = \inf\{n : m + A_n - D_n = 0\}$$

denote the length of a busy period initiated by $m > 0$ customers.

Then $n_i \geq 0$, $i = 0, \ldots, r$:

$$P_m(\mathbf{A}_n = \mathbf{n}, D_n = n_0, T_m > n)$$
$$= R(m, n_0, \mathbf{n}, n - \sum_{i=0}^{r} n_i) \left(\prod_{i=1}^{r} \alpha_i^{n_i}\right) \gamma^{n_0} \beta^{n - \sum_{i=0}^{r} n_i}. \tag{9.32}$$

The subscript m of P indicates that this probability is conditional on the event that there are m customers waiting initially.

Particularly simple is the joint probability function of $T_m, \mathbf{A}_n$ and D_n. Using (9.29), we immediately obtain

$$\begin{aligned} P_m(\mathbf{A}_n = \mathbf{n}, D_n = n_0, T_m = n) &= f(m, n_0, \mathbf{n}, n) \\ &= Q(m, n_0, \mathbf{n}, n - \sum_{i=0}^{r} n_i) \left(\prod_{i=1}^{r} \alpha_i^{n_i}\right) \gamma^{n_0} \beta^{n - \sum_{i=0}^{r} n_i}, \end{aligned} \tag{9.33}$$

where n_0 given by

$$n_0 = m + \sum_{i=1}^{r} b_i n_i,$$

and again $m > 0$ and $n_i \geq 0$, $i = 1, \ldots, r$.

9.2.5. A simple Jackson network. In the previous section we encountered the first instance of lattice path counting in higher dimensional space. A similar situation will arise in this section.

The last model we are going to discuss is an example of a simple Jackson network (see Section 2.8 in Chapter 2), an r-node tandem system in discrete time. Such models play an important role in various applications, notably in the study of assembly lines.

In particular we consider a service facility in which $r > 1$ servers are arranged in a service line. Customers enter the system at the first node and have to pass through all subsequent nodes.

Let us first introduce some basic assumptions and fix the necessary notation:

- There are $r > 1$ server arranged in a line.
- The probability that a customer arrives during a slot equals α.
- The probability that a customer finishes service at node i during a particular slot equals γ_i, if node i is not empty, $i = 1, 2 \ldots, r$.
- Service times at the various nodes are mutually independent.
- The service times of a particular customer are also independent at the various nodes.
- During a particular slot not more than one arrival or one service completion can occur.

- Initially there are $m_i > 0$ customers waiting at node i.

Let $Q_i(n)$ denote the number of customers waiting at node i at time n. It will be convenient to introduce a vector notation, so we define r-dimensional random vectors

$$\mathbf{Q}(n) = (Q_1(n), Q_2(n), \ldots, Q_r(n)).$$

Next let us introduce random stopping times T_i:

$$T_i = \inf\{n : Q_i(n) = 0\},$$

and let

$$T = \min(T_1, T_2, \ldots, T_r).$$

Finally we define two r-dimensional vectors

$$\mathbf{m} = (m_1, m_2, \ldots, m_r),$$

and

$$\mathbf{k} = (k_1, k_2, \ldots, k_r), \quad k_i > 0,$$

where k_i equals the number of customers waiting at node i at time n.

Again we focus on zero-avoiding transition probabilities, which are defined as the probability that there is a transition from a (now multivariate) state $\mathbf{m}$ to a state $\mathbf{k}$ in such a way that none of the servers becomes idle during the time interval $(0, n]$, which we express simply by requiring $T > n$. Thus let

$$^0P(\mathbf{m}, \mathbf{k}, n) = P(\mathbf{Q}(n) = \mathbf{k}, T > n | \mathbf{Q}(0) = \mathbf{m}). \tag{9.34}$$

In order to find a suitable lattice path representation, we have to go another way now. The following multivariate generalization of the ballot theorem will be important:

Let $N(r; \mathbf{a}, \mathbf{b})$ denote the number of r-dimensional lattice paths from the point $\mathbf{a} = (a_1, a_2, \ldots, a_r)$ to the point $\mathbf{b} = (b_1, b_2, \ldots, b_r)$ such that every lattice point $(x_1, x_2, \ldots, x_r)$ on the paths satisfies the condition $x_1 \geq x_2 \geq \ldots \geq x_r$. Then

$$N(r; \mathbf{a}, \mathbf{b}) = \left(\sum_{\nu=1}^{r} (b_\nu - a_\nu) \right)! \|c_{ij}\|_{r \times r}, \tag{9.35}$$

where $\|c_{ij}\|_{r \times r}$ is the $r \times r$ determinant with $(i, j)-$th element given by

$$c_{ij} = \frac{1}{(b_i - a_j - i + j)!} \qquad i, j = 1, 2, \ldots, r.$$

Consider now the system first only at those times where either an arrival at the first node or a service completion at any of the nodes occurs. Such time points will be called *instants*. Let α_{0j} denote the total number of arrivals at the first node at instant j, and similarly define α_{ij} as the total number of service completions at node i at instant j, $i = 1, 2, \ldots, r$. Obviously the random variables α_{ij} can assume only the values 0 and 1 and $\sum_{i=0}^{r} \alpha_{ij} = 1$ for all instances j. The total number of instants in $(0, n]$ will be denoted by N, and we assume that $N \leq n$.

The combinatorial problem may be stated as follows: enumerate the number of ways in which we can arrange arrivals and departures at the various nodes, such that none of the servers becomes idle, or in other words that the event

$$\{\mathbf{Q}(t) = \mathbf{k}, T > t | \mathbf{Q}(0) = \mathbf{m}\} \tag{9.36}$$

occurs. Some reflection shows that this event can occur if and only if for all nodes $i = 1, \ldots, r$ and all instances $\nu = 1, 2, \ldots, N$, the following set of fundamental inequalities is satisfied:

$$m_i + \sum_{\ell=1}^{\nu} \alpha_{i-1,\ell} > \sum_{\ell=1}^{\nu} \alpha_{i,\ell}. \tag{9.37}$$

Thus at any node i the number of customers waiting there initially plus the accumulated number of arrivals at this node should always exceed the accumulated number of departures.

To simplify notation we set $n_0 = \sum_{\ell=1}^{N} \alpha_{0,\ell}$, thus n_0 equals the total number of arrivals at the first node in $(0, n]$, the system's total inflow.

Since we require that there are k_i customers waiting at node i at time n, we have also a set of balancing equations:

$$m_i + \sum_{\ell=1}^{N} \alpha_{i-1,\ell} - \sum_{\ell=1}^{N} \alpha_{i,\ell} = k_i. \tag{9.38}$$

In fact, (9.38) is a recurrence relation in the sum of the α's, where recurrence is with respect to the first index. We may resolve this recurrence to find the following representation of the total number of service completions at node i:

$$\sum_{\ell=1}^{N} \alpha_{i,\ell} = n_0 + \sum_{j=1}^{i} (m_j - k_j) \qquad \text{for } i = 1, 2, \ldots, r. \tag{9.39}$$

Next let us introduce $(r+1)$-dimensional vectors $\mathbf{x}_\nu$:

$$\begin{aligned} x_{i,\nu} &= \sum_{j=i+1}^{r} (m_j - 1) + \sum_{\ell=1}^{\nu} \alpha_{i,\ell} \\ x_{r,\nu} &= \sum_{\ell=1}^{\nu} \alpha_{r,\ell} \\ & \qquad i = 0, 1, \ldots, r-1; \quad \nu = 1, 2, \ldots, N \end{aligned} \tag{9.40}$$

and if we agree that an empty summation in (9.40) has to be interpreted as zero, then we have also

$$\begin{aligned} \mathbf{x}_0 &= (x_{00}, x_{10}, \ldots, x_{r0}) \\ &= \left(\sum_{j=1}^{r} (m_j - 1), \sum_{j=2}^{r} (m_j - 1), \ldots, m_r - 1, 0 \right). \end{aligned}$$

The inequalities (9.37) together with the balancing conditions (9.38) may be restated now in terms of the vectors $\mathbf{x}_\nu$, in particular, (9.37) and (9.38) are equivalent to

$$x_{0\nu} \geq x_{1\nu} \geq \ldots \geq x_{r\nu}, \qquad \nu = 0, 1, \ldots, N \tag{9.41}$$

with $\mathbf{x}_N = (x_{0,N}, x_{1,N}, \ldots, x_{r,N})$ given by

$$\begin{aligned} x_{i,N} &= \sum_{j=i+1}^{r} (m_j - 1) + \sum_{\ell=1}^{N} \alpha_{i,\ell} \\ &= n_0 - r + i + \sum_{j=1}^{r} m_j - \sum_{j=1}^{i} k_j, \qquad i = 0, 1, \ldots, r \end{aligned}$$

where an empty sum has to be interpreted again as zero. It follows that the number of ways in which we can arrange arrivals and departures at the nodes in such a way that all servers are continuously busy equals the number of integer sequences $\mathbf{x}_\nu$, which satisfy (9.41), and this number is equal to $N(r+1, \mathbf{x}_0, \mathbf{x}_N)$.

Now let us assign probabilities to these integer sequences. First we observe that the total number of instances equals

$$N = \sum_{i=0}^{r} \sum_{\ell=1}^{N} \alpha_{i,\ell} = (r+1)n_0 + \sum_{i=1}^{r} (r - i + 1)(m_i - k_i), \tag{9.42}$$

which follows immediately from (9.39). Since we have N instances and $n \geq N$ time slots, there are $n_1 = n - N$ slots, where the queues do not change, this event happens with probability $\beta = 1 - \alpha - \sum_{i=1}^{r} \gamma_i$. The number of ways in which we can select slots, where the queues do not change equals $\binom{n}{n_1}$ which in turn equals $\binom{n}{N}$. Thus we arrive at the following representation of the zero avoiding transition probabilities:

$$P(\mathbf{m}, \mathbf{k}, n) = \sum_{n_1 + N = n} \binom{n}{N} \beta^{n-N} N(r+1, \mathbf{x}_0, \mathbf{x}_N) \alpha^{\sum \alpha_{0,\ell}} \prod_{i=1}^{r} \gamma_i^{\sum \alpha_{i,\ell}}. \tag{9.43}$$

Further simplification is possible by observing that

$$N(r+1, \mathbf{x}_0, \mathbf{x}_N) = N! C(n_0, r, \mathbf{m}, \mathbf{k}),$$

where $C(n_0, r, \mathbf{m}, \mathbf{k})$ is a determinant of order $(r+1) \times (r+1)$ with entries given by

$$c_{ij} = \frac{1}{(n_0 + \sum_{\nu=1}^{j} m_\nu - \sum_{\nu=1}^{i} k_\nu)!}, \qquad i, j = 0, 1, \ldots, r. \tag{9.44}$$

If we replace the sums $\sum_{\ell=1}^{N} \alpha_{i,\ell}$ in (9.43) by their values given in (9.39) and carry out summation on all $n_0 \geq 0$, then we obtain finally:

$$P(\mathbf{m}, \mathbf{k}, n) = \delta_r \sum_{n_0 \geq 0} (n)_N \beta^{n-N} \eta_r^{n_0} C(n_0, r, \mathbf{m}, \mathbf{k}), \tag{9.45}$$

where C is given in (9.44), $(n)_N = n(n-1)\ldots(n-N+1)$ and

$$\delta_r = \prod_{i=1}^{r} \gamma_i^{\sum_{j=1}^{i}(m_j-k_j)}, \qquad \eta_r = \alpha \prod_{i=1}^{r} \gamma_i,$$

and empty sums are zero again.

9.2.6. A limiting procedure. It is well known that if a random variable X has a binomial distribution with parameters n and α, then, as $n \to \infty$ in such a way that $\alpha = \frac{\lambda t}{n}$, for t and $\lambda > 0$ fixed, X will converge in distribution to a Poisson distribution with expectation λt (see Section A.4 in Appendix A).

In principle the same idea of convergence applies also to Markovian queueing models in discrete time, and it is best explained at the simple $M/M/1$ system.

Let us rescale time in such a way that the width of a slot is equal to $\Delta = \frac{t}{n}$, for some $t > 0$. Now set

$$\alpha = \frac{\lambda t}{n}, \quad \gamma = \frac{\mu t}{n}, \quad \text{and} \quad \beta = 1 - \frac{t}{n}(\lambda + \mu), \tag{9.46}$$

where $\lambda, \mu, t > 0$ and fixed. If we let $n \to \infty$, then the distributions of Q_n, S_n and also of various functionals, like the maximum queue length or stopping times, converge to certain limiting distributions, which turn out to be the distributions associated with the corresponding continuous time $M/M/1$ model. Actually, what is happening behind the scene is that the process Q_n when scaled in the above manner converges in the weak sense to the corresponding process Q_t.

As a particular example let us consider the limit of (9.5). Using the scaling (9.46), the right-hand side of (9.5) now reads as

$$\left[\binom{n_1+n_2}{n_1} - \binom{n_1+n_2}{m+n_1}\right]\binom{n}{n-n_1-n_2}\frac{\lambda^{n_1}\mu^{n_2}t^{n_1+n_2}}{n^{n_1+n_2}} \times \left(1 - \frac{t}{n}(\lambda+\mu)\right)^{n-n_1-n_2}.$$

If $n \to \infty$, while keeping n_1 and n_2 fixed,

$$\left(1 - \frac{t}{n}(\lambda+\mu)\right)^{n-n_1-n_2} \to e^{-(\lambda+\mu)t}.$$

It remains to determine the limit of $\frac{1}{n^{n_1+n_2}}\binom{n}{n-n_1-n_2}$. For this we find

$$\begin{aligned}
&\frac{1}{n^{n_1+n_2}}\binom{n}{n-n_1-n_2} \\
&\quad = \frac{1}{n^{n_1+n_2}} \frac{n!}{(n_1+n_2)!(n-n_1-n_2)!} \\
&\quad = \frac{1}{n^{n_1+n_2}} \frac{n(n-1)\dots(n-n_1-n_2+1)}{(n_1+n_2)!} \\
&\quad = \frac{1}{(n_1+n_2)!} 1 \left(1-\frac{1}{n}\right)\left(1-\frac{2}{n}\right)\cdots\left(1-\frac{n_1+n_2-1}{n}\right) \\
&\quad \to \frac{1}{(n_1+n_2)!}.
\end{aligned}$$

Thus the right-hand side of (9.5) converges to

$$\left[\binom{n_1+n_2}{n_1} - \binom{n_1+n_2}{n_1+m}\right] \frac{\lambda^{n_1}\mu^{n_2}t^{n_1+n_2}}{(n_1+n_2)!} e^{-(\lambda+\mu)t}.$$

This, however, is the joint distribution of the processes Q_t and D_t, the continuous time analogues of Q_n and D_n. Thus

$$\begin{aligned}
P(Q_t = m+&n_1+n_2, D_t = n_2, \tau_m > t | Q_0 = m) \\
&= e^{-(\lambda+\mu)t}\lambda^{n_1}\mu^{n_2}t^{n_1+n_2}\left[\frac{1}{n_1!n_2!} - \frac{1}{(n_1+m)!(n_2-m)!}\right],
\end{aligned} \tag{9.47}$$

where $\tau_m = \inf\{t : Q_t = 0 | Q_0 = m)$, the length of a busy period initiated by $m > 0$ customers in a $M/M/1$ system in continuous time.

Using the power series expansion of the modified Bessel functions

$$I_k(2t\sqrt{\lambda\mu}) = (\lambda\mu)^{k/2} \sum_{\nu \geq 0} \frac{(\lambda\mu)^\nu t^{k+2\nu}}{\nu!(k+\nu)!},$$

and summing over n_1 and n_2 such that $n_1 - n_2 = k - m$, we get the well-known formula

$$P_{m,k}^{(0)}(t) = e^{-(\lambda+\mu)t}\rho^{\frac{k-m}{2}}[I_{k-m}(2t\sqrt{\lambda\mu}) - I_{k+m}(2t\sqrt{\lambda\mu})] \tag{9.48}$$

for the zero-avoiding transition probabilities of the $M/M/1$ system in continuous time, which agrees with the first term in (2.61). By substituting (9.48) in the limit of (9.9), we can obtain the expression (2.50) for $P_n(t)$ (see Exercise 4). When performing this limiting procedure some care is needed whenever stopping time densities are involved. Again let us demonstrate the technicalities by an example and consider the limit of (9.8), which we expect to be

$$P(t - dt < \tau_m \leq t, D_{\tau_m} = k).$$

In fact we find

$$
\begin{aligned}
&\frac{m}{k}\binom{2k-1-m}{k-1}\binom{n-1}{2k-1-m}\frac{\lambda^{k-m}\mu^k t^{2k-m}}{n^{2k-m}}\left(1-\frac{t}{n}(\lambda+\mu)\right)^{n-2k+m} \\
&= \frac{m}{k}\frac{1}{(k-1)!(k-m)!}\frac{(n-1)!}{(n-2k+m)!}\frac{\lambda^{k-m}\mu^k t^{2k-m}}{n^{2k-m}}\left(1-\frac{t}{n}(\lambda+\mu)\right)^{n-2k+m} \\
&= \frac{m}{k!(k-m)!}1\left(1-\frac{1}{n}\right)\cdots\left(1-\frac{2k-m-1}{n}\right)\lambda^{k-m}\mu^k t^{2k-m-1} \\
&\qquad\times\left(1-\frac{t}{n}(\lambda+\mu)\right)^{n-2k+m}\frac{t}{n} \\
&\to \frac{m}{k!(k-m)!}\lambda^{k-m}\mu^k t^{2k-m-1}e^{-(\lambda+\mu)t}\,\mathrm{d}t.
\end{aligned}
$$

Thus if $h(t,k)$ denotes the joint density of τ_m and D_{τ_m}, we get

$$h(t,k) = \frac{m}{k!(k-m)!}e^{-(\lambda+\mu)t}\lambda^{k-m}\mu^k t^{2k-m-1}. \tag{9.49}$$

Summing on all $k \geq m$, we obtain the formula for the density $f_m(t)$ of τ_m:

$$f_m(t) = \frac{m}{t}\rho^{-m/2}e^{-(\lambda+\mu)t}I_m(2t\sqrt{\lambda\mu}). \tag{9.50}$$

Particularly interesting is the limit of the batch system discussed in Section 9.2.2. We use the same scaling as above. Then it is not difficult to show that as $\to \infty$ the limit of (9.16) is

$$\frac{k}{n_1!(m+Rn_!)!}e^{-(\lambda+\mu)t}\lambda^{n_1}\mu^{m+Rn_1}t^{m+n_1(R+1)-1}\,\mathrm{d}t.$$

Now let again τ_m denote the length of a busy period initiated by $m > 0$ customers and let $f_m(t)$ denote its density function. By summing on n_1, we get

$$f_m(t) = \frac{m}{t}e^{-(\lambda+\mu)t}\sum_{\nu\geq 0}\frac{\lambda^\nu\mu^{m+r\nu}t^{m+\nu(R+1)}}{\nu!(m+R\nu)!}.$$

This formula can be conveniently expressed in terms of generalized Bessel functions, which are defined by

$$I_m^R(x) = \sum_{\nu\geq 0}\frac{(x/2)^{m+\nu(R+1)}}{\nu!(m+R\nu)!}. \tag{9.51}$$

Setting $w = \lambda^{\frac{1}{R+1}}\mu^{\frac{R}{R+1}}$, get after some transformations the formula

$$f_m(t) = \frac{m}{t}e^{-(\lambda+\mu)t}\rho^{-\frac{m}{R+1}}I_m^R(2wt). \tag{9.52}$$

9.3. Recurrence Relations

We have repeatedly stressed the fact that transient analysis by analytical methods is not at all easy, even for Markovian models. Therefore we preferred to take advantage of some numerical techniques, as done in the last section. If, however, the models are non-Markovian, we do not expect any simple explicit solution by analytical methods (see Chapter 6). It is therefore natural to look for some numerical procedure to provide an approximate solution.

Recurrence is one of the oldest procedures (due to L. Euler) to solve numerically hard problems in mathematical analysis. The basic idea is to discretize continuous variables which enter equations via derivatives. Through the discretization, partial differential equations are approximated by partial difference equations which give rise to recurrence relations. If the necessary starting values are given, then these difference equations can be solved recursively.

In queueing models the continuous variable is nothing but the time. The above method suggests to discretize the time axis, dividing it into segments of some fixed duration and observe the queueing process at the end of each slot. It is known from Section 3.7 in Chapter 3 that such a model is called a discrete-time model and in the same section the steady-state solutions have been studied. In this section we first formulate the recurrence relations for the state probabilities of discrete time $Geo/G/1$ and $G/Geo/1$ models and then demonstrate how to implement these recurrences on a computer and how to calculate other interesting queueing characteristics, such as moments of the queue length, expectations of sojourn times and busy period analysis. The parameterizations of these models will be as general as possible. The numerical solutions so obtained may be used as approximations to the corresponding continuous-time models. Of course, as any numerical technique, this approach has its limitations, a point we will discuss briefly as a remark at the end of this section.

9.3.1. The $Geo/G/1$ model. Suppose we have chosen some time unit, whatever it will be, then instead of continuous time, we have now a slotted time axis, which consists of a sequence of contiguous slots:

$$(0,1], (1,2], \ldots, (n-1, n], \ldots,$$

and all events, like arrivals and departures will occur during particular slots. Our interest is in the state of the system at the end of a slot, i.e., at epochs $1, 2, \ldots, n$. To be more specific, the model we are going to analyze is characterized by the following assumptions:

1. The numbers of arrivals in successive time slots are mutually independent and identically distributed random variables. Let p_ν, $\nu = 0, 1, \ldots, K$ denote the probability that ν customers join the system during a given slot. Observe that we assume K, the maximum number of arrivals in a slot, to be finite. This is a natural

assumption and it is difficult to imagine a practical situation where it is not satisfied.

2. The system has a waiting room with finite capacity $C > K$. Whenever an arriving customer finds the waiting room exhausted, he will be lost. Again this is a natural assumption. In fact, the prevalence of queueing systems with infinite capacity in theoretical work is generally due to mathematical convenience. Finite capacity introduces one more boundary condition to the equations governing the state probabilities, and this makes the analysis much more intricate.
3. The service times are i.i.d. integer-valued random variables. Let r_j denote the probability that a customer will require j units of service time $j = 1, 2, \ldots, \sigma$. We assume that the service times are bounded from above with maximum service time equal to some constant σ.
4. The queue discipline is immaterial except when we are interested in waiting times. In such cases it will be always assumed that the discipline is FCFS.
5. The mechanism determining the order in which arrivals and departures occurring during the same slot affect the queue is: arrivals (if there are any) join the queue immediately at the beginning of a slot, departures (if there are any) are recorded immediately before the end of a slot:

6. At time $n = 0$ there are $0 \leq i_0 \leq C$ customers already waiting. If $i_0 > 0$, the customer in service has residual service time $1 \leq j_0 \leq \sigma$, otherwise the residual service time is equal to zero by definition.

One final remark to these assumptions: we did not make any parametric assumption about the service time distribution or the distribution of the number of arrivals during a slot. This is a point of considerable practical relevance. Because in practice it will rarely happen that these distributions are known. In general they have to be estimated by taking samples. Simple histogram estimates, for instance, of the probabilities r_j may be used directly in the recurrence relations below. The same is true for the arrival probabilities p_k.

Let $Q(n)$ denote the number of customers in the system at time n, i.e., exactly at the end of a slot. Furthermore, denote by $R(n)$ the residual service time of the customer being in service at time n. Then due to our assumptions the bivariate process $(Q(n), R(n))$ is a Markov chain with state space

$$\mathcal{S} = \{(0,0)\} \cup \{1, 2, \ldots, C\} \times \{1, 2, \ldots, \sigma\}.$$

Let

$$P_n(i,j) = P(Q(n)=i,\ R(n)=j|Q(0)=i_0,\ R(0)=j_0).$$

Note that the introduction of $R(n)$ allows us to bring back the Markov property and $R(n)$ is called a *supplementary variable.* The approach of using supplementary variables was done in Chapter 6.

By considering the various possibilities by which a particular state can be entered in one step, we will derive now the recurrence relations for the probabilities $P_n(i,j)$.

The state $(0,0)$ can be entered in one step only from state $(0,0)$ or from state $(1,1)$. In any case such a transition requires that there are no arrivals during that time slot. Hence

$$P_n(0,0) = p_0[P_{n-1}(0,0) + P_{n-1}(1,1)]. \tag{9.53}$$

Consider now one-step transitions into state (i,j), where $i = 1,2,\ldots,K$ and $j = 1,2,\ldots,\sigma-1$. This state may be entered from

1. State $(i,j+1)$, if there is no arrival,
2. State $(i-\nu,j+1)$, $\nu = 1,2,\ldots,i-1$, if exactly ν customers arrive,
3. State $(0,0)$, if i customers arrive and the first requires j units of service,
4. State $(i-\nu+1,1)$, $\nu = 0,1,\ldots,i$, if ν customers arrive, the customer in service finishes necessarily and the next customer requires j units of service time.

Thus for $i = 1,2,\ldots,K$ and $j = 1,2,\ldots,\sigma-1$, we have the difference equation

$$\begin{aligned} P_n(i,j) &= p_0 P_{n-1}(i,j+1) + \sum_{\nu=1}^{i-1} p_\nu P_{n-1}(i-\nu,j+1) \\ &+ r_j p_i P_{n-1}(0,0) + r_j \sum_{\nu=0}^{i} p_\nu P_{n-1}(i-\nu+1,1). \end{aligned} \tag{9.54}$$

If $i = K+1,\ldots,C-1$ and $j = 1,2,\ldots,\sigma-1$, state (i,j) can be entered from

1. State $(i,j+1)$, if there is no arrival,
2. State $(i-\nu,j+1)$, $\nu = 1,2,\ldots,K$, if ν customers arrive,
3. State $(i-\nu+1,1)$, $\nu = 0,1,\ldots,K$, if ν customers arrive and the next customer commencing service requires j units of service.

Therefore we have for $i = K+1,\ldots,C-1$ and $j = 1,2,\ldots,\sigma-1$:

$$\begin{aligned} P_n(i,j) &= p_0 P_{n-1}(i,j+1) + \sum_{\nu=1}^{K} p_\nu P_{n-1}(i-\nu,j+1) \\ &+ r_j \sum_{\nu=0}^{K} p_\nu P_{n-1}(i-\nu+1,1). \end{aligned} \tag{9.55}$$

Next consider one-step transitions into state (C,j), $j = 1,2,\ldots,\sigma-1$. This state can be entered from

1. State $(C-\nu,j+1)$, $\nu = 0,1,\ldots,K$, if at least ν customers arrive,

2. State $(C-\nu+1,1)$, $\nu=1,2,\ldots,K$, if at least ν customers arrive and the next customer to be served requires j units of service time.

Let q_ν, $\nu=0,1,\ldots,K$ denote the probability that there are at least ν arrivals in a slot, i.e., $q_\nu=\sum_{m=\nu}^{K}p_\nu$. Then for $j=1,2,\ldots,\sigma-1$:

$$P_n(C,j)=\sum_{\nu=0}^{K}q_\nu P_{n-1}(C-\nu,j+1)+r_j\sum_{\nu=1}^{K}q_\nu P_{n-1}(C-\nu+1,1). \quad (9.56)$$

The state (i,σ), $i=1,2,\ldots,K$ can be entered from

1. State $(0,0)$, if i customers arrive and the first customer requires σ units of service,
2. State $(i-\nu+1,1)$, $\nu=0,1,\ldots,i$, if ν customers arrive and the next customer requires σ units of service.

Thus for $i=1,2,\ldots,K$:

$$P_n(i,\sigma)=r_\sigma p_i P_{n-1}(0,0)+r_\sigma\sum_{\nu=0}^{i}p_\nu P_{n-1}(i-\nu+1,1). \quad (9.57)$$

If $i=K+1,\ldots,C-1$, then state (i,σ) can be entered from state $(i-\nu+1,1)$, $\nu=0,1,\ldots,K$, if ν customers arrive and the next customer requires maximal service time σ. Therefore for $i=K+1,\ldots,C-1$:

$$P_n(i,\sigma)=r_\sigma\sum_{\nu=0}^{K}p_\nu P_{n-1}(i-\nu+1,1). \quad (9.58)$$

Finally the state (C,σ) can be entered from state $(C-\nu+1,1)$, $\nu=1,2,\ldots,K$, if at least ν customers arrive and the next customer requires maximal service time:

$$P_n(C,\sigma)=r_\sigma\sum_{\nu=1}^{K}q_\nu P_{n-1}(C-\nu+1,1). \quad (9.59)$$

The recurrence relations (9.53) to (9.59) may be written much more compactly. For this purpose let $1(A)$ denote the indicator function of event A. Then the following set of partial difference equations determines the probabilities $P_n(i,j)$:

$$P_n(0,0)=p_0[P_{n-1}(0,0)+P_{n-1}(1,1)]. \quad (9.60)$$

For $i=1,2,\ldots,C-1$ and $j=1,2,\ldots,\sigma$:

$$\begin{aligned} P_n(i,j) &= 1(j<\sigma)\sum_{\nu=0}^{\min(i-1,K)} p_\nu P_{n-1}(i-\nu,j+1) \\ &\quad +r_j\left[1(i\le K)p_i P_{n-1}(0,0)+\sum_{\nu=0}^{\min(i,K)} p_\nu P_{n-1}(i-\nu+1,1)\right], \end{aligned} \quad (9.61)$$

from Equations (9.54), (9.55), (9.57) and (9.58).

For $j = 1, 2, \ldots, \sigma$ we get from (9.56) and (9.59):

$$P_n(C, j) = 1(j < \sigma) \sum_{\nu=0}^{K} q_\nu P_{n-1}(C - \nu, j + 1) + r_j \sum_{\nu=1}^{K} q_\nu P_{n-1}(C - \nu + 1, 1). \tag{9.62}$$

These are the recurrence relations from which the probabilities $P_n(i, j)$ can be determined.

9.3.2. Computer implementation. To demonstrate how the recurrence relations derived above may be implemented on a computer, we decided to use the C programming language.

The following code example contains several comments and is therefore more or less self-explaining. The interested reader will have no difficulties in adapting this code fragment for his personal needs.

```
#include <stdio.h>
#include <malloc.h>
/* some defines, replace them by your favorite values...                 */
#define N        30       /* we want to calculate P_N(i,j)                */
#define K        10       /* maximum batch size                           */
#define SIGMA    10       /* maximum service time                         */
#define C       100       /* maximum capacity of waiting room             */
#define I0        0       /* these are the initial states                 */
#define J0        0
/* a useful macro:                                                        */
#define MIN(a,b) (((a)<(b))?(a):(b))
/* the arrival and service time probabilities are stored in vectors
   which are set up by these functions:                                   */
void arrival(float *x, int k);
void service(float *x, int k);
/* a simple error handler                                                 */
void Error(char *s);
int  main(void)
{
int  i, j, m, n;
float x;
/* the pointers p0 and p1 hold the vectors P(N-1) and P(N).
   swp is used for swapping, P_r and  P_p hold the probabilities r_j
   and p_j, P_q holds the probs. q_j .                                    */
    float *p0, *p1, *swp, *P_r, *P_p, *P_q;
/* First we allocate memory and check for memory exhaustion:              */
   if (NULL == (p0 = (float *) calloc(1 + C * SIGMA, sizeof(float))))
      Error("Out of Memory!\n");
   if (NULL == (p1 = (float *) calloc(1 + C * SIGMA, sizeof(float))))
      Error("Out of Memory!\n");
   if (NULL == (P_r = (float *) calloc(1 + SIGMA, sizeof(float))))
      Error("Out of Memory!\n");
   if (NULL == (P_p = (float *) calloc(1 + K, sizeof(float))))
```

```
      Error("Out of Memory!\n");
   if (NULL == (P_q = (float *) calloc(1 + K, sizeof(float))))
      Error("Out of Memory!\n");
   /* Now we set up the vectors P_r, P_p and P_q:                   */
   service(P_r, 1 + SIGMA);
   arrival(P_p, 1 + K);
   P_q[0] = 1.0;
   for (i = 1; i <= K; i++)
      for (j = i; j <= K; j++)
         P_q[i] += P_p[j];
/* Set up the inital state:
   the state (0,0) is element 0 of p0 and p1, the other states (i,j)
   will be located at index (i-1)*SIGMA+j of p0 and p1.             */
   if (I0 == 0 && J0 == 0)
      p0[0] = 1.0;
   else
      p0[(I0 - 1) * SIGMA + J0] = 1.0;
/* Here comes the recurrence:                                       */
   for (n = 1; n <= N; n++)
   {
      p1[0] = P_p[0] * (p0[0] + p0[1]);
      for (i = 1; i < C; i++)
         for (j = 1; j <= SIGMA; j++)
         {
            x = 0.0;
            for (m = 0; m <= MIN(i, K); m++)
               x += P_p[m] * p0[1 + SIGMA * (i - m)];
            if (i <= K)
               x += P_p[i] * p0[0];
            x *= P_r[j];
            if (j < SIGMA)
               for (m = 0; m <= MIN(i - 1, K); m++)
                  x += P_p[m] * p0[1 + j + SIGMA * (i - m - 1)];
            p1[j + SIGMA * (i - 1)] = x;
         }
      for (j = 1; j <= SIGMA; j++)
      {
         x = 0.0;
         for (m = 1; m <= K; m++)
            x += P_q[m] * p0[1 + SIGMA * (C - m)];
         x *= P_r[j];
         if (j < SIGMA)
            for (m = 0; m <= K; m++)
               x += P_q[m] * p0[1 + j + SIGMA * (C - m - 1)];
         p1[j + SIGMA * (i - 1)] = x;
      }
      /* Now swap pointers:                                         */
      swp = p0;
      p0 = p1;
```

```
        p1 = swp;
    }

     /* The probabilities P_n(i,j) are now ready in p0.
    Here follows optional code to calculate various statistics,
    report printing and graphics etc.
    ..........................................................     */
    return 0;
}
/* This function generates the arrival probabilities                */
void arrival(float *x, int k)
{
/* replace this code by your batch distribution                     */
    x[0] = 0.9;
    x[K] = 0.1;
}

/* This function generates the service time probabilities. In this
    example we have a discrete uniform distribution                 */
void service(float *x, int k)
{
/* replace this code by your service time distribution              */
int  i;
    for (i = 1; i < k; i++)
        x[i] = 1 / (float) SIGMA;
}

/* This is a simple error handler                                   */
     void Error(char *s)
{
    fprintf(stderr, "%s", s);
    exit(1);                    /* exit gracefully                  */
}
```

In the above code example we used single precision arithmetic because this results in shorter execution times. In the experiments we carried out we did not find any evidence of rounding errors. Therefore even if one uses single precision the program seems to be numerically stable, which is due to the fact that only additions and multiplications of probabilities are necessary.

An important issue is computational complexity measured by the number of floating point operations (flops) a program has to carry out. For the example program given above this complexity is roughly proportional to $KC\sigma n$.

The storage requirements are determined primarily by the pointers `p0` and `p1`, which use $8(C+1)\sigma$ bytes in single precision arithmetic.

Remarks:

1. Once the vector $P_n(i,j)$ has been found, several other interesting quantities may be computed very easily.

   For instance the moments of the queue length are simply given by

$$E(Q(n)^k) = \sum_{i=1}^{C} i^k \sum_{j=1}^{\sigma} P_n(i,j).$$

   Furthermore, since recursive computation of $P_n(i,j)$ yields as intermediate result all $P_k(i,j)$, $0 \leq k \leq n$, we may use these results to compute, e.g., the expected local time or sojourn time in a particular state. More precisely, let $\tau_\ell(n)$ denote the expected time the process $Q(n)$ spends in state ℓ, $0 \leq \ell \leq C$. Then

$$E(\tau_\ell(n)) = \sum_{i=0}^{n} \sum_{j=1}^{\sigma} P_i(\ell, j).$$

   Also the computation of the waiting time distribution is easy. Let $W(n)$ denote the waiting time of a customer arriving immediately after time n. Then

$$P(W(n) = 0) = P_n(0,0),$$

   and for $k > 0$:

$$\begin{aligned} &P(W(n) = k) \\ &\quad = P_n(1,k) + \sum_{i+j=k} P_n(2,i) r_j + \sum_{i+j=k} P_n(3,i) r_j^{(2)} \\ &\qquad + \cdots + \sum_{i+j=k} P_n(C,i) r_j^{(C-1)}, \end{aligned}$$

   where $r_j^{(m)}$ is the m-fold convolution of the service time distribution r_j. The above formula suggests recursive computation of $P(W(n) = k)$. For this purpose let

$$W_1(k) = P_n(C,k),$$

   and compute recursively

$$\begin{aligned} W_2(k) &= \sum_{i+j=k} W_1(k) r_j + P_n(C-1,k) \\ W_3(k) &= \sum_{i+j=k} W_2(k) r_j + P_n(C-2,k) \\ &\cdots \end{aligned}$$

   Then of course $P(W(n) = k) = W_C(k)$.

2. Another important issue is busy period analysis. Let T_{i_0,j_0} denote the duration of a busy period initiated by $i_0 > 0$ customers, the first customer having residual service time j_0. To compute $P(T_{i_0,j_0} \leq n)$ only a slight modification of the recurrence equations (9.60) – (9.62) is necessary. All we have to do is to make the state $(0,0)$ absorbing; therefore, we have to delete all possibilities for transitions out of state $(0,0)$, and this yields the system of difference equations:

$$P_n(0,0) = p_0 P_{n-1}(1,1). \tag{9.63}$$

For $i = 1, 2, \ldots, C-1$:

$$\begin{aligned} P_n(i,j) = 1(j < \sigma) & \sum_{\nu=0}^{\min(i-1,K)} p_\nu P_{n-1}(i-\nu, j+1) \\ & + r_j \sum_{\nu=0}^{\min(i,K)} p_\nu P_{n-1}(i-\nu+1, 1), \end{aligned} \tag{9.64}$$

and

$$\begin{aligned} P_n(C,j) = 1(j < \sigma) & \sum_{\nu=0}^{K} q_\nu P_{n-1}(C-\nu, j+1) \\ & r_j \sum_{\nu=1}^{K} q_\nu P_{n-1}(C-\nu+1, 1). \end{aligned} \tag{9.65}$$

The state probabilities $P_n(i,j)$, however, have now a different meaning, in particular

$$P_n(i,j) = P(Q(n) = i, R(n) = j, T_{i_0,j_0} > n | Q(0) = i_0, R(0) = j_0),$$

and

$$P_n(0,0) = P(T_{i_0,j_0} \leq n | Q(0) = i_0, R(0) = j_0).$$

Unfortunately, the computation of moments of T_{i_0,j_0} is considerably more difficult.

9.3.3. The $G/Geo/1$ model. To derive recurrence relations for the state probabilities in a discrete time $G/M/1$ system we will use a somewhat different approach than has been used for the $M/G/1$ system. There, by the method of supplementary variables, we have kept track of the residual service time of a customer in order to retain the Markov property. Here we will use as supplementary variable the time to the next arrival, which has the advantage of yielding simpler equations.

The basic assumptions for the discrete time $G/Geo/1$ model are

1. The interarrival times are i.i.d. integer valued random variables having values $1, 2, \ldots, \sigma$ with probabilities r_j. As in the discrete $Geo/G/1$ case we assume that these times are bounded from above.

2. The number of customers leaving the system during a particular slot has probability function $p_0, p_1, \ldots, p_K$, and also here we assume that the number of service completions in a slot is bounded.
3. The system has finite capacity $C > K$, excess customers are lost.
4. We assume that departures occur immediately after the beginning of a slot, and an arrival joins the system immediately before the end of a slot:

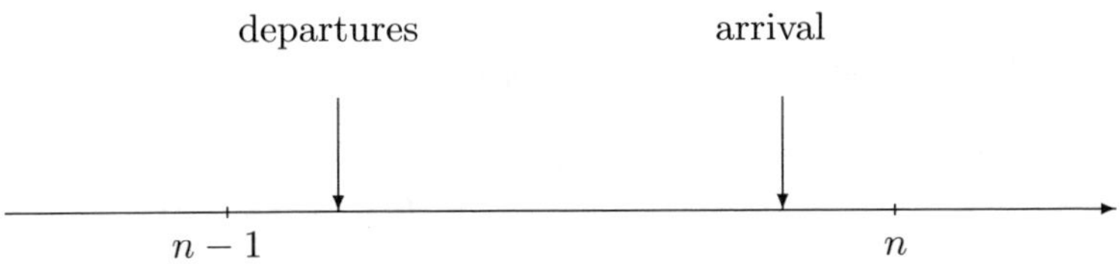

Let $Q(n)$ denote the number of customers in the system at time n and let $R(n)$ denote the residual time to the next arrival. Then the bivariate process $(Q(n), R(n))$ is easily seen to be a Markov chain with state space

$$\{0, 1, \ldots, C\} \times \{1, 2, \ldots, \sigma\} - (0, \sigma).$$

Initially there are $0 \leq i_0 \leq C$ customers waiting and the residual time to the next arrival is j_0, $1 \leq j_0 \leq \sigma$. If $i_0 = 0$, then we assume that $j_0 < \sigma$. Let $P_n(i, j) = P(Q(n) = i,\, R(n) = j | Q(0) = i_0,\, R(0) = j_0)$.

To derive recurrence relations for the state probabilities $P_n(i, j)$ we will argue as in the case of the $Geo/G/1$ system and consider all possible one-step transitions into a particular state.

The state $(0, j)$ can be entered only from state $(\nu, j + 1)$ with $\nu = 0, 1, \ldots, K$ and $j = 1, 2, \ldots, \sigma - 1$, if at least ν customers leave the system and the residual time to the next arrival is equal to $j + 1$. Hence for $j = 1, 2, \ldots, \sigma - 1$:

$$P_n(0, j) = \sum_{\nu=0}^{K} q_\nu P_{n-1}(\nu, j + 1), \tag{9.66}$$

where $q_\nu = \sum_{m=\nu}^{K} p_m$ is the probability that at least ν customers leave in a slot. Note that $P_n(0, \sigma) = 0$ by definition for all n.

The state $(1, j)$, $j = 1, 2, \ldots, \sigma$, can be entered from

1. State $(\nu, 1)$, if at least $\nu = 0, 1, \ldots, K$ customers leave, a new customer arrives necessarily and the time to the next arrival is equal to j,
2. State $(\nu + 1, j + 1)$, if $\nu = 0, 1, \ldots, K$ customers leave and $j < \sigma$.

Therefore for $j = 1, 2, \ldots, \sigma$:

$$P_n(1, j) = r_j \sum_{\nu=0}^{K} q_\nu P_{n-1}(\nu, 1) + 1(j < \sigma) \sum_{\nu=0}^{K} p_\nu P_{n-1}(\nu + 1, j + 1). \tag{9.67}$$

The state (i, j), $i = 2, 3, \ldots, C-1$, $j = 1, 2, \ldots, \sigma$, can be entered from

1. State $(\nu + i - 1, 1)$, $\nu = 0, 1, \ldots, \min(K, C - i + 1)$, if ν customers leave the system, a new customer arrives necessarily and the time to the next arrival is equal to j.
2. State $(\nu + i, j + 1)$, $\nu = 0, 1, \ldots, \min(K, C - i)$, if ν customers leave and $j < \sigma$.

Thus for $i = 2, 3, \ldots, C-1$ and $j = 1, 2, \ldots, \sigma$:

$$\begin{aligned} P_n(i,j) &= r_j \sum_{\nu=0}^{\min(K,C-i+1)} p_\nu P_{n-1}(\nu + i - 1, 1) \\ &+ 1(j < \sigma) \sum_{\nu=0}^{\min(K,C-i)} p_\nu P_{n-1}(\nu + i, j+1). \end{aligned} \tag{9.68}$$

Finally the state (C, j), $j = 1, 2, \ldots, \sigma$ may be entered from

1. State $(C-1, 1)$, if no customer leaves, a new one joins and the time to the next arrival equals j,
2. State $(C, j+1)$, if no customer leaves and $j < \sigma$,
3. State $(C, 1)$; here we have two possibilities: if no customer leaves (with probability p_0), the new arriving customer finds the waiting room full. In this case the customer is lost and the time to the next arrival is equal to j with probability r_j. Otherwise, if exactly one customer leaves, then the next arriving customer can join the system and the time to the next arrival is equal to j.

Hence for $j = 1, 2, \ldots, \sigma$:

$$\begin{aligned} P_n(C,j) = &\, p_0 r_j P_{n-1}(C-1, 1) + (p_0 + p_1) r_j P_{n-1}(C, 1) \\ &+ 1(j < \sigma) p_0 P_{n-1}(C, j+1). \end{aligned} \tag{9.69}$$

Similar to the previous case, a computer implementation program can be written to calculate the probabilities $P_n(i, j)$. Using its output, other queueing characteristics can be evaluated as done earlier. In addition to those, we will be able to obtain the waiting time distribution for this model.

Let $W(n)$ denote the waiting time of a customer arriving in $(n, n+1]$ and let $P_n(i)$ denote the probability that there are i customers present at time n:

$$P_n(i) = \sum_{j=1}^{\sigma} P_n(i, j).$$

Furthermore define $a_m(k)$, the probability that the time to serve k customers does not exceed $m, m \geq 1$. Then

$$P(W(n) = 0) = P_n(0),$$

and

$$P(W(n) \leq m) = P_n(1) a_m(1) + P_n(2) a_m(2) + \ldots + P_n(C) a_m(C).$$

The probabilities $a_m(k)$ can be computed recursively. To obtain the recurrence we may argue as follows:

Of course, $a_m(0) = 1$ for all $m \geq 1$. Now consider $a_m(1)$, the probability that the service time of one customer will not exceed m. We have

$$a_m(1) = q_1 a_{m-1}(0) + p_0 a_{m-1}(1),$$

because either this customer has been served in $(0, m-1]$, then there cannot be any service in $(m-1, m]$ which happens with probability p_0, or there has been no service in $(0, m-1]$ and in $(m-1, m]$ the customer leaves. Continuing this argument we find that

$$a_m(i) = q_i + \sum_{\nu=1}^{i} p_{i-\nu} a_{m-1}(\nu), \qquad i = 1, 2, \ldots, K \tag{9.70}$$

and

$$a_m(i) = \sum_{\nu=0}^{K} p_{k-\nu} a_{m-1}(\nu + 1). \tag{9.71}$$

Remarks:

1. Of course, any method in numerical analysis has its limitations and so is the case for the recursive techniques presented in this section. The most serious point is that of dimensionality. Using large values of C and/or of σ increases the computational complexity considerably. Furthermore there is no simple generalization to systems with several parallel servers (but see the exercises below). In the $Geo/G/c$ case one has to keep track of the residual service times at any of the servers, thus for each additional server we have one additional variable in our difference equations. In such cases one has to resort to alternative methods like simulation, which was dealt with in Section 5.3 in Chapter 5.

2. A serious problem is the discretization error. Generally it is very difficult to analyze and it is even more difficult to get information about the propagation of this error in the process of recurrence. For a general discussion of this point see Collatz (1966).

3. To achieve satisfactory accuracy, it is often necessary to use extremely fine grids when discretizing the time variable. As a result the computational complexity (measured by the number of floating point operations) will be increased considerably. This is especially true if we are faced with a higher dimensional problem in several state variables, as it is the case for systems with several parallel servers.

9.4. Algebraic Methods

Let us revisit Section 4.4.1 in Chapter 4, the eigenvalue method which is an algebraic method (see Section 4.5). A parallel approach can be adapted for the three-step random walk of discrete-time birth-death model with state space $\{0, 1, \ldots, N\}$ in which

$$P(X_i = 1) = \alpha_i, \;\; P(X_i = -1) = \gamma_i \text{ and } P(X_i = 0) = 1 - \alpha_i - \gamma_i = \beta_i$$

(in Section 9.2.1, $\alpha_i = \alpha$ and $\gamma_i = \gamma$), $\gamma_0 = 0$ and $\alpha_N = 0$. Denote by $P_n(k)$ the probability $P(Q_n = k | Q_0 = m)$. Then $P_n(k)$ satisfies the following equations:

$$\begin{aligned} P_{n+1}(0) &= (1-\alpha_0)P_n(0) + \gamma_1 P_n(1), \\ P_{n+1}(k) &= \alpha_{k-1}P_n(k-1) + \beta_k P_n(k) + \gamma_{k+1}P_n(k+1), \; k = 1, \ldots, N-1, \end{aligned}$$

and

$$P_{n+1}(N) = \alpha_{N-1}P_n(N-1) + (1-\gamma_N)P_n(N), \tag{9.72}$$

with $P_0(k) = \delta_{mk}$, $0 \le k \le N$. Let $G_k(z)$ be the generating function of $P_n(k)$, i.e.,

$$G_k(z) = \sum_{n=0}^{\infty} P_n(k) z^n, \quad |z| < 1, \; k = 0, 1, \ldots, N.$$

Then taking the generating function of (9.72), we have the following matrix equation:

$$\begin{bmatrix} 1-(1-\alpha_0)z & -\gamma_1 z & & & & \\ -\alpha_0 z & 1-\beta_1 z & -\gamma_2 z & & & \\ & -\alpha_1 z & 1-\beta_2 z & -\gamma_3 z & & \\ & & \ddots & & & \\ & & & -\alpha_{N-2}z & 1-\beta_{N-1}z & -\gamma_N z \\ & & & & -\alpha_{N-1}z & 1-(1-\gamma_N)z \end{bmatrix}$$

$$\times \begin{bmatrix} G_0(z) \\ G_1(z) \\ \vdots \\ \\ G_N(z) \end{bmatrix} = \begin{bmatrix} \delta_{m0} \\ \delta_{m1} \\ \vdots \\ \\ \delta_{mN} \end{bmatrix}. \tag{9.73}$$

Denote by $A_N(z)$ the determinant of the coefficient matrix on the left side of (9.73). By applying Cramer's rule, we can get a solution of $G_k(z)$, the denominator of which is $A_N(z)$.

For a partial fraction expansion of $G_k(z)$, we may examine zeroes of $A_N(z)$. From (9.73), it is evident that $z = 0$ is not a zero but $z = 1$ is. Divide each row of $A_N(z)$ by $-z$ to get

$$\begin{vmatrix} 1-\alpha_0-\frac{1}{z} & \gamma_1 & & & \\ \alpha_0 & \beta_1-\frac{1}{z} & \gamma_2 & & \\ & & \ddots & & \\ & & \alpha_{N-1} & \beta_{N-1}-\frac{1}{z} & \gamma_N \\ & & & \alpha_{N-1} & 1-\gamma_N-\frac{1}{z} \end{vmatrix}$$

which can be checked to be the same as

$$\begin{vmatrix} 1-\alpha_0-\frac{1}{z} & \sqrt{\alpha_0\gamma_1} & & & \\ \sqrt{\alpha_0\gamma_1} & \beta_1-\frac{1}{z} & \sqrt{\alpha_1\gamma_2} & & \\ & & \ddots & & \\ & & \sqrt{\alpha_{N-2}\gamma_{N-1}} & \beta_{N-1}-\frac{1}{z} & \sqrt{\alpha_{N-1}\gamma_N} \\ & & & \sqrt{\alpha_{N-1}\gamma_N} & 1-\gamma_N-\frac{1}{z} \end{vmatrix}. \tag{9.74}$$

Clearly z is a zero of $A_N(z)$ if and only if $\frac{1}{z}$ is an eigenvalue of

$$\mathbf{C}_N = \begin{bmatrix} 1-\alpha_0 & \sqrt{\alpha_0\gamma_1} & & & \\ \sqrt{\alpha_0\gamma_1} & \beta_1 & \sqrt{\alpha_1\gamma_2} & & \\ & & \ddots & & \\ & & \sqrt{\alpha_{N-2}\gamma_{N-1}} & \beta_{N-1} & \sqrt{\alpha_{N-1}\gamma_N} \\ & & & \sqrt{\alpha_{N-1}\gamma_N} & 1-\gamma_N \end{bmatrix} \tag{9.75}$$

(Compare (9.75) with (4.38) for similarity.)

$\mathbf{C}_N$ is a positive definite real symmetric tridiagonal matrix with non-zero sub-diagonal elements. Therefore it follows that (see Section A.10 in Appendix A) all eigenvalues of $\mathbf{C}_N$ are real, distinct and non-negative. Hence all zeroes of $A_N(z)$ are real, distinct and positive since $z = 0$ is not a zero. Let the zeroes be $z_0, z_1, \ldots, z_N$ with $z_0 = 1$. Then $G_k(z)$ can be written as

$$\begin{aligned} G_k(z) &= \frac{U_k(z)}{\prod_{j=0}^{N}(1-\frac{z}{z_j})} \\ &= \sum_{j=0}^{N} \frac{v_{j,k}}{(1-\frac{z}{z_j})}, \quad k = 0, 1, \ldots, N, \end{aligned} \tag{9.76}$$

by partial fraction expansion, where

$$v_{j,k} = \frac{U_k(z_j)}{\prod_{\substack{r=0 \\ r\neq j}}^{N}(1-\frac{z_j}{z_r})} \quad j = 0, 1, \ldots, N. \tag{9.77}$$

The power series expansion of (9.76) leads to

$$P_n(k) = P(Q_n = k|Q_0 = m) = v_{0,k} + \sum_{j=1}^{N} v_{j,k} z_j^{-n}, \quad k = 0, 1, \ldots, N. \tag{9.78}$$

As $n \to \infty$, the steady-state distribution is given by

$$\lim_{n\to\infty} P_n(k) = v_{0,k}, \qquad k = 0, 1, \ldots, N. \tag{9.79}$$

Remarks:

1. The eigenvalues of $\mathbf{C}_N$ or roots of $A_N(z) = 0$ may be numerically evaluated by using MATLAB or QROOT and thus we can obtain the distributions numerically.

2. The transient solution for large N may be used as an approximation for infinite state space.

Now what about the continued fraction method in Section 4.4.2 in Chapter 4? Consider the infinite number of equations

$$P_{n+1}(0) = (1-\alpha_0)P_n(0) + \gamma_1 P_n(1)$$

and

$$P_{n+1}(k) = \alpha_{k-1}P_n(k-1) + \beta_k P_n(k) + \gamma_{n+1}P_n(k+1), \ k = 1, 2, \dots,$$

with

$$P_0(k) = \delta_{0k}, \qquad k = 0, 1, \dots . \tag{9.80}$$

The last condition implies that initially the system is at state 0. Similar to Section 4.4.2 in Chapter 4 let us set

$$f_k = (-1)^k M_k G_k(z)$$

where

$$f_0 = G_0(z), \ M_k = \prod_{j=1}^{k}(\gamma_j z)$$

and $G_k(z)$ is the p.g.f. of $P_n(k)$. Then we find that

$$f_1 = 1 - (1-(1-\alpha_0)z)f_0.$$

and

$$f_{k+1} = -(1-\beta_k z)f_k - \alpha_{k-1}\gamma_k z^2 f_{k-1} \tag{9.81}$$

which has the general form

$$f_1 = a_1 - b_1 f_0,$$

and

$$f_{k+1} = a_{k+1}f_{k-1} - b_{k+1}f_k.$$

This is exactly the same as (4.50) except that in this case

$$a_1 = 1, \ b_1 = 1-(1-\alpha_0)z$$

and

$$a_{k+1} = -\alpha_{k-1}\gamma_k z^2, \ b_{k+1} = 1 - \beta_k z, \qquad k \geq 1. \tag{9.82}$$

Because of the exact structural setup as in Section 4.4.2 in Chapter 4, we should have (4.61) as

$$G_r(z) = \frac{(-1)^r}{M_r} f_r,$$

f_r being given by (4.60) with appropriate a's and b's from (9.82). Note that $M_r = z^r \prod_{j=1}^{r} \gamma_j$ and the factor α_{r+1} in the numerator of f_r has the expression

$$\alpha_{r+1} = \prod_{j=1}^{r+1} a_j = (-1)^{r+1} z^{2(r+1)} \prod_{j=2}^{r+1} \alpha_{j-2}\nu_j$$

and thus z^r from M_r in the denominator gets cancelled with those in α_{r+1} in the numerator. Therefore, the sth convergent $G_{k,s}(z)$ of $G_k(z)$ may be written as (see (4.62))

$$G_{k,s}(z) = \frac{(-1)^k A_s^{(r)}/M_k}{B_{s+k}}, \tag{9.83}$$

in which B_{s+k} is exclusively in the denominator. The reader is reminded that without any confusion we are using the same notations here. In the present case, it can be checked that

$$B_{s+k} = \begin{vmatrix} 1-(1-\alpha_0)z & 1 & & & \\ \alpha_0\gamma_1 z^2 & 1-\beta_1 z & 1 & & \\ & & \ddots & & \\ & & & 1-\beta_{s+k-2}z & 1 \\ & & & \alpha_{s+k-2}\gamma_{s+k-1}z^2 & 1-\beta_{s+k-1}z \end{vmatrix}.$$

Observe that $z=0$ is not a zero of B_{s+k}. After dividing each row by $-z$, we can show that the resulting determinant can be expressed as

$$\begin{vmatrix} 1-\alpha_0-\frac{1}{z} & \sqrt{\alpha_0\gamma_1} & & & \\ \sqrt{\alpha_0\gamma_1} & \beta_1-\frac{1}{z} & \sqrt{\alpha_1\gamma_2} & & \\ & & \ddots & & \\ & & & \beta_{s+k-2}-\frac{1}{z} & \sqrt{\alpha_{s+k-2}\gamma_{x+k-1}} \\ & & & \sqrt{\alpha_{s+k-2}\gamma_{x+k-1}} & \beta_{s+k-1}-\frac{1}{z} \end{vmatrix} \tag{9.84}$$

which is similar to (4.63). Following the argument used to arrive at (9.76), we should be able to write $G_{k,s}(z)$ as

$$G_{k,s}(z) = \sum_{j=1}^{s+k} \frac{H_{k,j}}{(1-z/z_j)}, \qquad k=0,1,\ldots, \tag{9.85}$$

where $z_1,\ldots,z_{s+k}$ are zeroes of B_{s+k} and $H_{k,j}$ is determined by the usual partial fraction expansion. Finally, an expansion of (9.85) in powers of z gives an approximation of $P_n(k)$ as

$$\sum_{j=1}^{s+k} H_{k,j} z_j^{-n}, \qquad k=0,1,\ldots, \tag{9.86}$$

which is obtained by the continued fraction method. It is of interest to derive an approximation when the system is initially at state m.

9.5. Discussion

One can easily be convinced that the study of transient behaviour shows more complications than its counterpart, namely, the study of steady-state behaviour. The steady-state behaviour suggests the process has to reach a well-behaved stage or we can say it is 'tamed.' In contrast, the transient nature of the process remains 'wild' until it can reach its well-behaved stage and thus contributes to complications. Therefore, its study deserves enough attention. Most of Chapter 6 and this chapter exclusively are devoted to that topic.

The transient nature in a discrete-time process is nothing but taking into account its movement in a fixed number of steps. If we restrict the movement only to certain types of steps, say an arrival, a service completion and none of this to happen in a unit time, then these types of movement can be represented as steps in a lattice path. Instead of rushing to use the method of transforms, one may wish to examine the nature of these paths in the context of a given model, in order to find out the existence of any pattern or structure which would facilitate in forming a solution. This may be termed as the bottom-to-top approach, as has been described in Section 6.5 in reference to Section 6.3.

The methods presented in Section 9.2 are particularly appealing because of their conceptual simplicity, which is highly due to the lattice path representation of Markovian queueing processes. Once such a representation has been found, many problems become more transparent and it is easy to see the structure of the processes involved. Exploiting these structures essentially means to look out for decompositions of paths in terms of certain classes of lattice paths which can be enumerated easily.

A natural question is, how far can one go using these methods?

One possible generalization concerns the slot mechanism. We have definitely excluded the possibility of more than one event in a slot. This assumption can be relaxed easily: suppose that now the occurrence of an arrival and a departure in the same slot is allowed, with the convention that in a slot arrivals are recorded before departures. In this new setting the one-step transition probabilities of the process Q_n change; however, these changes can be easily taken care of.

Control policies in queues are introduced in Section 2.9.2 in Chapter 2. Lattice path counting techniques are also capable of being utilized in discrete-time $(0, K)$-control policy and server vacations (see Böhm and Mohanty (1993, 1994b) and Kanwar Sen, Jain and Gupta (1993)). Here the idea is to split the lattice paths representing the sample paths of the queueing process into segments which correspond to busy and idle periods, where now idleness also includes the possibility of the server being on vacation. More on control policies appears in Section 10.4 in Chapter 10.

Queues involving batches can be dealt with in a far more general setting than we have discussed here (see Böhm and Mohanty (1994a)). For instance we may allow for batches of fixed size in both, arrivals and departures. Again counting results are available, although generally in the form of determinants (see Mohanty (1972) and consider the discrete-time analogue). It is also possible to consider batches of random size. For instance, if customers arrive in batches of random size and are served one after the other, the lattice paths representing the sample paths of this queueing process now have step set $(1, 0)$ for departures and $(0, k), k \geq 1$ for arrivals. Such problems may be treated combinatorially by either a discrete analogue of the Wiener-Hopf factorization technique (see Grassmann and Jain (1989)) or by the generalized ballot theorem of Takács (1967).

The Markovian structure is crucial in bringing the paths into picture. In analyzing the $Geo^R/Geo/1$ model, we realize that it is essentially the discretized version of the $M/E^R/1$ model. Here the Markovian property is achieved when each phase in the Erlangian distribution is considered as a state. Let us consider another situation in which the Markovian property is achieved. Suppose the distribution is Coxian with k independent exponential phases and service rate μ_j at phase j, $j = 1, \ldots, k$. This means that after completing service in phase j, the customer either enters phase $j+1$ with probability α_j or completes service with probability $1 - \alpha_j$. If we set $\alpha_0 = 1$, $\alpha_k = 0$ and let $A_i = \alpha_0 \alpha_1 \cdots \alpha_{i-1}$ be the probability that the customer enters state i, then the Coxian probability density $f(t)$ has Laplace transform

$$f^*(\theta) = \sum_{i=1}^{k} A_i(1-\alpha_i) \prod_{j=1}^{i} \frac{\mu_j}{\mu_j + \theta}.$$

Its expectation is given by $\sum_{j=1}^{k} A_j/\mu_j$. The importance of *Coxian distributions* is due to their universality, any distribution function can be approximated arbitrarily closely by a Coxian distribution by choosing suitable parameter values. Thus such a model may be thought of to replace the distribution G in a queueing context. Note that if $\alpha_k = 1$ for every j, the distribution is Erlangian with k phases. Again, if we take phases as states, the Markovian property is established. In discrete time, one can represent the process as a lattice path. For $k = 2$, the transient solution has been analysed by the lattice path approach in Kanwar Sen and Agarwal (1997). Both authors have considered other models with Coxian distribution. Their solution involves summation over several sets of variables which could be computationally challenging.

At this point, we may reexamine (9.4) and (9.6). In (9.4), note that the first term is

$$\binom{n_1+n_2}{n_1}\binom{n_1+n_2+n_3}{n_3} = \frac{(n_1+n_2+n_3)!}{n_1!n_2!n_3!} = \binom{n_1+n_2+n_3}{n_1, n_2, n_3},$$

which is the unusual trinomial coefficient. It gives the number of lattice paths from (0,0) to (n_1+n_3, n_2+n_3) such that there are n_1 horizontal steps, n_2 vertical steps and n_3 diagonal steps. Similarly the second term in (9.4) is also a trinomial coefficient. When the first term is substituted in (9.6), a sum like

$$\sum_{\substack{a,b,c\geq 0\\ a+b+c=n\\ a-b=k}} \binom{n}{a,b,c} \alpha^a \gamma^b \beta^c \tag{9.87}$$

is produced. But this happens to be a constrained sum of weighted lattice paths from (0,0) to $(a+c, b+c)$ having a, b and c horizontal, vertical and diagonal steps, respectively, such that each horizontal step is of weight α, vertical step of γ and diagonal step of β. Thus, the sum (9.87) is called a generalized trinomial coefficient in literature, for example see Böhm (1993), Böhm and Mohanty (1993, 1994a), and Mohanty and Panny (1990a,b). Letting $w_k(n) = P(S_n = k|S_0 = 0)$, we can easily see that $w_k(n)$ is nothing but this sum. Again for a good reason, the sum is also known as the Green function of the basic random walk $\{S_n\}$ (see Böhm (1993, p. 6)). This coefficient or function has many interesting properties; for instance, it can be expressed as a hyper-geometric function and can have various integral representations which are sometimes helpful in numerical computation (see Section 4.4.5 in Chapter 4 and Section 8.4 in Chapter 8). Such properties may be used to derive new results of queues. Although we have restricted ourselves only to lattice path approach for simplicity, an inquisitive reader may like to explore the implication of using the generalized trinomial coefficient in queues, from most of the references in Section 9.2.

In Sections 6.3 in Chapter 6 and 9.2, the application of lattice path combinatorics is conspicuously similar which leads us to think a link between the two exists somehow. Let us examine the first part of (6.36) and (9.48) which are arrived to apparently from two different points of view but lead to the same solution. Both represent the contribution from the zero-avoiding probabilities. Noting the observation in parentheses just after (6.43), we may say that the starting point in Chapter 6 is the randomization method, although our presentation does not begin that way. The randomization formula (6.42) is established by an algebraic method which exploits the Markovian nature of the queueing process (see Section 4.4.4 in Chapter 4). In the formula, the right side has the r-step transition probability in a random walk, the computation of which involves lattice path combinatorics. In contrast, here in this chapter we start with a discrete-time queueing model leading to a random walk which allows stay in any state. Then, in order to compute its transient solution which is the same as the r-step transition probability, we employ lattice path combinatorics. By a limiting process, we arrive at the continuous time solution. Although, randomization and discretization are equivalent from a purely formal point of view, they are conceptually quite different (see Böhm (1993, Section 1.6)). We refer to Böhm et al. (1997) for obtaining the transient solution of discrete-time birth-death processes by a combinatorial approach.

Looking back at the initial Kolmogorov differential equations, we remember that in Chapter 2 these have been obtained by establishing recurrence relations from the flow diagram. Because the time is continuous, we are unable to generate solutions in a recursive manner and rather become dependent on some analytic approach for solution. Sometimes, such a solution may not be easy to derive. Alternatively, the time can be discretized and the usual recurrence relations be written down. Using them, behaviour of queueing characteristics can be studied recursively. This has been the topic of Section 9.3, where the recurrence relation approach has been discussed in $Geo/G/1$ and $G/Geo/1$ models. For pedagogical reasons, we have included a computer program. However, improved programs are available. Evidently, there are problems of error accumulation and convergence as time increases. Nevertheless, it is one method which is elementary and can be easily implemented.

Use of recurrence relations is a numerical technique to study transient behaviour. Immediately we are reminded of a few more in Section 4.4 in Chapter 4 which are algebraic in nature. It is but natural to look for their applicability to the discrete-time models. We have found in Section 9.4 that extensions of eigenvalue and continued fraction methods can be developed in the present scenario. Furthermore, observe that deriving eigenvalues is related to root finding and therefore MATLAB or QROOT (see Chapter 4) may be effectively utilized for numerical computation. In fact, for a few applications of QROOT in discrete-time queues, one may refer to Chaudhry and Zhao (1994). Finally, we may ask whether there is any discrete-time analogue of power series method or of randomization method (see Sections 4.4.3 and 4.4.4 in Chapter 4) and the obvious answer is no.

The material in Section 9.3 is based on Dafermos and Neuts (1971). Heimann and Neuts (1973), Klimko and Neuts (1973a,b) and Neuts (1973) and those in Section 9.4 are based on Chaudhry and Zhao (1994) and Mohanty (1991). The references Böhm, Jain and Mohanty (1993), Böhm and Mohanty (1994c), Kanwar Sen and Jain (1993) and Mohanty (1979) are on Section 9.2 and related use in queues.

9.6. Exercises

1. For the $Geo/Geo/1$ system in discrete time show that

$$E(D_{T_m}) = \frac{m\gamma}{\gamma - \alpha}, \quad \text{and} \quad \text{var}(D_{T_m}) = m\frac{\alpha\gamma(\alpha+\gamma)}{(\gamma-\alpha)^3}.$$

2. Consider a $Geo/Geo^R/1$ system in discrete time in which customers are served in batches of fixed size $R \geq 1$. If at any time the number of customers in the system is less than R, the server is temporarily switched off and resumes service again only if there are R customers in the system again. Derive the distribution of the busy period as well as the transient distribution of the number of customers in the system using path combinatorics.

3. Find the general transient distribution of the model discussed in Section 9.2.4. Hint: determine first the number of lattice paths having a prescribed number of reflection points on the plane E_m. Then insert steps with probability $1 - \sum_{i=1}^{r} \alpha_i$ at those points.

4. By using (9.9) and (9.48), derive $P_n(t)$ as given in (2.50).

5. Determine the limits for the batch arrival model of Section 9.2.2. In particular show that the limit of (9.17) gives rise to a generalization of the modified Bessel functions. What are the properties of these functions?

6. Find the limit of (9.27). Hint: first keep n_1, n_2, n_3 and r fixed and consider the limit of the partial sum

$$\sum_{n_4+n_5+n_6=n-n_1-n_2-2n_3-r} \binom{n_1+n_2+n_4}{n_4}\binom{r+n_3+n_5-1}{n_5} \times \binom{n_3+n_6-1}{n_6}\beta_2^{n_4}\beta_1^{n_5}\beta_0^{n_6}.$$

7. Let $M_n = \max_{0\le\nu\le n} Q(n)$. How do we have to modify the recurrence relations for $G/Geo/1$ systems to get the joint distributions of $Q(n), M_n$ and $R(n)$?

8. Consider a $M/G/1$ system in continuous time with service time distribution $B(t)$, which is absolutely continuous and has a bounded derivative for all $t \geq 0$. Then $W(t)$, the waiting time of a customer arriving at time t, satisfies Takács's integro-differential equation

$$\frac{\partial F}{\partial t} - \frac{\partial F}{\partial x} = -\lambda F(x,t) + \lambda \int_0^x F(x-\nu,t)\,\mathrm{d}B(\nu),$$

where $F(x,t) = F(x_0;x,t) = P(W(t) \leq x|W(0) = x_0)$, λ is the arrival rate and $t > 0, x, x_0 \geq 0$. Find an analogue of this equation for the $Geo/G/1$ system.

9. Determine the recurrence relations for a $Geo/G/2$ system by taking as supplementary variables $R_n^{(1)}$ and $R_n^{(2)}$, the residual service times at counters 1 and 2.

10. Write a computer program to evaluate the waiting time distributions for the $Geo/G/1$ system.

11. Determine the recurrence for the $G/Geo/1$ system in discrete time by taking as supplementary variable the time since the last arrival. Compare this with the recurrence (9.66)–(9.69).

12. Consider a $G/G/1$ system in discrete time. Choose as supplementary variables the residual service time $R(n)$ and the time to the next arrival $A(n)$. Find the recurrence for the joint distribution of $Q(n), R(n)$ and $A(n)$.

13. Consider a $Geo/Geo/1$ model under $(0, K)$-control policy: the server waits until there are exactly K customers in the system. Upon the arrival of the K-th customer, service starts immediately and it continues until the system becomes empty. The server now enters the idle state and waits until the arrival of K new customers. Then service starts again and the process continues in the described way. Let $\zeta_n = 1$ if the server is busy at time n and zero otherwise. Determine $P(Q_n = k, \zeta_n = 1|Q_0 = m, \zeta_0 = 1)$.

Hint. Let $i \geq 0$ be the number of completed busy periods. Cut the sample paths into segments corresponding to idle and busy periods and join the segments of the idle and busy periods to two new paths. The lattice path (without diagonal steps) representing the transitions during busy periods has the following properties:

- It starts at the origin and terminates in the point $(n_1 - Ki, n_2)$.
- It does not touch or cross the line $y = x + m + Ki$, but if $i > 0$ it has to reach the line $y = x + m + K(i-1)$.

The number of such paths can be determined using (9.3). Then insert diagonal steps using the balls into boxes technique as suggested before (9.4).

14. *Continued.* Find $P(Q_n = k, \zeta_n = 0 | Q_0 = m, \zeta_0 = 0)$.

15. *Continued.* Let T_{00} denote the length of a complete cycle, i.e., an idle period followed by a busy period. Show that

$$E(T_{00}) = \frac{K}{\gamma - \alpha} + \frac{K}{\alpha},$$

where α and γ denote the probability of an arrival and a service completion.

16. *Continued.* Prove that for $\rho < 1$, $\rho = \alpha/\gamma$ a steady state exists and

$$\lim_{n\to\infty} P(Q_n = k, \zeta_n = 1 | Q_0 = m, \zeta_0 = 1)$$
$$= \begin{cases} \frac{1}{K}\rho^{k-K+1}(1-\rho^K) & k \geq K \\ \frac{1}{K}\rho(1-\rho)^k & 0 < k < K. \end{cases}$$

Hint. Use $E(T_{00})$ and the renewal theorem.

17. Prove that in the steady state the expected number of customers in the system is given by

$$E(Q) = \frac{K-1}{2} + \frac{\rho}{1-\rho}.$$

18. Using the $(N+1)$-state discrete-time birth death model as in Section 9.4, show that the length of a busy period is a mixture N geometric variables. (See Mohanty (1991).)

19. Consider a discrete-time queueing model with the following:
    a. The interarrival times are i.i.d. geometric random variables with λ_n ($n = 0, 1, , \ldots ; \lambda_N = 0$) being the probability that an arrival occurs in a time slot when there are n customers in the system ($\lambda_N = 0$ implies the finite system capacity).
    b. The service-times independent of interarrival times are i.i.d. geometric variables with μ_n ($n = 0, 1, \ldots ; \mu_0 = 0$) being the probability of service completion in a slot when there are n customers in the system.

c. Initially there are i customers. Find the distribution of the first-passage time to state N (i.e., the time needed for the system to reach state N for the first time). (See Chaudhry and Zhao (1994).)

References

Böhm, W. (1993). *Markovian Queueing Systems in discrete time*, Hain, Frankfurt/Main.

Böhm, W., Jain, J. L. and Mohanty, S. G. (1993). On zero avoiding transition probabilities of an r-node tandem queue. A combinatorial approach, *J. Appl. Prob.,* **30**, 737–741.

Böhm, W. and Mohanty, S. G. (1993). The transient solution of M/M/1 queues under $(M, N)-$policy. A combinatorial approach, *Journal of Statistical Planning and Inference*, **34**, 23–33.

Böhm, W. and Mohanty, S. G. (1994a). On discrete-time Markovian N-policy queues involving batches, *Sankhyā, Series A*, **56**, 144–163.

Böhm, W. and Mohanty, S. G. (1994b). Transient analysis of M/M/1 queues in discrete-time with general server vacations, *J. Appl. Prob.*, **31A**, 115–130,

Böhm, W. and Mohanty, S. G. (1994c). Transient analysis of queues with heterogeneous arrivals, *Queueing Systems*, **18**, 27–45.

Böhm, W., Krinik, A. and Mohanty, S. G. (1997). The combinatorics of birth-death processes and applications to queues, *Queueing Systems*, **26**, 255–267.

Chaudhry, M. L. and Zhao, Y. Q. (1994). First-passage-time and busy-period distributions of discrete-time Markovian queues: $Geom(n)/Geom(n)/1/N$, *Queueing Systems*, **18**, 5–26.

Collatz, L. (1966). *The Numerical Treatment of Differential Equations*, 2nd ed., Springer, Berlin, p. 268.

Dafermos, S. C. and Neuts, M. F. (1971). A single server queue in discrete-time, *Cahiers du Centre de Recherche Operationelle*, **13**, 23–40.

Grassmann, W. K. and Jain, J. L. (1989). Numerical solutions of the waiting time distribution and idle time distribution of the arithmetic $GI/G/1$, *Oper. Res.*, **37**, 141–150.

Heimann, D. and Neuts, M. F. (1973). The single server queue in discrete-time-numerical analysis IV, *Naval Research Log. Quart.*, **20**, 753–766.

Kanwar Sen and Agarwal, M. (1997). Transient busy period analysis of intially non-empty $M/G/1$ queues – lattice path approach, in *Advances in Combinatorial Methods and Applications to Probability and Statistics*, N. Balakrishnan (Ed.), Birkhäuser, Boston, 301–315.

Kanwar Sen, Jain, J. L. and Gupta, J. M. (1993). Lattice path approach to transient solution of $M/M/1$ with $(0, K)$ control policy, *Journal of Statistical Planning and Inference*, **34**, 259–268.

Kanwar Sen and Jain, J. L. (1993). Combinatorial approach to Markovian queues, *Journal of Statistical Planning and Inference*, **34**, 269–280.

Klimko, J. and Neuts, M. F. (1973a). The single server queue in discrete-time – numerical analysis II, *Naval Research Log. Quart.*, **20**, 305–319.

Klimko, J. and Neuts, M. F. (1973b). The single server queue in discrete-time – numerical analysis III, *Naval Research Log. Quart.*, **20**, 557–567.

Mohanty, S. G. (1972). On queues involving batches, *J. Appl. Prob*, **9**, 430–435.

Mohanty, S. G. (1979). *Lattice Path Counting and Applications*, Academic Press, New York.

Mohanty, S. G. (1991). On the transient behaviour of a finite discrete-time birth-death process, *Assam Statistical Review*, **5**, 1–7.

Mohanty, S. G. and Panny, W. (1990a). A discrete-time analogue of the $M/M/1$ queue and the transient solution: A geometric approach, *Sankhyā, Series A*, **52**, 364–370.

Mohanty, S. G. and Panny, W. (1990b). A discrete-time analogue of the $M/M/1$ queue and the transient solution, in *Proceedings of the 3rd. Hungarian Colloquium on Limit Theorems in Probability and Statistics*, P. Révész (Ed.), North-Holland, Amsterdam.

Neuts, M. F. (1973). The single server queue in discrete-time – numerical analysis I, *Naval Research Log. Quart.*, **20**, 297–304.

Takács, L. (1967). *Combinatorial Methods in the Theory of Stochastic Processes*, John Wiley & Sons, New York.

CHAPTER 10

Miscellaneous Topics

10.1. Introduction

Whereas each earlier chapter has a connected theme, the present chapter deals with topics of interest but cannot be a chapter by itself. Some topics such as design and control of queues and networks have been introduced elsewhere in Part I, mainly to expose these topics at a conceptual level to the readers. Their further developments are presented here for an advanced reader. Thus due to lack of continuity among topics, each section may be treated independently. Even one may sometimes find a fresh introduction of notations and definitions for the sake of completeness and better understanding.

10.2. Priority Queues

So far we have assumed that customers are taken for service on a first-come first-served (FCFS) basis in a queueing system. But in actual practice a queueing system may not behave like this and customers may also be served on a priority basis. A high priority customer enters for service before others, e.g., in a communication system urgent messages are sent first, or an emergency case gets higher priority in hospitals.

In this section we will examine the effects of various priority rules on the waiting time of customers from different classes, and in particular the effect of priorities on the average delay.

Customers with different priorities forming classes, say $1, \ldots, m$ arrive as input at a service station with the same or different arrival distributions, wait in their respective queues to be served on an FCFS basis within each priority class. Customers of class i ($i = 1, \ldots, m$) will receive service prior to those of class $i+1, \ldots, m$. It is standard practice to use a notation which attributes "higher" priority to a class with lower index i.

If at the time of arrival of a higher priority customer, a lower priority customer is getting service, then there are two possibilities. In the case of *non-preemptive priority*, (head-of-the-line) discipline, a unit at the service counter will be allowed to complete his service regardless of the priority, and at its completion instant, the unit with highest priority is taken for service. In the case of *preemptive priority*, however, any occupation of the service counter is interrupted as soon as a customer of higher priority arrives. The higher priority unit is taken for service and the interrupted unit goes back to the head of its queue. There are two main forms of preemptive priority discipline:

1. Preemptive-resume: the service is resumed where it was interrupted, i.e., no servicing time is lost.
2. Preemptive-repeat: the servicing time before the interruption is lost, i.e., the service of an ejected item begins from the starting point.

Note: It may be noted that in many cases the change in queue discipline does not change the overall behaviour of the system. The distribution of $X(t)$ is the same, the distribution of the busy period is the same and so on. It is only the individual behaviour of customers which is affected.

We shall examine single server queueing systems under non-preemptive and preemptive-resume queue disciplines. The main objective is to determine the steady-state average waiting time and total response time (i.e., average time spent in the system) for customers of different classes.

The following assumptions are made for the ith-priority class customers, $i = 1, 2, \ldots, m$:

1. The ith class customer arrives according to a Poisson process with rate λ_i;
2. $F_{S_i}(t)$ is the distribution function (d.f.) of the service time for customers of class i with mean $1/\mu_i$;
3. Customers are served on a first-come first-served basis within the ith priority class.

Regarding notations, we will use all earlier ones with a superscript i for the ith class, except that in this case let $\lambda = \sum_{i=1}^{n} \lambda_i$, which represents the overall arrival rate.

First we consider the queueing system under non-preemptive queue discipline. For the stationary distribution to exist, $\rho = \Sigma \rho_i$ should be less than 1.

10.2.1. Non-preemptive priority. Consider a customer of ith priority class ($i = 1, 2, \ldots, m$) and its waiting time W_{qi}. This waiting time consists of three components:

1. T_0, the time required to finish the service of a unit already in service;
2. The sum of the service times of customers of priorities $1, 2, \ldots, i$ that were present in the queue at the arrival epoch of the customer under consideration;
3. The sum of the service times of the customers of priorities $1, 2, \ldots, i-1$ that arrive during the waiting time W_{qi}.

The average waiting time $E(W_{qi})$ equals the sum of averages of those three components averages. We will consider them in turn.

The probability of the service counter being occupied by a customer of class k is ρ_k at the arrival instant of a customer of class i and its expected residual service duration is $\frac{1}{2}\mu_k E(S_k^2)$. Hence the average duration under (i) is

$$\frac{1}{2}\sum_{k=1}^{m}\rho_k\mu_k E(S_k^2) = \frac{1}{2}\sum_{k=1}^{m}\lambda_k E(S_k^2). \tag{10.1}$$

The average number of class k customers in the queue, met by the virtual unit on arrival is $\lambda_k E(W_{qk})$, $k = 1, 2, \ldots, i$. The average service times of those customers are independent of their indices. Hence, the expected total duration of (ii) is:

$$\sum_{k=1}^{i} E(S_k)\lambda_k E(W_{qk}) = \sum_{k=1}^{i}\rho_k E(W_{qk}). \tag{10.2}$$

The average number of class k customers arising during the waiting time W_{qi} is $\lambda_k W_{qi}$. The average of the ensuring service time is $\lambda_k W_{qi} E(S_k) = \rho_k W_{qi}$. When we take the classes $k = 1, 2, \ldots, i-1$ together and take the average of W_{qi}, we obtain for (iii)

$$E(W_{qi})\sum_{k=1}^{i-1}\rho_k. \tag{10.3}$$

Putting together (10.1), (10.2) and (10.3) we obtain a set of equations for the unknown average waiting times $E(W_{qi})$:

$$E(W_{qi}) = \frac{1}{2}\sum_{k=1}^{m}\lambda_k E(S_k^2) + \sum_{k=1}^{i}\rho_k E(W_{qk}) + E(W_{qi})\sum_{k=1}^{i-1}\rho_k, \tag{10.4}$$
$$i = 1, 2, \ldots, m.$$

An empty sum is equal to 0 by definition. The solution is easily found to be

$$E(W_{qi}) = \frac{\frac{1}{2}\sum_{k=1}^{m}\lambda_k E(S_k^2)}{\left(1 - \sum_{k=1}^{i-1}\rho_k\right)\left(1 - \sum_{k=1}^{i}\rho_k\right)}. \tag{10.5}$$

This is the average waiting time for priority i customers. The overall average waiting time of a customer for all classes is defined as

$$E(W_{\text{overall}}) = \sum_{i=1}^{m}\frac{\lambda_i E(W_{qi})}{\lambda}. \tag{10.6}$$

10.2.2. Preemptive-resume priority. Let us now consider the steady-state performance of the queueing model under the preemptive-resume priority discipline. The service of a lower priority customer may be interrupted by the subsequent arrivals of higher priority customers. This may happen many times before the service is completed.

On arrival, customers may go through the following phases within the system:

(i) There may be a time lag between the arrival and the instant of being taken for service, the so-called *initial waiting time*.
(ii) This is followed by what we shall call *attendance time*, which is the period between the start and the completion of the service.

Let

I_{qi} : initial waiting time of the ith priority class customer
V_i : attendance time of the ith priority class customer.

To determine the expected value of I_{qi}, the following two observations will be used. First, the customer of class $i+1, i+2, \ldots, m$ may be ignored when considering the performance of a class i customer. This is because a class i customer can preempt customers of classes $i+1, i+2, \ldots, m$ and so cannot be delayed by them. Second, during the initial waiting time of a class i customer, it does not matter whether the priorities of classes $1, 2, \ldots, i-1$ are preemptive or non-preemptive.

Thus the average of I_{qi} (initial waiting time of a class i customer) is equal to the average waiting time of a class i customer under non-preemptive queue discipline in a system where classes $i+1, i+2, \ldots, m$ do not exist. Using formula (10.5), we have for $i = 1, 2, \ldots, m$:

$$E(I_{qi}) = \frac{\frac{1}{2}\sum_{k=1}^{m} \lambda_k E(S_k^2)}{\left(1 - \sum_{k=1}^{i-1} \rho_k\right)\left(1 - \sum_{k=1}^{i} \rho_k\right)}. \tag{10.7}$$

Next we consider the attendance time of a class i customer, denoted by V_i. This consists of the customer's own service time, plus the service times of all higher priority class customers that arrive during the attendance time. During period V_i an average of $\lambda_j V_i$ customers of class j arrives, each of them takes an average of $\frac{1}{\mu_j}$ time units to be served. Hence we get

$$E(V_i) = \frac{1}{\mu_j} + \sum_{j=1}^{i-1} \frac{\lambda_j E(V_i)}{\mu_j}, \quad i = 1, 2, \ldots, m.$$

Solving this for $E(V_i)$ yields

$$E(V_i) = \frac{1}{\mu_i\left(1 - \sum_{j=1}^{i-1} \rho_j\right)}, \qquad i = 1, 2, \ldots, m. \tag{10.8}$$

The average time spent in the system (response time) and the total waiting time in the queue of class i customers are given by

$$E(W_i) = E(I_{qi}) + E(V_i), \tag{10.9}$$

$$E(W_{qi}) = E(I_{qi}) + E(V_i) - \frac{1}{\mu_i}. \tag{10.10}$$

$$i = 1, 2, \ldots, m.$$

The average number of customers of different classes in the system and in queues can be easily obtained by applying Little's formula.

10.3. Queues with Infinite Servers

So far we have dealt with queueing models having one server or at the most c (a finite number) servers. In such a system occasionally a queue is formed and the theory of Markov processes or its modification has been used to study the steady-state or time-dependent behaviour of the queue length and the waiting time processes.

In many real-life problems, customers may simply enter to receive service without interfering with the service of others, e.g., cars parked in a big lot which is never full, or a telephone exchange with a large number of telephone lines and a connection is made immediately after the arrival of a call. Such a system may be studied as a queueing system with an infinite number of servers. Every incoming customer starts being served immediately and there is never a queue.

In this section we would like to study two queueing systems with an infinite number of servers, namely $M/G/\infty$ and $G/M/\infty$.

Now we consider the $M/G/\infty$ model. As the notation suggests it is an infinitely many server system with the arrival and service times having the following properties:

(i) The arrival process is Poisson with rate λ;

(ii) Service times are i.i.d. with an arbitrary distribution function F_S.

Property (i) implies that the interarrival times are i.i.d. exponential (λ) random variables.

It is our intention to derive the transient distribution of the number of customers in the system at time t. Let the system-size process be called $X(t)$ and suppose that at time $t = 0$ there are i customers in the system. Furthermore, let $\gamma(t)$ be the number of customers arrived during the time $(0, t)$. By the law of conditional probability we find that

$$\begin{aligned} P_n(t) &= \sum_{k=0}^{n} P(k \text{ customers out of } i \text{ still being in service}) \times \\ &\times \sum_{m=0}^{\infty} P(n-k \text{ customers out of the new arrivals} \\ &\qquad \text{still being in service} | \gamma(t) = m) P(\gamma(t) = m). \end{aligned} \tag{10.11}$$

The probability that any customer among the initial ones will still be in the system at time t is given by $1 - F_S(t)$. Therefore,

$$\begin{aligned} &P(k \text{ customers out of } i \text{ still being in service}) \\ &\qquad = \binom{i}{k} [1 - F_S(t)]^k [F_S(t)]^{i-k}. \end{aligned} \tag{10.12}$$

The probability that a customer who arrives at time x will still be present at time t is given by $1 - F_S(t - x)$. Let the probability of an arbitrary one of the new arrivals still being in service given that $\gamma(t) = m$ be $q(t)$. Then

$$\begin{aligned} q(t) &= \int_0^t P(\text{service time} > t - x | \text{ an arrival at } x) \times \\ &\quad \times (\text{p.f. of an arrival at } x)\, \mathrm{d}x. \end{aligned}$$

Since the arrivals are Poisson and $\gamma(t) = m$, the m arrival instants $\tau_1, \ldots, \tau_m$ have the same distribution as the order statistics corresponding to m i.i.d. random variables uniformly distributed on the interval $(0, t]$. Hence

$$\text{p.f. of an arrival at } x = \frac{1}{t}.$$

Thus

$$\begin{aligned} q(t) &= \frac{1}{t} \int_0^t [1 - F_S(t - x)]\, \mathrm{d}x \\ &= \frac{1}{t} \int_0^t [1 - F_S(x)]\, \mathrm{d}x, \end{aligned} \tag{10.13}$$

and is independent of any other arrival.

Therefore,

$$\begin{aligned} &P(n - k \text{ customers out of new arrivals still being in service} | \gamma(t) = m) \\ &\qquad = \binom{m}{n-k} [q(t)]^{n-k} [1 - q(t)]^{m-n+k}. \end{aligned} \tag{10.14}$$

Since the arrival process is Poisson, we have

$$P(\gamma(t) = m) = e^{-\lambda t} \frac{(\lambda t)^m}{m!}. \tag{10.15}$$

Thus using (10.12), (10.14) and (10.15) in (10.11) we have

$$\begin{aligned} P_n(t) = \sum_{k=0}^{n} \binom{i}{k} [1 - F_S(t)]^k [F_S(t)]^{i-k} \\ \times \sum_{m=n-k}^{\infty} \binom{m}{n-k} [q(t)]^{n-k} [1 - q(t)]^{m-n+k} e^{-\lambda t} \frac{(\lambda t)^m}{m!}. \end{aligned}$$

On simplification the transient distribution is

$$\begin{aligned} P_n(t) &= \sum_{k=0}^{n} \binom{i}{k} [1 - F_S(t)]^k [F_S(t)]^{i-k} e^{-\lambda \int_0^t [1 - F_S(x)]\, \mathrm{d}x} \\ &\qquad \times \frac{\left[\lambda \int_0^t [1 - F_S(x)]\, \mathrm{d}x\right]^{n-k}}{(n-k)!}, \end{aligned} \tag{10.16}$$

which is the convolution of a binomial and a nonhomogeneous Poisson distribution. Hence

$$P(z, t) = [F_S(t) + z(1 - F_S(t))]^i \exp\left(-\lambda(1 - z) \int_0^t F_S(x)\, \mathrm{d}x\right). \tag{10.17}$$

One can easily check that the limiting distribution of the state of the system will follow a Poisson distribution with mean λ/μ.

For the model $M/M/\infty$ the transient solution for the number of customers in the system at time t will be a particular case of $M/G/\infty$ and can be obtained by substituting $F_S(t) = 1 - e^{-\mu t}$, and thus

$$P(z,t) = (q + pz)^i e^{-\frac{\lambda}{\mu} q(1-z)}, \tag{10.18}$$

where $p = e^{-\mu t}$ and $q = 1 - p$ (see Exercise 25, Chapter 2).

Next we consider the dual of the $M/G/\infty$ model, namely $G/M/\infty$. As the notation suggests, it is an infinitely many server system with the arrival and service times having the following properties:

(i) The interarrival times T are i.i.d. with an arbitrary distribution function F_T;

(ii) Service times are exponential with mean $1/\mu$.

Let Y_n denote the number of customers in the system just before the arrival of C_n (i.e., the nth customer). Now our objective is to determine the transient behaviour of the stochastic sequence $\{Y_n\}$. We shall also determine the asymptotic behaviour of the distribution of Y_n as $n \to \infty$.

Let

$$\begin{aligned} P_k^{(n)} &= P(Y_n = k), \qquad k = 0, 1, 2, \ldots\ , \\ B_\nu^{(n)} &= E[\binom{Y_n}{\nu}], \qquad \nu = 0, 1, 2, \ldots\ . \end{aligned}$$

$B_\nu^{(n)}$ represents the νth binomial moment of the distribution $\{P_k^{(n)}, k = 0, 1, 2, \ldots\}$. The sequence of random variables $\{Y_n\}$ forms an embedded Markov chain with transition probabilities

$$\begin{aligned} p_{jk} &= P(Y_{n+1} = k | Y_n = j), \qquad n = 1, 2, \ldots \\ &= \int_0^\infty \binom{j+1}{k} e^{-k\mu t}(1 - e^{-\mu t})^{j+1-k}\, \mathrm{d}F_T(t). \end{aligned} \tag{10.19}$$

Given that i customers are in the system (i.e., $X(0) = i$), we have

$$P(Y_1 = j | X(0) = i) = \int_0^\infty \binom{i}{j} e^{-j\mu t}(1 - e^{-\mu})^{i-j}\, \mathrm{d}F_T(t). \tag{10.20}$$

If X has a binomial distribution with parameters n and p, then one can check that

$$\begin{aligned} E\left[\binom{X}{\nu}\right] &= \sum_{k=\nu}^{n} \binom{k}{\nu}\binom{n}{k} p^k (1-p)^{n-k} \\ &= \binom{n}{\nu} p^\nu, \qquad \nu = 0, 1, 2, \ldots, n. \end{aligned}$$

Thus the νth binomial moment of a binomial distribution with parameters n and p is equal to $\binom{n}{\nu}p^\nu$. This will be of help in determining the $B_\nu^{(n)}$,

($\nu = 0, 1, 2, \ldots$), the νth binomial moment of the random variables Y_n for $n = 1, 2, \ldots$.

Given $X(0) = i$, we get from the above

$$E\left[\binom{Y_1}{\nu} | T_1 = t\right] = \binom{i}{\nu} e^{-\nu\mu t}, \tag{10.21}$$

since Y_1 is conditionally a binomial random variable with parameters i and $e^{-\mu t}$. Hence unconditionally the νth binomial moment of Y_1 is given by

$$B_\nu^{(1)} = E\left[\binom{Y_1}{\nu}\right] = \int_0^\infty \binom{i}{\nu} e^{-\nu\mu t}\, \mathrm{d}F_T(t) = \binom{i}{\nu} \Phi_T(\nu\mu), \qquad \nu = 1, 2, 3, \ldots \tag{10.22}$$

by recalling that $\Phi_T(\theta) = \int_0^\infty e^{-\theta t}\, \mathrm{d}F_T(t)$ is the L.S.T. of the distribution function of T.

Similarly, for $n = 1, 2, \ldots$

$$E\left[\binom{Y_{n+1}}{\nu} | Y_n = j, T_{n+1} = t\right] = \binom{j+1}{\nu} e^{-\nu\mu t}.$$

Unconditioning with respect to T_{n+1}, we have

$$\begin{aligned} E\left[\binom{Y_{n+1}}{\nu} | Y_n = j\right] &= \Phi_T(\nu\mu)\binom{j+1}{\nu} \\ &= \Phi_T(\nu\mu)\left[\binom{j}{\nu} + \binom{j}{\nu-1}\right] \end{aligned} \tag{10.23}$$

Multiplying (10.23) on both sides by $P(Y_n = j)$ and summing over j, we get for $n = 1, 2, \ldots$:

$$B_\nu^{(n+1)} = \Phi_T(\nu\mu)(B_\nu^{(n)} + B_{\nu-1}^{(n)}), \qquad \nu = 1, 2, \ldots . \tag{10.24}$$

One can determine $B_\nu^{(n)}$ recursively for every n and ν from (10.22) and (10.24).

Knowing $B_\nu^{(n)}$, one can determine the distribution of Y_n, by using the relationship

$$P(Y_n = k) = \sum_{\nu=k}^{\infty} (-1)^{\nu-k} \binom{\nu}{k} B_\nu^{(n)}. \tag{10.25}$$

The limiting distribution for the number of customers in the system will exist, if $\lambda < \infty$ and $\mu \neq 0$, and it is independent of i. Let

$$P_k = \lim_{n\to\infty} P_k^{(n)}, \qquad k = 0, 1, 2, \ldots .$$

Thus

$$P_k = \sum_{\nu=k}^{\infty} (-1)^{\nu-k} \binom{\nu}{k} B_\nu, \tag{10.26}$$

where B_ν is the νth binomial moment and is given by

$$B_\nu = \prod_{i=1}^{\nu} \frac{\Phi_T(i\mu)}{1 - \Phi_T(i\mu)}, \qquad \nu = 1, 2, \ldots, \tag{10.27}$$

and $B_0 = 1$ by definition.

If T follows an exponential distribution with mean $1/\lambda$, then we have the model $M/M/\infty$ and one can easily check that

$$P_k = e^{-\lambda/\mu}\frac{(\lambda/\mu)^k}{k!}, \qquad k = 0, 1, 2, \ldots \tag{10.28}$$

10.4. Design and Control of Queues

In Section 2.9 in Chapter 2, the concept of design problem in queueing theory has been introduced. The optimal values of the parameters c and μ for the multichannel Markovian queue have been derived. We also have investigated control policies in a single-server Markovian queueing model, where the server is activated and deactivated according to (1) $(0, K)$-policy and (2) random vacation policy. To continue our investigation further, in this section we would like to extend the study of control policies to the queueing model $M/G/1$. The control policies from the customer's point of view will be taken up later on in this section.

10.4.1. $M/G/1$ models operating under $(0, K)$-policy. We consider a model which may be called $M/G/1$ with removable server operating under $(0, K)$-policy. As the notation suggests, it is a single-server system with the interarrival times, service times and server process control having the following properties:

(i) The arrival process is Poisson with rate λ;
(ii) The service times are i.i.d. with an arbitrary distribution function F_S and mean $1/\mu$.

In addition, properties (iii), (iv) and (v) of $M/M/1$ are also valid.

The server continues to serve as long as there are customers in the system and the server is removed (i.e., deactivated) whenever the system becomes empty. The server is made available (i.e., activated) to customers only when the number of customers in the queue reaches K. Such a policy is known as $(0, K)$ server process control policy. When $K = 1$, this model is the same as the ordinary $M/G/1$ model of Section 3.2.

For the steady-state situation we assume that $E(S) < E(T)$, i.e., $\rho = \lambda/\mu < 1$.

Our system undergoes two phases: the idle period, during which no service is being rendered and the busy period during which the server is continuously busy.

Let $T_c = \Delta + I$ which represents the length of a busy cycle. Under the $(0, K)$ control policy, the expected duration of an idle period is equal to the average time taken for K customers to arrive and since the arrivals are according to a Poisson process with rate λ, we have

$$E(I) = \frac{K}{\lambda}, \tag{10.29}$$

which is the expected time the system is in the first phase.

The second phase starts with K customers in the system and the server starts serving them one by one and ends when the system becomes empty. Thus the expected duration of the second phase is equal to the expected length of a busy period with initially K customers and is

$$E(\Delta_K) = \frac{KE(S)}{1-\rho}. \tag{10.30}$$

Thus

$$E(T_c) = \frac{K}{\lambda} + \frac{KE(S)}{1-\rho} = \frac{K}{\lambda(1-\rho)}. \tag{10.31}$$

Therefore, the probability that the system is in the first phase is

$$\frac{K/\lambda}{K/(\lambda(1-\rho))} = 1-\rho, \tag{10.32}$$

and the probability that the system is in the second phase is

$$P(\text{system in 2nd phase}) = \rho. \tag{10.33}$$

Let E_i^0 denote the state that the service process is closed and i customers are in the system and let $p_i(0)$ be the probability associated with this event for $i = 0, 1, \ldots, K-1$. Similarly E_i^1 denotes the state that the server is busy and there are i customers in the system, and $p_i(1)$ denotes the probability associated with this event.

Thus $\bigcup_{i=0}^{K-1} E_i^0$ is the collection of states through which the system passes in the first phase. Since arrivals are according to a Poisson process, all states of this union are equi-probable. Hence by (10.33) we have

$$p_i(0) = \frac{1-\rho}{K}, \qquad i = 0, 1, \ldots, K-1. \tag{10.34}$$

Furthermore, $\bigcup_{i=0}^{\infty} E_i^1$ is the collection of states during which the server is busy and therefore

$$\sum_{i=1}^{\infty} p_i(1) = \rho. \tag{10.35}$$

The expectation of the residual service time of a customer in service at the time of the arrival of a customer is (see (A.70))

$$\frac{E(S^2)}{2E(S)} = \frac{\sigma_S^2 + \mu_S^2}{2\mu_S}. \tag{10.36}$$

The average time spent in the system by a customer if he arrives during the state E_i^0 $(0 \le i \le K-1)$ is

$$[K-(i+1)]\frac{1}{\lambda} + (i+1)\frac{1}{\mu}. \tag{10.37}$$

The first term is the expected time for the system to start the servicing process (i.e., the system reaches the level K), and the second term is the expected duration of service time devoted to the i customers in the system.

For an arrival during the state E_i^1 $(i = 1, 2, \ldots)$, its expected time to spend in the system is the sum of the expected residual service time (i.e., the expectation of the remaining service time of a customer in service at the time of the arrival of the customer), and the service time of i customers ($(i-1)$ customers in the queue and the arriving customer himself), i.e.,

$$\frac{\sigma_S^2 + \mu_S^2}{2\mu_S} + i\mu_s = \frac{1}{2\mu}(1 + \mu^2\sigma_S^2) + \frac{i}{\mu}, \tag{10.38}$$

where we have used $\mu_s = 1/\mu$.

Hence the expected time spent by a customer in the system is given by

$$\begin{aligned} E(W^{(0,K)}) &= \sum_{i=0}^{K-1} p_i(0)\left[\frac{K-(i+1)}{\lambda} + \frac{(i+1)}{\mu}\right] \\ &\quad + \sum_{i=1}^{\infty} p_i(1)\left[\frac{\mu^2\sigma_s^2 + 1}{2\mu} + \frac{i}{\mu}\right] \\ &= \frac{1}{\lambda}\frac{(1-\rho)}{K}\sum_{i=0}^{K-1}[(K-i+1+\rho] + \frac{\rho^2}{2\lambda}(1+\mu^2\sigma_s^2) \\ &\quad + \frac{\rho}{\lambda}\left[\sum_{i=0}^{K-1} ip_i(0) + \sum_{i=1}^{\infty} ip_i(1)\right], \end{aligned}$$

where $W^{(0,K)}$ denotes the limiting average waiting time over all customers in the system.

Thus

$$\lambda E(W^{(0,K)}) = (1-\rho)\frac{K-1}{2} + \rho(1-\rho) + \frac{\rho^2}{2}(1-\mu^2\sigma_S^2) + \rho L^{(0,K)}, \tag{10.39}$$

where $L^{(0,K)}$ denote the number in the system. Using Little's formula, $\lambda E(W) = L$, (10.39) can be simplified to give an expression for the expected number of customers in the system for the queueing system $M/G/1$ operating under the $(0, K)$ policy:

$$L^{(0,K)} = \rho + \frac{\rho^2}{2(1-\rho)}(1+\mu^2\sigma_S^2) + \frac{K-1}{2}. \tag{10.40}$$

The first two terms on the right-hand side of (10.40) constitute the P-K mean value formula without any restriction.

In order to find an optimal K, we consider the following costs already introduced in Section 2.9 in Chapter 2:

A : the cost of setting up and dismantling the server

C_1 : the cost of waiting per customer per unit of time

The costs are assumed to be linear with their average values

$$E(TC(K)) = \frac{A}{E(T_c)} + C_1 L^{(0,K)}, \tag{10.41}$$

where $TC(K)$ is the total cost under $(0, K)$-policy. Substituting (10.31) and (10.40) in (10.41) yields

$$E(TC(K)) = \frac{A\lambda(1-\rho)}{K} + C_1\left[\rho + \frac{\rho^2}{2(1-\rho)}(1+\mu^2\sigma_s^2) + \frac{K-1}{2}\right]. \quad (10.42)$$

When (10.42) is minimized with respect to K, the optimal value K^* is obtained as

$$K^* = \sqrt{\frac{2A\lambda(1-\rho)}{C_1}}, \quad (10.43)$$

and since

$$\frac{d^2TC(K)}{dK^2} = \frac{2A\lambda(1-\rho)}{K^3} > 0,$$

K^* is the minimizer of $TC(K)$.

10.4.2. ***M/G/*1 model with server vacations.** For the $M/G/1$ model operating under $(0, K)$ policy, it is necessary to continuously observe the growth of the queue whenever the server is not active. In many practical situations, however, this may be prohibitive due to the heavy costs to continuously observe the queue building up. So we look for another operating policy called T-policy. This is different from random vacation considered in Section 2.9 in Chapter 2.

Under this server control policy, the server leaves the system for a fixed duration V, called a vacation whenever the system becomes empty. On returning from the vacation, the server starts serving if there are customers in the system, otherwise he leaves immediately for another vacation of duration V.

A vacation period V_v is defined as the sum of a number of vacations after the end of the last busy period such that at least one customer arrives in the last of the vacations and none arriving in the earlier ones. The system goes from one busy cycle of length V_c to another one of random duration.

It is easy to see that

$$V_c = V_v + \Delta. \quad (10.44)$$

Let the vacation period V_v consist of N vacations of a fixed length V. The probability of the event $\{N = n\}$ is given by:

$$P(N = n) = (e^{-\lambda V})^{n-1}(1 - e^{-\lambda V}), \qquad n = 1, 2, \dots . \quad (10.45)$$

Thus

$$E(N) = \sum_{n=1}^{\infty} n(e^{-\lambda V})^{n-1}(1 - e^{-\lambda V}) = \frac{1}{1 - e^{-\lambda V}}. \quad (10.46)$$

By conditioning on N, we have

$$\begin{aligned} E(V_v) &= E\left[\sum_{i=1}^{N} V\right] = E\left[E\left(\sum_{i=1}^{N} V|N\right)\right] \\ &= VE(N) = V(1 - e^{-\lambda V})^{-1}. \end{aligned} \quad (10.47)$$

The probability that a busy period will start with $K > 0$ customers in the system is given by

$$\frac{e^{-\lambda V}(\lambda V)^K}{K!(1-e^{-\lambda V})}, \qquad k = 1, 2, \ldots. \tag{10.48}$$

Using the result (2.73) in Chapter 2 that the expected duration of a busy period beginning with K customers is $\frac{K\mu_S}{1-\rho}$, we have

$$\begin{aligned} E(\Delta) &= \sum_{k=1}^{\infty} \frac{K\mu_s}{(1-\rho)} \frac{e^{-\lambda V}(\lambda V)^K}{K!(1-e^{-\lambda V})} \\ &= \frac{\rho}{(1-\rho)} \frac{V}{(1-e^{-\lambda V})}. \end{aligned} \tag{10.49}$$

Using (10.44), (10.47) and (10.49) we get

$$\begin{aligned} E(V_c) &= \frac{V}{1-e^{-\lambda V}} + \frac{\rho}{(1-\rho)} \frac{V}{(1-e^{-\lambda V})} \\ &= \frac{V}{(1-\rho)(1-e^{-\lambda V})}. \end{aligned} \tag{10.50}$$

The probability that an arrival finds the server on vacation is

$$\frac{E(V_v)}{E(V_c)} = 1 - \rho, \tag{10.51}$$

and the probability that an arrival finds the server busy is

$$\frac{E(\Delta)}{E(V_c)} = \rho. \tag{10.52}$$

Let E_i^0 denote the state that the server is on vacation with i customers in the system and let $p_i(0)$ be the probability associated with this event, $i = 0, 1, 2, \ldots$. Similarly E_i^1 denotes the state that the server is busy with i customers in the system and let $p_i(1)$ be its probability, $i = 1, 2, \ldots$.

Equations (10.51) and (10.52) can be rewritten as

$$\sum_{i=0}^{\infty} p_i(0) = 1 - \rho \tag{10.53}$$

and

$$\sum_{i=1}^{\infty} p_i(1) = \rho \tag{10.54}$$

Using the concept of *residual waiting time* from renewal theory (see Section A.6 in Appendix A), the expectation of the remaining vacation time of the server is

$$\frac{E(V^2)}{2E(V)} = \frac{V^2}{2V} = \frac{V}{2}, \tag{10.55}$$

since the vacation time is of a fixed duration V. Similarly the expectation of the residual service time of a customer in service is

$$\frac{E(S^2)}{2E(S)} = \frac{\sigma_S^2 + \mu_S^2}{2\mu_S}. \tag{10.56}$$

Thus the average time spent in the system by a customer if he arrives during the state E_i^0 is

$$\frac{V}{2} + (i+1)\mu_S, i \geq 0. \tag{10.57}$$

The first term is the expected remaining vacation of the server and the second term is the expected duration of the service time devoted to i customers in the system.

If the arrival is taking place during state E_i^1, then the expected time spent in the system is the sum of the expected residual service time (i.e., the expectation of the remaining service time of a customer in service at the time of the arrival of the customer) and the service times of i customers ($i-1$ customers in the queue at his arrival instant and himself) and is

$$\frac{\sigma_S^2 + \mu_S^2}{2\mu_S} + i\mu_S. \tag{10.58}$$

Hence the expected time spent by a customer in the system is given by

$$\begin{aligned} E(W^{(T)}) &= \sum_{i=0}^{\infty} p_i(0) \left[\frac{V}{2} + (i+1)\mu_S\right] + \sum_{i=1}^{\infty} p_i(1) \left[\frac{\sigma_S^2 + \mu_S^2}{2\mu_S} + i\mu_S\right] \\ &= \frac{1}{\mu}\left[\sum_{i=0}^{\infty} ip_i(0) + \sum_{i=1}^{\infty} ip_i(1)\right] + (1-\rho)\frac{V}{2} + (1-\rho)\frac{1}{\mu} \\ &\quad + \frac{1}{2\mu}\rho(1+\mu^2\sigma_S^2). \end{aligned} \tag{10.59}$$

where $W^{(T)}$ is the waiting time under the T-policy. On multiplying both sides of (10.59) by λ and using the fact that

$$\sum_{i=0}^{\infty} ip_i(0) + \sum_{i=1}^{\infty} ip_i(1) = L,$$

and Little's formula, we have on simplifying

$$L^{(T)} = \rho + \frac{\rho^2}{2(1-\rho)}(1+\mu^2\sigma_S^2) + \lambda\frac{V}{2}, \tag{10.60}$$

where $L^{(T)}$ is the expected number of customers in the system for the queueing model $M/G/1$ operating under the T-policy. The first two terms on the right-hand side of (10.60) constitute the P-K mean value formula.

In order to find an optimal V, we consider the following costs:

A : the cost of going on one vacation

C_1 : the cost of waiting in the system per customer per unit of time

The costs are assumed to be linear with their average values and the busy cycle forms a renewal process, the total expected cost per unit time is

$$E(TC(V)) = A\frac{E(N)}{E(V_c)} + C_1 L^{(T)}.$$

Substituting the values of $E(N), E(V_c)$ and $L^{(T)}$ from (10.46), (10.50) and (10.60) yields

$$E(TC(V)) = A\frac{(1-\rho)}{V} + C_1\left[\rho + \frac{\rho^2}{2(1-\rho)}(1+\mu\sigma_S^2) + \lambda\frac{V}{2}\right]. \qquad (10.61)$$

Thus V^*, the optimal value of V, that minimizes (10.61) is given by

$$V^* = \sqrt{\frac{2A(1-\rho)}{\lambda C_1}}. \qquad (10.62)$$

Consider $M/G/1$ with general vacations. In a queueing system the server besides servicing the customers, called primary customers, may have to do some secondary job; a situation which arises naturally in many computer, communication and production systems. From the primary customer's point of view, the server working on secondary jobs is equivalent to the situation that the server is on vacation. Thus there is a natural interest to study queueing systems with server vacation, vacation time having a general distribution.

We consider an $M/G/1$ system in which the server begins a vacation of random length each time a busy period ends and the system becomes empty. If on return from a vacation the server finds no customers waiting, it waits for the arrival of a customer. This is called a *single vacation system* and is denoted by V_s. On the other hand, if it finds no customers waiting, it goes on taking vacations until on return from a vacation it finds at least one customer waiting then this will be called a *multi-vacation system*, denoted by V_m. We further assume that the lengths of vacations are i.i.d. and independent of the arrival process and the service times of customers. Let

$F_v(.)$: d.f. of the length of a vacation,

N : Number of customers in the system at the start of a busy period following a vacation,

$\alpha(z)$: p.g.f. of N,

$\Pi_v(z)$: p.g.f. of the number of customers at a departure epoch of a $M/G/1$ system with vacations.

Clearly,

$$\alpha(z) = \sum_{n=1}^{\infty} P(N=n)z^n.$$

Then for an $M/G/1$ queue with V_m server vacation, we have

$$\Pi_v(z) = \frac{1-\alpha(z)}{\alpha'(1)(1-z)}\Pi(z). \qquad (10.63)$$

(10.63) can be derived in a simple, direct and intuitive manner, see Fuhrmann (1984).

In the special case of $(0, K)$ policy $\alpha(z) = z^K$ for some fixed positive integer K and we have

$$\Pi_v(z) = \frac{1 - z^K}{K(1 - z)}\Pi(z). \tag{10.64}$$

10.4.3. Individual and social optimization. Another domain concerning design and control in a queueing system is some sort of control on the customers, in particular the determination of optimal joining rules. We would like to discuss the concept of *individual and social optimization* for the $M/M/c$ system and start with individual optimization.

We will be considering the standard $M/M/c$ model to obtain optimal balking and joining rules for the individual customer. By optimal, we mean optimal according to some economic criterion, and for this reason let us impose a cost structure on the queueing system.

Cost structure:

1. A customer who joins the system and gets served obtains a reward or benefit of R monetary units. The existence of such a reward attracts customers to the system.
2. Let $h(t)$ be the waiting cost of a customer who spends t units of time in the system.
3. If a customer decides to balk, that is, not to join the system, then a penalty of ℓ monetary unit is incurred ($\ell < \infty$).

Under a given cost structure one will be concerned with the following problems:

- When should an individual customer join the system?
- What is the structure of his optimal joining or balking rules?
- How are these rules affected by the cost parameters?

In this case each customer decides whether to join the system, with the objective of maximizing his own net benefit. The expected waiting cost of an arriving customer depends on the number of customers already in the system, and the assumption is that an arriving customer will join the queue if and only if the benefit R is large enough to cover the expected waiting cost. For the individual customer the optimal strategy is therefore to join the queue if and only if the number of customers already present in the system at the instant of his arrival is less than some critical number, say n_I. We shall make the obviously reasonable assumption that such an n_I exists.

Arrival-control policy is of the critical number form: admit a customer in state K if and only if $K < n_I$, where the critical number n_I is some fixed non-negative integer. Thus such an arrival control policy for a Markovian queueing system gives rise to a finite capacity model.

The queueing model underlying the present study is the $M/M/c$ with capacity n and FCFS as queue discipline (see Exercise 9, Chapter 2). Let $\rho = \lambda/\mu$. The stochastic process $\{X(t); t \geq 0\}$ is a continuous time Markov chain with state space $\{0, 1, \ldots, n\}$. It is ergodic, irrespective of the value of ρ. Let

$$d_k = \begin{cases} \frac{\rho^k}{k!}, & k = 0, 1, \ldots, c-1 \\ \frac{\rho^c}{c!}\left(\frac{\rho}{c}\right)^{k-c}, & k = c, c+1, \ldots, n, \end{cases}$$

and

$$D_n = \sum_{k=0}^{n} d_k, \qquad n \geq 0.$$

Then the steady-state probability of having k customers in the system with finite capacity n is given by

$$P_k(n) = \frac{d_k}{D_n}, \qquad k = 0, 1, \ldots, n.$$

Furthermore

$$L(n) = \sum_{k=0}^{n} k P_k(n)$$

is the expected number of customers in the system.

As to the cost-benefit structure of the model, let γ_k be the expected waiting cost for a customer who decided to join the queueing system when k customers are already in the system. A possible interpretation of γ_k is the following one: let T_k be a random variable that measures the time spent in the system by a customer who finds the system in state k at the instant of his arrival. Denoting the p.d.f. of T_k by $f_{T_k}(\cdot)$, we have

$$f_{T_k}(t) = \begin{cases} f_S(t) & 0 \leq k \leq c-1 \\ \int_0^t g_k(u) f_S(t-u)\,du & k \geq c, \end{cases}$$

where

$$g_k(u) = \frac{(c\mu)^{k-c+1}}{(k-c)!} u^{k-c} e^{-c\mu u}.$$

The expected waiting cost γ_k for the customer is given by

$$\gamma_k = \int_0^\infty h(t) f_{T_k}(t)\,dt,$$

and α_k, the expected net benefit of such a customer is given by

$$\alpha_k = R - \gamma_k - \ell.$$

If the cost function $h(t)$ is a monotonically increasing function, then it can be shown that

$$\alpha_0 = \alpha_1 = \ldots = \alpha_{c-1} > \alpha_c > \alpha_{c+1} \cdots.$$

Under the assumption that each individual customer adopts a strategy to maximize his expected benefit, a customer who upon his arrival finds k

customers present in the system will join the queue if $\alpha_k \geq 0$. Thus the balking level is determined by the condition

$$\alpha_{n_I-1} \geq 0 > \alpha_{n_I}.$$

Now we consider social optimization. Suppose that the customers form a cooperative and their joint objective is to find a policy that will maximize the long-run average net benefit per customer (or equivalently per unit of time) for all customers in the cooperative. Such on optimization problem is known as *social optimization.* Under social optimization questions that obviously arise are as follows:

1. Does an optimal policy exist?
2. If it does, is it a control-limit rule?

It can be shown using decision processes that this policy is a stationary Markovian policy and among all rules $R \in C_S$, there exists a deterministic control-limit rule with finite control limit denoted by n_S, that is optimal for social optimization. Moreover it can also be shown that, for fixed R, ℓ, λ, μ and c, the optimal control limit n_S is not necessarily the same as n_I, the optimal control limit for individual optimization, and in fact, $n_S \leq n_I$.

The problem of finding n_S can be formulated as a linear programming problem.

10.5. Networks of Queues II

Some of the most important applications of probabilistic modeling techniques are in the area of distributed computing. The term *distributed* means that various computational tasks that are somehow related can be carried out by different processors. To study the behaviour of such a system one normally needs a model involving a number of service centers with units arriving and circulating among different service centers according to a routing pattern. This leads in a natural way to the concept of a network of queues.

In order to define a queueing network completely, one has to make assumptions concerning the nature of the external arrival process, the routing of units among nodes, the number of servers and the service time distributions at different nodes. Sometimes we have to characterize each unit in the system.

We begin (see Section 2.8 in Chapter 2) by considering a queueing network, in which all units (customers) have identical characteristics, and such a network will be called a *single class queueing network.* Jackson network is known as a single class network. In contrast to Jackson's open network (considered in Section 2.8), we analyze in this section closed Markovian queueing networks, which have no external arrivals or departures. They are often more useful than open networks in modeling computer and communication systems. We are considering a single class network.

10.5.1. Closed Jackson network. We consider a queueing network having J service nodes (or systems), labeled $i = 1, 2, \ldots, J$, with the following properties:

1. The service times at node i are i.i.d. exponential random variables with parameter $\mu_i, i = 1, 2, \ldots, J$.
2. On completion of a service at node i, a unit goes instantaneously to node j with probability $p_{ij}, i, j = 1, 2, \ldots, J$, for its next service. These routing probabilities are independent of the history of the system.
3. There are no external arrivals or departures from the network. A fixed number of N units circulate through the network.

A network with these properties is called a *closed Jackson network* or *Gordon-Newell network.*

Now we derive the steady-state probability distribution of the number of units at the various nodes of the network. Similar notations as for the open Jackson network are used in this section (see Section 2.8 in Chapter 2).

Since there are J nodes and the network population is a fixed positive integer $N > 0$, the state space denoted by $S(J, N)$ is finite and given by

$$S(J, N) = \{(n_1, n_2, \ldots, n_J) | \sum_{i=1}^{J} n_i = N, n_i \geq 0, i = 1, 2, \ldots, J\}.$$

The number of elements (number of possible states) in the set $S(J, N)$ can be determined by simple combinatorial arguments and is the same as the number of ways of putting N balls into J cells (i.e., there are N indistinguishable units that must be placed in J different nodes), which is given by

$$\binom{N + J - 1}{J - 1} \quad \text{or} \quad \binom{N + J - 1}{N}.$$

Since there are no external arrivals or departures, the routing probabilities satisfy

$$\sum_{j=1}^{J} p_{ij} = 1, \qquad i = 1, 2, \ldots, J.$$

The traffic equations take the form

$$\alpha_i = \sum_{j=1}^{J} \alpha_j p_{ji}, \qquad i = 1, 2, \ldots, J. \tag{10.65}$$

Interestingly these traffic equations do not have a unique solution. Any solution $(\alpha_1, \alpha_2, \ldots, \alpha_J)$ multiplied by a real number c, i.e., $(c\alpha_1, c\alpha_2, \ldots, c\alpha_J)$ is also a solution.

Suppose we let $(\alpha_1, \alpha_2, \ldots, \alpha_J)$ be any nonzero solution, the α_i is proportional to the arrival rate at node i.

Typically, $(\alpha_1, \alpha_2, \ldots, \alpha_J)$ is chosen by fixing one component to a convenient value, such as $\alpha_1 = 1$. This means that for every visit to node 1, a customer makes α_i visits to node i on average. Thus α_i is called the average visitation rate of node i.

That the closed network $\mathbf{X}(t) = (X_1(t), \ldots, X_J(t))$ is an irreducible continuous time Markov chain with finite state space, implies the existence of a steady-state distribution. As in the open network, we obtain the steady-state distribution by using the network's steady-state global balance equations. Equating the net flow out of state $\mathbf{n}$ with the net flow into state $\mathbf{n}$, one has the global balance equations at state $\mathbf{n}$ as:

$$\sum_{i=1}^{J} \mu_i I_{(n_i>0)} P(\mathbf{n}) = \sum_{j=1}^{J} \sum_{j=1}^{J} \mu_j p_{ji} p(\mathbf{n} + \mathbf{e}(j) - \mathbf{e}(i)). \tag{10.66}$$

See Section 2.8 in Chapter 2 for notation. To solve the global balance equations we write the traffic equations as

$$1 = \sum_{j=1}^{J} p_{ji} \frac{\alpha_j}{\alpha_i}, \qquad i = 1, 2, \ldots, J. \tag{10.67}$$

Recall that $(\alpha_1, \alpha_2, \ldots, \alpha_J)$ is a solution of the traffic equations. Now multiply the left side of the global balance equation by this identity. Then one has

$$\sum_{i=1}^{J} \sum_{j=1}^{J} \mu_i I_{(n_i>0)} p_{ji} \frac{\alpha_j}{\alpha_i} P(\mathbf{n}) = \sum_{i=1}^{J} \sum_{j=1}^{J} \mu_i p_{ji} P(\mathbf{n} + \mathbf{e}(j) - \mathbf{e}(i)), \tag{10.68}$$

which can be rewritten as

$$\sum_{i=1}^{J} \sum_{j=1}^{J} p_{ji} \left[\mu_i \frac{\alpha_j}{\alpha_i} P(\mathbf{n} + \mathbf{e}(i)) - \mu_j P(\mathbf{n} + \mathbf{e}(j)) \right] = 0. \tag{10.69}$$

Clearly this equation will be satisfied if the following local balance equations hold, $n_i \geq 1$:

$$\mu_i \frac{\alpha_j}{\alpha_i} P(\mathbf{n} + \mathbf{e}(i)) = \mu_j P(\mathbf{n} + \mathbf{e}(j)), \qquad i, j = 1, 2, \ldots, J. \tag{10.70}$$

On rearranging, one has

$$P(\mathbf{n} + \mathbf{e}(i)) = \left(\frac{\alpha_i}{\mu_i} \right) \left(\frac{\alpha_j}{\mu_i} \right)^{-1} P(\mathbf{n} + \mathbf{e}(j)), \qquad i, j = 1, 2, \ldots, J. \tag{10.71}$$

This is equivalent to

$$P(n_1, n_2, \ldots, n_i, \ldots, n_J) = \frac{\alpha_i}{\mu_i} P(n_1, n_2, \ldots, n_i - 1, \ldots, n_J). \tag{10.72}$$

Continuing the recursion to zero on the ith term, one obtains

$$P(n_1, n_2, \ldots, n_i, \ldots, n_J) = \left(\frac{\alpha_i}{\mu_i} \right)^{n_i} P(n_1, n_2, \ldots, 0, \ldots, n_J). \tag{10.73}$$

Doing this for each node yields

$$P(\mathbf{n}) = \frac{1}{G} \prod_{i=1}^{J} \left(\frac{\alpha_i}{\mu_i} \right)^{n_i}, \tag{10.74}$$

where G is the normalization constant.

The steady-state distribution of the number of units in the network, $P(\mathbf{n})$, turns out to have a product form, similar to the one specified by the open Jackson network. The idea is to treat node i as an isolated $M/M/1$ queueing system.

Thus there are obvious similarities between the main result of closed and open Jackson networks, but also some important differences. First the fact that $\sum_{i=1}^{J} n_i = N$ holds, implies that the number of units present at the different J nodes are not independent random variables. Second, from a computational point of view, there is one difficult step. That step consists of determining the normalization constant over the state space of the network. The normalization constant is important in its own right for calculating performance measures such as mean queue length, throughputs implying the average departure rate and mean response times (derivation of performance measures as dealt with in Section 10.5.2). Since the number of states of the network grows exponentially with the number of customers and the number of service centers (or nodes), it is not feasible to evaluate the normalization constant by direct summation, because this would be too expensive and perhaps numerically unstable.

Even for small values of J and N, say $J = 8, N = 20$, the possible number of states is

$$\frac{27!}{20!7!} = 888030.$$

Thus calculation of the normalization constant by direct summation would require the summation of 888030 terms, each of which is the product of 8 factors.

Now we compute the normalization constant. Initially we consider a closed network with service rate of different nodes to be independent of the state of the nodes. To make the number of nodes (J) and the population (N) explicit in the normalization constant, let $G = G(J, N)$.

Let

$$S(\ell, n) = \{(n_1, \ldots, n_\ell | n_i \geq 0 \text{ for every } i), \sum_{i=1}^{\ell} n_i = n\},$$

and

$$G(\ell, n) = \sum_{\mathbf{n} \in S(\ell,n)} \prod_{i=1}^{\ell} \rho_i^{n_i}, \quad \text{where} \quad \rho_i = \frac{\alpha_i}{\mu_i}. \tag{10.75}$$

For $n, \ell > 0$,

$$\begin{aligned} G(\ell, n) &= \sum_{\substack{\mathbf{n} \in S(\ell, n) \\ n_\ell = 0}} \prod_{i=1}^{\ell} \rho_i^{n_i} + \sum_{\substack{\mathbf{n} \in S(\ell, n) \\ n_\ell > 0}} \prod_{i=1}^{\ell} \rho_i^{n_i} \\ &= \sum_{\mathbf{n} \in S(\ell-1, n)} \prod_{i=1}^{\ell-1} \rho_i^{n_i} + \rho_\ell \sum_{\substack{\mathbf{n} \in S(\ell, n) \\ k_i = n_i (i \neq \ell) \\ k_\ell = n_\ell - 1}} \prod_{i=1}^{\ell} \rho_i^{k_i}. \end{aligned} \tag{10.76}$$

The domain of the second summation is

$$\{\mathbf{k} | k_i \geq 0, \sum_{i=1}^{\ell} k_i = n - 1\} = S(\ell, n-1).$$

Thus we have the following recurrence:

$$G(\ell, n) = G(\ell - 1, n) + \rho_\ell G(\ell, n-1) \qquad \ell, n > 0. \tag{10.77}$$

The boundary conditions can be determined from the conditions

$$\begin{aligned} G(\ell, 0) &= 1, \qquad \ell > 0 \\ G(0, n) &= 0, \qquad n \geq 0. \end{aligned} \tag{10.78}$$

The recursive method to compute the normalization constant is called the convolution algorithm (the underlying difference equation is of the form of a discrete convolution) and is also known as *Buzen's Algorithm.*

An alternative derivation of the computational algorithm to obtain the value of the normalization constant $G(J, N)$ is discussed. Consider the following polynomial in z:

$$\begin{aligned} G(z) &= \prod_{i=1}^{J} \frac{1}{1 - \rho_i z} \\ &= (1 + \rho_1 z + \rho_1^2 z^2 + \ldots)(1 + \rho_2 z + \rho_2^2 z^2 + \ldots) \ldots \\ &\quad \times (1 + \rho_J z + \rho_J^2 z^2 + \ldots). \end{aligned} \tag{10.79}$$

It is clear that the coefficient of z^N in $G(z)$ is equal to the normalization constant $G(J, N)$, since the coefficient is just the sum of all terms of the form

$$\rho_1^{n_1} \rho_2^{n_2} \ldots \rho_J^{n_J} \quad \text{with} \quad \sum_{i=1}^{J} n_i = N.$$

In other words, $G(z)$ is the generating function of the sequence $G(J, 1)$, $G(J, 2), \ldots$,

$$G(z) = \sum_{n=0}^{\infty} G(J, n) z^n, \tag{10.80}$$

where $G(J, 0)$ is defined to be equal to unity.

In order to derive a recursive relation for computing $G(J, N)$, let us define:

$$G_\ell(z) = \prod_{i=1}^{\ell} \frac{1}{1 - \rho_i z}, \quad \ell = 1, 2, \ldots, N, \tag{10.81}$$

so that $G_N(z) = G(z)$. Also define

$$G_\ell(z) = \sum_{k=0}^{\infty} G(\ell, k) z^k, \quad \ell = 1, 2, \ldots, N. \tag{10.82}$$

Observe that

$$G_1(z) = \frac{1}{1 - \rho_1 z}$$

and

$$G_\ell(z) = G_{\ell-1}(z) \frac{1}{(1 - \rho_\ell z)}, \quad \ell = 2, 3, \ldots, N. \tag{10.83}$$

The last equation can be written as:

$$G_\ell(z) = \rho_\ell z G_\ell(z) + G_{\ell-1}(z),$$

or

$$\sum_{k=0}^{\infty} G(\ell, k) z^k = \sum_{k=0}^{\infty} \rho_\ell z G(\ell, k) z^k + \sum_{k=0}^{\infty} G(\ell - 1, k) z^k. \tag{10.84}$$

Equating the coefficient of z^n on both sides, we have a recursive formula for the computation of the normalization constant:

$$G(\ell, n) = G(\ell - 1, n) + \rho_\ell G(\ell, n - 1), \qquad \ell, n > 0. \tag{10.85}$$

Consider a cyclic queueing system with three nodes arranged in a circle. The service rates are $\mu_1 = \mu_2 = \mu_3 = \mu$. The mean throughput of each queue is equal and $\alpha_1 = \alpha_2 = \alpha_3 = 1.0$, as the queues are in series where the throughput is defined as the number of units passing through a node per unit of time. Let $N = 5$. Thus we have a closed network with $J = 3$ and a fixed population of 5 units.

Using the convolution algorithm, one can easily work out

$$G(3, 1) = \frac{3}{\mu}$$

$$G(3, 2) = \frac{6}{\mu^2}$$

$$G(3, 3) = \frac{10}{\mu^3}$$

$$G(3, 4) = \frac{15}{\mu^4}$$

$$G(3, 5) = \frac{21}{\mu^5}$$

Here is a simple C program for calculating the normalization constant.

```
/***************************************************************
Buzen's algorithm to calculate the normalizing constant in
```

```
closed queueing network.
****************************************************************/
#include <stdio.h>
#include <assert.h>
#define N    1000          /* size of customer population       */
#define J      50          /* number of nodes in network        */
                           /* replace this by your favorite     */
                           /* values                            */
double Rho[J];             /* vector for the traffic intensities */
double *G;                 /* vector of normalizing constants:  */
                           /* G[i]=G(J,i).                      */
void Intensities(void);    /* it is convenient to have a function */
                           /* for setting up the vector Rho      */
void Buzen(void);          /* this is the recurrence             */
    int main(void)
{
   int i;
       G=(double *) malloc((N+1)*sizeof(double));
       Intensities();
   Buzen();
   /* Here comes optional code to calculate various performance
      measures.
   ............................................................ */
return 0;
}
/* Set up vector Rho. Here we simply set alpha_i=1.0 and mu_i=2.0 */
void Intensities(void)
{
int i;
for(i=0;i<J;i++)
     Rho[i]=0.5;
}
/* This is Buzen's algorithm.                                      */
void Buzen(void)
{
int i,j;
double **g;
/* first allocate memory and check for errors.                     */
g=(double **) malloc((J+1)*sizeof(double*));
assert(g!=NULL);
for(i=0;i<=J;i++){
     g[i]=(double *)calloc(N+1,sizeof(double));
     assert(g[i]!=NULL);
     }
/* O.k. Now initialize the array g...                              */
for(i=1;i<=J;i++)
     g[i][0]=1.0;
/* Now go through recurrence ...                                   */
```

```
for(i=1;i<=J;i++)
     for(j=1;j<=N;j++)
          g[i][j]=g[i-1][j]+Rho[i-1]*g[i][j-1];
/* The last line of g contains now G(J,0), G(J,1),..., G(J,N)        */
/* copy the results to G... */
G=g[J];
}
```

10.5.2. Performance measures. The fact that the factors in the expression for $P(n_1, \ldots, n_J)$ are from the $M/M/1$ result is just a mathematical property and has no probabilistic interpretation. So, performance measures for each individual node in closed networks are not as easy to find as in open networks. Many performance measures can be determined by summing the product form solution over a subspace of the state space given by some restriction. Often this summation can be replaced by a normalization constant of a 'smaller' queueing network. Since nearly all performance measures can be expressed in terms of the normalization constant, the normalization constant $G(J, N)$ plays a central role in numerical evaluation of the queueing network model.

Probabilities of idleness. The probability that node J is idle can be determined directly by summing the joint probability over all states with $n_J = 0$, as follows:

$$P(N_J = 0) = \frac{1}{G(J, N)} \sum_{\substack{\mathbf{n} \in S(J, N) \\ n_J = 0}} \prod_{i=1}^{J-1} \rho_i^{n_i}$$

$$= \frac{G(J-1, N)}{G(J, N)}.$$

In general, for node i the idle probability is given by

$$P(N_i = 0) = \frac{G(J\backslash i, N)}{G(J, N)}, i = 1, 2, \ldots, J, \tag{10.86}$$

where $G(J\backslash i, N)$ is the normalization constant for a J-node network with node i being removed and the number of units still N.

Cumulative probabilities. Instead of considering the marginal system size probabilities, we may consider the cumulative marginal system size probabilities at different nodes, more precisely, its complements. Let N_i denote the steady-state number of customers at node i. For node i this is given by

$$P(N_i \geq n) = \frac{1}{G(J, N)} \sum_{\substack{\mathbf{n} \in S(J, N) \\ n_i \geq n}} \prod_{i=1}^{J} \rho_i^{n_i},$$

and it can be checked that for $n = 1, 2, \ldots, N$:

$$P(N_i \geq n) = \rho_i^n \frac{G(J, N-n)}{G(J, N)}, \qquad i = 1, 2, \ldots, J. \tag{10.87}$$

Utilizations. The utilization of node i is defined as the fraction of time the server at i is busy and is denoted by U_i. When $n = 1$, we obtain the node utilization of the ith node as

$$U_i = \rho_i \frac{G(J, N-1)}{G(J, N)}, \qquad i = 1, 2, \ldots, J. \tag{10.88}$$

Throughputs. The throughput of node i, i.e., the mean number of customers passing through node i per unit time, is given by

$$\mu_i U_i = \alpha_i \frac{G(J, N-1)}{G(J, N)}, \qquad i = 1, 2, \ldots, J. \tag{10.89}$$

As expected, the throughput at node i is proportional to its visitation rate.

System size probabilities. It is easy to determine the marginal steady-state system size probabilities at node i from the cumulative probabilities:

$$\begin{aligned} P_i(n) &= P(N_i \geq n) - P(N_i \geq n+1) \\ &= \rho_i^n \left[\frac{G(J, N-n) - \rho_i G(J, N-n-1)}{G(J, N)} \right], \quad n = 0, 1, \ldots, \\ &\qquad\qquad i = 1, 2, \ldots, J, \end{aligned} \tag{10.90}$$

where $G(J, -1) = 0$ by definition.

Mean system size. We denote the mean number of units at node i by $L_i(N)$, when the network population is N. One can easily verify that

$$\begin{aligned} L_i(N) &= \sum_{n=1}^{\infty} P(N_i \geq n) \\ &= \frac{1}{G(J, N)} \sum_{n=1}^{N} \rho_i^n G(J, N-n), \qquad i = 1, 2, \ldots, J. \end{aligned} \tag{10.91}$$

Mean value analysis. The traditional approach to the solution of Markovian queueing networks was to formulate a system of linear equations (balance equations) for the joint probability distribution of the vector-valued system state. For Jackson networks (open as well as closed), the solution of the balance equations is in the form of a product of simple terms. For closed networks we have discussed the convolution algorithm for computing the normalization constant and have expressed the different performance measures in terms of normalizing constants.

For practical purposes, however, the joint distributions contain far too much detail. Much simpler quantities such as the mean number of units at node i, L_i; the average time a unit spends at node i (on each visit to that node), W_i, utilizations, and throughputs were needed. Mean value analysis computes the standard performance measures of a queueing network, using only the arrival theorem and Little's Law. First we state the arrival theorem.

THEOREM 10.5.1. **[Arrival Theorem]** *In a closed queueing network, let* $\mathbf{P}(\mathbf{n}, N)$ *denote the steady-state probability that the network is in state* $\mathbf{n}$, *given the population size in the network is* N. *Then, when the network population* *is N, the equilibrium probability that the network is in state* $\mathbf{n}$ *just before an arrival at node i is*

$$\mathbf{A}(\mathbf{n}, N) = \mathbf{P}(\mathbf{n}, N-1).$$

Thus in a closed Markovian network, the system size distribution of any node observed by an arriving unit is the same as the distribution that will be observed at a randomly chosen instant if that particular unit were not contributing to the network load. In other words, if we denote by $\{A_i(n, N) : 0 \leq n \leq N-1\}$ the marginal distribution of units found by an arriving unit at node i, then

$$A_i(n, N) = P_i(n, N-1), \qquad 0 \leq n \leq N-1. \tag{10.92}$$

Mean value analysis (MVA) – Computational aspects. We will be studying the MVA approach for directly obtaining the different performance measures for the closed network models. Let

$$\begin{aligned} L_i(N) &: \text{ Mean number of units at node } i, \\ Y_i(N) &: \text{ Mean number of units at the } i\text{th node, found} \\ & \text{ by an arriving unit,} \\ E[W_i(N)] &: \text{ Mean time spent at node } i, \\ X_i(N) &: \text{ Throughput of node } i, \\ U_i(N) &: \text{ Utilization of node } i, \\ & \ i = 1, 2, \ldots, J; N = \text{population size.} \end{aligned}$$

The mean time spent. For the closed Jackson network with single servers we have

$$E[W_i(N)] = \frac{1}{\mu_i}[1 + Y_i(N)], \quad i = 1, 2, \ldots, J.$$

Since the arrival theorem implies that

$$Y_i(N) = L_i(N-1), \qquad i = 1, 2, \ldots, J.$$

Thus

$$E[W_i(N)] = \frac{1}{\mu_i}[1 + L_i(N-1)], \qquad i = 1, 2, \ldots, J. \tag{10.93}$$

Throughput. The key idea is to apply Little's result to the entire queueing network. That is, let the 'queueing system' be the entire network. Since we are considering a closed network, the average number of units in the system is just to be N. The mean time a unit spends in the system (during one cycle period) is simply the sum of the time it spends at each of the individual service nodes. But the average time a unit spends at node i is equal to $E[W_i(N)]X_i(N)$. Since the solution of the traffic equations is not unique, normally the α_i's are chosen, so that for one node it is equal to one, say for the first node. This node can be referred to as the reference node. By using Little's result to the entire queueing system, we have

$$N = \sum \alpha_i X_1(N) E[W_i(N)].$$

Hence

$$X_1(N) = \frac{N}{\sum_{i=1}^{J} \alpha_i E[W_i(N)]}. \tag{10.94}$$

The throughputs of the other nodes are computed from

$$X_i(N) = \alpha_i X_1(N), \qquad i = 2, 3, \ldots, J. \tag{10.95}$$

Utilization.

$$U_i(N) = \frac{1}{\mu_i} X_i(N), \qquad i = 1, 2, \ldots, J. \tag{10.96}$$

Mean number of units.

$$L_i(N) = X_i(N) E[W_i(N)], \qquad i = 1, 2, \ldots, J. \tag{10.97}$$

Thus the following algorithm may be used for computing different measures of the closed queueing network via MVA approach:

(i) Find the particular solution of the job flow balance equations (traffic equations) with $\alpha_1 = 1$;
(ii) Initialize $L_i(0) = 0, i = 1, 2, \ldots, J$;
(iii) Compute the performance measures of the network as follows:

```
    for n:=1 to N do
begin
```

$$\begin{aligned} E[W_i(n)] &= \frac{1}{\mu_i}[1 + L_i(n-1)], \quad i = 1, 2, \ldots, J \\ X_1(n) &= n / \sum \alpha_i E[W_i(n)] \\ X_i(n) &= \alpha_i X_1(n), \quad i = 2, 3, \ldots, J \\ U_i(n) &= \frac{1}{\mu_i} X_i(n), \quad i = 1, 2, \ldots, J \\ L_i(n) &= X_i(n) E[W_i(n)], \quad i = 1, 2, \ldots, J. \end{aligned}$$

```
end
```

Thus the performance measures can be computed using a simple algorithm.

Coded in C, this may look like the following:

```
/*******************************************************************
Mean value analysis
*******************************************************************/
#include <stdio.h>
#include <assert.h>
#define N     1000             /* size of customer population       */
#define J       50             /* number of nodes in network        */
                               /* replace these by your favorite    */
                               /* values                            */
double Alpha[J];               /* vector of the visitation rates    */
double Mu[J];                  /* vector of the service rates       */
double X[J];                   /* vector of throughputs             */
double W[J];                   /* vector of mean time spent         */
```

```
double L[J];                   /* Mean number of units              */
double U[J];                   /* Utilization                       */
void Parameters(void);         /* it is convenient to have a function */
                               /* for setting up the vectors Alpha    */
                               /* and Mu.                             */
void MeanValueAnalysis(void); /* this is the recurrence             */
int main(void)
{
   Parameters();
   MeanValueAnalysis();
       /* Here comes optional code to print measures etc.
       ...............................................................*/
return 0;
}

/* Set up parameters. Here we simply set alpha_i=1.0 and mu_i=2.0 */
void Parameters(void)
{
   int i;
       for(i=0;i<J;i++){
      Alpha[i]=1.0;
      Mu[i]=2.0;
      }
}
/* This is the MVA algorithm.                                        */
void MeanValueAnalysis(void)
{
   int i,n;
       /* Initial conditions...                                      */
       for(i=0;i<J;i++)
      L[i]=0.0;
       /* now go through recurrence...                               */
   for(n=1;n<=N;n++){
      for(i=0;i<J;i++)
         W[i]=(1.0+L[i])/Mu[i];
          X[0]=0.0;
      for(i=0;i<J;i++)
         X[0]+=Alpha[i]*W[i];
          X[0]=n/X[0];
      for(i=1;i<J;i++)
         X[i]=Alpha[i]*X[0];
          for(i=0;i<J;i++)
         U[i]=X[i]/Mu[i];
          for(i=0;i<J;i++)
         L[i]=X[i]*W[i];     }
}
```

Flexible manufacturing systems. Closed networks of queues play an important role in modelling of flexible manufacturing systems (FMSs). Typically an FMS consists of a load/unload station and several computer controlled machines which are connected by a material handling system, see Figure 10.1. The latter transports work pieces from machine to machine in any sequence, as it is required by the production process. When the processing of a work piece has been completed, it is immediately routed to the load/unload station, departs the systems and is replaced by a new work piece. Modelling an

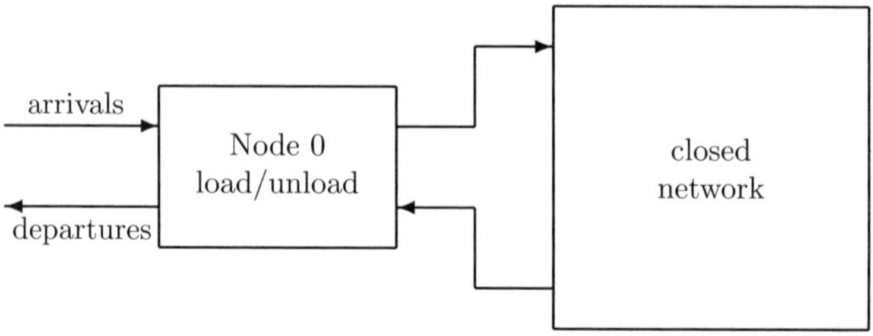

FIGURE 10.1. The schema of a flexible manufacturing system

FMS as a closed network requires a slight adaptation of the theory outlined above, since we have to take care of the special nature of the load/unload station. Let us label this station as node 0, so that our system now has nodes $0, 1, \ldots, J$. Furthermore assume that the mean service time $1/\mu_0$ at node 0 is equal to zero, since a work piece arriving at the load/unload station is immediately replaced by a new one. For the traffic equations (10.65) we enforce a unique solution by putting $\alpha_0 = 1$. As a result the solutions α_i attain a special meaning, α_i equals the mean number of visits to node i.

Now MVA may be carried out exactly along the same lines as for any closed network. The recurrence relations presented on page 370 remain unchanged except that $E[W_0(n)] = 0$ for all $n = 1, \ldots, N$, and the index i is running in range $0, 1, \ldots, J$ in each of the recurrence formulas.

Two important performance measures are directly connected to the load/ unload station:

1. The system throughput $X_0(N)$, the mean number of work pieces leaving the system per unit time, is given by (10.94):

$$X_0(N) = \frac{N}{\sum_{i=1}^{J} \alpha_i E[W_i(N)]}$$

2. The mean time $E[W(N)]$ spent by a work piece in the system or equivalently the mean time between successive returns to the load/ unload station. Since α_i now equals the mean number of visits to

node i, we have

$$E[W(N)] = \sum_{i=1}^{J} \alpha_i E[W_i(N)] = \frac{N}{X_0(N)}$$

Consider the following rather simplistic example of an FMS. It consists of a

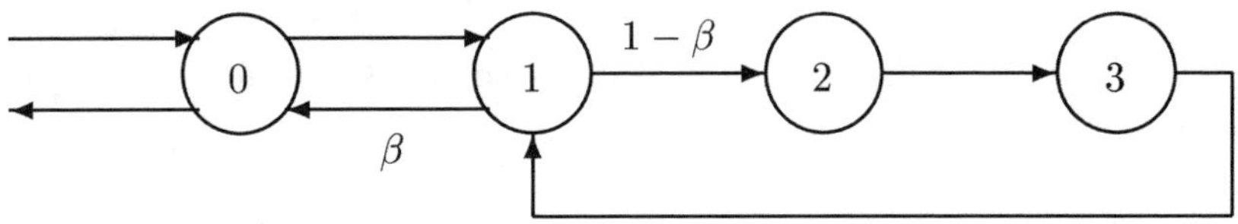

load/unload station and three processing units. The robot at node 1 does the major part of the work. If a work piece finishes processing at node 1 its quality is checked and if quality requirements are met, the work piece is sent to node 0 where it is replaced instantaneously by a new piece. This happens with probability β. With probability $1-\beta$ the work piece requires additional processing first on machine 2 and subsequently on machine 3. From there it is sent back to node 1 for quality inspection. Processing times at the machines are independent exponential random variables with means $1/\mu_i, i = 1, 2, 3$. To calculate the standard performance measures by MVA we first require the visitation rates $\alpha_i, i = 1, 2, 3$. These are the solutions of the traffic equation $\boldsymbol{\alpha} = \boldsymbol{\alpha}\mathbf{P}$, where the routing matrix is given by

$$\mathbf{P} = \begin{bmatrix} 0 & 1 & 0 & 0 \\ \beta & 0 & 1-\beta & 0 \\ 0 & 0 & 0 & 1 \\ 0 & 1 & 0 & 0 \end{bmatrix}.$$

Putting $\alpha_0 = 1$, the unique solution is given by

$$\alpha_0 = 1, \quad \alpha_1 = \frac{1}{\beta}, \quad \alpha_2 = \alpha_3 = \frac{1}{\beta} - 1.$$

These together with the mean service times are the input data for the MVA, bearing in mind that $E[W_0(n)] = 0$ for all n.

To illustrate the situation, suppose that $\beta = 0.8$, which means that a fraction of 20% of the work pieces require postprocessing. Suppose further that the mean processing times are

$$1/\mu_1 = 5, \quad 1/\mu_2 = 2, \quad 1/\mu_3 = 1 \qquad \text{(minutes)}$$

The MVA yields for a population size up to $n = 10$:

n	$E[W_1(n)]$	$E[W_2(n)]$	$E[W_3(n)]$	$L_1(n)$	$L_2(n)$	$L_3(n)$	$X_0(n)$	$E[W(n)]$
1	5.00	2.00	1.00	0.89	0.07	0.04	0.14	7.00
2	9.46	2.14	1.04	1.87	0.08	0.04	0.16	12.63
3	14.37	2.17	1.04	2.87	0.09	0.04	0.16	18.77
4	19.36	2.17	1.04	3.87	0.09	0.04	0.16	25.00
5	24.35	2.17	1.04	4.87	0.09	0.04	0.16	31.25
6	29.36	2.17	1.04	5.97	0.09	0.04	0.16	37.50
7	34.36	2.17	1.04	6.87	0.09	0.04	0.16	43.75
8	39.36	2.17	1.04	7.87	0.09	0.04	0.16	50.00
9	44.36	2.17	1.04	8.87	0.09	0.04	0.16	56.25
10	49.36	2.17	1.04	9.87	0.09	0.04	0.16	62.50

Observe that the system throughput stabilizes quickly at a value of 0.16 pieces per minute.

10.6. Discussion

This chapter is a potpourri of topics, a medley of notes on queues that ties down some of the loose ends which have not been accommodated elsewhere. On their own, they seem to be discordant, but when viewed along with the rest of the book, they do have a harmonizing effect. For example, the priority queues in Section 10.2 only suggest a different queueing discipline, and we ask what happens. The models $M/G/\infty$ and $G/M/\infty$ in Section 10.3 are, in a natural way, extensions of $M/M/c$. We are indeed looking for more of networks and optimization after seeing their introductory appearances in Sections 2.8 and 2.9 in Chapter 2. They have reappeared in the previous two sections. More specifically, the networks on queues has been substantially elaborated. Most of the materials are classical and the treatment and presentation of results are no doubt selective.

The references for Section 10.2 are Cobham (1954), Jaiswal (1968) and Kleinrock (1976), for Section 10.3, Liu, Kashyap and Templeton (1987) and Takács (1962), for Section 10.4 Heyman (1977), Knudsen (1972), Naor (1969), Teghem (1976, 1977), Yadin and Naor (1963), and Yechiali (1971, 1972) and finally for Section 10.5 are Bruell and Balbo (1980), Buzen (1973), Denning and Buzen (1989), Gordon and Newell (1967), Harrison and Patel (1993), Jackson (1963), Kobayashi (1978), Lavenberg (1982), Reiser and Lavenberg (1980) and Mitrani (1987).

It is time to stop! In Part I of the book, a reader has got a good taste of what the subject is about. There has been a horizontal spread to cover various ingredients such as, optimization, networks, discrete-time queues, statistical analysis and simulation, in order to make the flavour wholesome. In Part II, we delve into some topics further and introduce some more whose understanding needs experience. In this regard, combinatorial treatments in Chapter 6, computational methods in Chapter 8 and discrete-time queues in Chapter 9 stand out. These are very much new materials, not covered by other textbooks and their addition is expected to enrich the theoretical perceptibility and enhance the computational ability. We have stopped with the hope that the inquisitiveness of the reader is aroused sufficiently, so as not to stop learning more.

10.7. Exercises

1. Consider a $M/G/1$ system under processor sharing (PS) scheduling strategy. Let $W(x)$ denote the steady-state average response time for a job whose required service-time is x, then show that:
$$W(x) = \frac{x}{1-\rho}$$
and W the (unconditional) average response time is
$$W = \frac{1}{\mu(1-\rho)}.$$

2. For the $M/G/1$ system, let C_S^2 be the square of the coefficient of variation of service: Check that if $C_S^2 > 1$, then PS scheduling is better, otherwise FCFS scheduling is better.

3. Suppose that the job population consists of k types of jobs, arriving in independent Poisson streams with possibly different rates (λ_i for type i), and having different distributions of required service-time ($F_{S_i}(x)$ for type i with mean $\frac{1}{\mu_i}$).
Discuss the queueing system under PS queue discipline and show that:
(i)
$$\bar{W}_i = \int_{x=0}^{\infty} w_i(x) f_{S_i}(x)dx = \frac{1}{\mu_i(1-\rho)},$$
where $(P = \sum_{i=0}^{K} P_i)$, $i = 1, 2, \ldots, K$.
(ii)
$$L_i = \frac{\rho_i}{1-\rho}$$
where $P_i = \frac{\lambda_i}{\mu_i}$, $i = 1, 2, \ldots, K$.
(iii)
$$P(n_1, n_2, \ldots, n_k) = \rho^n (1-\rho) P(n_1, n_2, \ldots, n_k \mid \sum_{i=1}^{K} n_i = n)$$
where $P(n_1, n_2, \ldots, n_k \mid \sum_{i=1}^{K} n_i = n) = \frac{n!}{\rho^n} \Pi_{i=1}^{k} \frac{\rho_i}{(n_i)!}$
(iv)
$$P(m \text{ jobs of type } i \text{ in the system}) = (1-\sigma_i)\sigma_i^m$$
where $\sigma_i = \frac{\rho_i}{1-\rho+\rho_i}$, the marginal distribution of the number of type i jobs in the system.

4. For the $M/G/1$ queueing system under the last-come first-served (LCFS) queueing discipline, show that the Laplace transform of W_q (waiting time in the queue) is:

$$f^*_{W_q}(s) = (1-\rho) + \frac{\lambda[1 - f^*_{\Delta}(s)]}{s + \lambda - \lambda f^*_S(s)},$$

where $f^*_{\Delta}(s)$ is the Laplace transform of the busy period and ρ is the utilization of the system.

5. For the $M/G/1$ queueing system show that:

$$E[W_q(\text{under FCFS})] = E[W_q(\text{under LCFS})]$$

$$\sigma^2_{W_q(\text{under FCFS})} = (1-\rho)\sigma^2_{W_q(\text{under LCFS})} - \rho E(W_q).$$

6. For the queueing model $M/M/\infty$ let $X_1(t)$ be the number of units in the system at time t from the initial i units and $X_2(t)$ is the number of units in the system at time t from among the new arrivals in $(0,t]$. Show that $X_1(t)$ follows a binomial distribution with parameter $n = i$ and $p = e^{-\mu t}$, $X_2(t)$ follows a Poisson distribution with parameter $\frac{\lambda}{\mu}(1 - e^{-\mu t})$.

7. Show that the limiting distribution of the system for the model $M/G/\infty$ will follow a Poisson distribution with parameter $\frac{\lambda}{\mu}$.

8. Consider an $M/M/\infty$ queueing system, in which customers have a choice of individual service or batch service (batch service is more economical than individual service, such a system may arise in a transit terminal where a large number of taxis and buses are competing for customers) with R being the fixed size of the service batch. Let $\frac{n+1}{R}$ be the probability that an arriving customer chooses batch service when there are already n waiting customers prior to his arrival $(n = 0, 1, \ldots, R-1)$.
Show that the (marginal) distribution P_n for the customers waiting for batch service is given by:

$$P_n = \left(\sum_{K=1}^{R} \frac{1}{K}\right)^{-1} \frac{1}{n+1}, n = 0, 1, \ldots, R-1,$$

and

$$L_q = \sum_{n=0}^{R-1} nP_n = R^2 P_{R-1} - 1.$$

(See Liu, Kashyap and Templeton (1987).)

9. Consider the queueing system $M/G/1$ and let each busy period start with N^* number of customers. Let $R_{N^*}(z)$ be the pgf of N^*. Show that

(i)

$$\Pi(z) = \left[\frac{(1-\rho)(1-z)f_S^*(\lambda-\lambda z)}{f_S^*(\lambda-\lambda z)-z}\right]\frac{1-R_{N^*}(z)}{E(N^*)(1-z)}$$

(See Doshi (1986).)

(ii)

$$f^*{}_{W_q}(\theta) = \frac{(\theta/\lambda)(1-z)}{f_S^*(\theta)-z}\frac{1-f_v^*(\theta)}{\theta E(v)}$$

10. Show that for the model $M/G/1$ operation under multivacation with $f_v(\cdot)$ as the p.d.f. of the single vacation;

    (i)

$$\Pi(z) = \frac{(1-\rho)(1-z)f_S^*(\lambda-\lambda z)}{f_S^*(\lambda-\lambda z)}\frac{1-f_v^*(\lambda-\lambda z)}{\lambda E(v)(1-z)}$$

(ii)

$$f_{W_q}^*(\theta) = \frac{(\theta/\lambda)(1-z)}{f_S^*(\theta)-z}\frac{f_v^*(\lambda)+\frac{\lambda}{\theta}[1-f_v^*(\theta)]}{\lambda E(v)+f_v^*(\lambda)}$$

where $f^*(\theta)$ is the Laplace transform of the p.d.f. for vacation. Interpret the result. (See Fuhrmann (1984).)

11. For the $M/G/1$ model operating under single vacation with $F_v(\cdot)$ as the d.f. Check that

$$\Pi(z)_{\text{with vacation}} = \Pi(z)\frac{1-f_v^*(\lambda-\lambda z)+(1-z)f_v^*(\lambda)}{(1-z)[\lambda E(V)+f_v^*(\lambda)]}.$$

12. Verify that $\Pi(z)$ the pgf for the number of units at service completion epoch for the $M/G/1$ queueing system operating under $(0,K)$ operating policy,

    (i)

$$\Pi(z) = \Pi(z)_{M/G/1}\frac{1-z^K}{K(1-z)}$$

(See Heyman (1968).)
where $\Pi(z)_{M/G/1}$ means the usual $M/G/1$ queueing system.

(ii)

$$L^{(0,K)} = \rho + \frac{\rho^2}{2(1-\rho)}(1+\mu^2\sigma_S^2) + \frac{K-1}{2}$$

13. Consider an open Jackson queueing network with J nodes and with the routing matrix P is given below

$$P = \begin{bmatrix} p_1 & 1-p_1 & 0 & 0 & & & 0 \\ p_2 & 0 & 1-p_2 & 0 & \cdots & & 0 \\ p_3 & 0 & 0 & 1-p_3 & & \cdots & 0 \\ \cdots & \cdots & \cdots & \cdots & & \cdots & \cdots \\ \cdots & \cdots & \cdots & \cdots & & & \cdots \\ \cdots & \cdots & \cdots & \cdots & & & \cdots \\ p_{J-1} & 0 & 0 & 0 & & & 1-p_{J-1} \\ p_J & 0 & 0 & 0 & & & 0 \end{bmatrix}$$

where $0 \le p_i < 1$. Let the external input node be 1 with rate λ.
   (i) Find α_i explicitly for $i = 1, 2, \ldots, J$.
   (ii) Let μ_i be the service rate for node i $(i = 1, 2, \ldots, J)$ and each node has a single server. What relationship must hold among the $\{\mu_i\}$ in order that each node has the same utilization.
   (iii) Suppose we convert this to a closed Jackson network by letting the last row of P become $[p_J, p_1, p_2, \ldots, p_{J-1}]$
      (i) What must the first column sum up to ?
      (ii) What is the only sensible value for λ ?

14. Consider a closed Jackson queueing network with five nodes. Let nodes 1,2,3 and 4 have a single server, μ_i $(i = 1, 2, 3, 4)$ as their rate of service. Node 5 has c identical servers each with μ_5 as the rate of service. The number of customers in the network are J. Assuming

$$P = \begin{bmatrix} 0 & \frac{1}{2} & \frac{1}{2} & 0 & 0 \\ 0 & 0 & \frac{1}{2} & \frac{1}{2} & 0 \\ 0 & \frac{1}{2} & 0 & \frac{1}{2} & 0 \\ 0 & 0 & 0 & 0 & 1 \\ 1 & 0 & 0 & 0 & 0 \end{bmatrix}$$

   (i) Draw a fully labeled graph of the network.
   (ii) Formulate the traffic equations and find a solution assuming $\alpha_5 = 1$.
   (iii) Determine $P(n_1, n_2, n_3, n_4, n_5)$, the joint steady-state probability function.

15. Consider the following closed Jackson network with one customer $(N = 1)$ and four service nodes $(J = 4)$ with μ_1, μ_2, μ_3, and μ_4 as the rate of service at node 1, 2, 3 and 4, respectively (see Figure 10.2). Find $P(n_1, n_2, n_3, n_4)$ explicitly in terms of μ_i, $(i = 1, 2, 3, 4)$ and p.

16. Show that the number of distinguishable states of a closed queueing network with N customers and J service nodes is equal to the

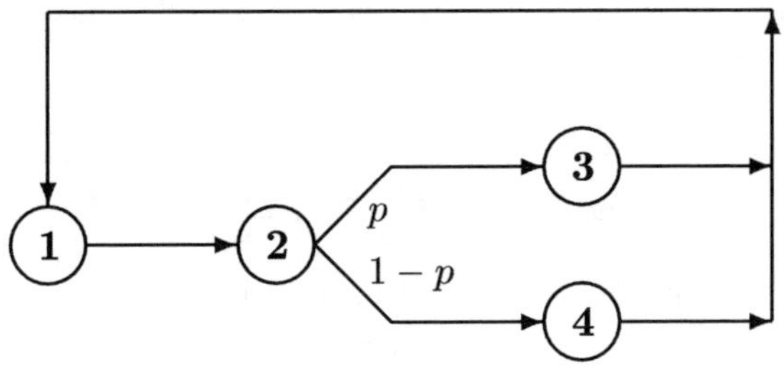

FIGURE 10.2. Closed Jackson network

binomial coefficient

$$\binom{J+N-1}{J-1}.$$

Interpret the expression.

17. Show that the output process of a $M/M/1$ queueing system, in which the initial distribution is the stationary distribution, is a Poisson process. Moreover, the state of the system $X(t)$ and the number of departures in $(0,t]$ are independent. (See Burke (1956).)

18. If $X(t)$ is a birth-death process satisfying (1) $\lambda_i = \lambda$, (2) $\sum \pi_j < \infty$, and (3) the initial distribution is given by $P_j = \frac{\pi_j}{\sum_j \pi_j}$, then the sequence of death times forms a Poisson process of rate λ, and the number of deaths in $(0,t]$ and the state of the system at time t are independent.

19. Show that for a $M/M/1$ queueing system with Bernoulli feedback with parameter p
    (i) The steady-state distribution is a geometric distribution with parameter $\frac{\lambda}{\mu(1-p)}$.
    (ii) In the steady state, let T be an epoch where a customer (exogenous or feedback) joins the queue and let $T+U$ be the next such epoch. Then U is independent of T, and

    $$\bar{F}_u(t) = \beta e^{-\mu t} + (1-\beta)e^{-\lambda t}$$

    where $\beta = \frac{p\mu}{\mu-\lambda}$.

20. Show that for the closed Jackson networks with J nodes and population size N,

$$G(\rho_1, \rho_2, \ldots, \rho_J; N) = \sum_{j=1}^{J} \frac{\rho_j^{N+J-1}}{\Pi_{i=1, i\neq j}^{J}(\rho_j - \rho_i)}$$

where $\rho_i = \frac{\alpha_i}{\mu_i}$, $i = 1, 2, \ldots, J$ and all are distinct. (See Harrison (1985).)

21. For the closed Jackson networks with J nodes and the population is of size N, show that the mean queue length of the load independent service node j is given by

$$L_j(N) = \sum_{n=1}^{N} \rho^n \frac{G(J, N-n)}{G(J,N)}$$

and using the fact that its utilization is given by

$$U_j(N) = \rho_j \frac{G(J, N-1)}{G(J,N)}$$

$$L_j(N) = U_j(N)[1 + L_j(N-1)]$$

a recursive relation with utilization with the initial condition $L_j(0) = 0$ for all j.

22. Using the fact that

$$U_j(N) = 1 - \frac{G(J/j, N)}{G(J,N)}$$

is another expression for the utilization of the jth node and equating it with $U_j(N) = \rho_j \frac{G(J,N-1)}{G(J,N)}$, yields the following recurrence relation

$$G(J,N) = G\left(\frac{J}{j}, N\right) + \rho_j G(J, N-1)$$

23. For the open Jackson network with J nodes and $\mathbf{P} = (p_{ij})$ as the routing matrix, $W_{i,0}$ the average interval between a customer arrival at node i and its departure from the network ($i = 1, 2, \ldots, J$) satisfies the following set of equations:

$$W_{i,0} = W_j + \sum_{j=1}^{J} p_{ij} W_{j,0}, i = 1, 2, \ldots, J$$

where W_i is the sojourn time at node i, ($i = 1, 2, \ldots, J$).

24. For the Jackson network (with μ_i as the service rate and α_i as the composite arrival at node i ($i = 1, 2, \ldots, J$)) suppose we have a control over the service rates $\mu_1, \mu_2, \ldots, \mu_J$ but with a constant $\sum_{i=1}^{J} \mu_i = c$. Using Lagrangian multipliers, show that the optimal set $\{\mu_i\}$ that minimizes the average network population $L = \sum \left(\frac{\alpha_i}{\mu_i - \alpha_i}\right)$ is given by:

$$\mu_i = \alpha_i + \frac{\sqrt{\alpha_i}}{\sum_{i=1}^{J} \sqrt{\alpha_i}} \left(c - \sum_{i=1}^{J} \alpha_i\right), i = 1, 2, \ldots, J.$$

25. Consider a system with three nodes, which we refer to as U (operating base station), B (base repair facility) and D (depot repair facility). Holding times at all nodes are assumed to be independent exponentially distributed random variables. At node U, the holding

time is the time to failure of a component (under operation), with the mean failure rate μ. At nodes B and D, the holding times are repair times and the mean repair rates are denoted by μ_2 and μ_3, respectively.

Let C_1 be the number of repair channels at the base node, C_2 the number of repair channels at the depot node. M (maximum) number of components at the operating base station node are in operation mode and the rest, if any, are in the queue (as space components at the base station). The total number of components in the system is N, α is the fraction of failed components that are diagnosed as base repairable and sent directly to base repair and $1-\alpha$ is the fraction sent directly to depot repair. Of those that are sent to base repair a further fraction β, after undergoing service, cannot be fixed and are sent to depot repair. The routing matrix for our problem is the following:

$$P = \{p_{ij}\} = \begin{matrix} \\ U \\ B \\ D \end{matrix} \begin{matrix} \begin{matrix} U & B & D \end{matrix} \\ \begin{bmatrix} 0 & \alpha & 1-\alpha \\ 1-\beta & 0 & \beta \\ 1 & 0 & 0 \end{bmatrix} \end{matrix}$$

Formulate the given repair system as closed Jackson network and obtain its solution.

26. The parameters $\rho_i = \alpha_i/\mu_i$ can be interpreted as the total mean service time a job requires from node i. Prove that for an FMS modelled as a closed network with single exponential servers the system throughput always satisfies

$$X_0(N) \le \min_i \frac{1}{\rho_i}.$$

27. Consider a computer system consisting of a central processing unit (CPU) modelled as node 1 and various input/output devices, like terminals and printers, modelled as nodes $2, 3, \ldots, J$. When a job is completed at the CPU it is routed to device $i = 2, 3, \ldots, J$ with probability β_i. With probability $\beta_1 = 1 - \beta_2 - \cdots - \beta_J$ the job leaves the network and is immediately replaced at the CPU by a new job. All nodes are assumed to have a single server with exponential service times having mean $1/\mu_i$. Set up the traffic equations and calculate performance measures of the system for different parameterizations (number of jobs being simultaneously in the system, number of devices, mean service times).

28. By the arrival theorem $Y_i(N) = L_i(N-1)$. Now consider the approximation

$$Y_i(N) \approx \frac{N-1}{N} L_i(N), \qquad i = 1, 2, \ldots, J.$$

This approximation, which is clearly exact for $N = 1$, is known to be asymptotically exact for large values of N. Check how the recurrence relations of the MVA change when introducing this approximation and design an efficient computational procedure to calculate the performance measures approximately.

References

Bruell, S. C. and Balbo, G. (1980). *Computational Algorithms for Closed Queueing Networks*, North-Holland, Amsterdam.

Burke, P. (1956). The output of a queueing system, *Opns. Res.*, **4**, 699–704.

Buzen, J. P. (1973). Computational algorithm for closed networks with exponential servers, *Comm. Assoc. Comput. Mach.*, **16**, 527–531.

Cobham, A. (1954). Priority assignment in waiting line problems, *Opn. Res.*, **2**, 70–76.

Denning, P. J. and Buzen, J. P. (1989). The operational analysis of queueing networks, *ACM Computing Surveys*, **10**, 225–241.

Doshi, B. T. (1986). Queueing systems with vacation survey, *Queuing Systems*, **1**, 29–66.

Fuhrmann, S. (1984). A note on the $M/G/1$ queue with server vacations, *Opns. Res.*, **32**, 1368–1373.

Gordon, W. J. and Newell, G. F. (1967). Closed queueing networks with exponential servers, *Oper. Res.*, **15**, 254–265.

Harrison, P. G. and Patel, N. M. (1993). *Performance Modelling of Communication Networks and Computer Architectures*, Addison-Wesley, Reading, MA.

Harrison, P. G. (1985). On normalizing constants in queueing networks, *Oper. Res.*, **33**, 464–468.

Heyman, D. P. (1977). The T-policy for the $M/G/1$ queue, *Management Sci.*, **23**, 775–778.

Heyman, D. P. (1968). Optimal operating policies for the $M/G/1$ queueing systems, *Opns. Res.*, **16**, 362–382.

Jackson J. R. (1963). Jobshop-like queueing systems, *Management Sci.*, **10**, 131–142.

Jaiswal, N. K. (1968). *Priority Queues*, Academic Press, New York.

Kleinrock, L. (1976). *Queueing Systems (Computer Applications)*, Vol. 2, John Wiley & Sons, New York.

Knudsen, N. C. (1972). Individual and social optimization in a multiserver queue with a general cost benefit structure, *Econometrica*, **40**, 515–528.

Kobayashi, H. (1978). *Modeling and Analysis: An Introduction to System Performance Evaluation Methodology*, Addison-Wesley, Reading, MA. (For the derivation of different performance measures of queueing networks in terms of the normalization constant).

Lavenberg, S. S. (1982). *Computer Performance Modelling Handbook*, Academic Press.

Liu, L., Kashyap, B. R. K. and Templeton, J. G. C. (1987). The service system $M/M^R/\infty$ with impatient customers, *Queueing Systems*, **2**, 363–372.

Mitrani, I. (1987). *Modelling Computer and Communication Systems*, Cambridge University Press, Cambridge, New York.

Naor, P. (1969). The regulation of queue size by levying tolls, *Econometrica*, **37**, 15–24.

Reiser, M. and Lavenberg, S. S. (1980). Mean value analysis of closed multichain queueing networks, *J. ACM*, **22**, 313–322.

Takács, L. (1962). *Introduction to the Theory of Queues*, Oxford Univ. Press, New York.

Teghem, J. Jr. (1976). Properties of $(0, K)$ policies in a $M/G/1$ queue and optimal joining rules in a $M/M/1$ queue with removable server, Haley, K. B. (Ed.), *Operational Research*, North-Holland, Amsterdam, p. 75.

Teghem, J. Jr. (1977). Optimal pricing and operating policies in a queueing system, in *Advances in O.R.*, Proc. of the Second European Conference on O.R., Roubens, M., Ed., North-Holland, Amsterdam, 489–496.

Yadin, M. and Naor, P. (1963). Queueing systems with removable service stations, *Oper. Res. Quart.*, **14**, 393–405.

Yechiali, U. (1971). On optimal balking rules and toll charges in the $GI/M/1$ queueing process, *Opns. Res.*, **19**, 349–370.

Yechiali, U. (1972). Customer's optimal joining rules for the $GI/M/s$ queue, *Management Sci.*, **18**, 434–443.

Appendices

APPENDIX A

A.1. Probability

The aim of this section is to review some basic definitions and results in probability theory which are pertinent for the development of queueing theory.

The uncertainty nature of many real world events denies the use of the usual deterministic mathematical models to describe them. Yet, there are situations which are not characterized by total uncertainty but rather display the property of so-called statistical regularity. For example, if a coin is tossed once, we cannot predict beforehand whether it will turn up heads or tails. However, if it is tossed long enough, the proportion of time it turns up heads will settle down to some number. This long run 'settling down' property is known as the statistical regularity. Events possessing this attribute can be described by what we call the probabilistic models and the study relating to such events is called the probability theory.

To each event under study we may associate an experiment (say, tossing a coin) which leads to a well-defined set of possible outcomes (heads, tails) and which at least in principle can be repeated under identical conditions. The last part is essential to ensure the property of statistical regularity. Our concern being the events under uncertainty, the probability of an event may be in some way the measure of uncertainty. From the assumption of statistical regularity, a natural measure of uncertainty may be defined as the limit of the relative frequency of the event under consideration in many independent trials of the experiment. In spite of its simplicity and naturalness, this definition has a limited scope as a mathematical entity.

In order to provide the theory of probability a sound mathematical structure, some elementary set theoretic notions and operations are useful and are described hereafter. An outcome of an experiment is called a *sample point* and may be denoted by ω, and the set of possible outcomes is called the *sample space* which is denoted by S. The membership of ω in S is denoted by $\omega \in S$. An event is a subset of S, i.e., if A is an event then in notation $A \subset S$. Usually notation for an event is a capital letter such as A, B or C and for the impossible event ϕ. An event may be written by listing the sample points belonging to it or by its defining property. For instance the event that the service time is less than 5 can be represented by A as $A = \{\omega : \omega < 5\}$. Let us define the following operations on events A and B in a given S:

$A \cup B$ (union of A and B): either A or B or both to occur,
$A \cap B$ (intersection of A and B): both A and B to occur,
A^c (complement of A): A not to occur.

We say that two events A and B are *mutually exclusive* if $A \cap B = \phi$ (i.e., no sample point is common to both) and that a finite or infinite set of events $\{A_1, A_2, \ldots\}$ is *exhaustive* if $A_1 \cup A_2 \cup \cdots = S$.

When the sample space S contains an infinite number of sample points, it can be seen that neither all subsets of S are of interest as events nor are technically easy to deal with. Therefore we consider a family of subsets which are meaningful as events and are mathematically manageable. A family of events ε is a class of subsets of S such that (i) S is an event, (ii) if A is an event then A^c is an event, (iii) if $A_1, A_2, \ldots$ are any finite or infinite set of events then both $A_1 \cup A_2 \cup \cdots$ and $A_1 \cap A_2 \cap \cdots$ are events. The *probability* P as a measure is a real-valued function defined on the family of events ε which satisfies the following properties (or axioms):

(a) For any event A in ε, $0 \leq P(A) \leq 1$.
(b) $P(S) = 1$.
(c) If $A_1, A_2, \ldots$ are a finite or infinite set of mutually exclusive events in ε then $P(A_1 \cup A_2 \cup \cdots) = P(A_1) + P(A_2) + \cdots$.

The triplet (S, ε, P) with (a), (b) and (c) is called a probability model. It can be seen that the definition based on relative frequency satisfies all these properties. From the definition it follows that

(i) $P(A^c) = 1 - P(A)$,
(ii) $P(\phi) = 0$,
(iii) $P(A) \leq P(B)$ if $A \subset B$,
(iv) $P(A \cap B^c) = P(A) - P(A \cap B)$,
(v)

$$\begin{aligned} P\left[\bigcup_{i=1}^{n} A_i\right] &= \sum_{i=1}^{n} P(A_i) - \sum_{\substack{i=1 \\ i<j}}^{n} \sum_{j=1}^{n} P(A_i \cap A_j) \\ &\quad + \sum_{\substack{i=1 \\ i<j<k}}^{n} \sum_{j=1}^{n} \sum_{k=1}^{n} P(A_i \cap A_j \cap A_k) - \cdots \\ &\quad + (-1)^{n-1} P(A_1 \cap A_2 \cap \cdots \cap A_n). \end{aligned}$$

The above definition does not help us in assigning probabilities to various events. However, if the sample space S is finite consisting of n sample points $\{\omega_1, \omega_2, \ldots, \omega_n\}$ and if $P\{\omega_i\} = 1/n$ for every i, then for any event A

$$P(A) = \frac{\text{number of sample points in } A}{n}.$$

When these assumptions are valid, we say that the probability model is an equally likely or uniform model. Since in this model counting plays a vital role, three counting rules which are frequently used are given below.

1. If the number of choices at the first stage is m and for each of the first stage choices there are n second stage choices, then the total number of choices in mn.
2. The number of permutations (or arrangements) of n objects taken r at a time is

$$n(n-1)\cdots(n-r+1) = \frac{n!}{(n-r)!}\ .$$

3. The number of combinations (or selections) of n objects taken r at a time is

$$\binom{n}{r} = \frac{n!}{r!(n-r)!}\ .$$

Next, we introduce two other related and important concepts, namely, conditional probability and independence. Let A and B be two events in a given sample space. The *conditional probability* of A given that the event B has occured is denoted by $P(A|B)$ and is defined as

$$P(A|B) = \frac{P(A \cap B)}{P(B)} \tag{A.1}$$

provided $P(B) \neq 0$. The conditional probability can be seen as the unconditional probability with event B as the sample space. Two events A and B are said to be *independent* when

$$P(A \cap B) = P(A)P(B). \tag{A.2}$$

A consequence of (A.1) is

$$P(A \cap B) = P(A|B)P(B) \tag{A.3}$$

which is often used in computing $P(A \cap B)$ when in general the occurrence of A follows the occurrence of B.

To conclude this section, two well-known theorems are presented. First, we state the *theorem of total probability* which is the most useful theorem in the theory of queues. If B is an event and $\{A_1, A_2, \ldots\}$ is a set of mutually exclusive and exhaustive events, then

$$P(B) = \sum_i P(B|A_i)P(A_i) \tag{A.4}$$

which is simply a generalization of (A.3). The second theorem is called *Bayes' theorem* which is given by

$$P(A_i|B) = \frac{P(B|A_i)P(A_i)}{\sum_j P(B|A_j)P(A_j)}\ . \tag{A.5}$$

As an application of (A.4), consider B to be the event that there have k customers arrived during a two-hour period and A_i to be the event that there have i customers arrived during the first hour. Then we can express $P(B)$ in terms of the right side of (A.4) provided $P(B|A_i)$ is computable or is given, which is so in many situations.

A.2. Random Variables

Although we have described an event in general terms as a collection of outcomes of an experiment, a numerical description of an event is often natural and sometimes practical. For instance, in a queueing system the number of customers waiting for service directly expresses the events as numerical measurements, whereas in an experiment of tossing a coin twice with $S = \{HH, HT, TH, TT\}$, H referring to heads and T to tails, the number of heads in two trials as an event is of practical interest. Observe that in the second example, there is a probability attached to each number of heads, such as the probability of getting heads once is equal to $P\{HT, TH\}$. This motivates us to introduce the concept of a random variable which takes on different values depending on which outcome is observed.

A *random variable* (briefly, r.v.) is a real-valued function defined on S with the associated probability based on P in (S, ε, P). The r.v.'s are generally denoted by capital letters such as X, Y, Z. The associated probability rule for a r.v. X is that the probability of the values of X belonging to a set A on the real line (denoted by $(X \in A)$) equals $P\{\omega : X(\omega) \in A\}$. Since X is real valued, we may consider its sample space to be the real line and A is an event in the family of events constructed from the real line. Once the probability of events for X is specified, the original experiment and the corresponding sample space stay in the background. Thus the same notation P is used for the probability measure for X, although strictly speaking there should be a new notation.

In our coin tossing example, assume that the coin is unbiased and let the r.v. X represent the number of heads. Then

$$\begin{aligned} P(X=0) &= P\{\omega : X(\omega) = 0\} = P\{TT\} = 1/4, \\ P(X=1) &= P\{\omega : X(\omega) = 1\} = P\{HT, TH\} = 1/2, \\ P(X=2) &= P\{\omega : X(\omega) = 2\} = P\{HH\} = 1/4. \end{aligned}$$

The set of values of the r.v. X together with the associated probability measure on the family of events is called the *probability distribution* (in short distribution) of X. It is most conveniently described by the *distribution function* (briefly d.f.) of X which is denoted by $F_X(\cdot)$ and defined as

$$F_X(x) = P(X \le x) = P\{\omega : X(\omega) \le x\} \tag{A.6}$$

for every real number x. To see that the distribution is determined by the d.f. observe that

$$P(a < X \le b) = F_X(b) - F_X(a)$$

and

$$P(X = x) = F_X(x) - F_X(x-)$$

where $F_X(x-)$ is the limit of $F_X(b)$ when b approaches x from the left. The d.f. satisfies the following:

(i) $F_X(x) \ge 0$ for every x, $F_X(-\infty) = 0$, $F_X(\infty) = 1$.
(ii) $F_X(\cdot)$ is continuous from the right.

In fact any function having these properties can be treated as a d.f. of some random variable. We say that two r.v.'s are *equivalent* if their distributions are the same.

The most commonly met random variables are of two types (1) discrete r.v.'s (2) continuous r.v.'s. A *discrete random variable* is one which takes on only finite or a countably infinite number of values each with positive probability. If X is a discrete r.v. then $p_X(\cdot)$ is called its *probability mass function* (briefly p.m.f.) and is given by

$$p_X(x) = P(X = x) \tag{A.7}$$

for every x. Note that $p_X(x) \geq 0$ for every x and $\sum_x p_X(x) = 1$ and that any function satisfying these two properties is a p.m.f. of some r.v. Also it is easy to verify that given $p_X(\cdot)$, we can evaluate the probability of any event concerning X and in that sense the p.m.f. completely characterizes the distribution of X. Clearly, $p_X(x)$ and $F_X(x)$ are related by

$$F_X(x) = \sum_{y \leq x} p_X(y).$$

There is another class of r.v.'s namely, continuous r.v.'s for which the p.m.f. does not exist. A random variable is a *continuous random variable* for which there exists a non-negative function $f_X(\cdot)$, called its *probability density function* (briefly, p.d.f.) such that for any event E defined on X,

$$P(E) = \int_E f_X(x)\, dx.$$

Similar to the p.m.f., a function $f_X(\cdot)$ is a p.d.f. if $f_X(x) \geq 0$ for every x and $\int_{-\infty}^{\infty} f_X(s)\, dx = 1$. Also note that

$$F_X(x) = \int_{-\infty}^{x} f_X(y)\, dy$$

and

$$f_X(x) = \frac{dF_X(x)}{dx},$$

and the distribution of a continuous r.v. is completely specified by the p.d.f. Once the concept of p.m.f. and p.d.f. is clear, we call both the *probability functions* (briefly, p.f.) for simplicity and denote it by f_X without any confusion since it is clear by now that for computation of probabilities, f_X is summed if X is discrete and is integrated if X is continuous over the appropriate region.

Next we may think of having several r.v.'s defined over the same (S, ε, P). Let X and Y be two random variables. Their distribution is given by the joint d.f. $F_{X,Y}(x, y)$ defined for all real numbers x and y as

$$F_{X,Y}(x, y) = P(X \leq x, Y \leq y). \tag{A.8}$$

Note that

$$\begin{aligned} P(a < X \leq a+b, c < Y \leq c+d) &= F_{X,Y}(a+b, c+d) - F_{X,Y}(a+b, c) \\ &\quad - F_{X,Y}(a, c+d) + F_{X,Y}(a, c). \end{aligned}$$

The joint p.f. is defined as

$$f_{X,Y}(x,y) = \begin{cases} P(X=x, Y=y) & \text{if } X \text{ and } Y \text{ are discrete,} \\ \frac{\partial^2 F_{X,Y}(x,y)}{\partial x \partial y} & \text{if } X \text{ and } Y \text{ are continuous.} \end{cases} \tag{A.9}$$

It satisfies similar properties as in the case of one variable. The probability of a region $\{(x,y) : (x,y) \in B\}$ is determined as follows:

$$P((X,Y) \in B) \quad = \quad \begin{cases} \sum\sum_B f_{X,Y}(x,y) & \text{in the discrete case,} \\ \int\int_B f_{X,Y}(x,y)\,dx dy & \text{in the continuous case.} \end{cases}$$

From the joint distribution, one can get the distribution of any individual r.v. For example, the distribution of X for every x is given by

$$\begin{aligned} F_X(x) &= F_{X,Y}(x,\infty), \\ f_X(x) &= \begin{cases} \sum_y f_{X,Y}(x,y) & \text{in the discrete case,} \\ \int_{-\infty}^{\infty} f_{X,Y}(x,y)\,dy & \text{in the continuous case,} \end{cases} \end{aligned}$$

where $f_X(x)$ is generally known as the *marginal p.f.* of X. Similar to the definition of conditional probability, the *conditional p.f.* of X at x given $Y = y$ is defined as

$$f_{X|Y}(x,y) = \frac{f_{X,Y}(x,y)}{f_Y(y)} \tag{A.10}$$

provided $f_Y(y) > 0$. Two variables X and Y are said to be *independent* if

$$F_{X,Y}(x,y) = F_X(x)F_Y(y) \tag{A.11}$$

or in terms of the p.f. if

$$f_{X,Y}(x,y) = f_X(x)f_Y(y). \tag{A.12}$$

All the definitions of two variables extend to more variables in a routine manner. In case of several continuous variables, say $X_1, \ldots, X_n$, the marginal p.f. of r variables $X_1, \ldots, X_r$ $(r < n)$ is given by

$$f(x_1,\ldots,x_r) = \int_{-\infty}^{\infty} \cdots \int_{-\infty}^{\infty} f(x_1,\ldots,x_n)\,dx_{r+1}\cdots dx_n \tag{A.13}$$

(omitting the subscripts) and the conditional p.f. of $(X_{r+1},\ldots,X_n)$ at $(x_{r+1},\ldots,x_n)$ given $X_1 = x_1, \ldots, X_r = x_r$ is given by

$$f(x_{r+1},\ldots,x_n|x_1,\ldots,x_r) = \frac{f(x_1,\ldots,x_n)}{f(x_1,\ldots,x_r)}. \tag{A.14}$$

The condition for independence of n variables is

$$f(x_1,\ldots,x_n) = \prod_{i=1}^{n} f(x_i). \tag{A.15}$$

A set of independent and identically distributed r.v.'s is briefly referred to as i.i.d. r.v.'s. Also the distribution of several r.v.'s is called the *multivariate distribution* as opposed to the univariate distribution of a single r.v. Among the class of several r.v.'s, two classes are of interest to us. We say that $X_1, \ldots, X_n$ are *cyclically interchangeable* r.v.'s if all n cyclic permutations of $(X_1, \ldots, X_n)$ have the same distribution. They are said to be *interchangeable* if all $n!$ permutations of $(X_1, \ldots, X_n)$ have the same distribution. Note that i.i.d. r.v.'s are interchangeable and interchangeable r.v.'s are cyclically interchangeable.

In all our discussion, we are assuming that variables are either discrete or continuous. But in a real-life situation this need not be so. For instance, we are interested in the number of customers served during a busy period and its length, the first random variable being discrete and the second continuous. In such situations, the rule of the game is to write the joint distribution either in terms of the d.f. and get the remaining derivations from it or to simply follow the basic principle of putting the p.f. (p.m.f. for a discrete r.v. and p.d.f. for a continuous r.v.) and use summation or integration appropriately to determine the probability of an event.

Sometimes an r.v. as a function of one or several r.v.'s is of interest. For simplicity, let us consider $Y = g(X)$. Since the distribution is completely determined by the d.f. we can write

$$F_Y(y) = P(Y \leq y) = P(g(X) \leq Y) = P\{x : g(x) \leq y\}. \tag{A.16}$$

In some cases, the solution set of $g(x) \leq y$ is readily available and the corresponding probability may be found.

The same general procedure can be followed even in the case of Y as a function of several r.v.'s, say $Y = g(X_1, \ldots, X_n)$ and get the d.f. of Y as

$$F_Y(y) = P\{(x_1, \ldots, x_n) : g(x_1, \ldots, x_n) \leq y\}.$$

For example, let $Y = X_1 + X_2$. Then

$$\begin{aligned} F_Y(y) &= P(X_1 + X_2 \leq y) \\ &= \int_{-\infty}^{\infty} \int_{-\infty}^{y-x_2} f(x_1, x_2)\, dx_1 dx_2, \quad \text{if } x_1 \text{ and } x_2 \text{ are continuous} \\ &= \int_{-\infty}^{\infty} \left[\int_{-\infty}^{y-x_2} f_{X_1}(x_1)\, dx_1 \right] f_{X_2}(x_2)\, dx_2, \quad \text{if } x_1 \text{ and } x_2 \text{ are independent} \\ &= \int_{-\infty}^{\infty} f_{X_1}(y - x_2) f_{X_2}(x_2)\, dx_2 \end{aligned}$$

and

$$f_Y(y) = \int_{-\infty}^{\infty} f_{X_1}(y - x_2) f_{X_2}(x_2)\, dx_2$$

which is called the *convolution* of the density functions of X_1 and X_2, The convolution operator is denoted by

$$f_Y(y) = f_{X_1}(y) \star f_{X_2}(y). \tag{A.17}$$

The convolution operator $\star$ may be used on the d.f. as

$$F_Y(y) = F_{X_1}(y) \star F_{X_2}(y) \tag{A.18}$$

which expresses the d.f. of $Y = X_1 + X_2$ in terms of the d.f.'s of X_1 and X_2. If $X_1, \ldots, X_n$ are i.i.d. r.v.'s each having the distribution of X and $Y = X_1 + \cdots + X_n$ then we adopt the following notation:

$$F_Y(y) = F_{X_1}(y) \star \cdots \star F_{X_n}(y) = F_X^{(n)}(y). \tag{A.19}$$

There is another method for determining the distribution of functions of r.v.'s, more particularly the sum of r.v.'s which will be discussed in the next section.

In $Y = g(X)$, if X is continuous and g is strictly monotonic, then the p.f. of Y can be expressed as

$$f_Y(y) = f_X(g^{-1}(y))\frac{dg^{-1}(y)}{dy}$$

in the appropriate region. Note that the strict monotonicity guarantees the existence of the inverse. This procedure may be generalized to determine the joint distribution of several functions $u_1(X_1, \ldots, X_n), \ldots, u_r(X_1, \ldots, X_n)$ $(r \leq n)$ of continuous r.v.'s. Put $u_i(x_1, \ldots, x_n) = u_i \; i = 1, \ldots, r$ and introduce $u_{r+1} = x_{r+1}, \ldots, u_n = x_n$. Solve for $x_1, \ldots, x_n$ in terms of $u_1, \ldots, u_n$. The solution may not be unique and there may be several sets of solutions (of course, $x_{r+1} = u_{r+1}, \ldots, x_n = u_n$ for each set). Denoting by $(x_{1j}^*, \ldots, x_{nj}^*)$ the jth set of solutions, the joint p.f. is

$$f_{U_1,\ldots,U_n}(u_1, \ldots, u_n) = \sum_j f_{X_1,\ldots,X_n}(x_{1j}^*, \ldots, x_{nj}^*)|J_j| \tag{A.20}$$

where $|J_j|$ is the absolute value of the Jacobian of transformation J_j which is the determinant given by

$$J_j = \begin{vmatrix} \frac{\partial x_{1j}^*}{\partial u_1} & \frac{\partial x_{1j}^*}{\partial u_2} & \cdots & \frac{\partial x_{1j}^*}{\partial u_n} \\ \frac{\partial x_{2j}^*}{\partial u_1} & \frac{\partial x_{2j}^*}{\partial u_2} & \cdots & \frac{\partial x_{2j}^*}{\partial u_n} \\ & & \vdots & \\ \frac{\partial x_{nj}^*}{\partial u_1} & \frac{\partial x_{nj}^*}{\partial u_2} & \cdots & \frac{\partial x_{nj}^*}{\partial u_n} \end{vmatrix}.$$

The marginal p.f. of $U_1, \ldots, U_r$ is obtained by integrating the joint p.f. over $u_{r+1}, \ldots, u_n$.

An application of the formula is in order. Given a set of r.v.'s $X_1, \ldots, X_n$ the set $X_{(1)}, \ldots, X_{(n)}$ where $X_{(i)}$ is the ith smallest r.v. $i = 1, \ldots, n$ is called its order statistics. Let us find the joint distribution of $(X_{(1)}, \ldots, X_{(n)})$ when

$X_1, \ldots, X_n$ are continuous and i.i.d. r.v.'s. Observe that $(x_{(1)}, \ldots, x_{(n)})$ is a permutation of $(x_1, \ldots, x_n)$ and conversely. Therefore, there is a set of $n!$ solutions of $(x_1, \ldots, x_n)$ in terms of $(x_{(1)}, \ldots, x_{(n)})$ and $|J|$ of each transformation is 1. Using (A.20), the resulting joint p.f. is

$$f(x_{(1)}, \ldots, x_{(n)}) = n! \prod_{i=1}^{n} f(x_{(i)}). \tag{A.21}$$

Besides the sum of random variables $X_1, \ldots, X_n$, one may consider a *mixture* of these random variables. Let F_i be the d.f. of X_i, $i = 1, \ldots, n$ and let $a_i \geq 0$ be real numbers such that $\sum_{i=1}^n a_i = 1$. Then it can be shown that $\sum_{i=1}^n a_i F_i$ is a d.f. of a certain random variable, which is called the mixture of $X_1, \ldots, X_n$.

A.3. Expectation

In the application of probability models, the average or expectation of an r.v. is a useful concept. In order to define it with respect to any d.f. we may define a Stieltjes integral which is a generalization of a Riemann integral. Given a continuous function $g(\cdot)$ and a d.f. $F(\cdot)$, we define the Stieltjes integral of $g(\cdot)$ with respect to $F(\cdot)$ over $(a, b]$ written $\int_a^b g(x)dF(x)$ as

$$\int_a^b g(x)dF(x) = \lim_{n \to \infty} \sum_{i=1}^{n} g(y_i)(F(x_i) - F(x_{i-1}))$$

where $a = x_0 < x_1 < \cdots < x_n = b$, $x_{i-1} < y_i \leq x_i$ $i = 1, \ldots, n$ and the limit is being taken when $\max_i |x_i - x_{i-1}| \to 0$. As usual

$$\int_{-\infty}^{\infty} g(x)dF(x) = \lim_{\substack{a \to -\infty \\ b \to \infty}} \int_a^b g(x)dF(x).$$

The *expectation* of a continuous function $g(X)$ of an r.v. X denoted by $E(g(X))$ is given by

$$E(g(X)) = \int_{-\infty}^{\infty} g(x)dF(x). \tag{A.22}$$

Stieltjes integrals are only of theoretical interest. In practice, a d.f. F is expressed as

$$F(x) = c_1 \int_{-\infty}^{x} f(y)dy + c_2 \sum_{y \leq x} p(y),$$

the first term being the continuous part and the second term the discrete part and c_1 and c_2 being non-negative constants such that $c_1 + c_2 = 1$. In that case,

$$E(g(X)) = c_1 \int_{-\infty}^{\infty} g(x)f(x)dx + c_2 \sum_{x} g(x)p(x).$$

The properties of Stieltjes integrals being very much the same as those of Riemann integrals, we end in the following important results:

(i) $E(X_1 + \cdots + X_n) = E(X_1) + \cdots E(X_n)$. (A.23)

(ii) If X and Y are independent, then $E(XY) = E(X)E(Y)$(A.24)

(iii) $E(X) = E(E(X|Y))$. (A.25)

The *rth moment* of an r.v. X is defined by $E(X^r)$ and the *rth central* moment by $E(X - E(X))^r$. The first moment $E(X)$ is called the *mean* or the average or the expected value of X and is denoted by μ_X, whereas the second central moment of X is called the *variance* of X and is denoted by $\text{Var}(X)$ or σ_X^2. For computational purpose, we have

$$\sigma_X^2 = E(X^2) - \mu_X^2.$$

If $X_1, \ldots, X_n$ are independent, then

$$\text{Var}(X_1 + \cdots + X_n) = \text{Var}(X_1) + \cdots + \text{Var}(X_n). \tag{A.26}$$

The *standard deviation* and the *coefficient of variation* of X are respectively given by

$$\sigma_X = \sqrt{\text{Var}(X)} \qquad \text{and} \qquad \frac{\sigma_X}{\mu_X}.$$

The *covariance* of two r.v.'s X and Y is defined by $\text{Cov}(X, Y) = E[(X - \mu_X)(Y - \mu_Y)]$ and the *correlation coefficient* by $\text{Cov}(X, Y)/\sigma_X\sigma_X$.

The moments of an r.v. X can be obtained from its *moment generating function* (briefly m.g.f.) denoted by $m_X(\theta)$, which is defined as

$$m_X(\theta) = E(e^{\theta X}) \tag{A.27}$$

provided the expectation exists for some real θ, θ being the parameter. It can be easily verified that

$$E(X^r) = \left.\frac{\partial^r m_X(\theta)}{\partial \theta^r}\right|_{r=0}$$

or equal to the coefficient of $\frac{\theta^r}{r!}$ in the power services expansion of $m_X(\theta)$. Two properties of the m.g.f. are listed below.

(i) $m_{aX+b}(\theta) = e^{\theta b} m_X(a\theta)$. (A.28)

(ii) If X and Y are independent, then $m_{X+Y}(\theta) = m_X(\theta)m_Y(\theta)$.
(A.29)

The m.g.f. is useful in another way for finding the distribution of a function of r.v.'s. If the m.g.f. of a function Y exists and happens to be the m.g.f. of some known distribution, then it is known that Y has that distribution. This result is referred to as the *uniqueness theorem*. The method is often applied to derive the distribution of the sum of i.i.d. r.v.'s which will be illustrated in the next section. It may be noted that the m.g.f. method for derivation is often simpler than the direct approach.

We state a very useful result.

Weak Law of Large Numbers

Let $X_1, X_2, \ldots$ be a sequence i.i.d. r.v.'s with mean μ and σ^2 which are finite. Let

$$\bar{X}_n = \frac{X_1 + \cdots + X_n}{n} .$$

Then for any $\epsilon > 0$,

$$\lim_{n\to\infty} P(|\bar{X}_n - \mu| \geq \epsilon) = 0.$$

The implication of the law is that the arithmetic mean $\bar{X}_n$ in the limit approaches its expectation. It is easy to check that

$$E(\bar{X}_n) = \mu \qquad \text{and} \qquad \text{Var}(\bar{X}_n) = \frac{\sigma^2}{n} .$$

An r.v. X is said to have a *degenerate* distribution at a if

$$P(X = a) = 1.$$

Clearly $E(X) = a$ and $\text{Var}(X) = 0$. Thus, the weak law of large numbers says that $\bar{X}_n$ in the limit approaches a degenerate r.v. at μ. The distribution is also called *deterministic* as opposed to random and is denoted by D.

A.4. Special Distributions

We present a list of distributions along with some of their properties. The list starts with discrete distributions.

(i) Bernoulli distribution

An r.v. X is said to have the *Bernoulli distribution* if

$$P(X = 1) = p \quad \text{and} \quad P(X = 0) = 1 - p = q \text{ (say)}. \tag{A.30}$$

Here p is the parameter of the distribution. Any experiment (tossing a coin) which results in two outcomes (heads, tails) can be represented by a Bernoulli r.v. (heads by 1, tails by 0). Actually the values of X may be changed to any two arbitrary values depending on a specific problem. It can be seen that

$$\mu_X = p, \quad \sigma_X^2 = pq \quad \text{and} \quad m_X(\theta) = pe^\theta + q.$$

(ii) Binomial distribution

The p.f. of the *binomial distribution* is

$$f_X(x) = \binom{n}{x} p^x (1-p)^{n-x}, \qquad x = 0, 1, \ldots, n, \tag{A.31}$$

where n and p are the parameters of the distribution. The binomial r.v. arises as the sum of n i.i.d. Bernoulli r.v.'s. In other words, in a sequence of n independent experiments each having two outcomes, say a success with probability p or a failure with probability $1 - p$, the number of successes has the binomial distribution. Its mean, variance and m.g.f., respectively, are

$$\mu_X = np, \quad \sigma_X^2 = npq \quad \text{and} \quad m_X(\theta) = (pe^\theta + q)^n. \tag{A.32}$$

Although these expressions can be obtained directly from (A.31), one can also get these from the Bernoulli distribution by remembering that the binomial r.v. is the sum of n i.i.d. Bernoulli r.v.'s and by using (A.23), (A.26) and (A.29).

We present an application of the uniqueness theorem by considering $Y = X_1 + X_2$ where X_i $i = 1, 2$ has the binomial distribution with parameters n_i and p. Then

$$\begin{aligned} m_Y(\theta) &= E(e^{\theta Y}) = E[e^{\theta(X_1+X_2)}] \\ &= E(e^{\theta X_1})E(e^{\theta X_2}) \text{ since } X_1 \text{ and } X_2 \text{ are independent} \\ &= m_{X_1}(\theta)m_{X_2}(\theta) \\ &= (pe^{\theta} + q)^{n_1+n_2} \end{aligned}$$

which by the uniqueness theorem says that Y has the binomial distribution with parameters $n_1 + n_2$ and p.

(iii) Multinomial distribution

The *multinomial distribution* as the name suggests is an extension of the binomial distribution in the sense that instead of two resulting outcomes the experiment has r possible outcomes with respective probabilities $p_1, \ldots, p_r$ ($p_i \geq 0$ for all i, $p_1 + \cdots + p_r = 1$). Let X_i be the number of times the ith outcome occurs in n trials. Then the joint p.f. of the multivariate distribution is

$$f(x_1, \ldots, x_r) = \frac{n!}{x_1! \cdots x_r!}\, p_1^{x_1} \cdots p_r^{x_r} \tag{A.33}$$

for $x_i = 0, 1, \ldots, n$ for every i and $x_1 + \cdots + x_r = n$.

(iv) Geometric distribution

The p.f. of the *geometric distribution* is

$$f_X(x) = p(1-p)^x, \qquad x = 0, 1, \ldots \tag{A.34}$$

with p as the parameter. It arises in a sequence of i.i.d. Bernoulli trials with outcomes say, a success and a failure. Letting X represent the number of failures preceding the first success we can see that $P(X = x)$ is given by (A.34). In other words, the waiting time for the first success has the geometric distribution. By this time, one may have noticed that an r.v. or the corresponding distribution can be described either by its p.f. or by a random phenomenon of an appropriate experiment. The mean, variance and m.g.f. of the distribution, respectively, are

$$\frac{q}{p}, \ \frac{q}{p^2} \text{ and } \frac{p}{1-qe^{\theta}} \qquad \text{where } q = 1-p. \tag{A.35}$$

It can be checked that

$$\bar{F}_X(x) = P(X \geq x) = q^{x+1}.$$

Therefore

$$P(X \geq x + t | X \geq t) = P(X \geq x)$$

which is independent. This property of independence of the conditional probability is known as the *Markov property* or the *memoryless property.*

(v) Negative binomial distribution

In i.i.d. Bernoulli trials with a success and a failure as outcomes if X represents the number of failures preceding the rth success, then X has the *negative binomial distribution* with p.f.

$$f_X(x) = \binom{r+x-1}{x} p^r (1-p)^x, \qquad x = 0, 1, \ldots \tag{A.36}$$

r and p being the parameters. The fact that X is the sum of r independent geometric r.v.'s can be seen by treating the sequence of trials after each success a fresh one. Therefore by using (A.23), (A.26) and (A.29) we can get μ_X, σ_X^2 and $m_X(\theta)$ as

$$\frac{rq}{p}, \frac{rq}{p^2} \quad \text{and} \quad \left[\frac{p}{1-qe^\theta}\right]^r, \tag{A.37}$$

respectively.

(vi) Poisson distribution

The p.f. of the *Poisson distribution* is

$$f_X(x) = \frac{e^{-\lambda}\lambda^x}{x!}, \qquad x = 0, 1, \ldots \tag{A.38}$$

$\lambda(\lambda > 0)$ being the parameter of the distribution (briefly written to be Poisson (λ)). The binomial distribution can be approximated by the Poisson distribution as $n \to \infty$ and $p \to 0$ such that $np = \lambda$. Using the approximation one can visualize a practical situation leading to the Poisson distribution which will now be described. Suppose events occur during a unit of time interval. Divide the interval into n equal sub-intervals. Suppose the length of the sub-interval is so small that during that period at most one event can occur and the probability of the occurrence is p. Clearly this setting corresponds to n i.i.d. Bernoulli trials. Let the mean number of events during the unit interval be λ which is equal to np. The fact that at most one event can occur in any sub-interval implies n to be large. In addition if λ is fixed then p has to be small. Under this structure, the number of events occuring during a unit interval has the Poisson distribution.

The Poisson distribution is quite common in queueing theory and the random phenomenon that leads to the distribution will be discussed under Poisson processes and will have a structure similar to the one described above. The mean, variance and m.g.f., respectively, are

$$\lambda, \lambda \text{ and } e^{\lambda(e^\theta - 1)}. \tag{A.39}$$

By using the m.g.f. technique as was done in the case of the binomial distribution, we can show that the sum of two Poisson variables with parameters λ_1 and λ_2 has the Poisson distribution with parameter $\lambda_1 + \lambda_2$.

Listed below are some continuous distributions.

(i) Uniform distribution

The p.f. of the *uniform distribution* over the interval (a, b) is

$$f_X(x) = \frac{1}{b-a}\,, \qquad a < x < b. \tag{A.40}$$

Its d.f. is

$$F_X(x) = \begin{cases} 0 & x < a, \\ \dfrac{x}{b-a} & a \le x \le b, \\ 1 & x > b. \end{cases}$$

The mean, variance and m.g.f., respectively, are

$$\frac{a+b}{2}\,, \quad \frac{(b-a)^2}{12} \quad \text{and} \quad \frac{e^{\theta b} - e^{\theta a}}{\theta(b-a)}\,. \tag{A.41}$$

(ii) Exponential distribution

Besides the Poisson distribution, a continuous distribution used quite frequently in queueing theory is the *exponential distribution* which is defined by its p.f. with parameter μ $(\mu > 0)$ as

$$f_X(x) = \mu e^{-\mu x}, \qquad x \ge 0. \tag{A.42}$$

Its d.f. is

$$F_X(x) = 1 - e^{-\mu x}, \qquad x \ge 0, \tag{A.43}$$

and μ_X, σ_X^2 and $m_X(\theta)$, respectively, are

$$\frac{1}{\mu}\,, \quad \frac{1}{\mu^2} \quad \text{and} \quad \frac{\mu}{\mu - \theta}\,. \tag{A.44}$$

An r.v. having the exponential distribution with parameter μ is briefly written as exponential (μ). This distribution is the continuous version of the geometric distribution. Just as the Poisson distribution is derived as a limiting case of the binomial distribution, we get the exponential distribution as a limiting distribution of the geometric distribution. The situation is the same except that here we consider a time interval $(0, x)$ and the random phenomenon Y of interest is the number of nonoccurrences of the event before the first occurrence of the event. By the geometric distribution we know that

$$P(Y \ge k) = (1-p)^k.$$

Suppose the mean number of events per unit time (i.e., the rate) is fixed to be μ. Then

$$\mu = \frac{np}{x}\,.$$

Let $n \to \infty$, $p \to 0$ so that $\mu x = np$ and let X be the limiting random variable. We have

$$P(X > x) = \lim_{n\to\infty} P(Y \ge n) = \lim_{n\to\infty} \left(1 - \frac{\mu x}{n}\right)^n = e^{-\mu x}$$

which by comparing with (A.43) shows that X is an exponential random variable. Our discussion indicates that X can be considered as the duration between the occurrence of two consecutive events, such as the interarrival time or service time. The exponential variable X also satisfies the Markov or memoryless property since

$$P(X > x + t | x > t) = P(X > x) = e^{-\mu x} . \tag{A.45}$$

The existence of the Markov property simplifies a great deal of mathematical complexities that could arise in solving a problem. Thus there would be a great temptation to use the exponential distribution in queueing models. Indeed, observations have shown that this distribution provides a good approximation to some service-time distributions.

Let us further discuss the Markov property (A.45). Consider a random variable X which satisfies

$$P(X > x + t | X > t) = e^{-\mu x},$$

or equivalently

$$P(X > x + t | X > t) = 1 - \mu x + o(x).$$

Then it follows that

$$\frac{d}{dt} P(X > t) = -\mu P(X > t)$$

which has the unique solution

$$P(X > t) = e^{-\mu t}$$

by using the boundary condition $P(X > 0) = 1$. This proves that the Markov property implies the exponentiality of the random variable. In fact, it can be proved that the exponential distribution is the only distribution which satisfies

$$P(X > x + t | X > t) = P(X > x)$$

which is equivalent to

$$P(X > x + t) = P(X > t) P(X > x).$$

A mixture of exponential distributions is known as *hyper-exponential distribution.*

(iii) Erlangian(gamma) distribution

First, we define the gamma function $\Gamma(\cdot)$ as

$$\Gamma(t) = \int_0^\infty x^{t-1} e^{-x}\, dx, \qquad t > 0. \tag{A.46}$$

Clearly

$$\Gamma(t) = (t-1)\Gamma(t-1) \qquad \text{and} \qquad \Gamma(1) = 1$$

from which it follows that for any non-negative integer n

$$\Gamma(n+1) = n!$$

It can be proved that

$$\Gamma(1/2) = \sqrt{\pi} .$$

An r.v. X has the *gamma distribution* with parameters α and λ ($\alpha > 0, \lambda > 0$) if its p.f. is written as

$$f_X(x) = \frac{\lambda^\alpha}{\Gamma(\alpha)}\, e^{-\lambda x} x^{\alpha-1}, \qquad x \geq 0. \tag{A.47}$$

When α is a positive integer and equal to n, we get the *Erlangian distribution* having the p.f.

$$f_X(x) = \frac{\lambda^n}{(n-1)!}\, e^{-\lambda x} x^{n-1} \qquad x \geq 0. \tag{A.48}$$

Usually the parameter n in the Erlangian distribution is referred to as the number of stages or phases. The mean, variance and m.g.f. of the gamma distribution, respectively, are

$$\frac{\alpha}{\lambda},\ \frac{\alpha}{\lambda^2} \text{ and } \left(1 - \frac{\theta}{\lambda}\right)^{-\alpha}. \tag{A.49}$$

Using the m.g.f. and the uniqueness theorem, again it can be shown that the sum of two independent gamma r.v.'s with parameters (α_1, λ) and (α_2, λ) is a gamma r.v. with parameters $(\alpha_1 + \alpha_2, \lambda)$. From the point of queueing theory more important than this is the fact that the sum of n i.i.d. exponential r.v.'s each having the same parameter λ has the Erlangian distribution as given by (A.48). Obviously, the Erlangian distribution is a continuous analogue of the negative binomial distribution.

(iv) Normal distribution

The *normal distribution* is characterized by its p.f. which is given by

$$f_X(x) = \frac{1}{\sqrt{2\pi}\sigma}\, e^{-(x-\mu)/2\sigma^2}, \qquad x \geq 0 \tag{A.50}$$

with parameters μ ($-\infty < \mu < \infty$) and $\sigma(\sigma > 0)$. The mean, variance and m.g.f., respectively, are

$$\mu,\ \sigma^2 \text{ and } e^{\mu\theta + (1/2)\sigma^2\theta^2}. \tag{A.51}$$

The distribution is denoted by $N(\mu, \sigma^2)$. When $\mu = 0$ and $\sigma = 1$, the distribution is known as the *standard normal distribution* and is denoted by $N(0, 1)$. An r.v. X is *standardized* by a new variable $Z = \frac{X-\mu}{\sigma}$. Note that if X has normal distribution with mean μ and variance σ^2, then Z has the standard normal distribution.

From the m.g.f. it is obvious that the sum of n independent normal variables with mean and variance as $(\mu_i, \sigma_i^2), i = 1, \ldots, n$ is again a normal random variable with mean and variance as $\Sigma\mu_i$, $\Sigma\sigma_i^2$, respectively. In queueing theory the normal distribution may arise as a limiting distribution of sums of random variables through what is known as the central limit theorem which is stated next.

Central Limit Theorem:

Let $\{X_i\}$ be a set of i.i.d. r.v.'s with mean μ and variance σ^2 which are assumed to be finite. Then the standardized r.v. Z_n of $\sum_{i=1}^n X_i$ as given by

$$Z_n = \frac{\sum_{i=1}^n X_i - n\mu}{\sqrt{n}\sigma}$$

approaches the standard normal distribution as $n \to \infty$.

A useful result connecting the d.f. of a continuous r.v. and a uniform distribution is as follows:

Let X be a continuous r.v. and let $Y = F_X(X)$. Then Y has the uniform distribution over (0,1).

This is so simply because

$$F_Y(y) = P(F_X(X) \leq y) = P(X \leq F_X^{-1}(y)) = F_X F_X^{-1}(y) = y,$$

and F_X^{-1} exists.

The references for A.1–A.4 are given below.

References

Hogg, R. V. and Craig, A. T. (1978). *Introduction to Mathematical Statistics*, 4th ed., Macmillan Publishing Co., New York.

Ross, S. M. (1985). *Introduction to Probability Models*, 3rd ed., Academic Press, New York.

A.5. Complex Variables and Integral Transforms

Analytic Functions

A function $f(z)$ of a complex variable z is said to by *analytic* at the point $z_0 \in C$, C being an open subset of the complex plane, if it is complex differentiable at each point of a neighborhood of z_0. The terms *holomorphic* and *regular* are synonyms of analytic. An *entire function* is a function which is analytic at each point of the complex plane. In contrast to real analysis, a function which is complex differentiable has derivatives of any order and therefore can be expanded as a Taylor series around z_0:

$$f(z) = f(z_0) + f'(z_0)(z - z_0) + \frac{f''(z_0)}{2!}(z - z_0)^2 + \dots \tag{A.52}$$

This series converges for each z with $|z - z_0| < \rho$, ρ is called the *radius of convergence*.

One of the most fundamental results in complex variable theory is the *integral formula of Cauchy* (Henrici (1974)). Let $f(z)$ be analytic everywhere inside a region which is enclosed by a simple closed contour C with positive orientation. Then for any z_0 interior to C

$$f(z_0) = \frac{1}{2\pi i} \oint_C \frac{f(z)}{z - z_0} dz. \tag{A.53}$$

Now let $f(z)$ be analytic in the *annulus* $A : \rho_2 < |z - z_0| < \rho_1$, i.e., the region bounded by two concentric circles C_1 and C_2 centered at z_0, C_1 having radius ρ_1, C_2 radius ρ_2. Then $f(z)$ can still be expanded in powers of $z - z_0$, but the resulting series, which is called *Laurent series*, may contain positive and negative powers of $z - z_0$:

$$f(z) = \sum_{n=0}^{\infty} a_n (z - z_0)^n + \sum_{n=1}^{\infty} \frac{b_n}{(z - z_0)^n}, \qquad z_0 \in A, \tag{A.54}$$

where

$$a_n = \frac{1}{2\pi i} \oint_{C_1} \frac{f(z)}{(z - z_0)^{n+1}} dz, \quad n = 0, 1, \ldots$$

and

$$b_n = \frac{1}{2\pi i} \oint_{C_2} \frac{f(z)}{(z - z_0)^{-n+1}} dz, \quad n = 1, 2, \ldots$$

The coefficient b_1 is called the *residue* of $f(z)$ at z_0 and is denoted by $\mathrm{Res}(f, z_0)$.

Some Important Theorems on Analytic Functions

The following results are particularly useful when dealing with Laplace transforms and generating functions.

1. Residue Theorem. Suppose $f(z)$ is analytic inside and on a simple closed curve C except for isolated singularities $a_1, a_2, \ldots, a_m$ inside C. Then by the *residue theorem*

$$\oint_C f(z)dz = 2\pi i \sum_{j=1}^{m} \mathrm{Res}(f, a_j) \tag{A.55}$$

Observe that Cauchy's integral formula is a special case of the residue theorem.

2. Lagrange Series. Let $f(z)$ and $\Phi(z)$ be analytic inside and on the contour C surrounding a point a, and let t be such that the inequality $|t\Phi(z)| < |z - a|$ holds at all points on the perimeter of C. Then the equation $\xi = a + t\Phi(\xi)$ regarded as an equation in ξ has exactly one root inside C and any function $f(\xi)$ which is analytic inside C can be expanded as a power series in t by

$$f(\xi) = f(a) + \sum_{i=1}^{\infty} \frac{t^n}{n!} \frac{d^{n-1}}{da^{n-1}} \left[f'(a)\Phi^n(a) \right]. \tag{A.56}$$

The series (A.56) is called *Lagrange expansion* of f.

3. Rouché's Theorem. Let $f(z)$ and $g(z)$ be analytic inside a region C and $|g(z)| < |f(z)|$ in each point on the boundary of C. Then $f(z)$ and $f(z) \pm g(z)$ have the same number of zeroes inside C.

4. Liouville's Theorem. If $f(z)$ is an entire function and bounded for all values z in the complex plane, then $f(z)$ is constant.

Laplace Transforms

The *Laplace transform* (briefly, L.T.) of a function $f(t), t \geq 0$ will be denoted by $f^*(s)$ and is defined as (Henrici (1977))

$$f^*(s) = \int_0^\infty e^{-st} f(t)dt. \tag{A.57}$$

A sufficient condition for the existence of (A.57) is that $f(t)$ must be of *bounded exponential growth*, which means that there are numbers μ and γ, such that for all $t > 0$ $|f(t)| < \mu e^{\gamma t}$. If two functions f_1 and f_2 have the same L.T., then they are equivalent in the sense that f_1 and f_2 have the same values for all t except at most on a set of points with no finite point of accumulation. This property is quite significant since it guarantees the existence of an *inverse transform.*

The *Laplace-Stieltjes transform* (briefly, L.S.T.) is an extension of the L.T. It is the transform of the d.f. $F(t)$ of a non-negative random variable X to be denoted by $\Phi_X(s)$ and defined by

$$\Phi_X(s) = \int_0^\infty e^{-st} dF(t). \tag{A.58}$$

Clearly $\Phi_X(s) = f^*(s)$, if X has a density $f(t)$. Differentiation of the L.S.T. with respect to s is justified and yields the moments of X. The r-th moment of X is given by

$$E(X^r) = (-1)^r \left.\frac{d^r \Phi_X(s)}{ds^r}\right|_{s=0}. \tag{A.59}$$

Table A.1 summarizes the most important properties of Laplace transforms and Table A.2 lists important functions and their L.T.'s.

Finding the inverse $f(t)$ of a transform $f^*(s)$ is in general a quite difficult task. Sometimes this task can be accomplished by simple table lookup (see Table A.2) and the use of particular properties of L.T.'s. But in principle the inverse transform can always be determined by means of the *complex inversion formula:*

$$f(t) = \frac{1}{2\pi i} \int_{\gamma - i\infty}^{\gamma + i\infty} e^{st} f^*(s)ds, \qquad t > 0, \tag{A.60}$$

where γ is a real number to be chosen in such a way that all singularities of $f^*(s)$ lie left from the vertical line $s = \gamma + iy$.

If $f^*(s)$ happens to be a *rational function*, then formal inversion is possible by a *partial fraction expansion.* Suppose $f^*(s) = U(s)/V(s)$, where U and V are two polynomials in s, V being of degree n and U having degree $< n$. Assume further that V has n distinct roots $\alpha_i, i = 1, \ldots, n$, so that

$$V(s) = (s - \alpha_1)(s - \alpha_2) \cdots (s - \alpha_n)$$

TABLE A.1. Properties of Laplace transforms

$f^*(s)$	$f(t)$
$\alpha f_1^*(s) + \beta f_2^*(s)$	$\alpha f_1(t) + \beta f_2(t)$
$f^*(s/a)$	$af(at)$
$f^*(s-a)$	$e^{at}f(t)$
$sf^*(s) - f(0)$	$f'(t)$
$s^2 f^* - sf(0) - f'(0)$	$f''(t)$
$\frac{f^*(s)}{s}$	$\int_0^t f(u)du$
$f^*(s)g^*(s)$	$\int_0^t f(u)g(t-u)du$
$\frac{U(s)}{V(s)}$ $U(s)$ is a polynomial of degree $< n$ $V(s) = (s-\alpha_1)(s-\alpha_2)\cdots(s-\alpha_n)$, roots α_i distinct	$\sum_{i=1}^n \frac{U(\alpha_i)}{V'(\alpha_i)} e^{\alpha_i t}$

Then an expansion of $f^*(s)$ into *partial fractions* yields

$$f^*(s) = \sum_{i=1}^n \frac{A(\alpha_i)}{s-\alpha_i},$$

where $A(\alpha_i) = U(\alpha_i)/V'(\alpha_i)$. Since there are only finitely many terms and the inverse transform of $1/(s-\alpha_i)$ is $e^{\alpha_i t}$, a term by term inversion yields

$$f(t) = \sum_{i=1}^n \frac{U(\alpha_i)}{V'(\alpha_i)} e^{\alpha_i t}$$

The *Final Value Theorem* establishes an important relation between the behavior of $f^*(s)$ at zero and $f(t)$ at infinity. If both limits $\lim_{s\to 0} sf^*(s)$ and $\lim_{t\to\infty} f(t)$ exist, then

$$\lim_{s\to 0} sf^*(s) = \lim_{t\to\infty} f(t).$$

Generating Functions

Let $\{a_n, n = 0,1,\ldots\}$ be a sequence of real numbers. The *generating function* (briefly, g.f.) $G(z)$ of this sequence is defined as $G(z) = \sum_{i=0}^{\infty} a_n z^n$. If $\{a_n\}$ is the p.f. of a non-negative discrete random variable X, then $G(z)$ is called a *probability generating function* (briefly, p.g.f). Whereas a g.f. may be defined only as a formal power series, a p.g.f. is an analytic function, which converges for all $|z| < 1$. Moreover for a p.g.f. $G(1) = 1$. Repeated

TABLE A.2. Table of Laplace transforms

$f^*(s)$		$f(t)$
$\frac{1}{s}$		1
$\frac{1}{s^\nu}$	$\nu > 0$	$\frac{t^{\nu-1}}{\Gamma(\nu)}$
$\frac{1}{s-a}$		e^{at}
$\frac{1}{(s-a)^\nu}$	$\nu > 0$	$\frac{t^{\nu-1}e^{at}}{\Gamma(\nu)}$
$\frac{1}{\sqrt{s^2-a^2}}$		$I_0(at)$
$\frac{(s-\sqrt{s^2-a^2})^\nu}{\sqrt{s^2-a^2}}$	$\nu > -1$	$a^n I_n(at)$
$\frac{s}{(s^2-a^2)^{3/2}}$		$tI_0(at)$
$\frac{1}{(s^2-a^2)^{3/2}}$		$\frac{tI_1(at)}{a}$
$\frac{e^{-as}}{s}$		$f(t) = \begin{cases} 0 & t < a \\ 1 & t \geq a \end{cases}$

differentiation of $G(z)$ yields the *factorial moments* of X:

$$E[X(X-1)\cdots(X-r+1)] = \left.\frac{d^r G(z)}{dz^r}\right|_{z=1}$$

Given a g.f. $G(z)$ the original sequence $\{a_n\}$ can be recovered by expanding $G(z)$ into a power series. This can be done in various ways. In principle it is always possible to find a_n by repeated differentiation as

$$a_n = \frac{1}{n!} \left.\frac{d^n G(z)}{dz^n}\right|_{z=0}.$$

An alternative is provided by the *residue theorem*:

$$a_n = \frac{1}{2\pi i} \oint_C \frac{f(z)}{z^{n+1}} dz,$$

where C is a sufficiently small circle around the origin with positive orientation. If $G(z)$ is a *rational function*, then the sequence $\{a_n\}$ can be determined by partial fraction expansion.

Bessel Functions of Integer Order

Consider the *Bessel differential equation* (Abramowitz and Stegun (1965))

$$z^2 \frac{d^2 u}{dz^2} + z\frac{du}{dz} + (z^2 - n^2)u = 0,$$

where n is a non-negative integer. A family of solutions of this equations is given by the *Bessel functions of the first kind* of order n, denoted by $J_n(z)$. $J_n(z)$ is an entire function and has the following power series representation:

$$J_n(z) = \sum_{k=0}^{\infty} \frac{(-1)^k (z/2)^{n+2k}}{k!(n+k)!}.$$

The values of $J_n(z)$ on the imaginary axis $z = iy$ define a real valued function $I_n(y) = i^{-n} J_n(iy)$, which is called *modified Bessel function* of order n. Its power series representation is

$$I_n(y) = \sum_{k=0}^{\infty} \frac{(y/2)^{n+2k}}{k!(n+k)!}.$$

It can be shown that for large values of y

$$I_n(y) \sim \frac{e^y}{\sqrt{2\pi y}}.$$

A generalization of the functions $I_n(y)$ has been introduced into queueing theory by Luchak (1956), the *generalized modified Bessel functions* $I_n^\ell(y)$, defined by the power series expansion

$$I_n^\ell(y) = \sum_{k=0}^{\infty} \frac{(y/2)^{m+k(\ell+1)}}{k!(m+\ell k)!}.$$

Properties of modified Bessel functions. For all non-negative integers n $I_n(y) = I_{-n}(y)$ and $I_n(0) = \delta_{0n}$, the Kronecker delta. The functions $I_n(y)$ satisfy the recurrence relation

$$I_{n-1}(y) - I_{n+1}(y) = \frac{2n}{y} I_n(y).$$

The *generating function* of $I_n(y)$ is given by the Laurent series

$$\sum_{n=-\infty}^{\infty} x^n I_n(y) = \exp\left[\frac{y}{2}\left(x + \frac{1}{x}\right)\right].$$

Differentiating with respect to x and setting $y = 2t\sqrt{\lambda\mu}$ and $x = \sqrt{\mu/\lambda}$ yields

$$\frac{e^{-(\lambda+\mu)t}}{\mu t} \sum_{n=-\infty}^{\infty} n \left(\frac{\mu}{\lambda}\right)^{n/2} I_n(2t\sqrt{\lambda\mu}) = 1 - \rho,$$

where $\rho = \lambda/\mu$. This can be rewritten to give the important identity

$$(1-\rho)\rho^n = \frac{\rho^n e^{-(\lambda+\mu)t}}{\mu t} \sum_{n=-\infty}^{\infty} n\rho^{-n/2} I_n(2t\sqrt{\lambda\mu}).$$

Another important identity is

$$\sum_{k=-\infty}^{n+i} k\rho^{-k/2} I_k(2t\sqrt{\lambda\mu}) = \sum_{m=0}^{\infty} \frac{(\lambda t)^m}{m!} \sum_{k=0}^{m+n+i} (k-m) \frac{(\mu t)^k}{k!}.$$

References

Abramowitz, M. and Stegun, I. A. (1965). *Handbook of Mathematical Functions*, Dover, New York.

Henrici, P. (1974). *Applied and Computational Complex Analysis*, Vol. 1, John Wiley & Sons, New York.

Henrici, P. (1977). *Applied and Computational Complex Analysis*, Vol. 2, John Wiley & Sons, New York.

Luchak, G. (1965). The solution of the single channel queueing equations characterized by a time-dependent Poisson-distributed arrival rate and a general class of holding times. *Oper. Res.*, **4**, 711–732.

A.6. Stochastic Processes

The queueing system is a function of time and thus is a *stochastic process* which in general, denoted by $\{X(t), t \in T\}$, is a family of r.v.'s $X(t)$ indexed by the set T, being known as the *parameter space* of the process. Usually t represents the time and $X(t)$ is the r.v. to represent the state of the process at time t. Examples of processes are the number of crimes in a country in each year over a period, the number of customers in a queueing system at any time, the temperature at a city on successive days of a year, the amount of water in a dam at any time.

If T is countable (or finite), then the process is called a *discrete-time process* and usually written as $\{X_n,\ n = 0, 1, \ldots\}$. On the other hand, if it consists of a set of finite or infinite intervals on the time axis, then we say the process to be a *continuous-time process.* The set of possible values of $X(t)$ is known as its *state space.* By analogy with the parameter space, the behaviour of the state space makes the process to be either a *discrete-state space* or a *continuous-state space* depending on whether the state space is countable or not.

Besides the parameter space and the state space, a process is characterized by the dependency relation among random variables $X(t)$ at different values of t. On the basis of the relation, various classes of processes are defined. A continuous-time stochastic process $\{X(t), t \in T\}$ is said to have *independent increments* if for any $t_0 < t_1 < \cdots < t_n$, $X(t_1) - X(t_0), X(t_2) - X(t_1), \ldots, X(t_n) - X(t_{n-1})$ are independent. It is *stationary* if $X(t_2 + s) - X(t_1 + s)$ has the same distribution as of $X(t_2) - X(t_1)$ for all $t_1, t_2, t_1 + s, t_2 + s \in T$, and $s > 0$.

A very important class of discrete-state processes is referred to as Markov processes in which the future represented by the r.v. $X(s + t)$, $s > 0$ given the present variable represented by $X(t)$, and the past variables represented by $X(u)$, $u < t$ is independent of the past. A process $\{X(t),\ t \in T\}$ is called a *Markov process* if it is stationary and

$$\begin{aligned} P(X(t_{n+1}) &= x_{n+1} \mid X(t_1) = x_1, \ldots,\ X(t_n) = x_n) \\ &= P(X(t_{n+1}) = x_{n+1} \mid X(t_n) = x_n) \end{aligned} \tag{A.61}$$

for any $t_1 < t_2 < \cdots < t_{n+1}$ holds good. The definition suggests that a Markov process should have the memoryless property. In fact, it is known that for a continuous-time process the time spent in any given state (i.e., the sojourn time) has an exponential distribution. If the Markov process is a discrete-time process with similar property, then it is called a Markov chain. More precisely, a *Markov chain* is a discrete-time process $\{X_n,\ n = 0, 1, \ldots\}$ with discrete state space which is stationary and satisfies

$$P(X_{n+1} = x_{n+1} \mid X_1 = x_1, \ldots, X_n = x_n) = P(x_{n+1} = x_{n+1} \mid X_n = x_n). \tag{A.62}$$

Interestingly, but not surprisingly the time that the process spends in a given state has a geometric distribution which is the discrete version of the memoryless property. Letting P_{ij} to be the *one-step transition probability* of moving from state i to state j, it can be shown that a Markov process (chain) can be alternatively defined as a stochastic process with transition probabilities $\{P_{ij}\}$ and the sojourn time in a state or duration between any two consecutive events to be exponential (geometric). Realize that $P_{ii} = 0$ for a Markov process.

An important special case of Markov process is the birth-death process which plays a significant role in the development of queueing theory. A *birth-death process* has the state space $\{0, 1, 2, \ldots\}$ and is a Markov process in which the transition from any state k only to its two neighbouring states $k + 1$ and $k - 1$ (when $k = 0$, $k - 1$ does not exist) is permitted with rates (mean per unit time) depending on k. For example, in a queueing system the state may be considered to be the number of customers in the system. If the number of customers at a given time is k ($k \geq 1$) then the process can change its position from k to either $k + 1$ or $k - 1$. A *Poisson process* is again a special case of the birth-death process when the process from a given state k moves only to $k + 1$ with a fixed rate irrespective of the state. In this way, the Poisson process $\{X(t),\ t \geq 0\}$ is a *counting process* where $X(t)$ represents the number of events occurred during the interval $(0, t)$ and the duration between any two consecutive events is exponential. The birth-death process possesses the property of stationary independent increments. We will study both birth-death and Poisson processes in detail in a separate section.

We have observed that in a Markov process the time spent in a state has the distribution (exponential or geometric) with memoryless property. When the distribution is arbitrary, the ensuing process is called a *semi-Markov process* which is a generalization of the Markov process. An example of semi-Markov processes is a class of processes of interest known as random walks. Consider a set of i.i.d. r.v.'s $\{X_n\}$. Then the discrete-time process $\{S_n,\ n = 0, 1, \ldots\}$ where $S_0 = 0$ and $S_n = X_1 + \cdots + X_n$ is called a *random walk*. It may be thought of as a particle moving from a state S_{n-1} at stage $n-1$ to S_n at the next stage by an amount X_n. A random walk can also be a continuous-time process if the instants at which a new variable is added are points on a continuum. Note that it does not necessarily possess the Markov property because the transition interval has an arbitrary distribution and thus it is a semi-Markov process.

A counting process $\{N(t),\ t \geq 0\}$ which is a generalization of the Poisson process is called a *renewal process* if the durations between any two consecutive events called *interarrival times* are i.i.d. r.v.'s (not necessarily exponential). Clearly if S_n represents the duration for the nth event to occur then $\{S_n\}$ is a random walk. The following equivalence holds good:

$$\{N(t) \geq n\} \equiv \{S_n \leq t\} \quad \Longrightarrow \quad P(N(t) \geq n) = P(S_n \leq t). \tag{A.63}$$

A renewal process is completely characterized by the common distribution function $F(t)$ of the interarrival times X_n. Since $S_n = \sum_{i=1}^{n} X_i$ and the X_i are independent, S_n has distribution function $F_n(t)$, where $F_n(t)$ denotes the $n-$fold convolution of F with itself. Furthermore because of the relation (A.63)

$$\begin{aligned} P(N(t) = n) &= P(S_n \leq t) - P(S_{n+1} \leq t) \\ &= F_n(t) - F_{n+1}(t). \end{aligned} \tag{A.64}$$

The *renewal function* $H(t)$ is defined as the expected number of renewals in $(0, t)$. By (A.64) we have

$$H(t) = E[N(t)] = \sum_{n=1}^{\infty} n[F_n(t) - F_{n+1}(t)] = \sum_{n=1}^{\infty} F_n(t). \tag{A.65}$$

Conditioning on the time $s, 0 \leq s \leq t$, of the first renewal it follows that $H(t)$ satisfies the integral equation

$$H(t) = F(t) + \int_0^t H(t-s)dF(s), \tag{A.66}$$

which is called *renewal equation* (see Takács (1958)).

We state a few limit theorems. Let $\mu = E(X_n)$, which always exists since X_n is non-negative though μ need not be finite. The *elementary renewal theorem* states that

$$\frac{H(t)}{t} \to \frac{1}{\mu} \quad \text{as} \quad t \to \infty, \tag{A.67}$$

and this limit has to be interpreted as zero if $\mu = \infty$. A similar result is *Blackwell's Theorem* (Blackwell (1945)). If the interarrival times are continuous random variables, then for $a > 0$

$$(H(t+a) - H(t)) \to \frac{a}{\mu} \quad \text{as} \quad t \to \infty. \tag{A.68}$$

This is a special case of the *Key Renewal Theorem*, also called *Smith's Theorem* (Smith (1958)). Let $G(t)$ be Riemann integrable with $G(t) = 0$ for $t < 0$. If F is a continuous distribution, then

$$\lim_{t \to \infty} \int_0^t G(t-x)dH(x) = \frac{1}{\mu} \int_0^{\infty} G(t)dt. \tag{A.69}$$

This limit has to be interpreted as zero, if $\mu = \infty$. If we put $G(t) = 1$ for $0 \le t \le a$ and zero otherwise, then we immediately obtain (A.68).

Next we introduce residual waiting time and age. Let $Y(t)$ be the time until the next renewal and $Z(t)$ the time from the last renewal to t, i.e.,

$$Y(t) = S_{N(t)+1} - t, \quad Z(t) = t - S_{N(t)}.$$

$Y(t)$ is called the *residual waiting time* and $Z(t)$ the *age* of the renewal process. The distribution of $Y(t)$ is given by

$$P(Y(t) \le x) = F(t+x) - \int_0^t [1 - F(t+x-y)]\, dH(y).$$

From the equivalence

$$\{Z(t) > x\} \equiv \{\text{no renewal in } [t-x, t]\} \equiv \{Y(t-x) > x\}$$

it follows that

$$P(Z(t) \le x) = \begin{cases} F(t) - \int_0^{t-x} [1 - F(t-y)]\, dH(y) & x \le t \\ 1 & x > t \end{cases}.$$

By Smith's Theorem $Y(t)$ and $Z(t)$ have the same limiting distribution

$$\lim_{t\to\infty} P(Y(t) \le x) = \lim_{t\to\infty} P(Z(t) \le x) = \frac{1}{\mu}\int_0^x [1 - F(y)]\, dy, \tag{A.70}$$

with expectation $E(X_n^2)/(2E(X_n))$, provided F is continuous.

A simple process $\{S_n, n = 0, 1, \ldots\}$ of interest is called a *Bernoulli process*, if $S_0 = 0$ and $S_n = \sum_{i=1}^n X_i$, where the increments X_i are i.i.d. Bernoulli r.v.'s (A.30).

What we considered so far are Markov processes with discrete state space. Now we describe stochastic processes of interest with continuous state space.

A stochastic process $\{X(t), t \ge 0\}$ is said to be a *Brownian motion* or *Wiener-Lévy process*, if $\{X(t), t \ge 0\}$ has stationary and independent increments and for every time interval (s, t) the increments $X(t) - X(s)$ have a normal distribution with mean $\mu(t-s)$ and variance $\sigma^2(t-s)$ for all $t > 0$, μ being called the *drift* and σ^2 the *variance* parameter of $X(t)$. $X(t)$ is a Markov process in continuous time with continuous state space and with *transition function*

$$p(x; y, t) = P(y < X(t) < y + dy | X(0) = x).$$

It can be shown that $p(x; y, t)$ satisfies the *diffusion equation*

$$\frac{1}{2}\sigma^2 \frac{\partial^2}{\partial y^2} p(x; y, t) - \mu \frac{\partial}{\partial y} p(x; y, t) = \frac{\partial}{\partial t} p(x; y, t).$$

Since increments of $X(t)$ have a normal distribution, $p(x; y, t)$ is given by

$$p(x; y, t) = \frac{1}{\sigma\sqrt{2\pi t}} \exp\left[-\frac{(y - x - \mu t)^2}{2\sigma^2 t}\right].$$

By differentiation it is easily verified that $p(x; y, t)$ is indeed a solution of the diffusion equation.

A.7. Poisson Processes, Birth-Death Processes

One way of introducing the Poisson process may be to first discuss the Markov process and its properties and then bring out the Poisson process as a special case. However, because of the importance of the Poisson process particularly in queueing theory it is perhaps expedient to begin with it, followed by the birth-death process and finally come to the Markov process.

From the application point of view, a Poisson process $\{X(t),\ t \geq 0\}$ with rate $\lambda(\lambda > 0)$ is seen to be a counting process (i.e., counting events in the interval $(0, t)$) with the following properties:

(i) $X(0) = 0$.
(ii) The process has stationary independent increments.
(iii) $P(X(\Delta t) = 1) = \lambda \Delta t + o(\Delta t)$ and $P(X(\Delta t) \geq 2) = o(\Delta t)$.

[Note: A function $f(\cdot)$ is said to be $o(\Delta t)$ if $\lim_{\Delta t \to 0} \frac{f(\Delta t)}{\Delta t} = 0$.]

Denoting by $P_n(t) = P(X(t) = n)$, let us write the difference-differential equations for the process. The argument and the development being the same as in Section 2.2 in Chapter 2, we have for $n > 0$,

$$\begin{aligned} P_n(t + \Delta t) &= P_n(t)P_0(\Delta t) + P_{n-1}(t)P_1(\Delta t) + o(\Delta t) \\ &= (1 - \lambda \Delta t)P_n(t) + \lambda \Delta t P_{n-1}(t) + o(\Delta t) \end{aligned}$$

and similarly for $n = 0$,

$$P_0(t + \Delta t) = P_0(t)(1 - \lambda \Delta t) + o(\Delta t).$$

These lead to

$$\begin{aligned} \frac{dP_n(t)}{dt} &= -\lambda P_n(t) + \lambda P_{n-1}(t), \quad n > 0, \qquad \text{and} \\ \frac{dP_0(t)}{dt} &= -\lambda P_0(t). \end{aligned} \tag{A.71}$$

These equations are called *Kolmogorov's equations.* From the second one, we get

$$P_0(t) = e^{-\lambda t}$$

and from the first

$$e^{\lambda t}\left[\frac{dP_n(t)}{dt} + \lambda P_n(t)\right] = \lambda e^{\lambda t} P_{n-1}(t),$$

or

$$\frac{d}{dt}\left(e^{\lambda t} P_n(t)\right) = \lambda e^{\lambda t} P_{n-1}(t)$$

or

$$e^{\lambda t} P_n(t) = \int \lambda e^{\lambda t} P_{n-1}(t) dt + c.$$

By induction, it can be seen that

$$P_n(t) = \frac{e^{-\lambda t}(\lambda t)^n}{n!}, \qquad n = 0, 1, 2, \ldots . \tag{A.72}$$

In a simple language, a Poisson process arises when (i) events occur independently in nonintersecting intervals and with the same probability if the length is the same for intervals and (ii) in a small interval at most one event can occur at the rate of λ per unit of time. The r.v. representing the number of events occurring during the interval $(0, t)$ has the Poisson distribution with mean λt. This establishes the connection between the Poisson process and the Poisson distribution.

Denoting by T_n the elapsed times between the $(n-1)$st and the nth event, it can be shown that the process $\{X(t), t \geq 0\}$ is a Poisson process with rate λ if and only if (i) $X(0) = 0$ and (ii) T_n's are i.i.d. exponential (λ). It amounts to saying that the Poisson process possesses the Markov or the memoryless property and therefore is a Markov process.

Some results about Poisson processes are listed below.

(a) Let $S_n = T_1 + \cdots + T_n$ which represents the arrival time of the nth event. Given that $S_{k+1} = t$, the conditional joint distribution of $S_1, \ldots, S_k$ is that of the order statistics of k i.i.d. uniform r.v.'s on $(0, t)$, i.e., the conditional p.f. is

$$f(s_1, \ldots, s_k \mid S_{k+1} = t) = \frac{k!}{t^k}, \qquad 0 < t_1 < \cdots < t_k < t.$$

(See (A.21).)

(b) Given that $X(t) = n$, the n arrival times $S_1, \ldots, S_n$ have the same distribution as the order statistics of n i.i.d. uniform r.v.'s on $(0, t)$.

(c) We define the superimposition of two independent Poisson processes as the process resulting from pooling together the time points of events occurring in each separate Poisson process. The superimposition of two independent Poisson processes with rates λ_1 and λ_2 is a Poisson process with rate $\lambda_1 + \lambda_2$.

(d) Given the superimposition of two independent Poisson processes $\{X_1(t)\}$ and $\{X_2(t)\}$ with respective rates λ_1 and λ_2, each event that occurs will be an event of $\{X_1(t)\}$ with probability $\lambda_1/(\lambda_1 + \lambda_2)$ and of $\{X_2(t)\}$ with probability $\lambda_2/(\lambda_1 + \lambda_2)$, independent of what has previously happened. A sketch of the proof is as follows: It is not difficult to prove that the probability of the first event from $\{X_1(t)\}$ occurring before (after) the first event from $\{X_2(t)\}$ is $\lambda_1/(\lambda_1 + \lambda_2)(\lambda_2/(\lambda_1 + \lambda_2))$. Given the first event to be from $\{X_1(t)\}(\{X_2(t)\})$, the probability that the second event from $\{X_1(t)\}(\{X_2(t)\})$ precedes the first event from $\{X_2(t)\}(\{X_1(t)\})$ is again $\lambda_1/(\lambda_1 + \lambda_2)(\lambda_2/(\lambda_1 + \lambda_2))$ because of the memoryless property and property (ii) of the Poisson process. A continuation of this argument leads to the stated result.

Two generalizations of Poisson processes are presented. First, a process possessing the defining properties of the Poisson process except that it is not stationary and its λ is not a constant but depends on t is called a *nonhomogeneous Poisson process*. A second one is a *compound Poisson process*

$\{X(t),\ t \geq 0\}$ which by definition is represented by

$$X(t) = \sum_{i=1}^{N(t)} Y_i, \qquad t \geq 0$$

where $\{N(t),\ t \geq 0\}$ is a Poisson process and $\{Y_n,\ n = 1, 2, \ldots\}$ is a sequence of i.i.d. r.v.'s independent of the Poisson process.

Now let us turn to some discussion on the birth-death process which is also a generalization of the Poisson process. Let the state space be $\{0, 1, 2, \ldots\}$ which could be finite. In the birth-death process the transition takes place from state $k (k > 0)$ either to state $k+1$ (i.e., a birth occurs) at rate λ_k or to state $k-1$ (i.e., a death occurs) at rate μ_k and births and deaths occur independently of each other. When the process is at 0, only a birth can occur at rate λ_0. If no death occurs (i.e., $\mu_k = 0$), then the process is called a pure birth process and if in addition the birth rate is constant (i.e., $\lambda_k = \lambda$) the process becomes a Poisson process. We may be reminded that the birth-death process has the Markov property. For this process, we have

$$\begin{aligned}
P(X(t+\Delta t) &= k+1 \mid X(t) = k) = \lambda_k \Delta t + o(\Delta t), \\
P(X(t+\Delta t) &\geq k+2 \mid X(t) = k) = o(\Delta t), \\
P(X(t+\Delta t) &= k-1 \mid X(t) = k) = \mu_k \Delta t + o(\Delta t), \\
P(X(t+\Delta t) &\leq k-2 \mid X(t) = k) = o(\Delta t).
\end{aligned} \tag{A.73}$$

Many queueing systems are modeled as birth-death processes by simply representing an arrival by a birth and a departure by a death and assuming the Markov property for the process.

Let us derive the Kolmogorov equations for the birth-death process, even if the argument is being repeated. Consider the process to be in state n at time $t + \Delta t$.

$$\begin{aligned}
P(X(t+\Delta t) = n) &= P(X(t+\Delta t) = n \mid X(t) = n) P(X(t) = n) \\
&\quad + P(X(t+\Delta t) = n \mid X(t) = n+1) P(X(t) = n+1) \\
&\quad + P(X(t+\Delta t) = n \mid X(t) = n-1) P(X(t) = n-1).
\end{aligned}$$

Recalling $P_n(t) = P(X(t) = n)$ and using (A.73), the equations become

$$\begin{aligned}
P_n(t+\Delta t) &= (1 - \lambda_n \Delta t)(1 - \mu_n \Delta t) P_n(t) \\
&\quad + (1 - \lambda_{n+1} \Delta t) \mu_{n+1} \Delta t P_{n+1}(t) \\
&\quad + \lambda_{n-1} \Delta t (1 - \mu_{n-1} \Delta t) P_{n-1}(t) \\
&\quad + o(\Delta t).
\end{aligned}$$

Transfer $P_n(t)$ from the right side to the left, divide through by Δt and take the limit as $\Delta t \to 0$. Then the difference-differential equations for $n \geq 1$ are

$$\frac{dP_n(t)}{dt} = -(\lambda_n + \mu_n) P_n(t) + \mu_{n+1} P_{n+1}(t) + \lambda_{n-1} P_{n-1}(t), \tag{A.74}$$

and similarly for $n = 0$ is

$$\frac{dP_0(t)}{dt} = -\lambda_0 P_0(t) + \mu_1 P_1(t).$$

Unlike the Poisson process, no easy method of getting an explicit solution has been found yet. However, often in queueing theory the solution for $P_n(t)$ when $t \to \infty$ is of practical importance. If it exists, then we get the difference or the stationary equations as given in (2.11) of Chapter 2. In the next section, we will see that the limit of $P_n(t)$ for a Markov process of which the birth-death process is a particular case exists and the set $\{\lim_{t\to\infty} P_n(t)\}$ forms a probability distribution of practical significance.

A.8. Markov Chains, Markov Processes

Markov chains and processes are defined in Section A.6. What is intended here is to present two basic results which are more frequently referred to in queueing theory.

We begin with Markov chains since the results for the processes are analogous. Let P_{ij} be the one-step transition probability from state i to state j. In other words,

$$P_{ij} = P(X_{n+1} = j \mid X_n = i)$$

where P_{ij} depends only on i and j. The *transition probability matrix* $\mathbf{P} = (P_{ij})$ is one in which the (i, j)th element is P_{ij}. Let

$$P_{ij}^m = P(X_{n+m} = j \mid X_n = i)$$

be the *m-step transition probability* from i to j and let $\mathbf{P}^m$ be the corresponding matrix. In order to reach from state i to state j in $m + n$ steps, the chain has to go through some state k in m steps and then to j in n steps, it immediately follows that

$$P_{ij}^{m+n} = \sum_{k=0}^{\infty} P_{ik}^m P_{kj}^n \qquad \text{for all } m, n, i, j \geq 0 \tag{A.75}$$

which are the well-known *Chapman-Kolmogorov equations* and form the first basic result. In matrix formulation (A.75)

$$\mathbf{P}^{m+n} = \mathbf{P}^m \mathbf{P}^n \tag{A.76}$$

which by using step-by-step iteration is equal to the $(m + n)$th power of $\mathbf{P}$ and therefore our notation is justified without any ambiguity. Letting $\pi_0(i) = P(X_0 = i)$ we can compute

$$P(X_n = j) = \sum_i \pi_0(i) P_{ij}^n. \tag{A.77}$$

In order to develop the second result which deals with the asymptotic theory, we need a series of definitions on classification of states. A Markov chain is said to be *irreducible* if every state can be reached from every other one at some positive step. Let f_i be the probability that starting from state i, the chain will ever re-enter state i. State i is said to be *recurrent* or *transient* according as $f_i = 1$ or $f_i < 1$. Moreover it is *periodic* with period d if $P_{ii}^n = 0$ when n is not divisible by d, and d is the largest integer with such property. A state with period 1 is called *aperiodic*. Furthermore, any state is called *positive recurrent* if it is recurrent and starting from it the expected number of steps for the first return to itself is finite. Finally an *ergodic* state is one which is positive recurrent and aperiodic. An *ergodic Markov chain* is one if all its states are ergodic.

An important limit theorem is stated as follows:

In an irreducible ergodic Markov chain for every i and j $\lim_{n\to\infty} P_{ij}^n$ exists and is independent of i.

Because of this theorem, let

$$\pi_j = \lim_{n\to\infty} P_{ij}^n.$$

By using (A.75) we get

$$P_{ij}^{n+1} = \sum_{k=0}^{\infty} P_{ik}^n P_{kj}.$$

When we take limit on both sides, we obtain

$$\pi_j = \sum_{k=0}^{\infty} \pi_k P_{kj}, \qquad j = 0, 1, \ldots,$$

with

$$\sum_{j=0}^{\infty} \pi_j = 1. \tag{A.78}$$

Thus π_j's are obtained as the solution of (A.78). Letting the probability vector $\boldsymbol{\pi} = (\pi_0, \pi_1, \ldots)$, the first equation of (A.78) is simply

$$\boldsymbol{\pi} = \boldsymbol{\pi}\mathbf{P}$$

or

$$\boldsymbol{\pi}(\mathbf{I} - \mathbf{P}) = \mathbf{0}, \tag{A.79}$$

where $\mathbf{I}$ is the identity matrix. Moreover

$$\lim_{n\to\infty} P(X_n = j) = \pi_j \text{ for every } j.$$

The result states that the limiting probability that a chain will be in state j at time n is equal to the long-run proportion of time that the chain will be in state j and is determined by (A.78). The set $\{\pi_j\}$ is a probability distribution and is called the *steady-state* or *stationary distribution* of the chain.

The following two results are particularly important for $M/G/1$ and $G/M/1$ systems (Foster (1953)).

Result 1. An irreducible aperiodic Markov chain is ergodic if there exists a non-negative solution of the inequalities

$$\sum_{j\geq 0} P_{ij} y_i \leq y_i - 1, \qquad i \neq 0, \tag{A.80}$$

such that $\sum_{j\geq 0} P_{0j}y_j < \infty$.

Result 2. If the same Markov chain is ergodic, then the finite mean first-passage times d_j from the jth state to the zero-th state satisfy the equations

$$\sum_{j\geq 1} P_{ij}d_j = d_i - 1,\ i \neq 0,\ \text{and} \sum_{j\geq 1} P_{0j}d_j > \infty. \tag{A.81}$$

Next for the study of the Markov processes let

$$P_{ij}(t) = P(X(t+s) = j \mid X(s) = i).$$

The *Chapman-Kolmogorov equation* for the process is

$$P_{ij}(t+s) = \sum_{k=0}^{\infty} P_{ik}(t)P_{kj}(s) \tag{A.82}$$

which can be established in the same way as (A.75). Denoting by $\mathbf{P}(t)$ the matrix $(P_{ij}(t))$ (A.82) can be expressed as

$$\mathbf{P}(t+s) = \mathbf{P}(t)\mathbf{P}(s).$$

Regarding the limiting probabilities, since Markov processes do not have periodicities and other definitions are similar, it can be shown that for an irreducible positive recurrent process the stationary distribution $\{\pi_j\}$ exists and satisfies

$$\lim_{t\to\infty} P_{ij}(t) = \lim_{t\to\infty} P(X(t) = j) = \pi_j \text{ for every } j \tag{A.83}$$

and

$$\sum_j \pi_j = 1.$$

Let us develop the system of equations for determining $\{\pi_j\}$. Recall that for a Markov process the time spent in any given state has an exponential distribution. Let us assume that the exponential distribution has mean $1/\nu_i$ when the process is in state i. Also note that $1 - P_{ii}(\Delta t)$ is the probability that during interval Δt a transition from i has occurred. Because of the connection between the exponential distribution and the Poisson process we derive from property (iii) of the Poisson process that

$$1 - P_{ii}(\Delta t) = \nu_i \Delta t + o(\Delta t). \tag{A.84}$$

Similarly $P_{ij}(\Delta t)$ being the probability that during time interval Δt the state i changes to state j, has the expression

$$P_{ij}(\Delta t) = \nu_i \Delta t P_{ij} + o(\Delta t) \tag{A.85}$$

where P_{ij} is the probability that the process leaves state i and enters state j.

Now let us put $s = \Delta t$ in (A.82). Then we obtain

$$P_{ij}(t+\Delta t) = \sum_{k=0}^{\infty} P_{ik}(t)P_{kj}(\Delta t)$$

or

$$\lim_{\Delta t \to 0} \frac{P_{ij}(t+\Delta t) - P_{ij}(t)}{\Delta t} =$$

$$= \lim_{\Delta t \to 0} \left[\sum_{k \neq j} P_{ik}(t) \frac{P_{kj}(\Delta t)}{\Delta t} - P_{ij}(t) \frac{1 - P_{jj}(\Delta t)}{\Delta t} \right],$$

or

$$P'_{ij}(t) = \sum_{k \neq j} P_{ik}(t) \nu_k P_{kj} - P_{ij}(t) \nu_j \tag{A.86}$$

when (A.84) and (A.85) are used. As $t \to \infty$, $P_{ij}(t) \to \pi_j$ and $P'_{ij}(t) \to 0$. Therefore, we have

$$\pi_j \nu_j = \sum_{k \neq j} \pi_k \nu_k P_{kj} \text{ for all } j. \tag{A.87}$$

We recognize that the left side represents the rate at which the process leaves state j and the right side the rate at which the process enters state j. The principle of flow conservation is to make the rate of flow in equal to the rate of flow out and thus produces the stationary equations (sometimes called the balance equations) the solution of which will give $\{\pi_j\}$. One may note that matrix equation (A.79) can be directly written by using the principle of flow conservation. Also we recognize that (A.87) can be written in a matrix form similar to (A.79) as

$$\boldsymbol{\pi} \mathbf{Q} = \mathbf{0}, \tag{A.88}$$

where the (i, j)th element Q_{ij} of $\mathbf{Q}$ is given by

$$Q_{ij} = \begin{cases} -\nu_i & i = j, \\ \nu_i P_{ij} & i \neq j. \end{cases}$$

Remember that $\sum_{j=0}^{\infty} \pi_j = 1$.

Two results on positive recurrence are stated below.

(i) An irreducible Markov chain or process having finite number of states is positive recurrent.

(ii) An irreducible Markov chain or process is positive recurrent if and only if it has a stationary distribution.

In summary, we have presented the Chapman-Kolmogorov equations and two limit theorems. These lead us to the stationary equations for obtaining the steady-state probability distribution which is extremely relevant in queueing theory.

The references for Sections A.6–A.8 are Cramer (1946), Doob (1953), Feller (1971), Loéve (1960), Ross (1985) and Takács (1960).

References

Blackwell, D. (1945). A renewal theorem, *Duke Math. J.*, **15**, 145–150.

Cramer, H. (1946). *Mathematical Methods of Statistics*, Princeton University Press, Princeton.

Doob, J. L. (1953). *Stochastic Processes*, John Wiley & Sons, New York.

Feller, W. (1971). *An Introduction to Probability Theory and Its Applications*, Vol. 2., 2nd ed., John Wiley & Sons, New York.

Foster, F. G. (1953). On stochastic matrices associated with certain queueing processes, *Ann. Math. Statist.*, **24**, 355–360.

Loéve, M. (1960). *Probability Theory*, 2nd ed., Van Nostrand, New York.

Ross, S. M. (1985). *Introduction to Probability Models*, 3rd ed. Academic Press, New York.

Smith, W. L. (1958). Renewal theory and its ramifications. *J. Roy. Stat. Soc.*, Ser. B, **20**, 243–302.

Takács, L. (1958). On a general probability theorem and its applications in the theory of stochastic processes. *Proc. Camb. Phil. Soc.*, **54**, 219–224.

Takács, L. (1960). *Stochastic Processes*, Methuen, London.

A.9. Statistical Concepts

Statistical concepts are useful in model checking and in simulation experiments.

To begin with, we say a *population* consists of all objects to be studied which are considered as outcomes of a sample space (see A.1). A *random sample* is a sample chosen in a way that each member has the same chance of being picked up for the sample. Let us express it in a formal way. Suppose the experiment that results in an r.v. X with p.f. $f(x)$ is repeated n times independently. If $X_1, \ldots, X_n$ denote the random variables corresponding to X for each experiment, then the collection $(X_1, \ldots, X_n)$ of n mutually independent and identically distributed (in short i.i.d.) r.v., is called a random sample of *size* n from a population having p.f. $f(x)$.

Statistical methodologies are developed to make inferences on population characteristics based on characteristics of a random sample. Quite often inferences are made on $f(x)$ and its unknown parameters which are to be estimated. A function of a random sample that does not depend on any unknown parameter is called a *statistic*.

Point Estimation

A *point estimate* of a population parameter is a single value of a statistic used for the purpose of estimation. The statistic itself is called an *estimator*. There are some measures of quality of estimators. Any estimator whose expected value equals the parameter it is estimating is known to be *unbiased*. It is called an *unbiased minimum variance estimator* if its variance is not larger than the variance of any other unbiased estimator. While these are properties based on samples of any size, we think of a large sample desirable property.

Any statistic that converges probabilistically to the estimated parameter is called a *consistent* estimator of that parameter.

There are a few estimation procedures. Among the most commonly used is the method of *maximum likelihood estimation.* Denote by θ the parameter to be estimated that appears in $f(x)$. Thus we may write $f(x)$ as $f(x;\theta)$. The joint p.f. of the sample observations $X_1 = x_1, \ldots, X_n = x_n$ is $\prod_{i=1}^n f(x_i;\theta)$. Since θ is unknown, the joint p.f. may be regarded as a function of θ and is called the *likelihood function* and denoted by $L(\theta)$. The likelihood function represents how likely the observed sample $(x_1, \ldots, x_n)$ is for different values of θ. A *maximum likelihood estimator* (MLE) of θ is a statistic which maximizes $L(\theta)$. Since $L(\theta)$ and $\ln L(\theta)$ are maximized for the same value of θ, we consider $\ln L(\theta)$ as it is easier to handle for the purpose of maximization.

For example, suppose there is a random sample $x_1, \ldots, x_n$ from p.f.

$$\begin{aligned} f(x;\theta) &= \theta^x(1-\theta)^{1-x} \quad \text{for } x = 0, 1,\ 0 \le \theta \le 1 \\ &= 0 \qquad \text{otherwise.} \end{aligned}$$

The likelihood function is

$$L(\theta) = \theta^{\sum_1^n x_i}(1-\theta)^{n-\sum_1^n x_i}$$

and

$$\ln L(\theta) = \left(\sum_1^n x_i\right)\ln\theta + \left(n - \sum_1^n x_i\right)\ln(1-\theta).$$

In order to maximize $\ln L(\theta)$ with respect to θ, we have

$$\frac{d \ln L(\theta)}{d\theta} = \frac{\sum x_i}{\theta} - \frac{n - \sum x_i}{1-\theta} = 0,$$

leading to the ML estimate (we use MLE for both estimator and estimate) of θ to be $\frac{\sum_1^n x_i}{n} = \bar{x}$, the sample mean. It can be checked that the solution maximizes. If there are more parameters, say θ_1 and θ_2, one has to solve $\frac{\partial \ln L}{\partial \theta_1} = 0$ and $\frac{\partial \ln L}{\partial \theta_2} = 0$. A useful property of the MLE is that if $\hat{\theta}$ is the MLE of θ and $u(\theta)$ is a function with a single-valued inverse, then the MLE of $u(\theta)$ is $u(\hat{\theta})$. For example, let there be a random sample of size n from the population with normal distribution having mean μ and variance σ^2. Then by the procedure, the MLEs of μ and σ^2 are $\bar{X}$ and $\sum_{i=1}^n \frac{(X_i-\bar{X})^2}{n}$. A quick application of the property yields the MLE of σ as $\sqrt{\sum_{i=1}^n \frac{(X_i-\bar{X})^2}{n}}$.

An MLE is not necessarily unbiased. But it has some good large sample properties. Under quite general regularity conditions on p.f. which are satisfied most often, an MLE $\hat{\boldsymbol{\theta}}$ of θ is consistent. It is a best asymptotically normal estimator in the sense that $\sqrt{n}(\hat{\boldsymbol{\theta}} - \theta)$ has a limiting normal distribution with mean 0 and variance

$$\sigma^2 = \left[E\left\{\frac{\partial}{\partial\theta}\ln f(X;\theta)\right\}^2\right]^{-1}$$

and that if $\tilde{\boldsymbol{\theta}}$ is a consistent estimator of θ such that $\sqrt{n}(\tilde{\boldsymbol{\theta}} - \theta)$ has a limiting variance $\tilde{\sigma}^2$ then $\frac{\sigma^2}{\tilde{\sigma}^2} \leq 1$ (i.e., $\hat{\boldsymbol{\theta}}$ is efficient).

Sometimes it is impossible to find MLEs in closed form. Another procedure is to equate the first k moments of the distribution to the corresponding moments of the sample where k is the number of parameters in the distribution to be estimated. The solutions for these parameters are the estimates found by the *method of moments.*

Let us consider the gamma distribution as given in (A.47) in which we want to estimate both parameters α and λ. Because of the presence of $\Gamma(\alpha)$, the maximum likelihood technique is not convenient at all. Instead, we use the method of moments and equate

$$E(X) = \frac{\alpha}{\lambda} = \frac{\sum_{i=1}^{n} X_i}{n} \quad \text{and} \quad E(X^2) = \frac{\alpha}{\lambda^2} + \frac{\alpha^2}{\lambda^2} = \frac{\sum_{i=1}^{n} X_i^2}{n}$$

(see (A.49) and the expression before (A.26)). The solution gives the estimators of λ and α as $\frac{\bar{X}}{S^2}$ and $\frac{\bar{X}^2}{S^2}$, respectively, where $S^2 = \sum_{i=1}^{n} \frac{(X_i - \bar{X})^2}{n}$. The estimator by the method of moments is consistent and has a limiting normal distribution. However, it is not unbiased nor is it asymptotically efficient.

There is another method of estimation called *method of least squares* that arises when a random variable Y is linearly expressed as

$$Y = a + bx + e$$

with $E(Y) = a + bx$, where x is a known mathematical variable and a and b are unknown parameters to be estimated. Clearly e stands for error which is an r.v. such that $E(e) = 0$. Let $Y_1, \ldots, Y_n$ be a random sample. Let $Y_i = a + bx_i + e_i$, $i = 1, \ldots, n$, which means each Y_i as a random observation corresponding to x_i is expressible as $a + bx_i$ with a random error e_i. By the method least squares a and b are estimated by those values which will minimize the error sum of squares

$$\sum_{i=1}^{n} e_i^2 = \sum_{i=1}^{n} (Y_i - a - bx_i)^2.$$

It turns out that the least squares estimates of a and b are given by

$$\bar{y} - \bar{x}\frac{\sum_{i=1}^{n}(y_i - \bar{y})(x_i - \bar{x})}{\sum_{i=1}^{n}(x_i - \bar{x})^2} \quad \text{and} \quad \frac{\sum_{i=1}^{n}(y_i - \bar{y})(x_i - \bar{x})}{\sum_{i=1}^{n}(x_i - \bar{x})^2}, \tag{A.89}$$

respectively. Note that $\sum_{i=1}^{n}(y_i - \bar{y})(x_i - \bar{x}) = \sum_{i=1}^{n} y_i(x_i - \bar{x})$.

We observe that both estimates are linear functions of y_i's. It can be shown that these estimators are unbiased and are best linear unbiased estimators (BLUE) in the sense that among the class of linear estimators which are unbiased, estimators given by (A.89) has smaller variance than others under the assumption e_i's are uncorrelated with the same finite variance.

Interval Estimation

A point estimate of a parameter is not very meaningful without some measure of the possible error. It is preferable to give an interval around the estimate which is called an *interval estimate* with some measure of assurance that the unknown parameter lies within the interval. For example, consider a random sample $X_1, \ldots, X_n$ from normal population with mean μ and variance σ^2. Suppose σ^2 is known and we want an interval estimate of μ. It can be shown that $\bar{X}$ is the MLE of μ and $\bar{X}$ has a normal distribution with mean μ and variance $\frac{\sigma^2}{n}$. Then the standardized variable $Z = \frac{\bar{X}-\mu}{\sigma/\sqrt{n}}$ has standard normal distribution $N(0,1)$ (see Section A.4). From the Table B-1 in Appendix B of the normal distribution, we have

$$P\left(-2 < \frac{\bar{X}-\mu}{\sigma/\sqrt{n}} < 2\right) = 0.954$$

which can be rewritten as

$$P\left(\bar{X} - \frac{2\sigma}{\sqrt{n}} < \mu < \bar{X} + \frac{2\sigma}{\sqrt{n}}\right) = 0.954.$$

It implies that the probability of including μ in the random interval

$$\left(\bar{X} - \frac{2\sigma}{\sqrt{n}}, \bar{X} + \frac{2\sigma}{\sqrt{n}}\right)$$

is 0.954. In this case, we call the interval $(\bar{x} - 2\sigma/\sqrt{n}, \bar{x} + 2\sigma/\sqrt{n})$ a *confidence interval* estimate for μ with *confidence coefficient* 0.954. In general,

$$\left(\bar{x} - z_{\frac{\alpha}{2}}\frac{\sigma}{\sqrt{n}}, \bar{x} + z_{\frac{\alpha}{2}}\frac{\sigma}{\sqrt{n}}\right)$$

is a confidence interval estimate for μ with confidence coefficient $(1-\alpha)$ or we may say that the interval is a $100(1-\alpha)$ percent confidence interval, where $z_{\frac{\alpha}{2}}$ is given by

$$\int_{z_{\alpha/2}}^{\infty} \frac{1}{\sqrt{2\pi}} e^{-\frac{x^2}{2}}\, dx = \frac{\alpha}{2}.$$

If the sample size is large and σ is unknown, then it may be replaced by its estimate $\sum_{i=1}^{n} \frac{(x_i-\bar{x})^2}{n-1}$ in the interval estimate to get an approximation.

The same procedure may be used for the unknown parameter p in a Bernoulli distribution. We have checked in the previous section that $y = \frac{\sum_{i=1}^{n} x_i}{n}$ is the MLE of p. It can be seen that $E(Y) = p$ and therefore Y is unbiased and that $\sigma_Y^2 = \frac{p(1-p)}{n}$. Just as before standardize Y to have $\frac{Y-p}{\sqrt{\frac{p(1-p)}{n}}}$. For a large sample size, $\frac{Y-p}{\sqrt{\frac{p(1-p)}{n}}} \to N(0,1)$ due to the central limit theorem (Section A.4). Thus an approximate $100(1-\alpha)$ percent confidence interval for p is

$$\left(y - z_{\frac{\alpha}{2}}\sqrt{\frac{y(1-y)}{n}}, y + z_{\frac{\alpha}{2}}\sqrt{\frac{y(1-y)}{n}}\right) \tag{A.90}$$

when we replace p by its estimate y. The width of the interval is $2z_{\frac{\alpha}{2}}\sqrt{\frac{y(1-y)}{n}}$. Suppose we want the width of the interval to be limited to say, δ. In other words, we want

$$2z_{\frac{\alpha}{2}}\sqrt{\frac{y(1-y)}{n}} \leq \delta$$

leading to

$$n \geq 4z_{\frac{\alpha}{2}}^2 \frac{y(1-y)}{n}, \tag{A.91}$$

which means for a sample size at least as large as the right side expression, the width will be smaller than δ with confidence coefficient $1-\alpha$.

Hypothesis Testing

In queueing theory, one usually starts with an assumed model in which the probability distributions of interarrival times and of service times are hypothesized. Such hypotheses need to be checked.

Let us introduce some terminologies. A *statistical hypothesis* is a specification of the distribution. If it is a complete specification then the hypothesis is called *simple*, otherwise it is *composite*. For example, if we assume the distribution to be exponential with parameter λ and make a hypothesis $\lambda = \lambda_0$ (λ_0 as a known value), then this is a simple hypothesis. Note that the exponentiality is given whereas $\lambda = \lambda_0$ as a hypothesis is to be checked by some statistical testing procedure. Another hypothesis is to check exponentiality with a given λ. Then this is a simple hypothesis. A *test* is a rule based on the observed sample to decide whether to reject the hypothesis. Thus a test divides the sample space S of the random sample into two regions. The region in which if the observation falls leads to the rejection of the hypothesis is called the *critical* or *rejection region* and is denoted by C. The hypothesis to be tested is known as the *null hypothesis* and is denoted by H_0. It is tested against an *alternative hypothesis* and is usually denoted by H_1.

There are two types of error that arise in making the decision. An error occurs if H_0 is rejected when it is true and this is called *Type I error*. *Type II error* arises when H_0 is accepted given H_1 is true. A good test is one which will minimize the probabilities of both errors, but it happens that if one increases then the other decreases and vice versa. Thus it is reasonable to control or restrict one up to a given level and select a test which minimizes the other subject to this restriction. Usually the probability of Type I error is the one which is not to exceed a prescribed value, called the *significance level* or *size of the test* and denoted by α. Clearly, Type I error is considered to be of importance. Therefore it is important to set H_0 such that the consequence of rejecting it when it is true, is considered to be serious as against the other error. In our example, if we set $H_0 : \lambda = \lambda_0$ then we believe that accepting $\lambda \neq \lambda_0$ when $\lambda = \lambda_0$ is much less preferable than accepting $\lambda = \lambda_0$ given $\lambda \neq \lambda_0$.

After introducing several terminologies, we discuss how to set up a test. First, a test must be based on a statistic. The statistic selected is in general a good estimator and plays the role of a discriminator. The rule to reject

H_0 should be such that sample observations satisfying it look reasonable to confirm H_1. Going back to our example, we can show that the MLE of λ is $1/\bar{X}$. A reasonable test statistic is $|(1/\bar{X}) - \lambda_0|$ which measures the difference between the estimator and the hypothesized value $\lambda = \lambda_0$. Thus a good rejection rule based on this statistic is $|(1/\bar{X}) - \lambda_0|$ to be large, say $> c$, where c is called the *critical value* and the critical region C is $|(1/\bar{X}) - \lambda_0| > c$, once c is determined. The question remains: How to determine C? Here *P(Type I error)* comes into picture. Remember that if this probability is kept down then *P(Type II error)* goes up. Thus a good strategy is to let *P(Type I error)* $= \alpha$, the prescribed value and search for a test with the same size which will minimize *P(Type II error)*. In some situations this may be possible and if such a test exists we say it is a *uniformly most powerful* (UMP) test. In others, one is satisfied with a reasonably good statistic and determine C, the critical or rejection region. It is done by using *P(Type I error)* $=$ *P(Test statistics* $\in C|H_0) = \alpha$, which suggests that the *sampling distribution* (discussed at the end) of the statistic under H_0 is needed. It is a common practice to take $\alpha = 0.05$ or 0.01.

As for illustration, consider a random sample $X_1, \ldots, X_n$ of size n from a normal population $N(\mu, \sigma^2)$ with σ^2 known. We want to test $H_0 : \mu = \mu_0$ against $H_1 : \mu > \mu_0$, μ_0 being a known constant. Note H_0 is simple. Statistic $\bar{X}$ happens to be the MLE of μ. Then $\bar{X} - \mu_0$ seems to be a good test statistic. However, its largeness varies with different units of measurement. Thus if we consider the statistic to be $Z = \frac{\bar{X}-\mu_0}{\sigma/\sqrt{n}}$, then it is invariant of units of measurement and its distribution can be seen to be $N(0,1)$ under H_0, since $\bar{X}$ has variance σ^2/n. The rejection region $Z = \frac{\bar{X}-\mu_0}{\sigma/\sqrt{n}} > c$ looks reasonable and it makes the test a *one-sided* (*right-sided*) one. The natural rejection region for $H_1 : \mu < \mu_0$ is also one-sided (*left-sided*). In order to determine c, we use $P(Z > c) = \alpha$ and the table of the normal distribution. From Table B-1 in Appendix B, one gets $c = 1.645$ if $\alpha = 0.05$. This means that for size 0.05, we reject $\mu = \mu_0$ if the calculated value $\frac{\bar{x}-\mu_0}{\sigma/\sqrt{n}} > 1.645$. It can be seen that this is a UMP test. However, for $H_1 : \mu \neq \mu_0$ there does not exist a UMP test. Nevertheless, the *two-sided* test $|Z| > c$ is useful in practice where $c = 1.96$ if $\alpha = 0.05$.

Suppose σ^2 is unknown. Then replace σ in the test statistic by its well-known estimator

$$\sqrt{\sum_{i=1}^{n} \frac{(X_i - \bar{X})^2}{n-1}}$$

which is called the sample standard deviation. The changed statistic has a t-distribution with $n - 1$ degrees of freedom which is thoroughly studied and therefore tests can be developed following the procedure described above.

All our examples are to test hypotheses on parameters. However, our interest in theory of queues is generally to test the form of the distribution. Yet the procedure in developing such a test does not change. The steps are: (i) formulate the hypotheses, (ii) construct a test statistic for which the distribution can be determined under H_0, (iii) set an appropriate rejection region, and (iv) use *P(Type I error*$|H_0$*)* to completely determine the region. There are two tests of interest which are discussed in Section 5.2 in Chapter 5.

Sampling Distributions

As we have seen, we need sampling distributions of statistics which are either used as estimators or test statistics. A statistic being a function of a random sample, has its probability distribution called its *sampling distribution.*

Consider a random sample $X_1, \ldots, X_n$ from $N(\mu, \sigma^2)$. If both μ and σ are known, then the sampling distribution or simply the distribution of $\frac{\bar{X}-\mu}{\sigma/\sqrt{n}}$ is $N(0,1)$ for which a table of d.f. is given (Table B-1 in Appendix B). If σ^2 is unknown, then

$$(\bar{X}-\mu) \Bigg/ \left(\sqrt{\sum_{i=1}^{n} \frac{X_i - \bar{X})^2}{n-1}} \Bigg/ n \right)$$

has a t-distribution with $n-1$ degrees of freedom. The p.f. of t-distribution with k degrees of freedom is given by

$$f(t) = \frac{\Gamma(\frac{k+1}{2})}{\sqrt{k\pi}\Gamma(\frac{k}{2})} \frac{1}{(1+t^2/k)^{(k+1)/2}}, \quad -\infty < t < \infty \tag{A.92}$$

where $\Gamma(t)$ is defined in (A.46). In this text, we do not come across any application of t-distribution and therefore no table for the distribution is provided.

An important sampling distribution is chi-square distribution. Its p.f. is given as

$$f(x) = \frac{1}{\Gamma(k/2)2^{k/2}} x^{k/2-1} e^{-\frac{x}{2}}, \quad 0 < x < \infty \tag{A.93}$$

where k is the degrees of freedom of the distribution. Note that it is a special case of the gamma distribution (see (A.47)) with $\alpha = \frac{k}{2}$ and $\lambda = \frac{1}{2}$. As a sampling distribution it arises when a random sample $X_1, \ldots, X_n$ is taken from a population having the distribution $N(\mu, \sigma^2)$. We can prove that $\sum_{i=1}^{n}(X_i - \bar{X})^2/\sigma^2$ has the chi-square distribution with $n-1$ degrees of freedom. For our purpose, it suffices to state that the statistic defined in (5.20) has for large sample size, approximately the chi-square distribution with degrees of freedom explained in that section. The critical values of chi-square testing procedures for different confidence levels α are given in Table B-2 in Appendix B.

Another sampling distribution of interest is F-distribution. If U and V are two independent chi-square random variables with r_1 and r_2 degrees of freedom, respectively, then $(U/r_1)/(V/r_2)$ has a F distribution with (r_1, r_2) degrees of freedom. Since each chi-square arises as a sample statistic, the above ratio is also a statistic. In our case, we come across this distribution in (5.24). Some critical values are tabulated in Table B-3 in Appendix B. The reader is suggested to consult statistical tables in any standard textbook in Statistics.

References

Hogg, R. V. and Craig, A. T. (1978). *Introduction Mathematical Statistics*, 4th ed., McMillan Publishing Co., New York.

Rohatgi, V. K. (1976). *An Introduciton to Probability Theory and Mathematical Statistics*, John Wiley & Sons, New York.

A.10. Matrix Analysis

Definitions

In this part of the appendix we have summarized some basic facts and results from matrix analysis which are of significant importance in queueing theory.

To begin with, let us define a *matrix* as a two-dimensional arrangement of numbers. These numbers may be real or complex. They may be elements of any field of numbers, but in the theory of queues the real case is the most natural and important one.

Matrices are usually denoted by bold capital letters and the notation $\mathbf{A} \in \mathbb{R}^{m\times n}$ means that the matrix $\mathbf{A}$ has m rows, n columns and its components $a_{ij}, i = 1, \ldots m, j = 1, \ldots, n$ are real numbers. $\mathbf{A}$ is said to be of order $m \times n$ and is also represented by

$$\mathbf{A} = \begin{bmatrix} a_{11} & a_{12} & \ldots & a_{1n} \\ a_{21} & a_{22} & \ldots & a_{2n} \\ \vdots & & & \vdots \\ a_{m1} & a_{m2} & \ldots & a_{mn} \end{bmatrix} = [a_{ij}].$$

The special case $n = 1$ is called a *column vector* and it will be denoted by lowercase letters like for instance $\mathbf{a} \in \mathbb{R}^{m\times 1}$ or more compactly as $\mathbf{a} \in \mathbb{R}^m$ with $a_{i1} = a_i$. If $m = n$, then $\mathbf{A}$ is called a *square matrix*.

The *transpose* of a matrix $\mathbf{A}$ will be denoted by $\mathbf{A}'$. It is obtained from $\mathbf{A}$ by interchanging rows and columns:

$$\mathbf{A} = \left[a_{ij}\right] \in \mathbf{R}^{m\times n} \quad \Longrightarrow \quad \mathbf{A}' = \left[a_{ji}\right] \in \mathbf{R}^{n\times m},$$

and clearly $(\mathbf{A}')' = \mathbf{A}$. An important class of matrices are those which are invariant with respect to transposition, i.e., $\mathbf{A}' = \mathbf{A}$. Such matrices are called *symmetric* and they are necessarily square. Also vectors may be transposed. The transpose of a column vector $\mathbf{a}$ is called a *row vector* and denoted by $\mathbf{a}'$.

A square matrix $\mathbf{A}$ is *upper triangular*, if its components below the main diagonal are zero. The main diagonal is that from the northwest corner to the southeast corner in any square matrix. $\mathbf{A}$ is *lower triangular* if its transpose $\mathbf{A}'$ is upper triangular.

If all components of a square matrix $\mathbf{A}$ are zero except at least one component on the main diagonal, then $\mathbf{A}$ is called a *diagonal matrix*. A diagonal matrix is completely specified by its main diagonal components and often denoted in the abbreviated form $\mathbf{A} = \text{diag}(a_{11}, a_{22}, \ldots, a_{nn})$. If in a diagonal matrix $\mathbf{A}$, all main diagonal components are equal to some constant γ, then $\mathbf{A}$ is called a *scalar matrix*. The special case with $\gamma = 1$ is called the *identity matrix* and is always denoted by $\mathbf{I}$.

In the theory of birth-death processes *tridiagonal matrices* play an important role. In a tridiagonal matrix $\mathbf{A}$ only components on the main diagonal a_{ii}, the subdiagonal $a_{i,i-1}$ and the superdiagonal $a_{i,i+1}$ may be different from zero. All other components are zero. In main-, sub- and superdiagonal at least one component has to be non-null.

If in a matrix all components are equal to zero, then it is called the *zero matrix* and it is denoted by $\mathbf{0}$. The zero matrix need not be square.

Any matrix may be *partioned* to yield a matrix with *block structure*, the way the partitioning is done depends on the problem at hand. Consider the following example:

$$\mathbf{A} = \left[\begin{array}{cc|c} a_{11} & a_{12} & a_{13} \\ a_{21} & a_{22} & a_{23} \\ \hline a_{31} & a_{32} & a_{33} \end{array}\right] = \begin{bmatrix} \mathbf{A}_{11} & \mathbf{A}_{12} \\ \mathbf{A}_{21} & \mathbf{A}_{22} \end{bmatrix},$$

where $\mathbf{A}_{11} \in \mathbb{R}^{2\times 2}, \mathbf{A}_{12} \in \mathbb{R}^{2\times 1}, \mathbf{A}_{21} \in \mathbb{R}^{1\times 2}$ and $\mathbf{A}_{22} \in \mathbb{R}^{1\times 1}$.

A partioned matrix of shape

$$\mathbf{A} = \begin{bmatrix} \mathbf{A}_{11} & \mathbf{0} & \cdots \\ \mathbf{0} & \mathbf{A}_{22} & \cdots \\ \vdots & \vdots & \ddots \end{bmatrix}$$

is called *block diagonal*. The diagonal blocks $\mathbf{A}_{ii}$ have to be square matrices and at least one is not a zero matrix. Similarly one can define matrices with block-tridiagonal and block-triangular structure.

Operations and Properties

Having seen that there exists a large variety of matrices, we now introduce the basic arithmetical operations that can be performed. If $\mathbf{A}$ and $\mathbf{B}$ are two matrices of same order, then *addition* and *subtraction* are well defined and carried out component-wise:

$$\mathbf{A} \pm \mathbf{B} = \left[a_{ij} \pm b_{ij}\right].$$

Neither addition nor subtraction are defined if the orders of $\mathbf{A}$ and $\mathbf{B}$ do not coincide. A *scalar multiplication* by a scalar $\beta \in \mathbb{R}$ is performed component-wise as $\beta\mathbf{A} = \left[\beta a_{ij}\right]$.

Matrix multiplication is very different from addition or subtraction because it cannot be performed in a component-wise manner. Let $\mathbf{A}$ be a matrix of order $m \times n$ and $\mathbf{B}$ of order $n \times p$. The product $\mathbf{C} = \mathbf{AB}$ (in this order) is defined by

$$c_{ij} = (\mathbf{AB})_{ij} = \sum_{k=1}^{n} a_{ik} b_{kj}.$$

Note that each row of $\mathbf{A}$ is *multiplied* in this way by each column of $\mathbf{B}$. As a consequence $\mathbf{C}$ is of order $m \times p$. If the number of columns of $\mathbf{A}$ is different from the number of rows of $\mathbf{B}$, then the product $\mathbf{AB}$ is not defined. In general it is not commutative: $\mathbf{AB} \neq \mathbf{BA}$. Therefore in $\mathbf{AB}$ we say that $\mathbf{B}$ is premultiplied by $\mathbf{A}$ and $\mathbf{A}$ ist postmultiplied by $\mathbf{B}$. However,

$$\mathbf{IA} = \mathbf{AI} = \mathbf{A},$$

where if $\mathbf{A}$ is not square, then the identity matrices in this product will be of different order. Another exception is multiplication of two diagonal matrices. It is always commutative. Although matrix multiplication is not commutative it still is associative. Provided all products can be formed:

$$(\mathbf{AB})\mathbf{C} = \mathbf{A}(\mathbf{BC}) = \mathbf{ABC}.$$

If $\mathbf{A}$ is a square matrix, then by the associative rule the power of $\mathbf{A}^n$, n being a positive integer, is defined as $n-$fold product

$$\mathbf{A}^n = \underbrace{\mathbf{A} \cdot \mathbf{A} \ldots \cdot \mathbf{A}}_{n \text{ terms}}.$$

Matrix multiplication is also left and right distributive:

$$\mathbf{A}(\mathbf{B} + \mathbf{C}) = \mathbf{AB} + \mathbf{AC}$$
$$(\mathbf{B} + \mathbf{C})\mathbf{A} = \mathbf{BA} + \mathbf{CA},$$

provided products are defined. Transposing a matrix product changes the order of factors:

$$(\mathbf{AB})' = \mathbf{B}'\mathbf{A}'. \tag{A.94}$$

Multiplication of *partitioned* or *block structured* matrices is carried out in exactly the same way as ordinary matrix multiplication. Care has to be taken because the matrix products of the blocks is in general not commutative and blocks must be of suitable shape. The following example demonstrates multiplication of matrices with 2×2 block structure:

$$\begin{bmatrix} \mathbf{A}_{11} & \mathbf{A}_{12} \\ \mathbf{A}_{21} & \mathbf{A}_{22} \end{bmatrix} \begin{bmatrix} \mathbf{B}_{11} & \mathbf{B}_{12} \\ \mathbf{B}_{21} & \mathbf{B}_{22} \end{bmatrix} = \begin{bmatrix} \mathbf{A}_{11}\mathbf{B}_{11} + \mathbf{A}_{12}\mathbf{B}_{21} & \mathbf{A}_{11}\mathbf{B}_{12} + \mathbf{A}_{12}\mathbf{B}_{22} \\ \mathbf{A}_{21}\mathbf{B}_{11} + \mathbf{A}_{22}\mathbf{B}_{21} & \mathbf{A}_{21}\mathbf{B}_{12} + \mathbf{A}_{22}\mathbf{B}_{22} \end{bmatrix}.$$

Of particular importance is the product of a matrix $\mathbf{A} \in \mathbb{R}^{m \times n}$ and a column vector $\mathbf{x} \in \mathbb{R}^n$. This product is well defined and yields a column vector $\mathbf{Ax} = \mathbf{b} \in \mathbb{R}^m$. Note that in case $\mathbf{x}$ is unknown this is the matrix representation of a system of m linear equations in the n unknowns $x_1, x_2, \ldots, x_n$.

Vector Spaces and Linear Independence

To solve the system of linear equations $\mathbf{Ax} = \mathbf{b}$ some more requisits are required.

Let $\mathcal{A} = \{\mathbf{a}_1, \mathbf{a}_2, \ldots, \mathbf{a}_k\}$ denote a finite set of vectors, each $\mathbf{a}_i \in \mathbb{R}^n$. The vector

$$\mathbf{z} = \beta_1 \mathbf{a}_1 + \beta_2 \mathbf{a}_2 + \ldots + \beta_k \mathbf{a}_k, \tag{A.95}$$

where the scalars β_i are arbitrary real numbers, is called a *linear combination* of $\mathcal{A}$. Note that (A.95) can also be written as a matrix equation. For this purpose define a matrix $\mathbf{A}$ of order $n \times k$ whose columns are just the vectors $\mathbf{a}_i$ and let $\boldsymbol{\beta}$ denote the k-dimensional vector with components β_i. Then (A.95) simply reads as $\mathbf{A}\boldsymbol{\beta} = \mathbf{z}$.

The zero vector $\mathbf{0}$ is always a linear combination of any nonempty set $\mathcal{A}$. If $\mathbf{0}$ can be generated only by trivially putting $\beta_1 = \beta_2 = \ldots = \beta_k = 0$, then the set $\mathcal{A}$ is called *linearly independent*. Observe that a set of linearly independent vectors is minimal in the sense that no vector in this set can be represented as linear combination of the other vectors.

Suppose now that $\mathcal{A}$ is linearly independent. The set of all linear combinations of $\mathcal{A}$:

$$V(\mathcal{A}) = \{\mathbf{z} : \mathbf{z} = \sum_{i=1}^{k} \beta_i \mathbf{a}_i\}$$

is called a *vector space*, the space spanned by $\mathcal{A}$, and $\mathcal{A}$ is a *basis* of V. The *dimension* $\dim V$ of V is the number of vectors in $\mathcal{A}$. From a geometric point of view a vector space is a hyperplane passing through the origin, since the zero vector $\mathbf{0}$ is element of any vector space.

Vector spaces are always closed with respect to addition and scalar multiplication. If $\mathbf{z}_i$ and $\mathbf{z}_j$ are two arbitrary vectors belonging to a space V, then any linear combination $\alpha\mathbf{z}_i + \beta\mathbf{z}_j$ is also an element of V.

Let $\mathbf{A}$ be of order $m \times n$. The set of all vectors of the form

$$\mathcal{R} = \{\mathbf{z} : \mathbf{A}\mathbf{x} = \mathbf{z}, \mathbf{x} \in \mathbb{R}^n\},$$

is a special vector space, called the *range space* of $\mathbf{A}$. Its dimension is called the *rank* of $\mathbf{A}$ and denoted by $r(\mathbf{A})$. It can be shown that always $r(\mathbf{A}) \leq \min(m, n)$. If $r(\mathbf{A}) = min(m, n)$ then $\mathbf{A}$ is said to be of full rank. The *null space* $\mathcal{N}(\mathbf{A})$ is defined as the set of vectors

$$\mathcal{N}(\mathbf{A}) = \{\mathbf{x} : \mathbf{A}\mathbf{x} = \mathbf{0}\}.$$

It can be proved that $r(\mathbf{A}) = k$ if and only if $\dim \mathcal{N}(\mathbf{A}) = n - k$.

A necessary and sufficient condition for the existence of a unique solution of $\mathbf{A}\mathbf{x} = \mathbf{b}$ is $r(\mathbf{A}) = n$ or equivalently $\dim \mathcal{N}(\mathbf{A}) = 0$.

Inverse of a Matrix

Let $\mathbf{A}$ be a square matrix of order $n \times n$. If there exists a uniquely determined square matrix $\mathbf{X}$ which satisfies simultaneously

$$\mathbf{A}\mathbf{X} = \mathbf{X}\mathbf{A} = \mathbf{I}, \tag{A.96}$$

then $\mathbf{X}$ is called the *inverse* of $\mathbf{A}$ and it is denoted by $\mathbf{A}^{-1}$. If it exists, it is in fact uniquely determined.

Unfortunately the calculation of the inverse is a rather difficult process because it requires the solution of n systems of linear equations. A quite efficient algorithm to solve these equations is *Gaussian elimination*. Basically the algorithm applies linear row transformations on $\mathbf{A}$ in order to reduce it to the identity matrix $\mathbf{I}$. There are three types of allowed transformations: (i) interchange row i with row j, (ii) multiply row i by a scalar $\alpha \neq 0$ and (iii) add a scalar multiple of row i to row j.

If it is possible to reduce $\mathbf{A}$ to the identity $\mathbf{I}$, then

$$[\ \mathbf{A} \mid \mathbf{I}\ ] \quad \to \quad [\ \mathbf{I} \mid \mathbf{X}\ ],$$

and $\mathbf{X} = \mathbf{A}^{-1}$. In this case $\mathbf{A}$ is called a *regular* matrix, otherwise $\mathbf{A}$ does not have an inverse and it is called *singular*. Observe that a matrix $\mathbf{A}$ is regular if and only if $r(\mathbf{A}) = n$ or what amounts to the same, if its columns are linearly independent. If $\mathbf{A}^{-1}$ exists, then $\mathbf{Ax} = \mathbf{b}$ has the solution $\mathbf{x} = \mathbf{A}^{-1}\mathbf{b}$ by premultiplying $\mathbf{A}^{-1}$ on both sides.

The following properties of regular matrices are important:

$$(\mathbf{AB})^{-1} = \mathbf{B}^{-1}\mathbf{A}^{-1}, \tag{A.97}$$

and

$$(\mathbf{A}^{-1})^{-1} = \mathbf{A}, \tag{A.98}$$

which is an immediate consequence of (A.96).

We conclude with an interesting formula for the inverse of a partitioned matrix. Suppose that in (A.96)

$$\mathbf{A} = \begin{bmatrix} \mathbf{A}_{11} & \mathbf{A}_{12} \\ \mathbf{A}_{21} & \mathbf{A}_{22} \end{bmatrix}.$$

Since we already know how to multiply partitioned matrices, we may solve Equation (A.96) element by element to find

$$\mathbf{A}^{-1} = \begin{bmatrix} \mathbf{H}_1^{-1} & -\mathbf{A}_{11}^{-1}\mathbf{A}_{12}\mathbf{H}_2^{-1} \\ -\mathbf{A}_{22}^{-1}\mathbf{A}_{21}\mathbf{H}_1^{-1} & \mathbf{H}_2^{-1} \end{bmatrix}, \tag{A.99}$$

where $\mathbf{H}_1$ and $\mathbf{H}_2$ are given by

$$\begin{aligned} \mathbf{H}_1 &= \mathbf{A}_{11} - \mathbf{A}_{12}\mathbf{A}_{22}^{-1}\mathbf{A}_{21} \\ \mathbf{H}_2 &= \mathbf{A}_{22} - \mathbf{A}_{21}\mathbf{A}_{11}^{-1}\mathbf{A}_{12}. \end{aligned}$$

Determinants and Cramer's Rule

Let $\mathbf{A}$ be a square matrix of order $n \times n$. The *determinant* $|\mathbf{A}|$ of $\mathbf{A}$ is a linear function of the columns of $\mathbf{A}$ which may be defined recursively by the *expansion theorem of Laplace*.

If $\mathbf{A} = [a_{11}] \in \mathbb{R}^{1\times 1}$, then $|\mathbf{A}| = a_{11}$.

Now define the *minor* $|\mathbf{A}_{ij}|$ as the subdeterminant which is obtained from $\mathbf{A}$ by deleting row i and column j. Then

$$|\mathbf{A}| = \sum_{j=1}^{n}(-1)^{i+j}a_{ij}|\mathbf{A}_{ij}|, \qquad i = 1, \ldots, n \tag{A.100}$$

and i is an arbitrarily chosen row index. For instance

$$\mathbf{A} = \begin{bmatrix} a_{11} & a_{12} \\ a_{21} & a_{22} \end{bmatrix} \quad \Longrightarrow \quad |\mathbf{A}| = a_{11}a_{22} - a_{12}a_{21}.$$

The following rules are important in the calculus of determinants:

$$|\mathbf{AB}| = |\mathbf{A}||\mathbf{B}| \tag{A.101}$$

$$|\mathbf{A}^{-1}| = 1/|\mathbf{A}| \quad \text{if } |\mathbf{A}| \neq 0 \tag{A.102}$$

$$|\mathbf{A}'| = |\mathbf{A}|. \tag{A.103}$$

Determinants provide an alternative way to solve the system of linear equations $\mathbf{Ax} = \mathbf{b}$, provided $\mathbf{A}$ is a nonsingular square matrix. By *Cramer's rule*

$$x_i = \frac{|\mathbf{A}_i|}{|\mathbf{A}|}, \qquad i = 1, 2, \ldots n \tag{A.104}$$

where $\mathbf{A}_i$ is the matrix $\mathbf{A}$ with column i replaced by the column vector $\mathbf{b}$ of the right-hand side.

Eigenvalues and Eigenvectors

Throughout this section we assume that $\mathbf{A}$ is a square matrix of order $n \times n$. A vector $\mathbf{x} \neq \mathbf{0}$ is a *right eigenvector* of $\mathbf{A}$, if there exists a complex constant λ, such that

$$\mathbf{Ax} = \lambda\mathbf{x}. \tag{A.105}$$

The constant λ is called the *eigenvalue* associated with $\mathbf{x}$.

Eigenvectors are not unique, they are determined only up to a scalar multiple. Thus if $\mathbf{x}$ is an eigenvector, then so is $\alpha\mathbf{x}$ for any nonzero constant α, since

$$\mathbf{A}(\alpha\mathbf{x}) = \alpha\mathbf{Ax} = \lambda(\alpha\mathbf{x}).$$

By the Fix Point Theorem of Brouwer, it follows that any square matrix has at least one eigenvector.

In order to find these eigenvectors we rewrite (A.105) as

$$(\mathbf{A} - \lambda\mathbf{I})\mathbf{x} = \mathbf{0}. \tag{A.106}$$

From that equation it is immediately seen that $\mathbf{x} \in \mathcal{N}(\mathbf{A} - \lambda\mathbf{I})$. Thus nontrivial solutions $\mathbf{x} \neq \mathbf{0}$ exist if and only if $\dim \mathcal{N}(\mathbf{A} - \lambda\mathbf{I}) > 0$. This condition in turn is equivalent to

$$r(\mathbf{A} - \lambda\mathbf{I}) < n \quad \Leftrightarrow \quad |\mathbf{A} - \lambda\mathbf{I}| = 0.$$

Expanding the determinant, for instance by the formula of Laplace (A.100), we obtain a polynomial $p(\lambda)$ of degree n, the *characteristic polynomial* of $\mathbf{A}$. By the fundamental theorem of algebra this polynomial may be factored over the set of complex numbers $\mathbb{C}$ to yield the polynomial equation

$$(\lambda - \lambda_1)^{n_1}(\lambda - \lambda_2)^{n_2} \ldots (\lambda - \lambda_k)^{n_k} = 0, \qquad n_1 + n_2 + \ldots + n_k = n. \tag{A.107}$$

The roots λ_i are just the eigenvalues of $\mathbf{A}$ and n_i is called the *algebraic multiplicity* of λ_i. It follows that any $n \times n$ matrix has exactly n eigenvalues, which are in general complex numbers (even when $\mathbf{A}$ is a real matrix) and which are not necessarily distinct.

To determine the corresponding eigenvectors, we have to solve the systems (A.106). For this purpose define matrices $\mathbf{A}_i = \mathbf{A} - \lambda_i \mathbf{I}$ for each distinct eigenvalue λ_i. Then it is guaranteed that systems of linear equations $\mathbf{A}_i \mathbf{x}_i = \mathbf{0}$ have solutions $\mathbf{x}_i \neq \mathbf{0}$.

However, degeneracy may occur in case of multiple eigenvalues. For particular matrices the number of linearly independent solutions of $\mathbf{A}_i \mathbf{x} = \mathbf{0}$ may be less than the algebraic multiplicity of λ_i. This degeneracy cannot occur, if for instance:

- All eigenvalues of $\mathbf{A}$ are different.
- $\mathbf{A}$ is a symmetric matrix, even with multiple eigenvalues.

A matrix which is not degenerate in the above sense is said to be *semisimple*. It always has a complete set of eigenvectors $\mathbf{x}_1, \mathbf{x}_2, \ldots, \mathbf{x}_n$; moreover, these eigenvectors can be shown to be linearly independent.

In what follows we shall assume that $\mathbf{A}$ is semisimple. The most important case of semisimple matrices are those which are symmetric. For symmetric matrices the following statements are always true:

1. The eigenvalues of a symmetric matrix are real numbers, though not necessarily distinct.
2. If λ_i and λ_j are two distinct eigenvalues with eigenvectors $\mathbf{x}_i$ and $\mathbf{x}_j$, then these eigenvectors are orthogonal, which means that $\mathbf{x}_i'\mathbf{x}_j = 0$.

Let

$$\mathbf{X} = \begin{bmatrix} \mathbf{x}_1 & \mathbf{x}_2 & \ldots & \mathbf{x}_n \end{bmatrix}$$

denote the $n \times n$ matrix whose columns are the eigenvectors of $\mathbf{A}$. Since the latter are linearly independent, $\mathbf{X}$ has an inverse $\mathbf{X}^{-1}$. Multiplying $\mathbf{A}$ by $\mathbf{X}$ from the right yields:

$$\mathbf{AX} = \mathbf{X\Lambda}, \tag{A.108}$$

where $\mathbf{\Lambda}$ is the diagonal matrix of the eigenvalues:

$$\mathbf{\Lambda} = \text{diag}(\lambda_1, \lambda_2, \ldots, \lambda_n).$$

Multiplying (A.108) by $\mathbf{X}^{-1}$ from right yields a remarkable formula, called the *spectral decomposition* of $\mathbf{A}$:

$$\mathbf{A} = \mathbf{X}\boldsymbol{\Lambda}\mathbf{X}^{-1}. \tag{A.109}$$

Left eigenvectors $\mathbf{y}'$ are nontrivial solutions of the equation

$$\mathbf{y}'\mathbf{A} = \lambda\mathbf{y}'. \tag{A.110}$$

Since by (A.94) we have $(\mathbf{y}'\mathbf{A})' = \mathbf{A}'\mathbf{y}$, the left eigenvectors of $\mathbf{A}$ are just the right eigenvectors of $\mathbf{A}'$. Furthermore the characteristic polynomial $|\mathbf{A} - \lambda\mathbf{I}|$ does not change if $\mathbf{A}$ is replaced by $\mathbf{A}'$, thus $\mathbf{A}$ and $\mathbf{A}'$ have the same eigenvalues. Therefore no new theory is required for left eigenvectors.

Similarity Transforms

Let us rewrite the spectral formula (A.109) as

$$\boldsymbol{\Lambda} = \mathbf{X}^{-1}\mathbf{A}\mathbf{X}. \tag{A.111}$$

This formula tells us that any semisimple matrix can be transformed into a diagonal matrix by means of its eigenvectors. This is a special case of a *similarity transform.* Let $\mathbf{T}$ be any nonsingular matrix, then

$$\mathbf{B} = \mathbf{T}^{-1}\mathbf{A}\mathbf{T}$$

is called a similarity transform of $\mathbf{A}$ and $\mathbf{B}$ is said to be similar to $\mathbf{A}$. Thus any semisimple matrix is similar to a diagonal matrix.

It is important to note that a similarity transform preserves the structure of a matrix in the following sense:

- $\mathbf{A}$ and $\mathbf{B}$ have the same eigenvalues.
- If $\mathbf{X}$ is the matrix of eigenvectors of $\mathbf{A}$, then $\mathbf{T}^{-1}\mathbf{X}$ is the matrix of eigenvectors of $\mathbf{B}$.

Indeed:

$$\mathbf{A} = \mathbf{X}\boldsymbol{\Lambda}\mathbf{X}^{-1} \quad \Longrightarrow \quad \mathbf{B} = \mathbf{T}^{-1}\mathbf{X}\boldsymbol{\Lambda}\mathbf{X}^{-1}\mathbf{T}.$$

Tridiagonal Matrices

A tridiagonal matrix has the special structure

$$\mathbf{A} = \begin{bmatrix} b_0 & a_0 & 0 & 0 & \dots & 0 \\ c_1 & b_1 & a_1 & 0 & \dots & 0 \\ 0 & c_2 & b_2 & a_2 & \dots & 0 \\ \vdots & & & & \ddots & \vdots \\ 0 & \dots & & & c_N & b_N \end{bmatrix},$$

where the numbers a_i, b_i and c_i are nonzero and real.

Tridiagonal matrices have interesting properties, the two most important ones being

1. A tridiagonal matrix is always similar to a symmetric matrix. As a result, it is semisimple and has real eigenvalues only.
2. The eigenvalues of a tridiagonal matrix are distinct.

To prove the first statement, define a diagonal matrix $\mathbf{T} = \operatorname{diag}(t_0, t_1, \ldots, t_N)$ by

$$t_0 = 1, \quad t_k = \sqrt{\frac{c_1 c_2 \cdots c_k}{a_0 a_1 \cdots a_{k-1}}} \quad k = 1, 2, \ldots, N. \tag{A.112}$$

Hence the similarity transform of $\mathbf{A}$ yields

$$\mathbf{B} = \mathbf{T}^{-1}\mathbf{A}\mathbf{T} = \begin{bmatrix} b_0 & \sqrt{a_0 c_1} & 0 & \ldots & 0 \\ \sqrt{a_0 c_1} & b_1 & \sqrt{a_1 c_2} & \ldots & 0 \\ 0 & \sqrt{a_1 c_2} & b_2 & \ldots & 0 \\ \vdots & & & & \vdots \\ 0 & 0 & \ldots & \sqrt{a_{N-1} c_N} & b_N \end{bmatrix}.$$

To prove the second statement about tridiagonal matrices, namely that its eigenvalues are distinct, let $\mathbf{B}_n$ be the matrix obtained from $\mathbf{B}$ by taking the first $n+1$ rows and columns of $\mathbf{B}$ and let $p_k(\lambda)$ denote the characteristic polynomial of $\mathbf{B}_{k-1}, k = 1, 2, \ldots, n+1$. Applying Laplace's Theorem yields the three-term recurrence formula for the characteristic polynomials

$$p_k(\lambda) = (b_{k-1} - \lambda) p_{k-1}(\lambda) - a_{k-2} c_{k-1} p_{k-2}(\lambda), \qquad k = 2, 3, \ldots, N+1, \tag{A.113}$$

with boundary conditions $p_0(\lambda) = 1$ and $p_1(\lambda) = b_0 - \lambda$. The sequence of polynomials $p_n(\lambda)$ has a very special property, it forms a *Sturm chain.* By *Sturm's Theorem* it follows that $p_n(\lambda)$ has n distinct real roots in $(-\infty, +\infty)$.

In particular cases even sharper statements are possible. Suppose that $p_n(0) > 0$ for $n = 0, 1, \ldots$, this is precisely the situation encountered in Section 4.4 in Chapter 4. Then again by Sturm's Theorem the polynomial $p_n(\lambda)$ has n distinct roots which are now located in $(0, \infty)$ and as a result the matrix $\mathbf{B}_n, n = 1, 2, \ldots, N-1$ has n distinct positive eigenvalues.

Functions of a Matrix

The simplest functions of a square matrix $\mathbf{A}$ are the powers $\mathbf{A}^m$, where m is a positive integer. Assuming that $\mathbf{A}$ is semisimple and applying the spectral decomposition (A.109) we find:

$$\mathbf{A}^m = \mathbf{X}\mathbf{\Lambda}^m\mathbf{X}^{-1}. \tag{A.114}$$

But the m-th power of a diagonal matrix is easy to calculate:

$$\mathbf{\Lambda}^m = \operatorname{diag}(\lambda_1^m, \lambda_2^m, \ldots, \lambda_n^m).$$

Now let $f(z)$ be a function of a scalar complex variable z which is analytic inside the circle $|z| < \rho$. Then inside that circle $f(z)$ has a convergent power series expansion

$$f(z) = \sum_{n \geq 0} c_n z^n. \tag{A.115}$$

Under certain conditions this expansion can be given a meaning when z is replaced by a square matrix $\mathbf{A}$. Then we have using (A.114) and the convention $\mathbf{A}^0 = \mathbf{I}$:

$$
\begin{aligned}
f(\mathbf{A}) &= \sum_{n \geq 0} c_n \mathbf{A}^n = \sum_{n \geq 0} c_n \mathbf{X} \mathbf{\Lambda}^n \mathbf{X}^{-1} = \mathbf{X} \left[\sum_{n \geq 0} c_n \mathbf{\Lambda}^n \right] \mathbf{X}^{-1} \\
&= \mathbf{X} f(\mathbf{\Lambda}) \mathbf{X}^{-1},
\end{aligned} \tag{A.116}
$$

with

$$f(\mathbf{\Lambda}) = \mathrm{diag}(f(\lambda_1), f(\lambda_2), \ldots, f(\lambda_n)).$$

For this to make sense, it is necessary that $|\lambda_i| < \rho$ for all $i = 1, \ldots, n$ or, what amounts to the same, $\max |\lambda_i| < \rho$. The modulus of the largest eigenvalue $\max |\lambda_i|$ is also called the *spectral radius* $\mathrm{sp}(\mathbf{A})$.

From the representation (A.116) it follows that:

1. The eigenvalues of $f(\mathbf{A})$ are $f(\lambda_i)$.
2. $\mathbf{A}$ and $f(\mathbf{A})$ have the same eigenvectors.
3. $\mathbf{A}$ and $f(\mathbf{A})$ commute.

The most important example is $f(z) = e^z$. Its matrix analogue is the matrix exponential function

$$e^{\mathbf{A}} = \sum_{n \geq 0} \frac{1}{n!} \mathbf{A}^n = \mathbf{X} e^{\mathbf{\Lambda}} \mathbf{X}^{-1}, \tag{A.117}$$

where

$$e^{\mathbf{\Lambda}} = \mathrm{diag}(e^{\lambda_1}, e^{\lambda_2}, \ldots, e^{\lambda_n}). \tag{A.118}$$

Note that (A.117) is meaningful for any square matrix, since the Taylor series of e^z converges in the whole complex plane.

The multiplication formula for exponential functions does not hold for matrices in general: the functional relation

$$e^{\mathbf{A}} e^{\mathbf{B}} = e^{\mathbf{A}+\mathbf{B}} \tag{A.119}$$

is true if $\mathbf{A}$ and $\mathbf{B}$ commute.

The matrix exponential plays a central role in the theory of systems of linear differential equations with constant coefficients. In its most general form such a system is given by

$$
\begin{array}{ccccccccc}
\dfrac{dx_1}{dt} & = & a_{11}x_1 & + & a_{12}x_2 & + & \cdots & + & a_{1n}x_n \\
\dfrac{dx_2}{dt} & = & a_{21}x_1 & + & a_{22}x_2 & + & \cdots & + & a_{2n}x_n \\
\vdots & & & & & & & & \vdots \\
\dfrac{dx_n}{dt} & = & a_{n1}x_1 & + & a_{n2}x_2 & + & \cdots & + & a_{nn}x_n
\end{array} .
$$

The functions $x_i(t)$ are unknown and satisfy the initial conditions $x_i(0) = \xi_i$. Let $\mathbf{x}$ be a column vector with components $x_i(t), i = 1, \ldots, n$, and let $\mathbf{A}$

denote the matrix of coefficients a_{ij}. Then the system may be written in matrix form as:

$$\frac{d\mathbf{x}}{dt} = \mathbf{A}\mathbf{x}, \qquad \mathbf{x}(0) = \boldsymbol{\xi}. \tag{A.120}$$

The notation $d\mathbf{x}/dt$ means differentiation of each component, thus the components of $d\mathbf{x}/dt$ are the derivatives $dx_i(t)/dt$. The unique solution of (A.120) is given by the matrix exponential function

$$\mathbf{x} = e^{\mathbf{A}t}\boldsymbol{\xi}.$$

The proof is straightforward by inserting the Taylor series of the matrix exponential function in (A.120).

Another example of an important matrix function is the geometric series

$$\mathbf{G} = \mathbf{I} + \mathbf{A} + \mathbf{A}^2 + \cdots . \tag{A.121}$$

Since the scalar geometric series converges for $|z| < 1$, it follows that (A.121) converges, if all eigenvalues of $\mathbf{A}$ lie inside the unit circle, i.e., $\text{sp}(\mathbf{A}) < 1$. In that case:

$$\mathbf{G} = (\mathbf{I} - \mathbf{A})^{-1},$$

which may be verified directly by left or right multiplication of (A.121) with $\mathbf{I} - \mathbf{A}$. This result does not depend on $\mathbf{A}$ being semisimple.

Non-negative Matrices

The transition probability matrices we encounter in the theory of Markov chains have a special property: their components are non-negative. Non-negativity has some interesting and important consequences which we summarize in this section.

A matrix $\mathbf{A}$ is called *non-negative*, when all its components a_{ij} are non-negative and we write in this case $\mathbf{A} \geq \mathbf{0}$, where $\mathbf{0}$ denotes the zero matrix of order $n \times n$.

A non-negative matrix $\mathbf{A}$ is called *reducible*, if it can be transformed into a block structured matrix of form

$$\tilde{\mathbf{A}} = \begin{bmatrix} \mathbf{B} & \mathbf{0} \\ \mathbf{C} & \mathbf{D} \end{bmatrix}$$

by applying a particular permutation to its rows and columns. The submatrices $\mathbf{B}$ and $\mathbf{D}$ must be square. If such a transformation is not possible, then $\mathbf{A}$ is said to be *irreducible*.

Let $\mathbf{A}$ be an irreducible non-negative matrix. Then the classical *theorem of Perron and Frobenius* states that

- $\mathbf{A}$ has a real and positive eigenvalue ρ, the moduli of the other eigenvalues do not exceed ρ.
- The eigenvector corresponding to ρ has positive components.

- In case $\mathbf{A}$ has more than one eigenvalue with modulus ρ, that is $\lambda_1 = \rho, |\lambda_2| = |\lambda_3| = \ldots = |\lambda_h| = \rho$, then these eigenvalues are the roots of the equation

$$\lambda^h = \rho^h.$$

A non-negative matrix $\mathbf{A}$ is said to be *stochastic*, if its row sums are equal to one. Clearly the Perron-Frobenius theorem applies in case of irreducibility and the largest real eigenvalue equals $\rho = 1$ with corresponding right eigenvector $\mathbf{x} = \mathbf{e}$, $\mathbf{e}$ being a column vector with all components equal to one. In fact, right multiplication of any matrix by $\mathbf{e}$ yields the column vector of the row sums. Thus

$$\mathbf{Ae} = \mathbf{e},$$

which proves that 1 an eigenvalue and $\mathbf{e}$ the corresponding right eigenvector. The Perron-Frobenius theorem applies also to the transpose of $\mathbf{A}$. As a consequence, $\mathbf{A}$ has a left eigenvector $\mathbf{z}$ with eigenvalue 1 and positive components:

$$\mathbf{z}'\mathbf{A} = \mathbf{z}'$$

Since left and right eigenvectors are determined up to a scalar multiple, it is possible to *normalize* the vector $\mathbf{z}'$ in the last equation in such a way that the sum of its components equals one. Now it represents a probability distribution, the *steady-state vector* of the stochastic matrix $\mathbf{A}$.

References

Bellman, R. (1995). *Introduction to Matrix Analysis*, SIAM, Philadelphia.

Pease, M. C. (1965). *Methods of Matrix Algebra*, Academic Press, New York.

A.11. Numerical Analysis

Solution of Equations

In scientific and engineering work, a frequently occurring problem is to find the roots of equations of the form $f(x) = 0$. Here we describe some numerical methods for the solution of such equations, where $f(x)$ may be algebraic or transcendental or a combination of both.

The bisection method. It is based on *Bolzano's Theorem* that if a function is continuous in $[a, b]$ and $f(a)$ and $f(b)$ have different signs, then there must be at least one root in $[a, b]$. An approximation of this root is $x_0 = (a + b)/2$. If $f(x_0) = 0$ then x_0 is the sought root, otherwise the root lies between x_0 and b or between x_0 and a depending on whether $f(x_0)$ is positive or negative. We designate this new interval as $[a_1, b_1]$, its length is $(a + b)/2$. As before, the interval is bisected at x_1, say, and its length will be exactly half the length of the previous interval. The process is repeated until the interval $[a_n, b_n]$, which contains a root, has length less than some prescribed value $\epsilon > 0$. Since the width of the interval containing a root is reduced by a factor of $1/2$ at each iteration, we have $|b-a|/2^n \leq \epsilon$. Thus for a prescribed accuracy ϵ the number of iterations is about

$$n \geq \frac{\ln(|b-a|/\epsilon)}{\ln 2}.$$

Muller's Method. The idea of this method is to approximate $f(x)$ by a second degree polynomial in the vicinity of the root. The roots of the resulting quadratic equation are then approximations of the roots of $f(x) = 0$. Let x_1, x_2, x_3 be three arbitrary starting values and $g(x)$ be the unique parabola passing through the points $(x_i, f(x_i)), i = 1, 2, 3$; $g(x)$ is given by

$$g(x) = f(x_1) + f(x_1, x_2)(x - x_1) + f(x_1, x_2, x_3)(x - x_1)(x - x_2),$$

where

$$f(x_i, x_j) = \frac{f(x_i) - f(x_j)}{x_i - x_j}, \qquad i, j = 1, 2, 3$$

and

$$f(x_1, x_2, x_3) = \frac{f(x_1, x_2) - f(x_2, x_3)}{x_1 - x_3}.$$

Expanding $g(x)$ in powers of x yields $g(x) = a_0 + a_1 x + x_2^2$, with

$$a_0 = f(x_1) - x_1 f(x_1, x_2) + x_1 x_2 f(x_1, x_2, x_3)$$

$$a_1 = f(x_1, x_2) - (x_1 + x_2) f(x_1, x_2, x_3), \qquad a_2 = f(x_1, x_2, x_3).$$

The zeroes of $g(x)$ can be expressed as

$$x = \frac{2a_0}{-a_1 \pm \sqrt{a_1^2 - 4a_0 a_2}},$$

where the sign of the root is chosen such that the denominator is largest in magnitude. In case the root is a complex number, set it equal to zero. Put $x_1 \leftarrow x_2, x_2 \leftarrow x_3, x_3 \leftarrow x$ and continue until convergence is obtained.

Numerical Integration

The general problem of numerical integration may be stated as follows: given a set of points $(x_i, y_i), i = 0, \ldots, n$ of a function $y = f(x)$, it is required to compute the value of the definite integral $\int_a^b f(x)dx$. A successful strategy is to replace $f(x)$ by an interpolating polynomial $P(x)$ which can be integrated easily and its value is used as an approximation of $\int_a^b f(x)dx$. Different integration formulas can be obtained depending on the type of interpolation formula used. Here we derive a general formula for numerical integration using *Newton's forward difference formula.*

Let the interval $[a, b]$ be divided into n subintervals of length $h = (b-a)/n$, such that $a = x_0 < x_1 < \ldots < x_n = b$ and $x_n = x_0 + nh$. Writing $y = f(x)$ by means of Newton's forward difference formula we obtain

$$\int_{x_0}^{x_n} y dx = \int_{x_0}^{x_n} \left[y_0 + p\Delta y_0 + \frac{p(p-1)}{2}\Delta^2 y_0 + \frac{p(p-1)(p-2)}{6}\Delta^3 y_0 + \ldots \right] dx,$$

where $y_0 = f(x_0)$. Since $x = x_0 + ph$, $dx = hdp$, the above integral becomes

$$\int_{x_0}^{x_n} ydx =$$

$$= h\int_0^n \left[y_0 + p\Delta y_0 + \frac{p(p-1)}{2}\Delta^2 y_0 + \frac{p(p-1)(p-2)}{6}\Delta^3 y_0 + \ldots\right] dp.$$

Term by term integration yields

$$\int_{x_0}^{x_n} ydx = nh\left[y_0 + \frac{n}{2}\Delta y_0 + \frac{n(2n-3)}{12}\Delta^2 y_0 + \frac{n(n-2)^2}{24}\Delta^3 y_0 + \ldots\right]. \tag{A.122}$$

By putting $n = 1, 2, \ldots$, different integration rules can be obtained from this general formula.

The trapezoidal rule. Setting $n = 1$ in (A.122) and neglecting differences of order two and higher, we get

$$\int_{x_0}^{x_n} ydx = \frac{h}{2}\left[y_0 + 2(y_1 + y_2 + \ldots + y_{n-1}) + y_n\right] \tag{A.123}$$

Simpson's 1/3-rule. This rule follows, if we put $n = 2$ in (A.122) and neglect differences of order four and higher (observe that differences of order three cancel):

$$\int_{x_0}^{x_n} ydx = \frac{h}{3}\left[y_0 + 4y_1 + 2y_2 + 4y_3 + \ldots + 4y_{n-1} + y_n\right]. \tag{A.124}$$

Numerical Solution of Ordinary Differential Equations

Let

$$y' = f(x, y), \quad y(x_0) = y_0 \tag{A.125}$$

an ordinary differential equation which satisfies the conditions for existence and uniqueness of solution. Suppose we wish to solve (A.125) for values of y at $x = x_n = x_0 + nh$, where h is the stepsize sufficiently small. Integrating (A.125) with respect to x we obtain

$$y_1 = y_0 + \int_{x_0}^{x_1} f(x, y)dx. \tag{A.126}$$

Since the integrand contains the unknown function y, it is necessary to approximate the integral in a suitable way. Depending on how this approximation is done, various methods or *rules* result.

Euler's method. This method replaces the integrand by the known constant $f(x, y) = f(x_0, y_0), x_0 \le x \le x_1$. As a result we have the approximation

$$y_1 = y_0 + hf(x_0, y_0).$$

Similarly, in the range $x_1 \le x \le x_2$ we have

$$y_2 = y_1 + \int_{x_1}^{x_2} f(x, y)dx.$$

Within this range $f(x,y)$ is approximated by $f(x_1,y_1)$, thus

$$y_2 = y_1 + hf(x_1,y_1)$$

and proceeding in that way, we obtain the recurrence formula

$$y_{n+1} = y_n + hf(x_n,y_n), \qquad n = 0,1,\dots \tag{A.127}$$

Modified Euler's method. Instead of approximating $f(x,y)$ simply by $f(x_0,y_0)$ in (A.126) we approximate the integral by the trapezoidal rule to obtain

$$y_1 = y_0 + \frac{h}{2}\left[f(x_0,y_0) + f(x_1,y_1)\right].$$

A starting value $y_1^{(0)}$ is provided by Euler's method:

$$y_1^{(0)} = y_0 + hf(x_0,y_0)$$

to yield the iteration formula

$$y_1^{(r+1)} = y_0 + \frac{h}{2}\left[f(x_0,y_0) + f(x_1,y_1^{(r)})\right], \quad r = 0,1,2,\dots. \tag{A.128}$$

This recurrence is applied until convergence of the values of $y_1^{(r)}$ is obtained. Then we proceed to y_2 etc., until we arrive at y_n.

Runge-Kutta methods. Euler's method as well as its modification are not symmetrical in the sense that only information of the left endpoint of the interval $[x_0,x_1]$ is used. The idea of Runge-Kutta methods is to move inside the interval to obtain better approximations. For instance taking the midpoint of the interval one obtains:

$$\begin{aligned}
k_1 &= hf(x_0,y_0)\\
k_2 &= hf\left(x_0 + \frac{h}{2}, y_0 + \frac{k_1}{2}\right)
\end{aligned}$$

and as a result *Runge-Kutta's second order formula:*

$$y_1 = y_0 + k_2 + O(h^3), \tag{A.129}$$

whereas Euler's formula is of first order The error term can be improved further. The most commonly used approximation is the *fourth order Runge-Kutta formula:*

$$\begin{aligned}
k_1 &= hf(x_0,y_0)\\
k_2 &= hf\left(x_0 + \frac{h}{2}, y_0 + \frac{k_1}{2}\right)\\
k_3 &= hf\left(x_0 + \frac{h}{2}, y_0 + \frac{k_2}{2}\right)\\
k_4 &= hf(x_0 + h, y_0 + k_3)
\end{aligned}$$

with

$$y_1 = y_0 + \frac{1}{6}\left(k_1 + 2k_2 + 2k_3 + k_4\right) + O(h^5). \tag{A.130}$$

Iterative Solution of Systems of Equations

Let $\mathbf{Ax} = \mathbf{b}$ be a system of n linear equations in the n unknowns $x_1, x_2, \ldots, x_n$. If n is large but $\mathbf{A}$ is *sparse*, i.e., $\mathbf{A}$ contains many zeroes, then iterative algorithms are often an appropriate way to obtain a solution.

Consider equation number i in $\mathbf{Ax} = \mathbf{b}$:

$$\sum_{j=1}^{n} a_{ij} x_j = b_i$$

Solving for x_i while keeping the other variables fixed yields the recurrence relation

$$x_i^{(m)} = \frac{1}{a_{ii}} \left[b_i - \sum_{j \neq i} a_{ij} x_j^{(m-1)} \right], \quad m = 1, 2, \ldots \tag{A.131}$$

where $\mathbf{x}^{(0)} = (x_1^{(0)}, \ldots, x_n^{(0)})$ is an arbitrarily chosen starting vector. Recurrence (A.131) is known as the *Jacobi method.* Obviously it requires $a_{ii} \neq 0$ for all $i = 1, \ldots, n$. However, this condition does not ensure convergence of the sequence of Jacobi iterates $\mathbf{x}^{(m)}$. A sufficient condition for convergence is the *row sum criterion*

$$\sum_{j \neq i} |a_{ij}| < |a_{ii}|, \qquad i = 1, \ldots, n. \tag{A.132}$$

A square matrix satisfying (A.132) is called *diagonally dominant.*

Unfortunately convergence of Jacobi's method may be rather slow. Sometimes improved convergence is obtained by another iterative procedure, the *Gauss-Seidel method.* It uses the recurrence relation

$$x_i^{(m)} = \frac{1}{a_{ii}} \left[b_i - \sum_{j=1}^{i-1} a_{ij} x_j^{(m)} - \sum_{j=i+1}^{n} a_{ij} x_j^{(m-1)} \right], \qquad m = 1, 2, \ldots . \tag{A.133}$$

Observe that the calculation of $x_i^{(m)}$ requires the knowledge of $x_j^{(m)}$, $j = 1, \ldots, i-1$, in contrast to Jacobi iterations. As a consequence the Gauss-Seidel iterates $\mathbf{x}^{(m)}$ depend on the *ordering* of the equations. The row sum condition (A.132) is sufficient for convergence of this method.

The topics covered in this section can be found in most textbooks on numerical analysis, a good reference is the book by Sastri (2002).

Reference

Sastry, S. S. (2002). *Introductory Methods of Numerical Analysis*, 3rd ed., Prentice-Hall, India.

APPENDIX B

Table B-1: Distribution function of the standard normal distribution

$$F(z) = \frac{1}{\sqrt{2\pi}} \int_{-\infty}^{z} e^{-t^2/2} dt \text{ for } z \in [0, 3). \text{ Note that } F(-z) = 1 - F(z).$$

	0.00	0.01	0.02	0.03	0.04	0.05	0.06	0.07	0.08	0.09
0.0	0.500	0.504	0.508	0.512	0.516	0.520	0.524	0.528	0.532	0.536
0.1	0.540	0.544	0.548	0.552	0.556	0.560	0.564	0.567	0.571	0.575
0.2	0.579	0.583	0.587	0.591	0.595	0.599	0.603	0.606	0.610	0.614
0.3	0.618	0.622	0.626	0.629	0.633	0.637	0.641	0.644	0.648	0.652
0.4	0.655	0.659	0.663	0.666	0.670	0.674	0.677	0.681	0.684	0.688
0.5	0.691	0.695	0.698	0.702	0.705	0.709	0.712	0.716	0.719	0.722
0.6	0.726	0.729	0.732	0.736	0.739	0.742	0.745	0.749	0.752	0.755
0.7	0.758	0.761	0.764	0.767	0.770	0.773	0.776	0.779	0.782	0.785
0.8	0.788	0.791	0.794	0.797	0.800	0.802	0.805	0.808	0.811	0.813
0.9	0.816	0.819	0.821	0.824	0.826	0.829	0.831	0.834	0.836	0.839
1.0	0.841	0.844	0.846	0.848	0.851	0.853	0.855	0.858	0.860	0.862
1.1	0.864	0.867	0.869	0.871	0.873	0.875	0.877	0.879	0.881	0.883
1.2	0.885	0.887	0.889	0.891	0.893	0.894	0.896	0.898	0.900	0.901
1.3	0.903	0.905	0.907	0.908	0.910	0.911	0.913	0.915	0.916	0.918
1.4	0.919	0.921	0.922	0.924	0.925	0.926	0.928	0.929	0.931	0.932
1.5	0.933	0.934	0.936	0.937	0.938	0.939	0.941	0.942	0.943	0.944
1.6	0.945	0.946	0.947	0.948	0.949	0.951	0.952	0.953	0.954	0.954
1.7	0.955	0.956	0.957	0.958	0.959	0.960	0.961	0.962	0.962	0.963
1.8	0.964	0.965	0.966	0.966	0.967	0.968	0.969	0.969	0.970	0.971
1.9	0.971	0.972	0.973	0.973	0.974	0.974	0.975	0.976	0.976	0.977
2.0	0.977	0.978	0.978	0.979	0.979	0.980	0.980	0.981	0.981	0.982
2.1	0.982	0.983	0.983	0.983	0.984	0.984	0.985	0.985	0.985	0.986
2.2	0.986	0.986	0.987	0.987	0.987	0.988	0.988	0.988	0.989	0.989
2.3	0.989	0.990	0.990	0.990	0.990	0.991	0.991	0.991	0.991	0.992
2.4	0.992	0.992	0.992	0.992	0.993	0.993	0.993	0.993	0.993	0.994
2.5	0.994	0.994	0.994	0.994	0.994	0.995	0.995	0.995	0.995	0.995
2.6	0.995	0.995	0.996	0.996	0.996	0.996	0.996	0.996	0.996	0.996
2.7	0.997	0.997	0.997	0.997	0.997	0.997	0.997	0.997	0.997	0.997
2.8	0.997	0.998	0.998	0.998	0.998	0.998	0.998	0.998	0.998	0.998
2.9	0.998	0.998	0.998	0.998	0.998	0.998	0.998	0.999	0.999	0.999

Table B-2: Quantiles χ_p^2 of the chi-square distribution with m degrees of freedom

$$P(T \leq \chi_p^2) = p$$

$m\backslash p$	0.010	0.025	0.050	0.100	0.900	0.950	0.975	0.990
1	0.000	0.001	0.004	0.016	2.706	3.841	5.024	6.635
2	0.020	0.051	0.103	0.211	4.605	5.991	7.378	9.210
3	0.115	0.216	0.352	0.584	6.251	7.815	9.348	11.345
4	0.297	0.484	0.711	1.064	7.779	9.488	11.143	13.277
5	0.554	0.831	1.145	1.610	9.236	11.070	12.833	15.086
6	0.872	1.237	1.635	2.204	10.645	12.592	14.449	16.812
7	1.239	1.690	2.167	2.833	12.017	14.067	16.013	18.475
8	1.646	2.180	2.733	3.490	13.362	15.507	17.535	20.090
9	2.088	2.700	3.325	4.168	14.684	16.919	19.023	21.666
10	2.558	3.247	3.940	4.865	15.987	18.307	20.483	23.209
11	3.053	3.816	4.575	5.578	17.275	19.675	21.920	24.725
12	3.571	4.404	5.226	6.304	18.549	21.026	23.337	26.217
13	4.107	5.009	5.892	7.042	19.812	22.362	24.736	27.688
14	4.660	5.629	6.571	7.790	21.064	23.685	26.119	29.141
15	5.229	6.262	7.261	8.547	22.307	24.996	27.488	30.578
16	5.812	6.908	7.962	9.312	23.542	26.296	28.845	32.000
17	6.408	7.564	8.672	10.085	24.769	27.587	30.191	33.409
18	7.015	8.231	9.390	10.865	25.989	28.869	31.526	34.805
19	7.633	8.907	10.117	11.651	27.204	30.144	32.852	36.191
20	8.260	9.591	10.851	12.443	28.412	31.410	34.170	37.566
21	8.897	10.283	11.591	13.240	29.615	32.671	35.479	38.932
22	9.542	10.982	12.338	14.041	30.813	33.924	36.781	40.289
23	10.196	11.689	13.091	14.848	32.007	35.172	38.076	41.638
24	10.856	12.401	13.848	15.659	33.196	36.415	39.364	42.980
25	11.524	13.120	14.611	16.473	34.382	37.652	40.646	44.314
26	12.198	13.844	15.379	17.292	35.563	38.885	41.923	45.642
27	12.879	14.573	16.151	18.114	36.741	40.113	43.195	46.963
28	13.565	15.308	16.928	18.939	37.916	41.337	44.461	48.278
29	14.256	16.047	17.708	19.768	39.087	42.557	45.722	49.588
30	14.953	16.791	18.493	20.599	40.256	43.773	46.979	50.892
40	22.164	24.433	26.509	29.051	51.805	55.758	59.342	63.691
50	29.707	32.357	34.764	37.689	63.167	67.505	71.420	76.154
60	37.485	40.482	43.188	46.459	74.397	79.082	83.298	88.379
70	45.442	48.758	51.739	55.329	85.527	90.531	95.023	100.425
80	53.540	57.153	60.391	64.278	96.578	101.879	106.629	112.329
90	61.754	65.647	69.126	73.291	107.565	113.145	118.136	124.116
100	70.065	74.222	77.929	82.358	118.498	124.342	129.561	135.807

Table B-3: Quantiles F_p of the F-distribution with m and n degrees of freedom

$P(T \leq F_p) = 0.95$

$n\backslash m$	5	10	20	30	40	50	75	100
5	5.05	4.74	4.56	4.50	4.46	4.44	4.42	4.41
10	3.33	2.98	2.77	2.70	2.66	2.64	2.60	2.59
20	2.71	2.35	2.12	2.04	1.99	1.97	1.93	1.91
30	2.53	2.16	1.93	1.84	1.79	1.76	1.72	1.70
40	2.45	2.08	1.84	1.74	1.69	1.66	1.61	1.59
50	2.40	2.03	1.78	1.69	1.63	1.60	1.55	1.52
75	2.34	1.96	1.71	1.61	1.55	1.52	1.47	1.44
100	2.31	1.93	1.68	1.57	1.52	1.48	1.42	1.39

$P(T \leq F_p) = 0.99$

$n\backslash m$	5	10	20	30	40	50	75	100
5	10.97	10.05	9.55	9.38	9.29	9.24	9.17	9.13
10	5.64	4.85	4.41	4.25	4.17	4.12	4.05	4.01
20	4.10	3.37	2.94	2.78	2.69	2.64	2.57	2.54
30	3.70	2.98	2.55	2.39	2.30	2.25	2.17	2.13
40	3.51	2.80	2.37	2.20	2.11	2.06	1.98	1.94
50	3.41	2.70	2.27	2.10	2.01	1.95	1.87	1.82
75	3.27	2.57	2.13	1.96	1.87	1.81	1.72	1.67
100	3.21	2.50	2.07	1.89	1.80	1.74	1.65	1.60

Table B-4: Quantiles c_p of the 2-sided Kolmogorov-Smirnov statistic

$$P(D > c_p) = p$$

$n \backslash p$	0.2	0.1	0.05	0.02	0.01
1	0.9000	0.9500	0.9750	0.9900	0.9950
2	0.6838	0.7764	0.8419	0.9000	0.9293
3	0.5648	0.6360	0.7076	0.7846	0.8290
4	0.4927	0.5652	0.6239	0.6889	0.7342
5	0.4470	0.5094	0.5633	0.6272	0.6685
6	0.4104	0.4680	0.5193	0.5774	0.6166
7	0.3815	0.4361	0.4834	0.5384	0.5758
8	0.3583	0.4096	0.4543	0.5065	0.5418
9	0.3391	0.3875	0.4300	0.4796	0.5133
10	0.3226	0.3687	0.4092	0.4566	0.4889
11	0.3083	0.3524	0.3912	0.4367	0.4677
12	0.2957	0.3381	0.3754	0.4192	0.4490
13	0.2847	0.3255	0.3614	0.4036	0.4325
14	0.2748	0.3142	0.3489	0.3897	0.4176
15	0.2658	0.3040	0.3376	0.3771	0.4042
16	0.2577	0.2947	0.3273	0.3657	0.3920
17	0.2503	0.2863	0.3180	0.3553	0.3809
18	0.2436	0.2785	0.3094	0.3457	0.3706
19	0.2373	0.2714	0.3014	0.3369	0.3612
20	0.2315	0.2647	0.2941	0.3287	0.3524
25	0.2079	0.2377	0.2640	0.2952	0.3166
30	0.1903	0.2176	0.2417	0.2702	0.2899
35	0.1766	0.2018	0.2242	0.2507	0.2690
40	0.1654	0.1891	0.2101	0.2349	0.2521

APPENDIX C

Notations

Notations for Model Description

M	Markovian
E_k	k-stage Erlangian
H_R	R-stage Hyperexponential
D	Deterministic
G	General
PH	Phase type
$\bullet^k$	Model $\bullet$ with batches of k units

Notations for Queue Discipline

FCFS	First-Come, First-Served
LCFS	Last-Come, First-Served

General Notations

Notation	Description
$F_*(\bullet)$	Distribution function of the random variable $*$ at $\bullet$ (i.e., $P(* \leq \bullet)$)
$\bar{F}_*(\bullet)$	$1 - F_*(\bullet) = P(* > \bullet)$
$f_*(\bullet)$	Probability function of the random variable $*$ at $\bullet$
$\mu_* = E(*)$	Expected value or mean of the random variable $*$
$\sigma_*^2 = Var(*)$	Variance of the random variable $*$
L.T. $(\bullet)$	Laplace transform of $(\bullet)$
$a^*(\theta)$	Laplace transform of $a(t)$ (i.e., $\int_0^\infty e^{-\theta t} a(t)dt$)
$\Phi_*(\theta)$	Laplace-Stieltjes transform of $F_*(t)$ (i.e., $\int_0^\infty e^{-\theta t} dF_*(t)$)
P_{ij}	Transition probability from state i to state j in a Markov chain
C_n	nth customer to enter the system
τ_n	Arrival epoch of C_n
$T_n = \tau_n - \tau_{n-1}$	Interarrival time between C_{n-1} and C_n

T	Random variable representing interrarival times when these are i.i.d.
S_n	Service-time of C_n
S	Random variable representing service-times when these are i.i.d.
γ_n	Number of arrivals during S_n
γ_n^*	Number of customers being served during T_n
α_j	$P(\gamma_n = j)$
β_j	$P(\gamma_n^* = j)$
$X(t)$	Number of customers in the system at time t
$X_q(t)$	Number of customers in the queue at time t
$P_n(t)$	$P(X(t) = n)$
P_n	$\lim_{t\to\infty} P(X(t) = n)$, steady-state probability of n in the system at any instant
p_n	P(system in state n)
Π_n	Steady-state probability of n in the system at a departure instant
Q_n	Steady-state probability of n in the system at an arrival instant
L	Expected number of units in the system
L_q	Expected number of units in the queue
λ	Mean arrival rate
μ	Mean service rate
λ_n	Mean arrival rate, when n units in the system
μ_n	Mean service rate, when n units in the system
ρ	$\frac{\lambda}{\mu}$
W_n	Waiting time in the queue of C_n
W_q	Waiting time in the queue of a customer
W	Waiting time in the system of a customer
$V(t)$	Unfinished work in the system at time t
Δ	Length of a busy period
I	Length of an idle period
N	Number of units served during a busy period
$G(n,t)$	$P(N = n, \Delta \leq t)$
$P(z)$	Probability generating function of $\{P_n\}$
$\Pi(z)$	Probability generating function of $\{\Pi_n\}$
$Q(z)$	Probability generating function of $\{Q_n\}$
$\alpha(z)$	Probability generating function of $\{\alpha_n\}$
$\beta(z)$	Probability generating function of $\{\beta_n\}$
Δ_i	Length of a busy period with i initial customers
N_i	Number of customers served during a busy period with i initial customers
$\Delta(x)$	Length of a busy period when the initial workload is x

$N(x)$	Number of customers served during a busy period when the initial workload is x
X_n	Number of customers in the system at the departure of C_n
X_0	Initial number of customers waiting for service
Y_n	Number of customers in the system just before the arrival of C_n
$P(z,t)$	Probability generating function of $\{P_n(t)\}$
$P^*(z,\theta)$	Laplace transform of $P(z,t)$
$P_n^*(\theta)$	Laplace transform of $P_n(t)$

APPENDIX D

Formulas

Summary of models treated, with important results on them

Chapter 2

Model	Type of Result	Equation No.	Page
$M/M/1$	P_n	(2.7)	20
	L, L_q	(2.8), (2.9)	23
$M/M/c$	P_n, L_q	(2.17), (2.18)	25
$M/M/c$ finite population finite capacity	P_n	(2.20), (2.21)	26
$M/E_k/1$	$P(z) = \sum_{n=0}^{\infty} p_n z^n$	(2.25), (2.26)	29
$M/E_k/1$	$P(z)$	(2.32)	31
random bulk arrival	L_q, L	(2.33)	32
$E_k/M/1$	P_n	(2.38)	34
$M/M/1$	$P_n(t), n \geq 0$	(2.50)	37
	alter. formulas for $P_n(t)$	(2.61)	39
		(2.62), (2.63)	40
		(2.64), (2.65)	40
$M/M/1$	$E(W), E(W_q)$		42
$M/M/\infty$	$E(W)$		42
$M/M/1$	$F_{W_q}(t)$	(2.68), (2.69)	43
$M/M/1$	$f_\Delta(t), E(\Delta)$	(2.72), (2.73)	44
Open Jackson network	$P(\mathbf{n})$	(2.84)	48
$M/M/1$	$P_n[0], P_n[1]$	(2.97), (2.98)	53
$(0, K)$−policy	L, K^*	(2.99), (2.103)	53, 54
$M/M/1$	$P_n[0], P_n[1]$	(2.107), (2.108)	55
server vacation	L	(2.109)	55

$M/G/1$	transient solution	(6.13)	191
$G/M/1$	transient solution	(6.22)	194
$M/M/1$	$P_n(t)$ combinatorial solution	(6.35), (6.40)	201, 203
$M^m/M/1$	$G(n,t)$	(6.58)	211
$M^m/G/1$	$G(n,m;t)$	(6.62)	212
$M/G^k/1$	functional relations	(6.85) (6.95)	219 221
$G^k/M/1$	functional relations	(6.99)	222

Chapter 7

$G/G/1$	waiting time	(7.7) (7.17), (7.18)	233 235
	idle time	(7.32)	237
	bounds for average waiting time		
	upper bound	(7.39)	238
	lower bound	(7.48)	239
	heavy traffic approximation	(7.54), (7.55)	240
	diffusion approximation	(7.65)	243
Duality relation for $M/G/1$ and $G/M/1$		(7.91) (7.93)	249 250

Chapter 9

Geo/Geo/1	busy period $P(Q_n < k)$	(9.8) (9.10)	302 303
Geo^R/Geo/1	busy period $P(Q_n < k)$	(9.17) (9.20)	305 305
Geo/Geo/c	$P(Q_n = k)$	(9.27)	307
heterogeneous arrivals	transient solution	(9.33)	311
tandem queue	transient solution	(9.45)	314
Geo/G/1	recurrence for transient solution	(9.60)-(9.62)	321, 322
	waiting time		325

G/Geo/1	recurrence for transient solution	(9.66)-(9.69)	327, 328

Chapter 10

nonpreemptive priority	$E(W_{q_i})$	(10.5)	345
	$E(W_{\text{overall}})$	(10.6)	345
preemptive-resume priority	$E(I_{q_i})$	(10.7)	346
	$E(V_i)$	(10.8)	346
$M/G/\infty$	$P(z,t)$	(10.17)	348
$M/M/\infty$	$P(z,t)$	(10.18)	349
$M/G/1$ under $(0,K)$-policy	L	(10.40)	353
	K^*	(10.43)	354
$M/G/1$ with server vacation	$L^{(T)}$	(10.60)	356
	V^*	(10.62)	357
closed Jackson network	$P(\mathbf{n})$	(10.74)	362
	Buzen's algorithm	(10.77), (10.78)	364
	$P(N_i \geq n)$	(10.87)	367
	utilization U_i	(10.88)	368
	throughput	(10.89)	368
	$P_i(n)$	(10.90)	368
	$L_i(N)$	(10.91)	368

Index